Workbook to Accompany

Industrial Maintenance

Second Edition

Michael E. Brumbach
Jeffrey A. Clade

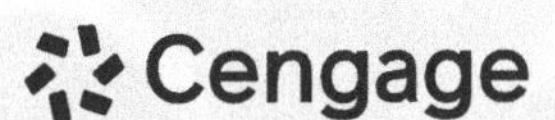

Australia • Brazil • Canada • Mexico • Singapore • United Kingdom • United States

***Workbook to Accompany: Industrial Maintenance*, 2E**
Michael E. Brumbach, Jeffrey A. Clade

Vice President, Editorial: Dave Garza

Director of Learning Solutions: Sandy Clark

Acquisitions Editor: Jim DeVoe

Managing Editor: Larry Main

Senior Product Manager: John Fisher

Editorial Assistant: Kaitlin Schlicht

Vice President, Marketing: Jennifer Baker

Director, Market Development Management: Debbie Yarnell

Marketing Development Manager: Erin Brennan

Director, Brand Management: Jason Sakos

Marketing Brand Manager: Erin McNary

Senior Production Director: Wendy Troeger

Production Manager: Mark Bernard

Content Project Manager: Barbara LeFleur

Senior Art Director: David Arsenault

Technology Project Manager: Joe Pliss

Library of Congress Control Number: 2012941115

ISBN-13: 978-1-133-13121-2
ISBN-10: 1-133-13121-2

Cengage
200 Pier 4 Boulevard
Boston, MA 02210
USA

Notice to the Reader

Publisher does not warrant or guarantee any of the products described herein or perform any independent analysis in connection with any of the product information contained herein. Publisher does not assume, and expressly disclaims, any obligation to obtain and include information other than that provided to it by the manufacturer. The reader is expressly warned to consider and adopt all safety precautions that might be indicated by the activities described herein and to avoid all potential hazards. By following the instructions contained herein, the reader willingly assumes all risks in connection with such instructions. The publisher makes no representations or warranties of any kind, including but not limited to, the warranties of fitness for particular purpose or merchantability, nor are any such representations implied with respect to the material set forth herein, and the publisher takes no responsibility with respect to such material. The publisher shall not be liable for any special, consequential, or exemplary damages resulting, in whole or part, from the readers' use of, or reliance upon, this material.

Printed in the United States of America
Print Number: 10 Print Year: 2022

CONTENTS

Name: ______________________ Date: ______________

ELECTRICAL SAFETY

1. List four general safety precautions that should be observed when working on or around electricity.

   a. ______________________

   b. ______________________

   c. ______________________

   d. ______________________

2. List five safety concerns to address before starting the work.

   a. ______________________

   b. ______________________

   c. ______________________

   d. ______________________

   e. ______________________

3. List three items relating to worker safety.

   a. ______________________

   b. ______________________

   c. ______________________

4. List two clothing-related safety issues.

   a. ______________________

   b. ______________________

5. List four suggested safety practices regarding personal protective equipment.

   a. ______________________

   b. ______________________

   c. ______________________

   d. ______________________

6. List three items of concern when dealing with scaffolds and ladders.

   a. ______________________________

   b. ______________________________

   c. ______________________________

7. List two safety suggestions related to the use and operation of portable hand and power tools.

   a. ______________________________

   b. ______________________________

8. List three items to remember when working on and around electrical circuits.

   a. ______________________________

   b. ______________________________

   c. ______________________________

9. List two grounding concerns.

   a. ______________________________

   b. ______________________________

10. List two items to remember when working on or operating electrical switches.

   a. ______________________________

   b. ______________________________

11. List three concerns when dealing with switches and circuit breakers.

   a. ______________________________

   b. ______________________________

   c. ______________________________

12. List three steps to minimize the risk of injury when working with motors.

   a. ______________________________

   b. ______________________________

   c. ______________________________

13. List two items of concern when working on or around capacitors.

    a. ______________________________

    b. ______________________________

14. List five items that will help ensure the safe operation of electrical test equipment.

    a. ______________________________

    b. ______________________________

    c. ______________________________

    d. ______________________________

    e. ______________________________

Name: ______________________ Date: ______________

NFPA STANDARD 704

1. Match each set of hazards to the correct NFPA 704 symbol.

   a. **Flammability**—The material has a flash point below 200°F but above 100°F. **Reactivity**—Shock and or heat may cause the material to detonate. **Health**—Slightly hazardous; irritation or minor residual injury may occur. **Special**—None.

   b. **Flammability**—The material will not burn. **Reactivity**—The material is very unstable and is capable of detonating itself or undergoes explosive decomposition or reaction at normal temperatures. **Health**—The material is considered to be hazardous and on intense or continued but not chronic exposure could cause temporary incapacitation or possible residual injury. **Special**—None.

   c. **Flammability**—The material has a flash point below 73°F. **Reactivity**—The material may undergo a violent chemical change at elevated temperatures and pressures, may react violently with water, or may form explosive mixtures with water. **Health**—An extreme danger exists. **Special**—The material will react with water.

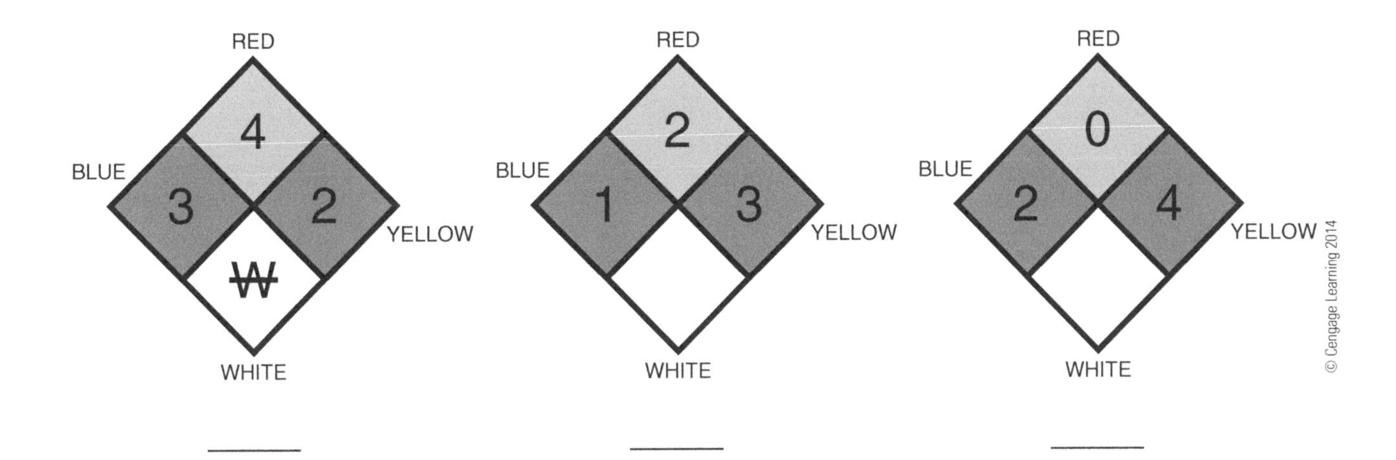

______ ______ ______

Name: ______________________________ Date: ____________________

PERSONAL PROTECTIVE EQUIPMENT

1. For each letter indicator as given in the Hazardous Material Information Guide (HMIG), select the correct group of symbols for recommended personal protective equipment.

a. HMIG indicator H ______

b. HMIG indicator F ______

c. HMIG indicator A ______

d. HMIG indicator J ______

e. HMIG indicator E ______

f. HMIG indicator C ______

g. HMIG indicator K ______

h. HMIG indicator I ______

1    

2     

3    

4    

5   

6    

7 

8    

Name: ______________________________ Date: ____________________

FIRE SAFETY—PART 1

1. Match the appropriate symbol to the correct fire classification.

   a. _______ A fire that burns on any electrical or electrically energized equipment

   b. _______ A fire that burns ordinary combustibles such as wood, paper, cloth, upholstery, trash, some plastics, or any other carbon-based solid materials that are not metals

   c. _______ A fire that burns combustible metals, such as magnesium, potassium, powdered aluminum, zinc, sodium, or titanium

   d. _______ A fire that burns flammable liquids or gases such as gasoline, oil, grease paints and thinners, propane, acetone, grease, or any other similar gas or liquid that is not metal based

1

2

3

4

Name: ______________________ Date: ______________

FIRE SAFETY—PART 2

1. Match each definition to the appropriate pictograph.

 ______

 ______

 ______

 ______

a. A fire that burns on any electrical or electrically energized equipment

b. A fire that burns ordinary combustibles such as wood, paper, cloth, upholstery, trash, some plastics, or any other carbon-based solid materials that are not metals

c. A fire that burns combustible metals, such as magnesium, potassium, powdered aluminum, zinc, sodium, or titanium

d. A fire that burns flammable liquids or gases such as gasoline, oil, grease paints and thinners, propane, acetone, grease, or any other similar gas or liquid that is not metal based

Name: ______________________________ Date: ______________________

PROPER FIRE EXTINGUISHER USE

1. Number the following procedures in the proper chronological order.

   a. _______ Use the PASS method to achieve the maximum firefighting capability of the extinguisher.

   b. _______ Make sure that the facility's emergency response or first responder team is notified.

   c. _______ Quickly identify the class of the fire and retrieve the appropriate extinguisher.

   d. _______ Get all nonessential personnel out of harm's way.

   e. _______ Deenergize all electrical sources that are in the immediate vicinity of the fire, if possible.

   f. _______ Locate and activate the fire alarm nearest to you. If no fire alarm is in the immediate area, then yell the word *fire* as loudly as possible to get the attention of a coworker or a supervisor.

   g. _______ Evacuate the facility and stand as far away from the fire as possible.

   h. _______ Continue to fight the fire as long as possible, preventing any spreading, until the local firefighters arrive.

   i. _______ Make certain that a call is placed to 911. It is most important to notify the local fire department of the fire. Remember that a loss of life due to a fire is not acceptable.

   j. _______ Always make sure that you have an escape route as you are trying to extinguish the fire.

Name: ____________________ Date: ____________________

IDENTIFYING EXTENSION LADDER PARTS

1. Label each part on the graphic below.

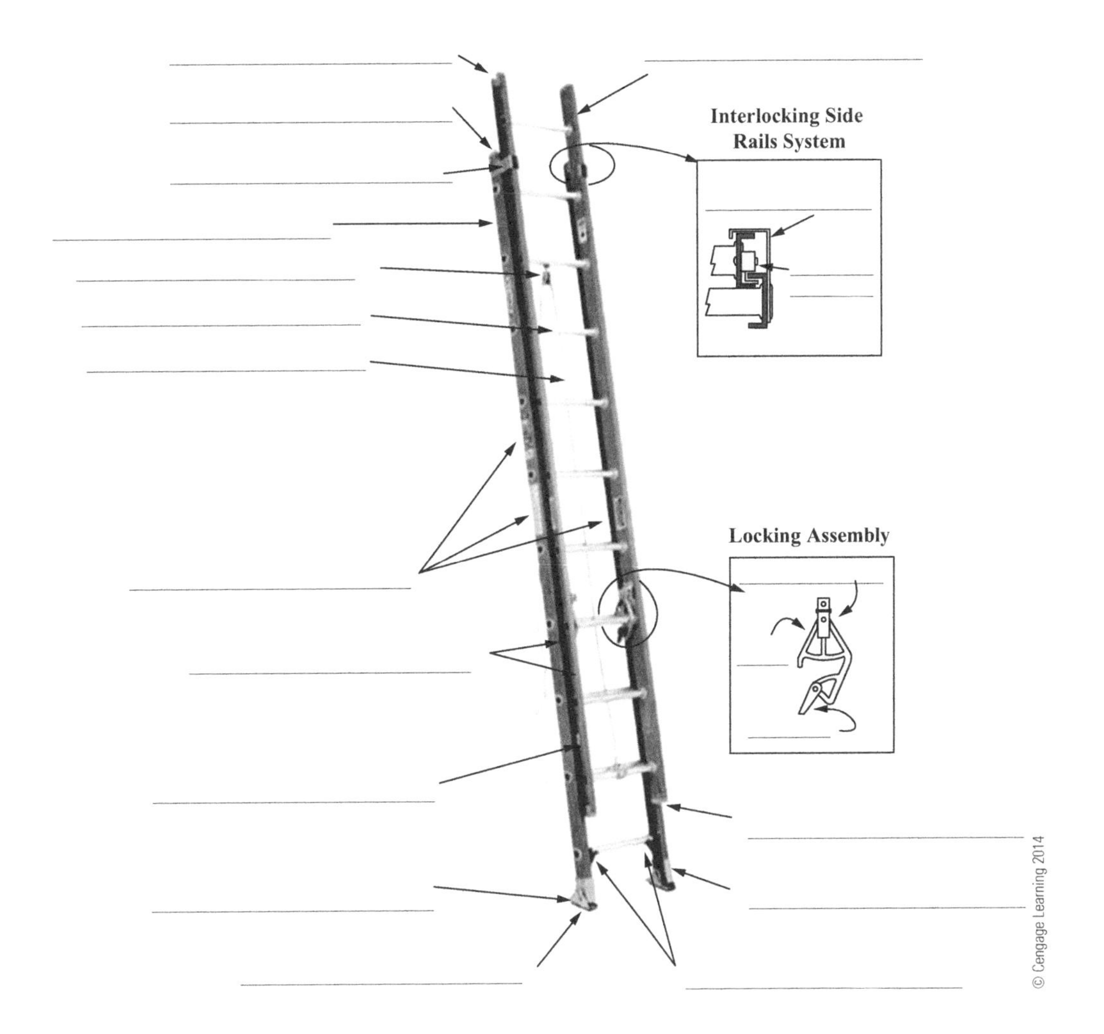

Name: ______________________ Date: ______________

LADDER ANSI CODES

1. Match the correct ANSI code with the correct type of ladder or scaffold.

   Rolling scaffold ________

   Wood ladders ________

   Fixed ladders ________

   Steel ladders ________

   Fiberglass ladders ________

   Metal ladders ________

a. ANSI A14.1

b. ANSI A14.2

c. ANSI A14.3

d. ANSI A14.5

e. ANSI A14.7

f. ANSI A10.8

Name: ______________________________ Date: ____________________

LADDER DUTY RATINGS

1. Match the correct rating with the correct rated use.

   Type I ________

   Type IA ________

   Type IAA ________

   Type II ________

   Type III ________

   a. Extra-heavy duty

   b. Light duty

   c. Special duty

   d. Medium duty

   e. Heavy duty

Name: ______________________________ Date: ____________________

HAND TOOLS: WRENCHES

1. Match the wrenches to the appropriate graphics.

   a. Open end

   b. Box end

   c. ⅜ in. ratchet

   d. Adjustable

   e. Flare nut

   f. Torque

 ______ ______ 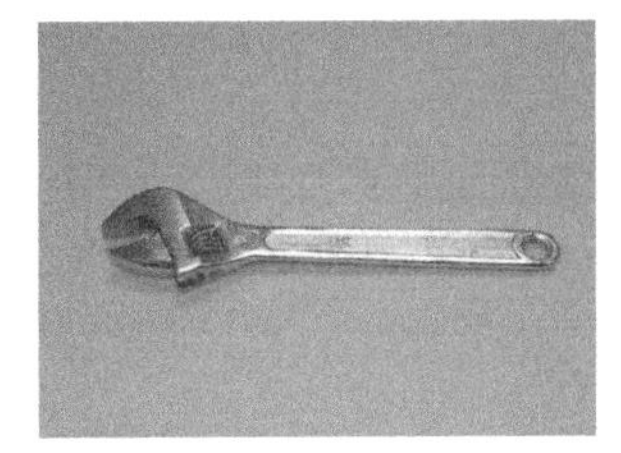

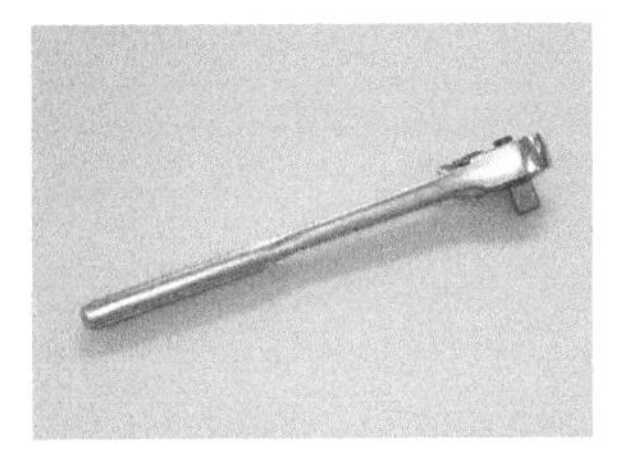 ______ ______ 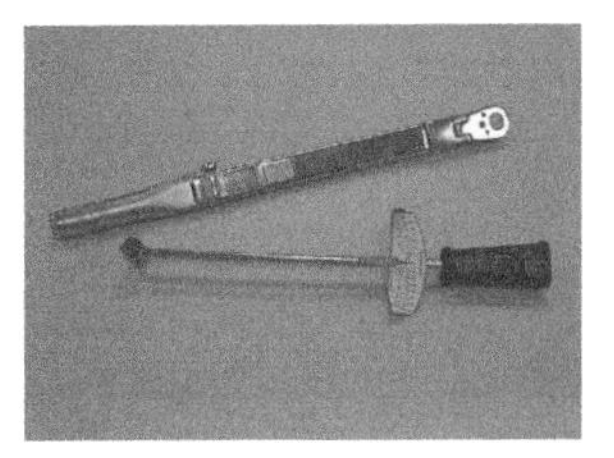

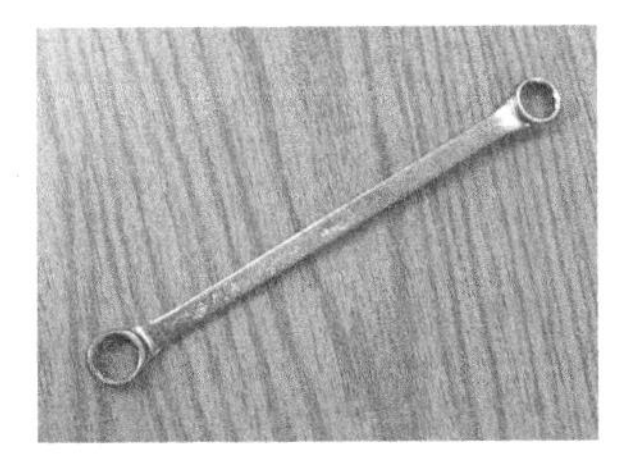 ______ ______ 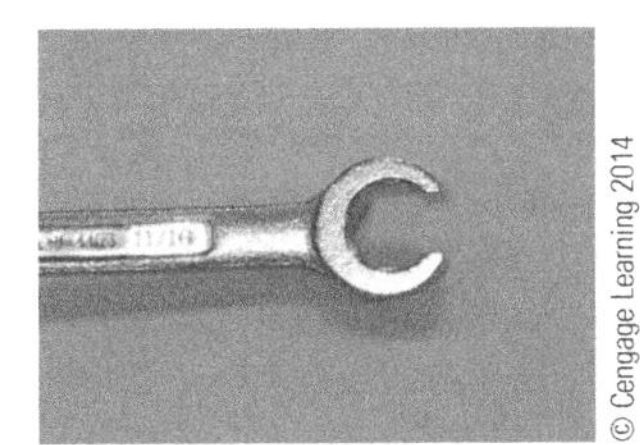

Name: ______________________________ Date: ____________________

HAND TOOLS: SCREWDRIVERS

1. The flukes of a Reed and Prince screwdriver are at what angle?
   a. 25°
   b. 30°
   c. 45°
   d. 60°

2. Match each screwdriver to its picture.

A

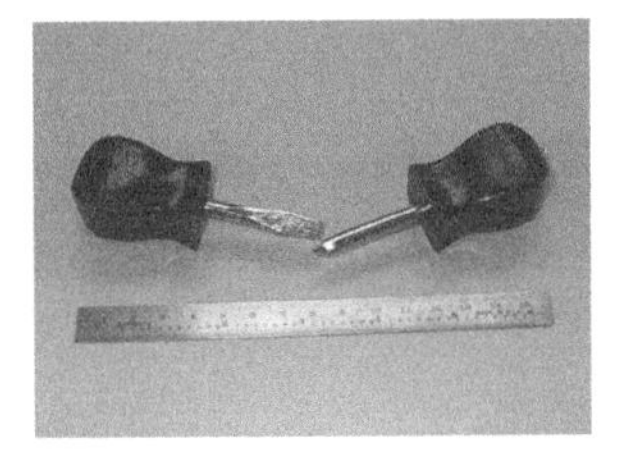

B

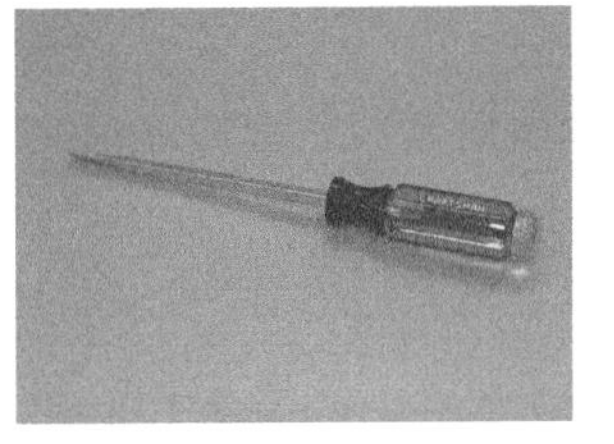

C

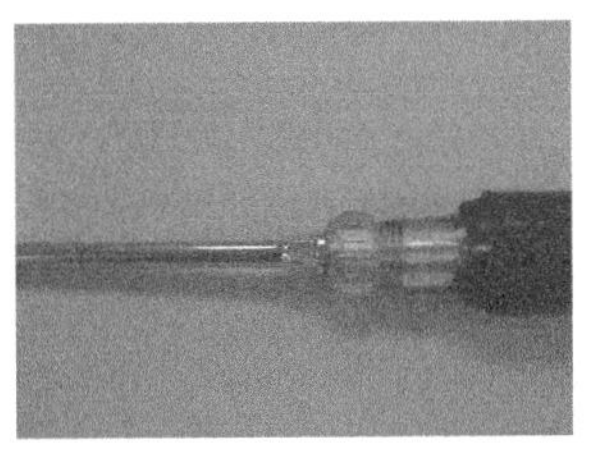

D

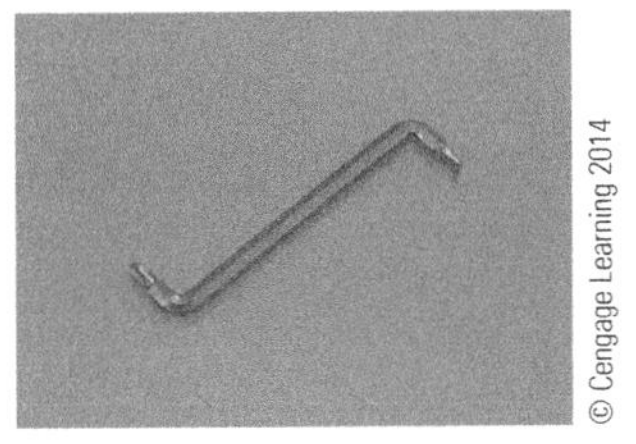

_______ Offset

_______ Round shank

_______ Stubby

_______ Square shank

Name: ______________________________ Date: ____________________

HAND TOOLS: PLIERS

1. Match each pliers to the appropriate graphic.
   a. Lineman's pliers
   b. Long-nose pliers
   c. Slip joint pliers
   d. Tongue and groove pliers
   e. Locking plier-wrench

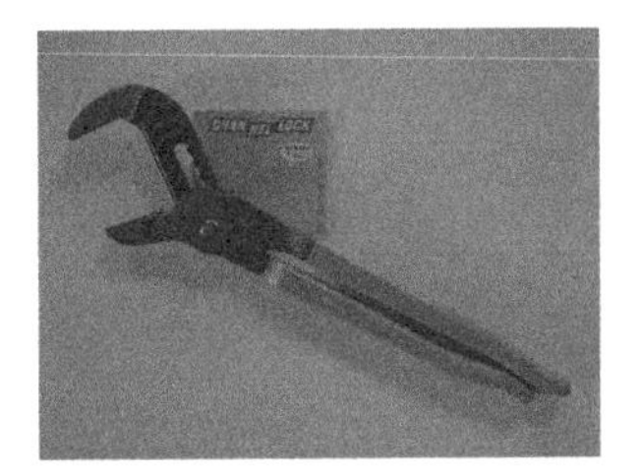

____

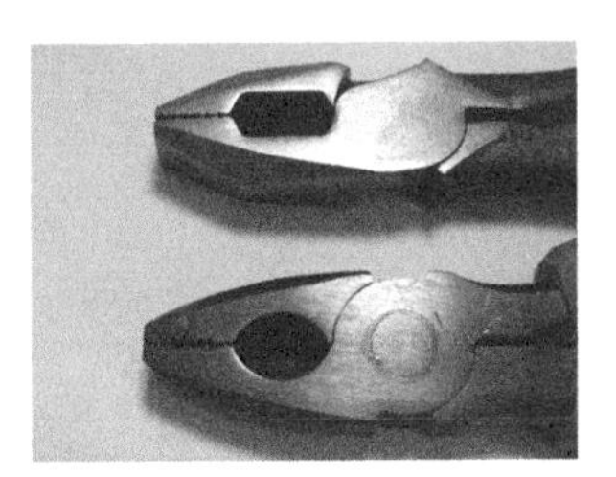

____

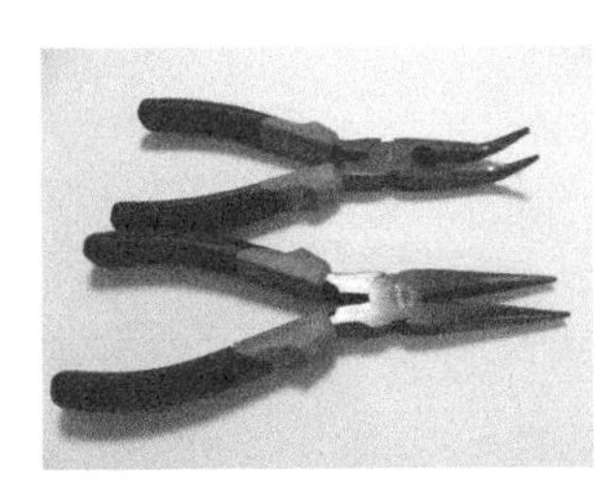

____

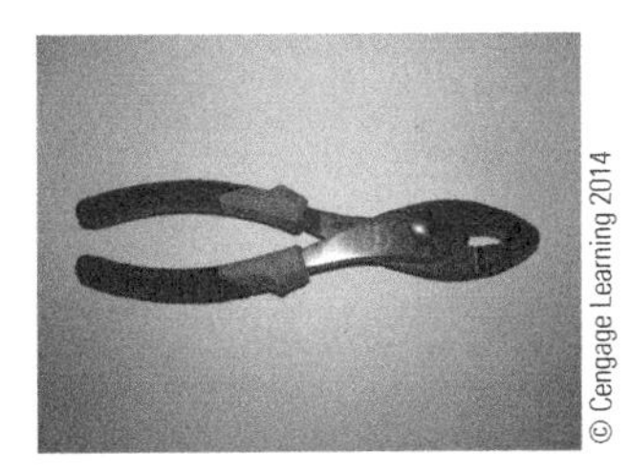

____

Name: ______________________________ Date: ____________________

HAND TOOLS: HAMMERS

1. Draw, to the best of your ability, each hammer that is listed below.

   a. Rubber mallet

   b. Ball-peen hammer

   c. Claw hammer

2. Match the correct hammer to the job that it would be used for.

   a. _______ Claw hammer

   b. _______ Rawhide hammer

   c. _______ Brass hammer

   d. _______ Ball-peen hammer

   1. Hammering on a finished surface that cannot be marred

   2. Hammering a nail in a wooden stud

   3. Hammering on a solid piece of iron

   4. Hammering on a solid piece of iron in a hazardous location

Name: ______________________ **Date:** ______________

HAND TOOLS: HACKSAWS

1. Match each tooth set to the appropriate graphic.

   a. _______ Raker set

   b. _______ Alternate set

   c. _______ Wave set

   d. _______ Double alternate set

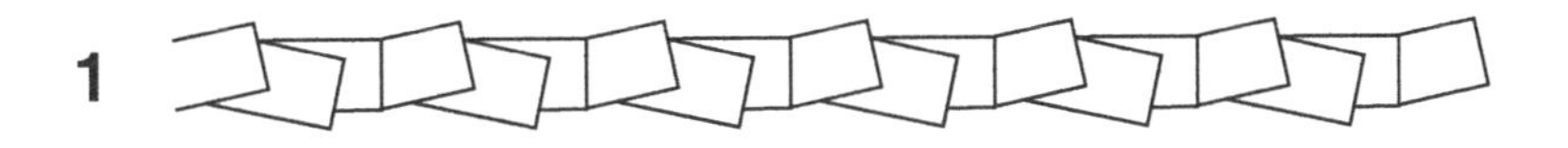

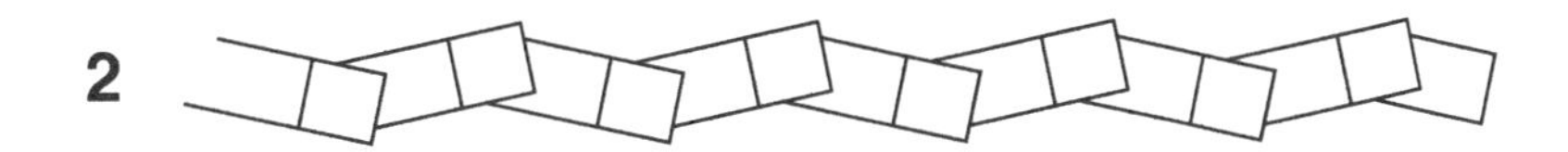

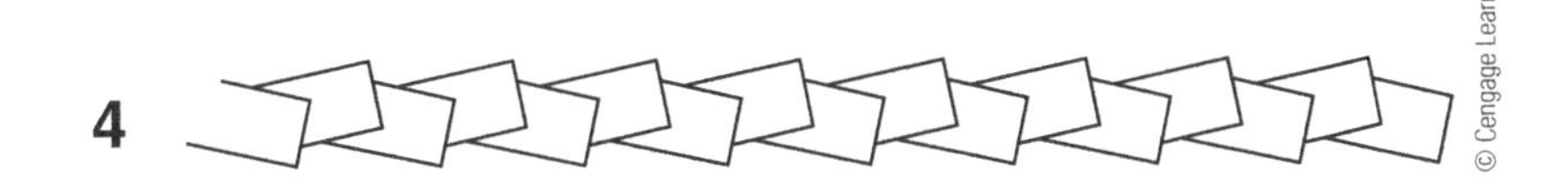

2. Match the proper hacksaw blade for the task at hand.

   a. _______ 14 teeth per inch

   b. _______ 18 teeth per inch

   c. _______ 24 teeth per inch

   d. _______ 32 teeth per inch

1. Cutting on a large section of mild material
2. Cutting on brass, thick wall pipe, angle iron, or copper
3. Cutting on a large section of hard material
4. Cutting on thin-wall tubing

Name: ______________________________ Date: ____________________

HAND TOOLS: POWER TOOLS

1. Describe, in your own words, what each power tool is used for.

   a. A drill

   ____

   ____

   ____

   b. A portable band saw

   ____

   ____

   ____

   c. A reciprocating saw

   ____

   ____

   ____

   d. A portable side grinder

   ____

   ____

   ____

Name: ______________________________ Date: ________________

FASTENERS: THREAD CONSTRUCTION—PART 1

1. Write the following thread specifications in their proper place on the graphic.

   a. Root

   b. Crest

   c. Thread depth

   d. Face

Name: ______________________ Date: ______________

FASTENERS: THREAD CONSTRUCTION—PART 2

1. Indicate on the graphic the pitch of the thread.

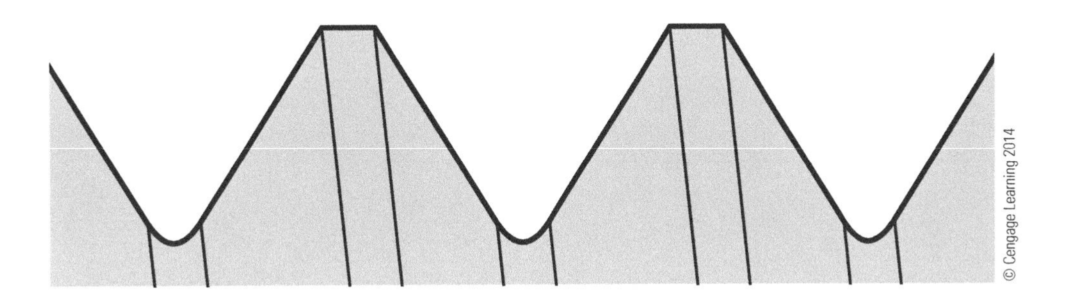

Name: ______________________________ Date: ______________________

FASTENERS: THREAD SERIES

1. Which thread series will be used in conditions where corrosion might occur?

   a. Constant pitch

   b. Extra-fine thread

   c. Fine thread

   d. Coarse thread

2. Which thread series will be used when the internal threads are being placed in thin-walled tubes, thin nuts, ferrules, or couplings that possess thin walls?

   a. Constant pitch

   b. Extra-fine thread

   c. Fine thread

   d. Coarse thread

3. Which thread series would usually be used for short-distance applications of a screw or bolt?

   a. Constant pitch

   b. Extra-fine thread

   c. Fine thread

   d. Coarse thread

4. In which thread series can two bolts have different diameters but still have the same thread pitch?

   a. Constant pitch

   b. Extra-fine thread

   c. Fine thread

   d. Coarse thread

Name: ______________________________ Date: ____________________

FASTENERS: THREAD DIAMETER

1. Draw a set of brackets on the internal thread that show its major diameter.

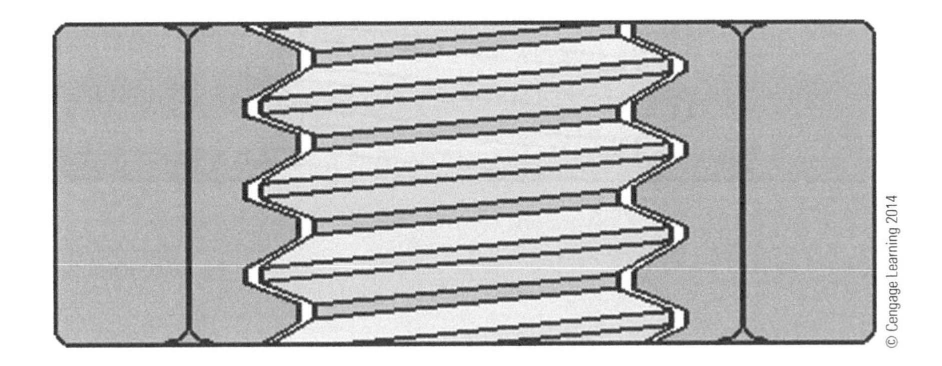

2. Draw a set of brackets on the internal thread that show its minor diameter.

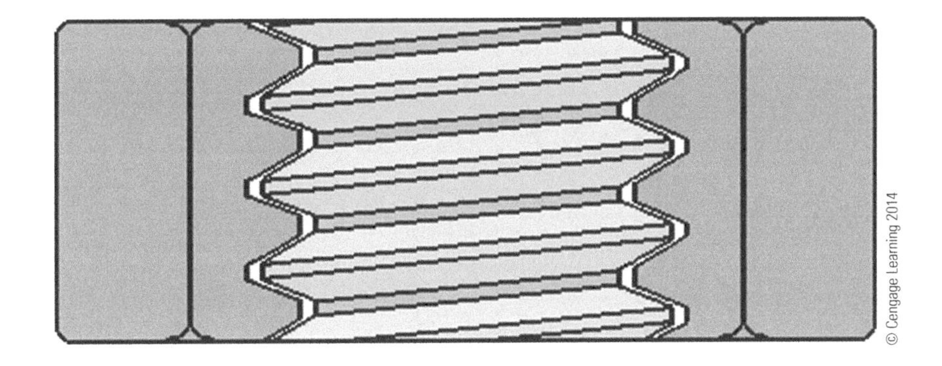

3. Draw a set of brackets on the external thread that show its major diameter.

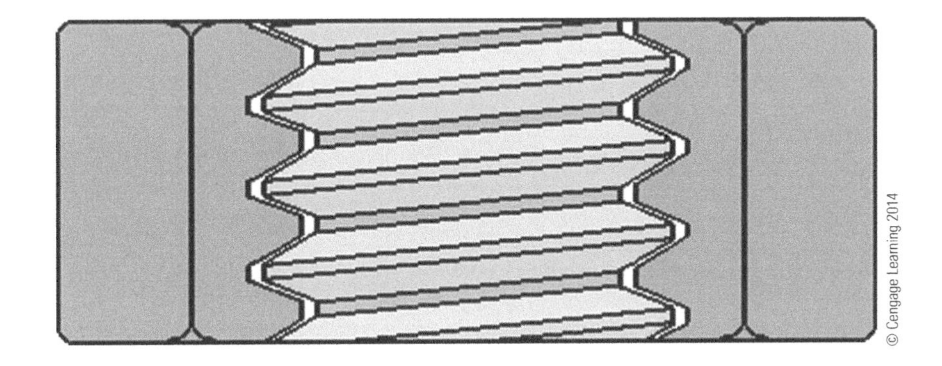

4. Draw a set of brackets on the external thread that show its minor diameter.

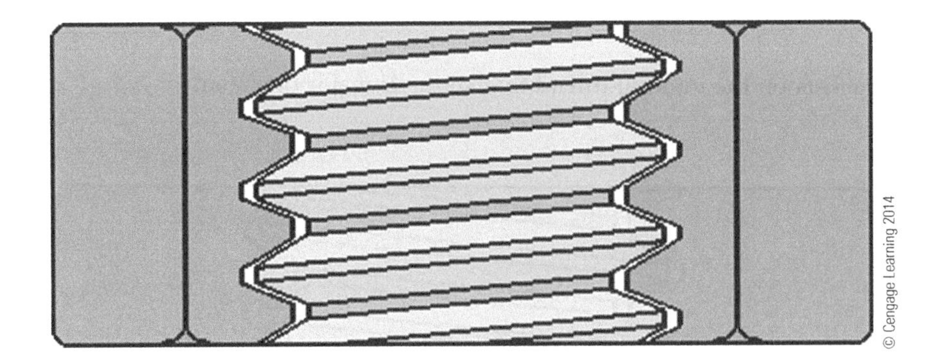

Name: ______________________________ Date: ____________________

FASTENERS: THREAD CLASS—PART 1

1. Match the following thread classes to the appropriate descriptions.

   a. ________ 1A

   b. ________ 3B

   c. ________ 2A

   d. ________ 3A

   e. ________ 1B

   f. ________ 2B

   1. A threaded nut with a relaxed tolerance

   2. A threaded bolt with a very close tolerance

   3. A threaded bolt with a medium tolerance

   4. A threaded nut with a tight tolerance

   5. A threaded nut with a medium tolerance

   6. A threaded bolt with a relaxed tolerance

Name: ______________________ Date: ______________

FASTENERS: THREAD CLASS—PART 2

1. Match the following sequences of letters and numbers to the appropriate descriptions.

   a. _______ ¼-20UNC-3A

   b. _______ ¼-28UNF-2B

   c. _______ ⅜-18UNC-1A

   d. _______ ⅜-24UNF-3B

   1. A ⅜-inch bolt with coarse threads and with a relaxed tolerance

   2. A ¼-inch bolt with coarse threads and with an extremely close tolerance

   3. A ⅜-inch nut with fine threads and with a medium tolerance

   4. A ⅜-inch nut with fine threads and with an extremely close tolerance

Name: ______________________ Date: ______________

FASTENERS: TAPS AND DIES—PART 1

1. Identify the following parts of the tap by writing the name of each in its proper place on the drawing.
   a. Flats
   b. Lands
   c. Flutes
   d. Shank

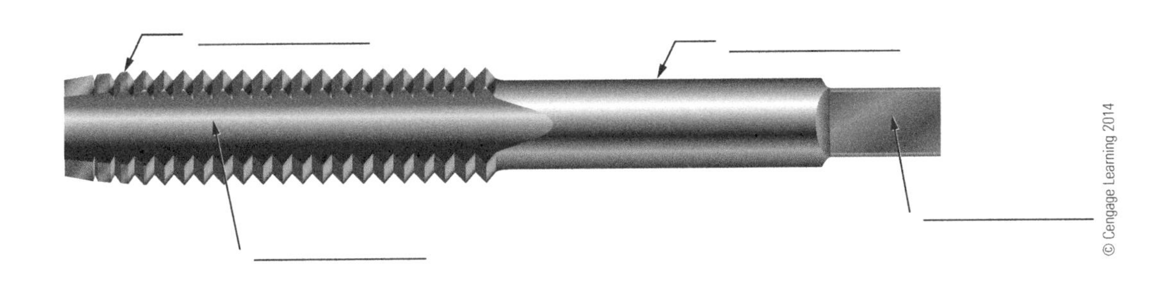

2. Identify the following parts of the die by writing the name of each in its proper place on the drawing.
   a. Lands
   b. Flutes

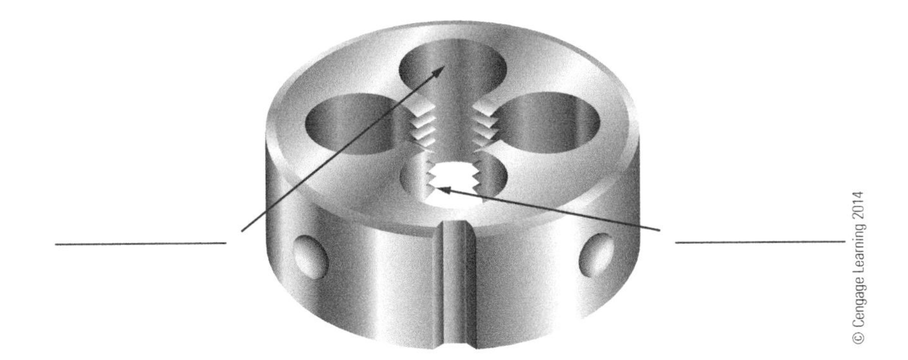

Name: ______________________________ Date: ____________________

FASTENERS: TAPS AND DIES—PART 2

1. Number the tapping procedures in the proper chronological order.

   a. _______ Insert tap into the predrilled hole.

   b. _______ Slowly rotate the tap clockwise, providing cutting oil as needed.

   c. _______ Determine the size of the drill bit needed.

   d. _______ Ensure that the tap is perpendicular to the surface of the work.

   e. _______ Insert the fastener.

   f. _______ Polish the threads with the plug tap and then the bottom tap.

   g. _______ Lubricate tap.

   h. _______ Put on safety glasses.

   i. _______ Rotate the tap counterclockwise to break away any burrs and chips.

   j. _______ Drill the hole.

   k. _______ Rotate the tap one revolution to cut the threads, as you add cutting oil.

   l. _______ Determine the location of the hole.

   m. _______ Repeat the cutting and cleaning steps until all the threads are cut.

   n. _______ Insert the tap into the tap handle.

   o. _______ Select the proper tap.

Name: ______________________ Date: ______________

FASTENERS: FASTENER TYPES

1. Match each type of threaded fastener to the correct graphic.

   a. _______ Bolt

   b. _______ Machine screw

   c. _______ Threaded stud

   d. _______ Double-ended stud

   e. _______ Masonry anchor

1

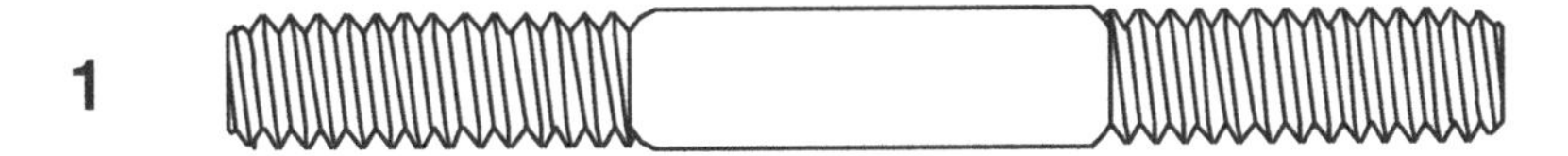

2

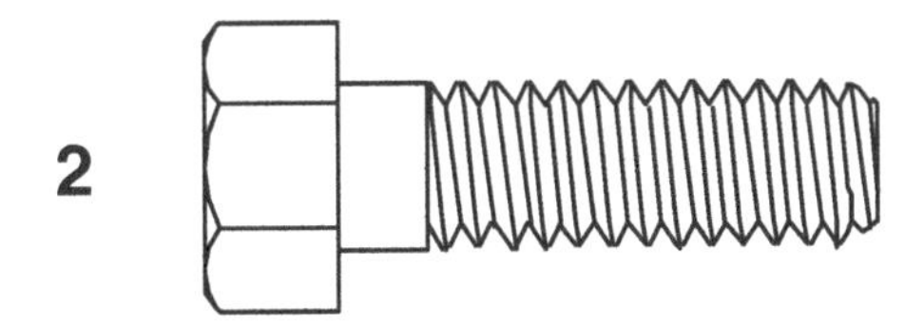

3

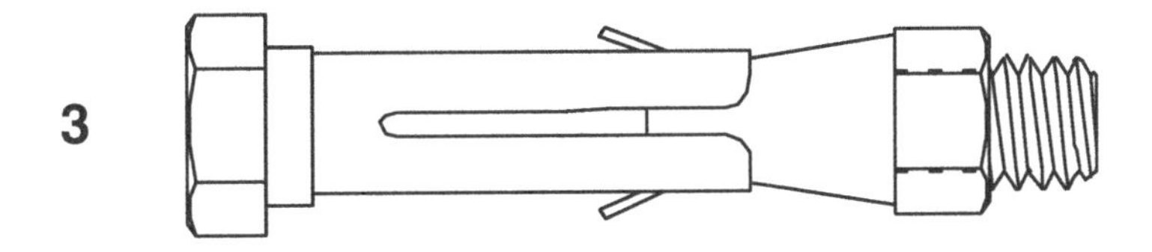

4

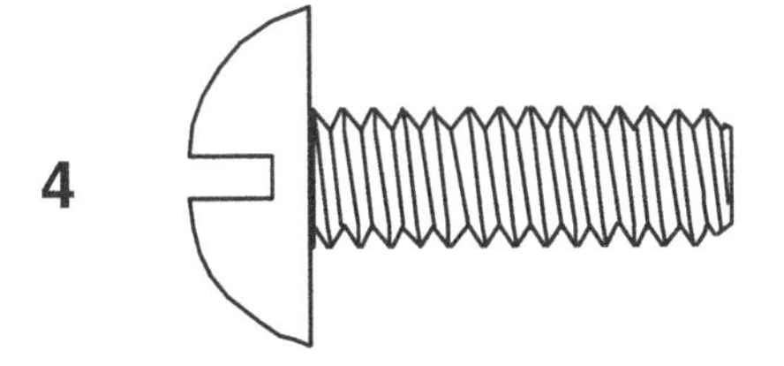

5

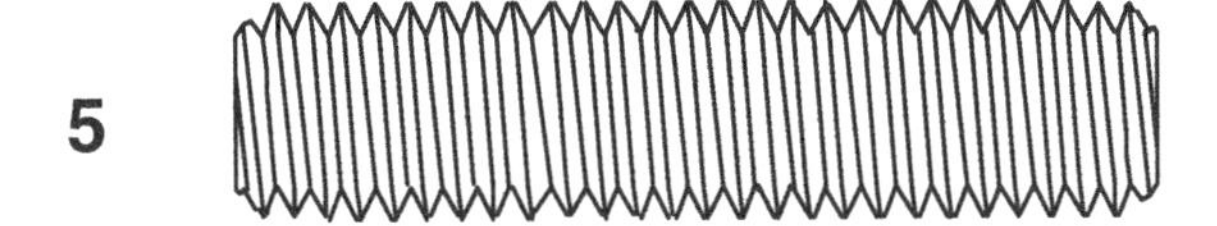

2. List at least five different types of retaining pins, and give a brief description of each.

a. ________________________________

b. ________________________________

c. ________________________________

d. ________________________________

e. ________________________________

Name: ______________________________ Date: ____________________

FASTENERS: FASTENER GRADES

1. Using the head markings, identify each bolt as ASTM, SAE, and/or Metric by writing ASTM, SAE, and/or Metric on the line with its graphic. There may be more than one answer per graphic. The first one is done as an example.

SAE

____

____

____

____

____

____

____

____

____

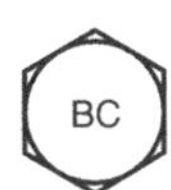

____

____

____

____

____

____

2. Draw a line from each standard fastener grade to the correct head marking for that fastener.

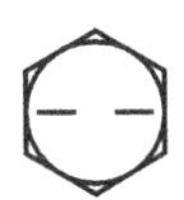

SAE Grade 3	ASTM A 490	SAE Grade 8	ASTM A 354 Grade BB	SAE Grade 5	ASTM A 354 Grade BC	SAE Grade 0-2	ASTM A 325	SAE Grade 7	SAE Grade 6
		ASTM A 354 Grade BD		ASTM A 449		ASTM A 307			

3. Draw a line from each metric fastener grade to the correct head marking for that fastener.

Property Class 9.8	Property Class 5.8	Property Class 8.8	Property Class 4.6	Property Class 10.9	Property Class 4.8

4. Match the standard fastener grade with the correct description by drawing a line from the head marking to its description.

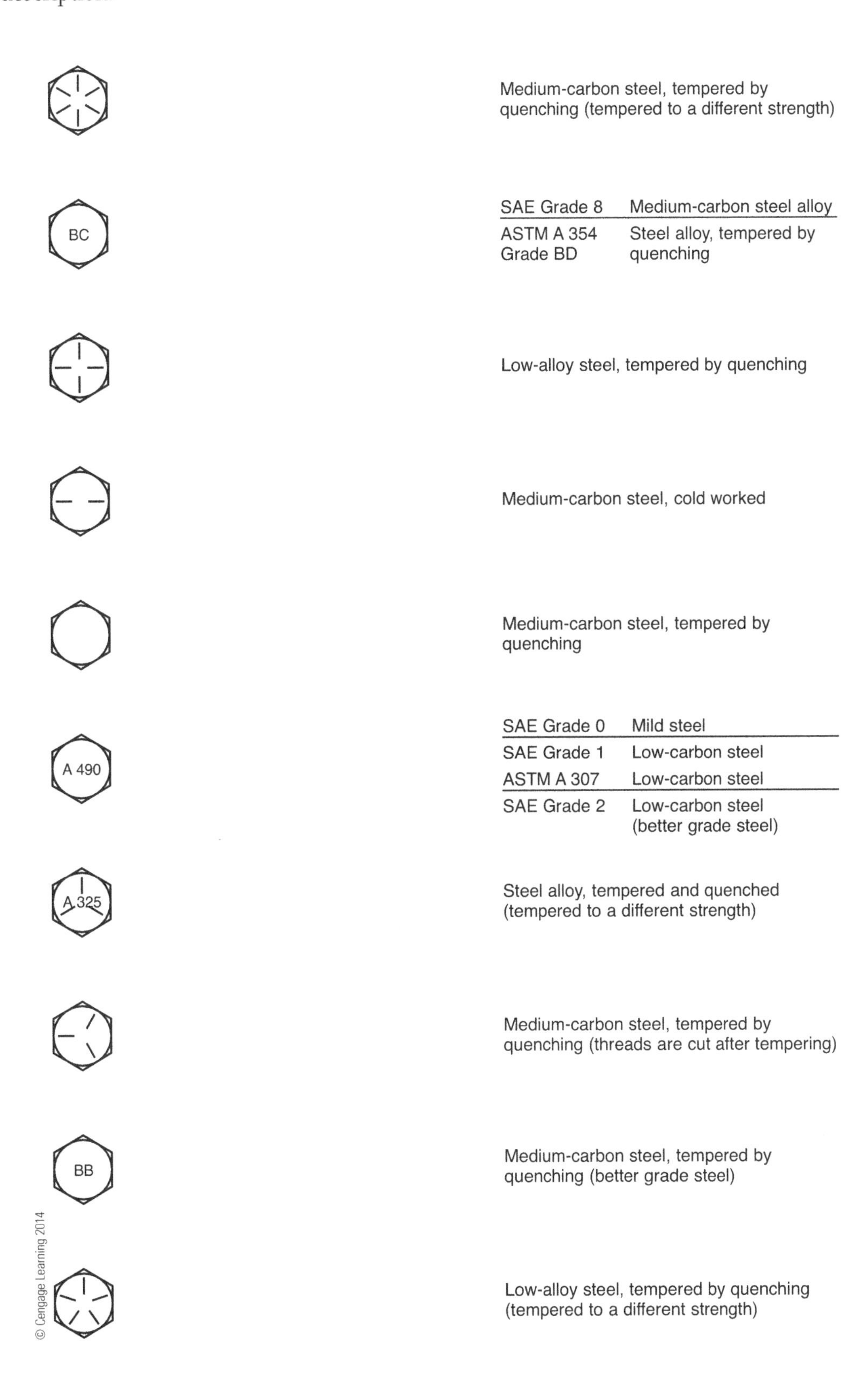

5. Match the metric fastener grades with the correct descriptions.

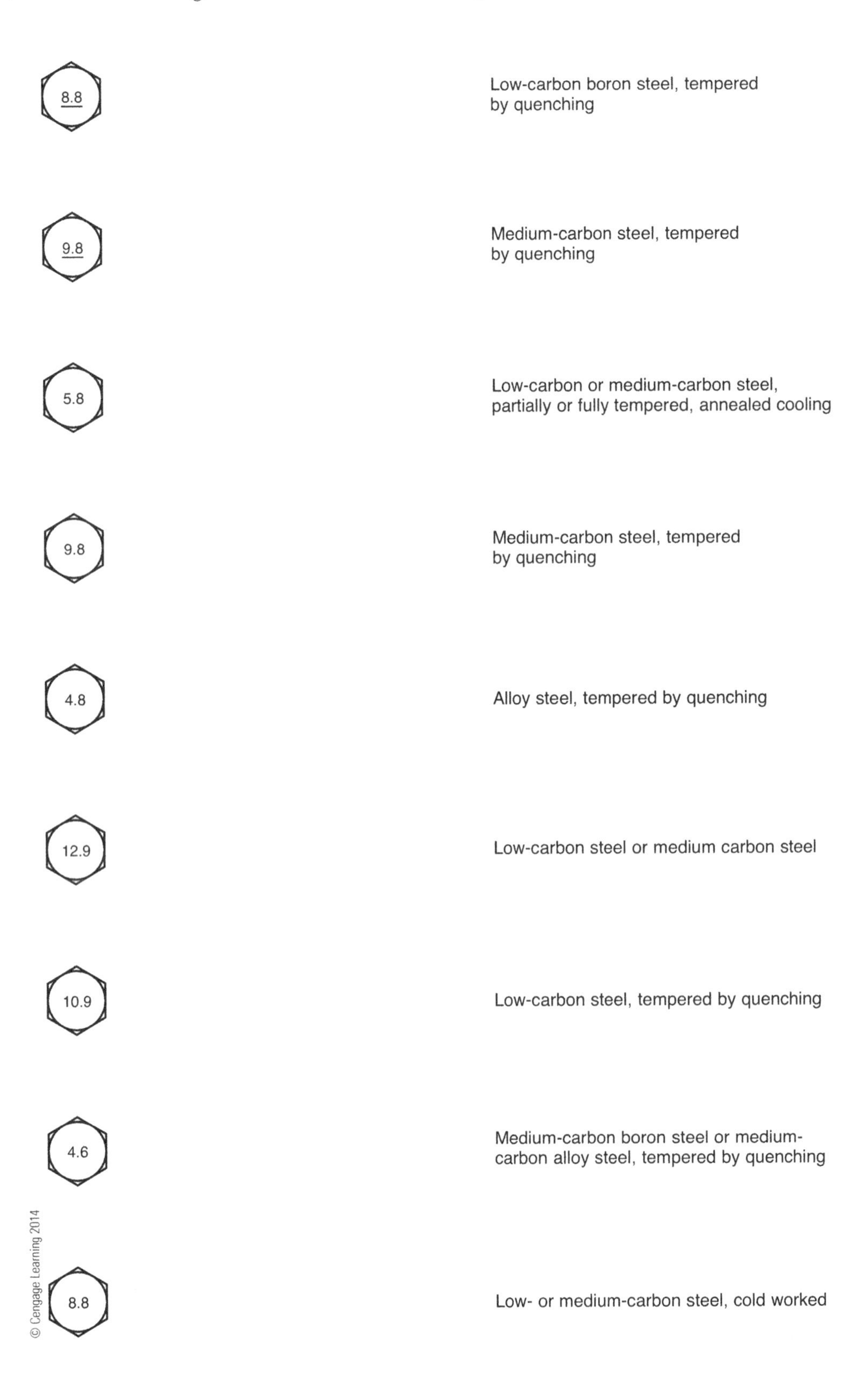

Name: ______________________ Date: ______________

FASTENERS: RETAINING RING FASTENERS

1. List at least four different types of retaining ring fasteners, and give a brief description of each.

   a. ______________________

   b. ______________________

   c. ______________________

   d. ______________________

Name: ______________________________ Date: ______________

MECHANICAL DRAWINGS: DIMENSIONAL

Refer to the drawing on page 58 to answer all of the questions on this worksheet.

1. What is the height of the bracket shown in the figure?

   ____

2. What is the distance between the two holes on the upright portion of this bracket?

   ____

3. What is the radius of both curved edges?

   ____

4. What is the thickness of the plate used to make the bracket?

   ____

5. What is the longest measurement of the bracket?

   ____

6. What is the shortest measurement of the bracket?

   ____

7. What is the center-to-center measurement between the two holes that are closest to the inside curve?

   ____

8. What size drill bit should be used if you were to drill the holes in this bracket?

   ____

9. Which of the four views—A, B, C, or D—would be used if you needed to know the measurements between the four holes that are drilled into the bracket base?

   ____

10. What is the name of this type of drawing?

    ____

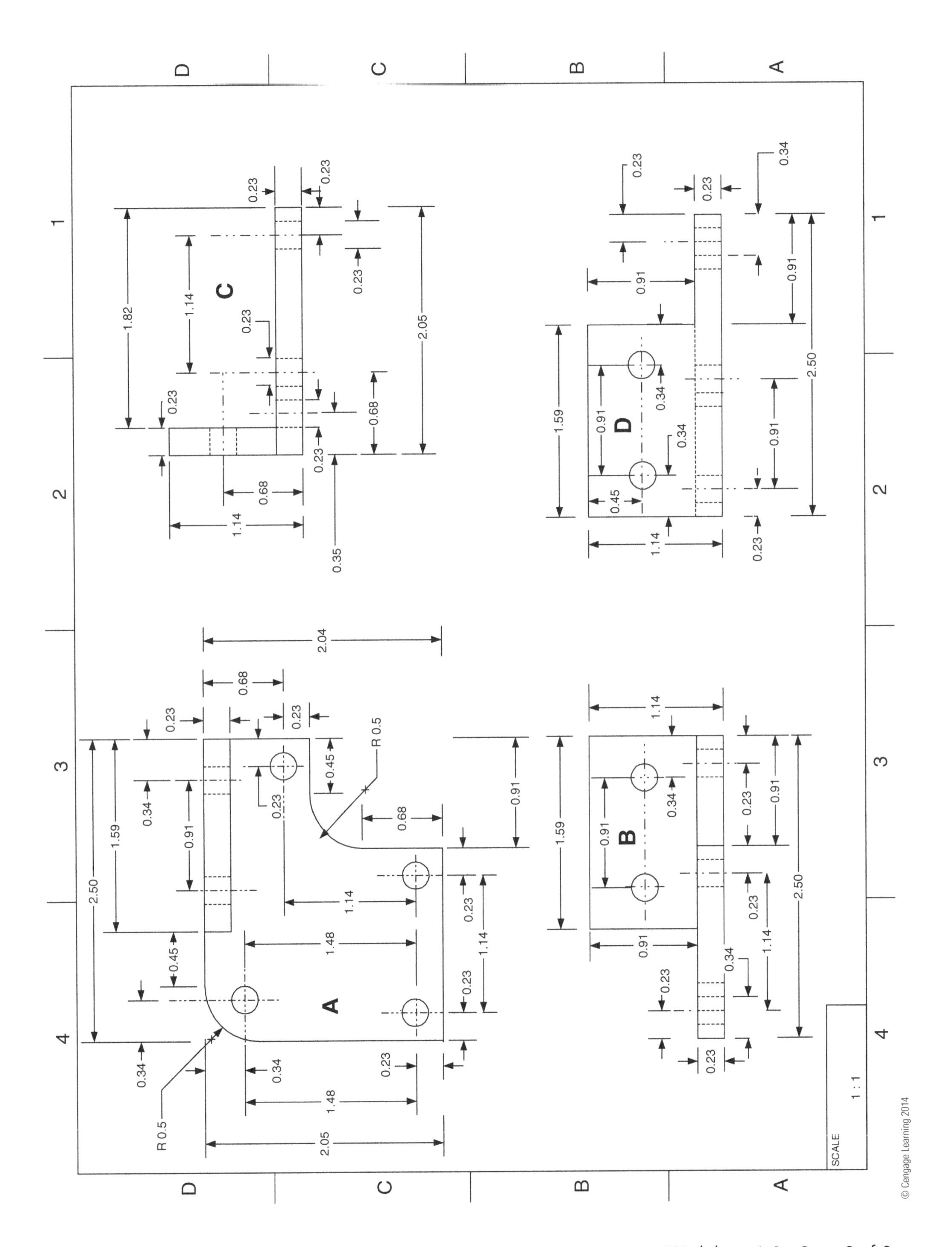
A
B
C
D
R 0.5
SCALE
1 : 1

Name: ______________________________ Date: ______________________

PIPING SYSTEMS—DRAWING TYPES

1. Name the type of single-line drawing that seems to match each description that is given.

   a. This view will be from the side in order to show all of the vertical runs.

   ____

   b. This view will be from the top of the piping system (looking down) showing all of the horizontal runs.

   ____

   c. This is a drawing that is made using two separate views.

   ____

   d. This drawing is considered to be a three-dimensional type of drawing. The drawing includes both the horizontal and the vertical runs of the piping system.

   ____

Name: ______________________________ **Date:** ____________________

PIPING SKETCHES—PART 1

1. Match each of the single-line pipe symbols to its name.

_______ 90° angle elbow, side view

_______ Reducers

_______ Wye, side view

_______ 45° angle elbow, side view

_______ 90° angle elbow, turned toward the viewer

_______ 45° angle elbow, turned toward the viewer

_______ Tee, side view

_______ 90° angle elbow, turned away from the viewer

_______ Union

_______ Tee, turned away from the viewer

_______ Wye, turned toward the viewer

_______ Wye, turned away from the viewer

_______ Tee, turned toward the viewer

_______ Valves

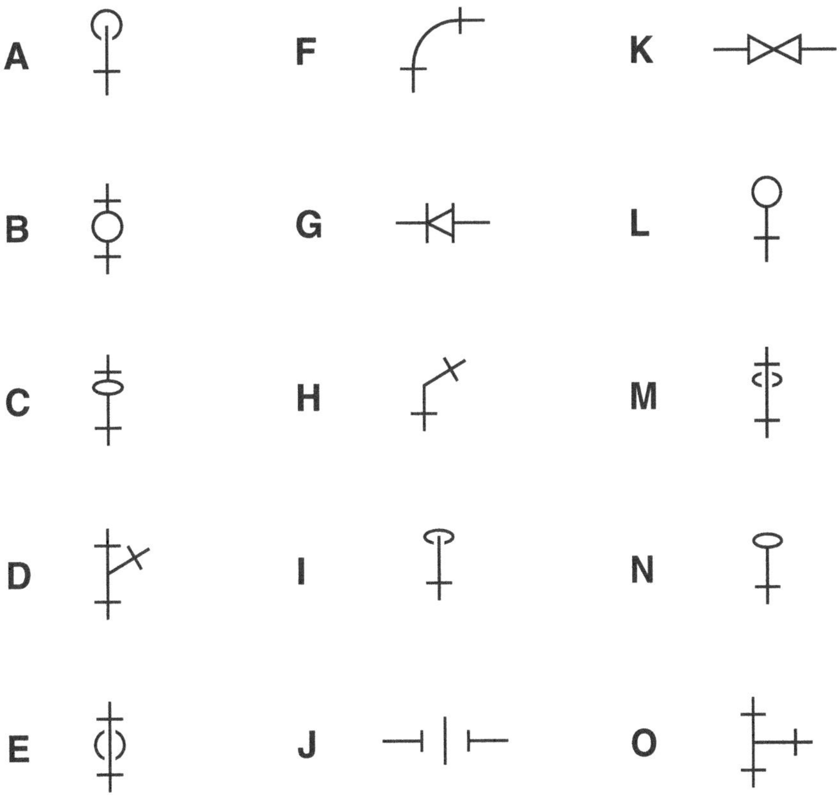

A F K

B G L

C H M

D I N

E J O

Name: ______________________ Date: ______________

PIPING SKETCHES—PART 2

1. Convert the orthographic sketch into an isometric sketch.

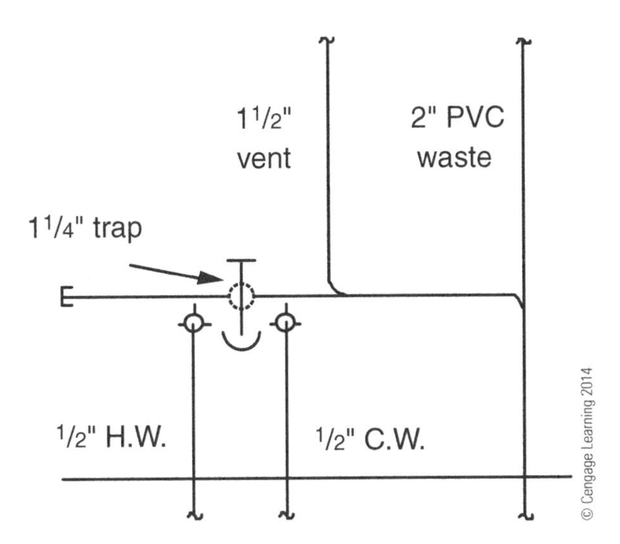

Name: ______________________ **Date:** ______________

PIPING SKETCHES—PART 3

1. Convert the isometric sketch into an elevated view orthographic sketch.

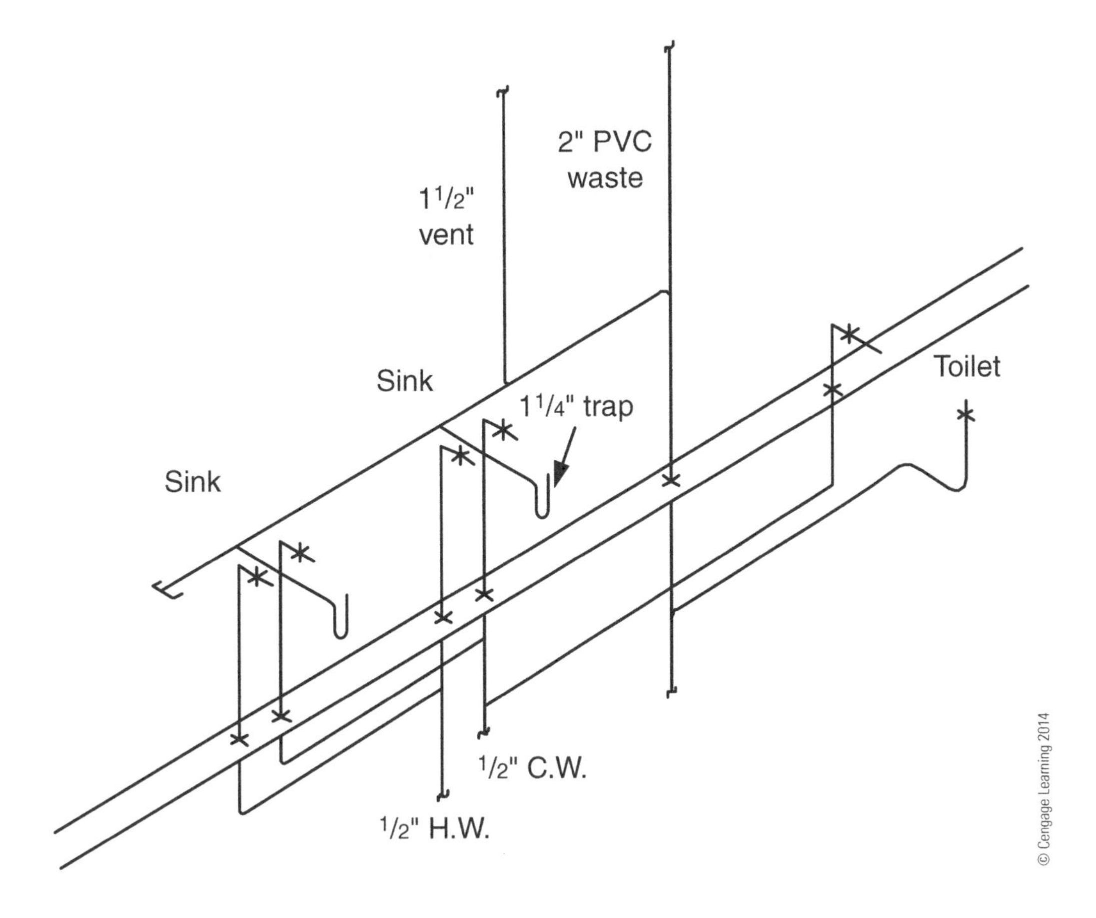

2. Convert the isometric sketch into a plan view orthographic sketch.

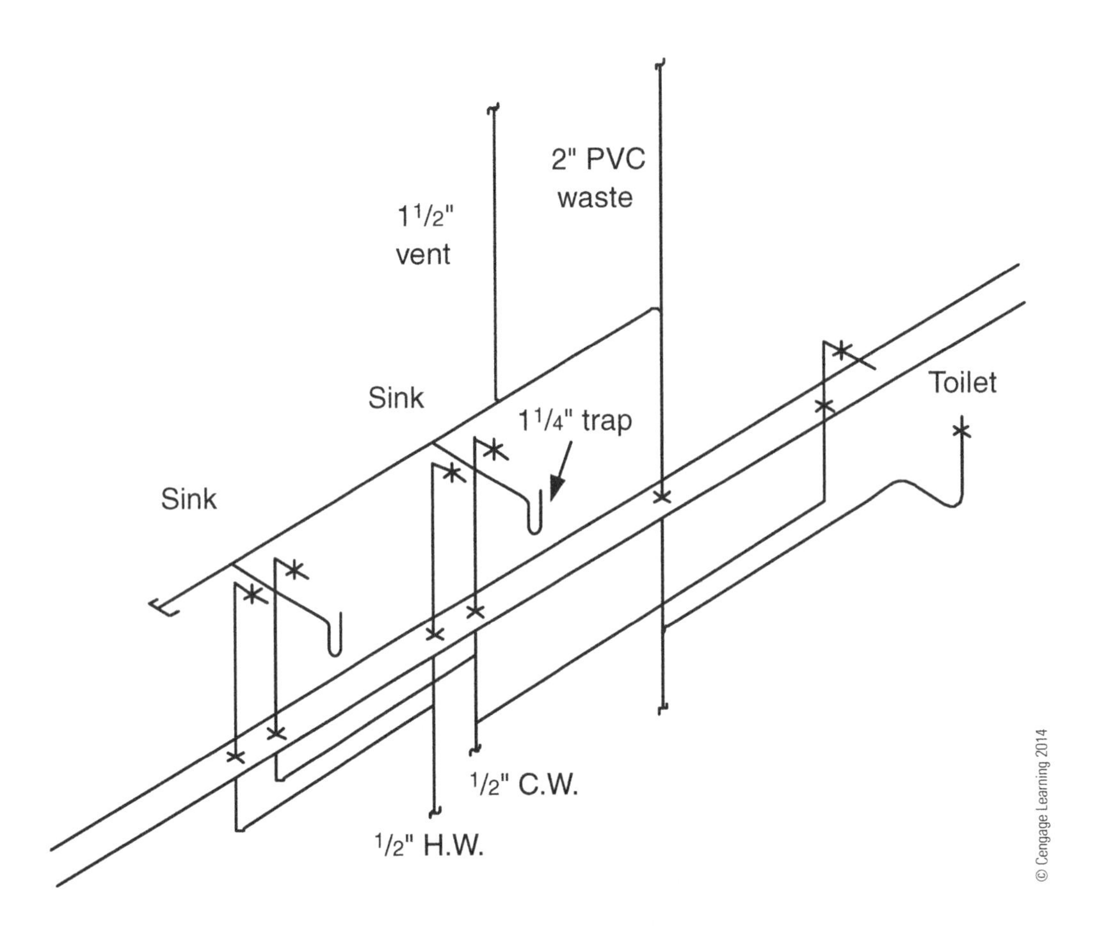

Name: ______________________ **Date:** ______________

PIPING SKETCHES—PART 4

1. Convert the bathroom builder's plan that is shown in the drawing into an isometric sketch.

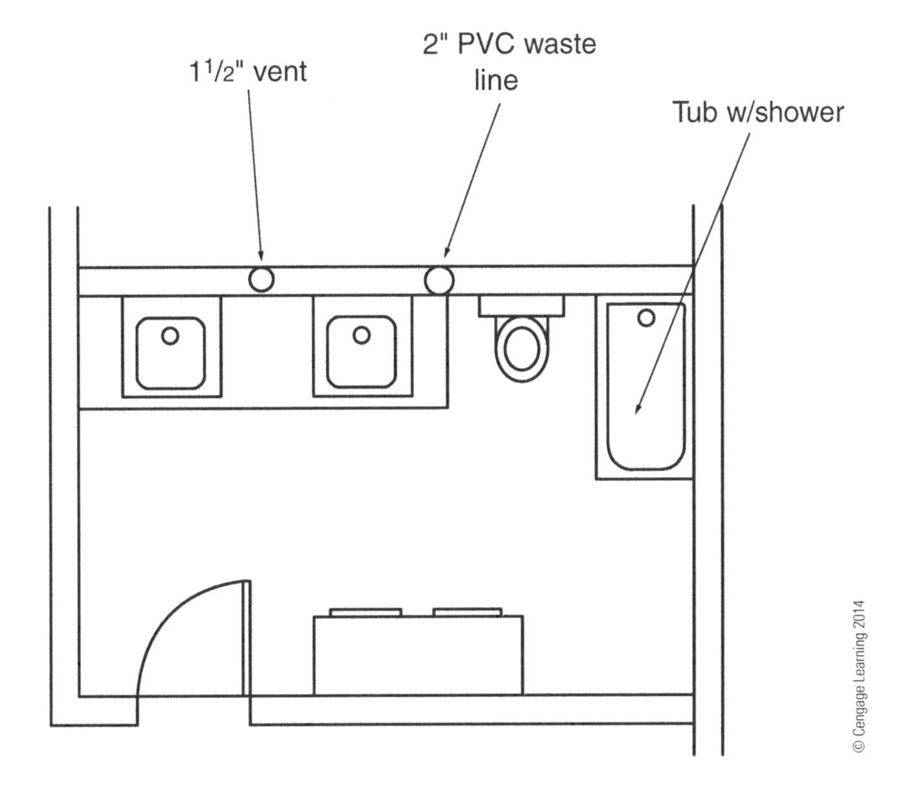

Name: ______________________ Date: ______________

PIPING SKETCHES—PART 5

1. Convert the same bathroom builder's plan that is shown in the drawing into an orthographic sketch including both the elevated view and the plan view.

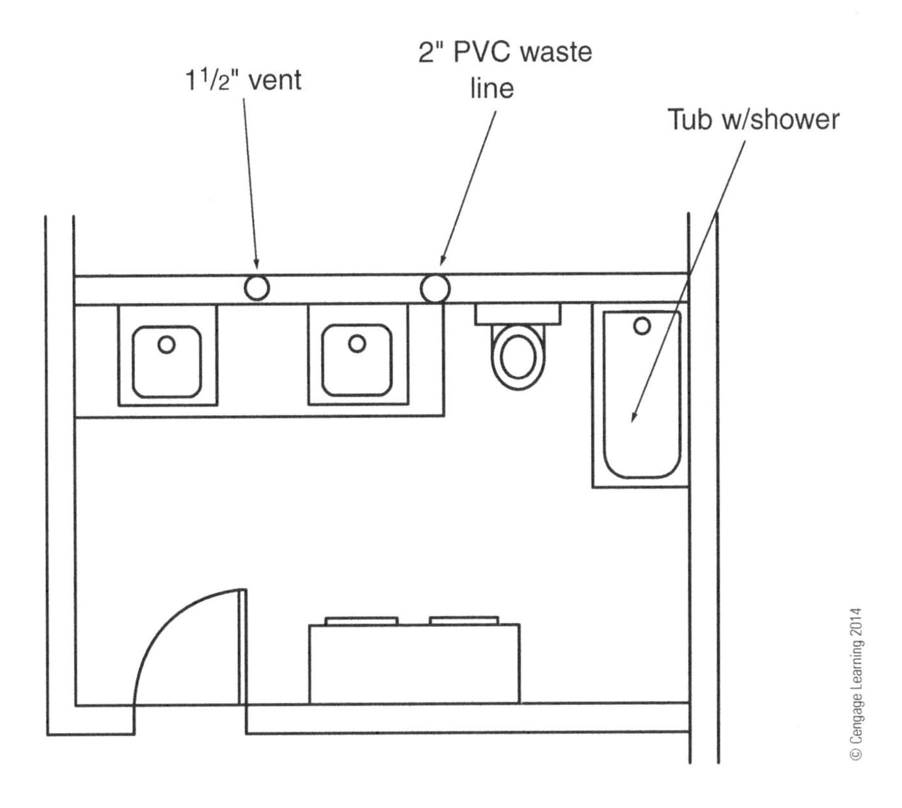

Name: ______________________________ Date: ______________________

SINGLE ACTUATORS

1. Match the actuator to the correct ANSI symbol in the drawing.

1. _______ Mechanical
2. _______ Lever
3. _______ Solenoid, spring return
4. _______ Foot pedal
5. _______ Pushbutton
6. _______ Detent
7. _______ Air pilot

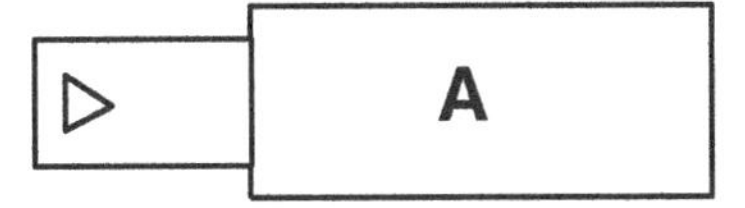

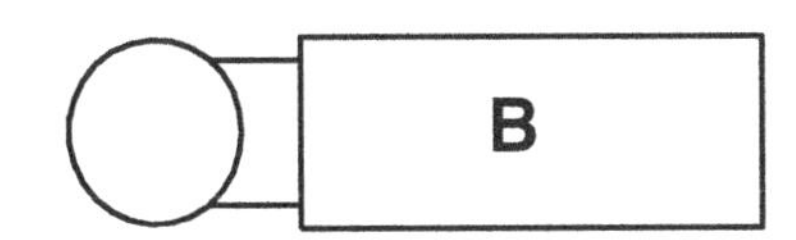

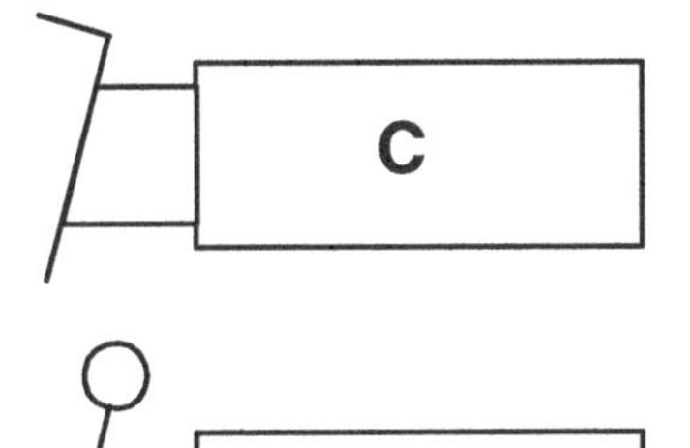

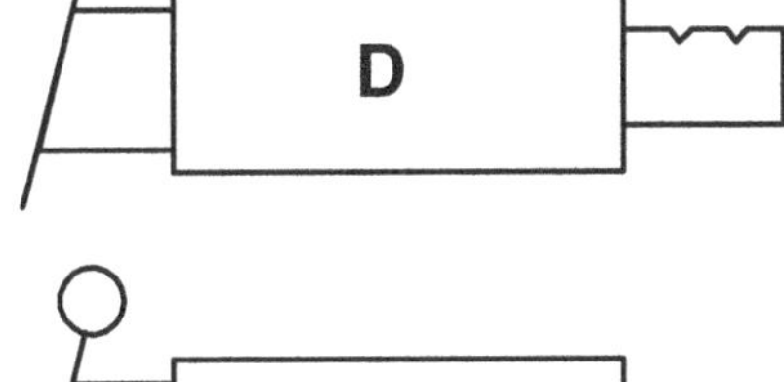

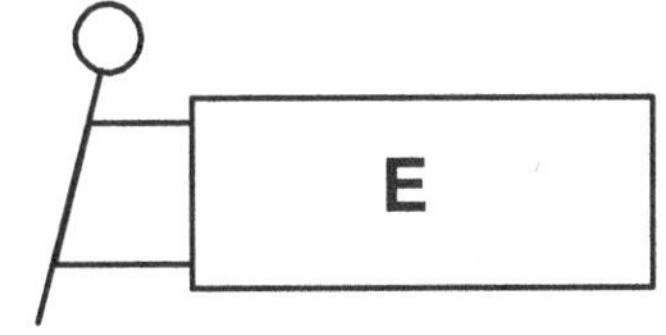

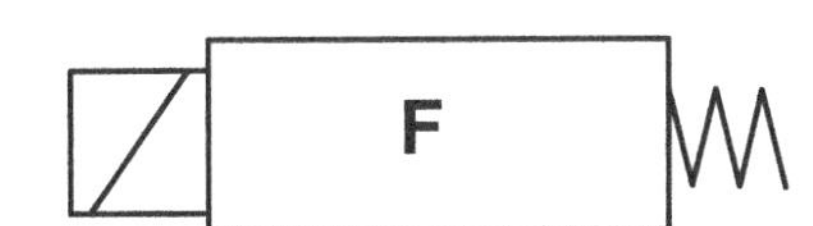

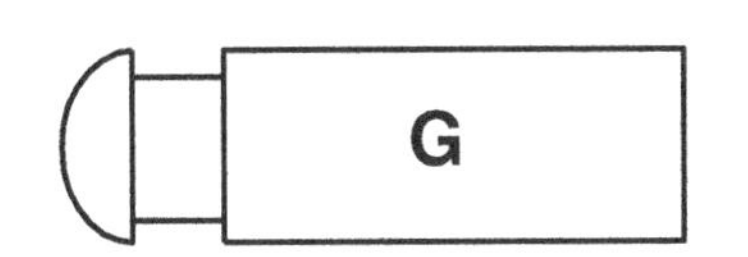

Name: ______________________________ **Date:** ____________________

ANSI SYMBOLS—PART 1

1. Select the correct symbol for each component in the list below.

_______ Check valve

_______ Double-acting, double-rod cylinder

_______ Pressure-compensated, unidirectional hydraulic pump

_______ Three-position, four-way valve

_______ Variable displacement, bidirectional pneumatic motor

_______ Unloading valve

_______ Single-acting cylinder

_______ Pressure gauge

_______ To tank

_______ Variable displacement, unidirectional hydraulic motor

_______ Flow control

_______ Double-acting cylinder

_______ Two-position, two-way, normally closed valve

_______ Variable displacement, unidirectional hydraulic pump

_______ Pressure-reducing valve

_______ Single-acting, spring-return cylinder

_______ One-way flow restrictor

_______ Two-position, four-way valve

_______ Pressure relief valve

_______ Fixed displacement, bidirectional hydraulic pump

_______ Two-position, three-way, normally opened valve

_______ Manually compensated, bidirectional hydraulic pump

_______ Variable displacement, pressure-compensated pneumatic motor

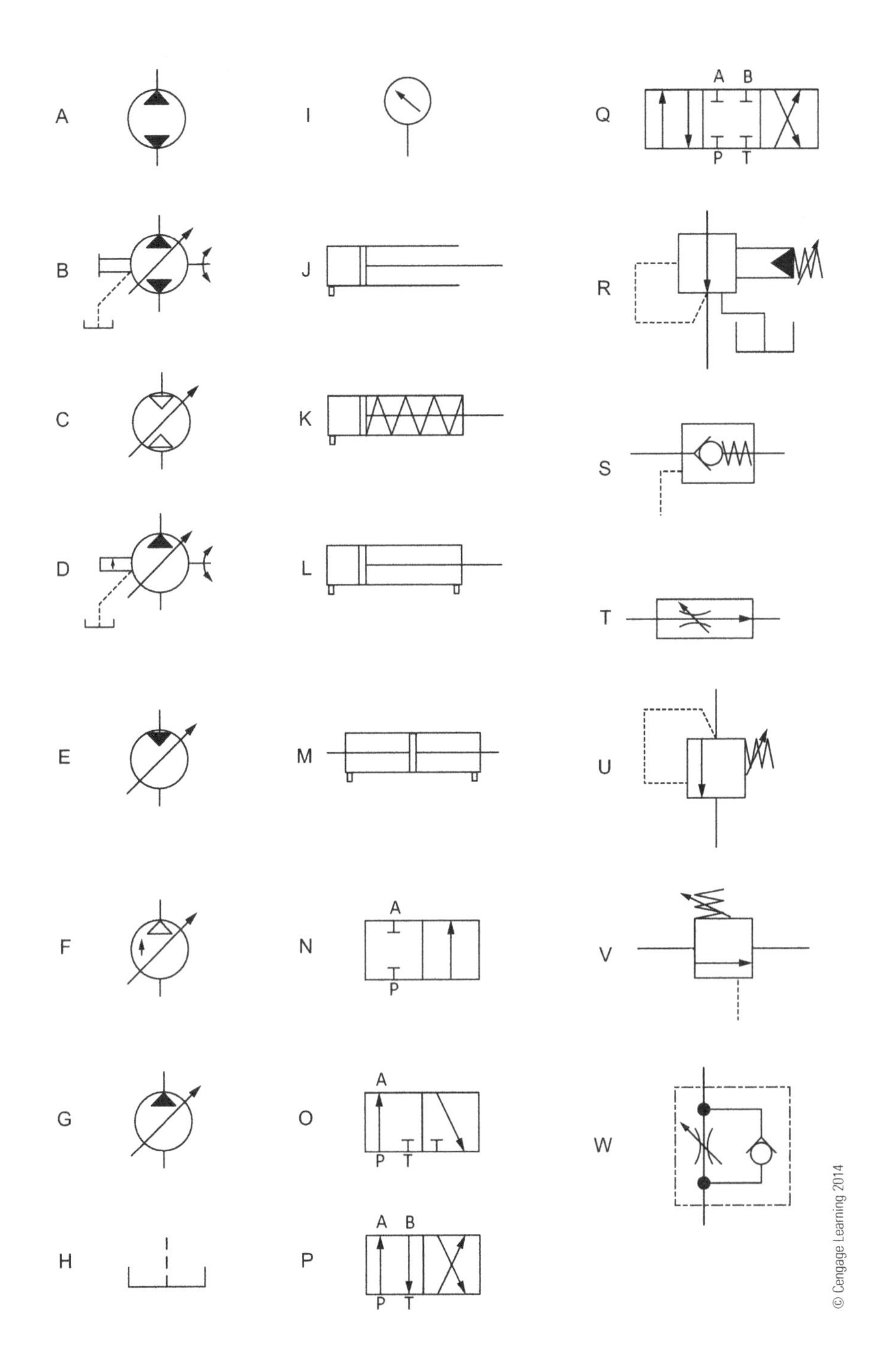
A
B
C
D
E
F
G
H
I
J
K
L
M
N
A
P
O
A
P T
P
A B
P T
Q
A B
P T
R
S
T
U
V
W
© Cengage Learning 2014

Name: ______________________________ Date: ____________________

ANSI SYMBOLS—PART 2

1. Match each ANSI symbol to the correct device.

1. _______ Pump
2. _______ Motor
3. _______ Check valve
4. _______ Air compressor
5. _______ Pressure-reducing valve
6. _______ Reservoir
7. _______ Pressure-relief valve
8. _______ Cylinder
9. _______ Bi-directional motor
10. _______ 3-way spool valve

A

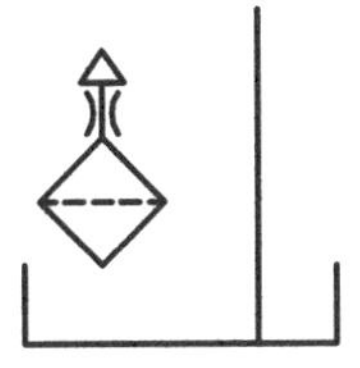

F

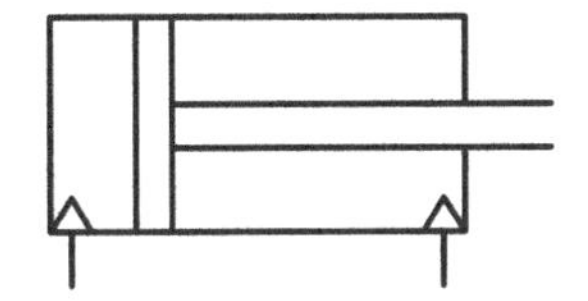

B

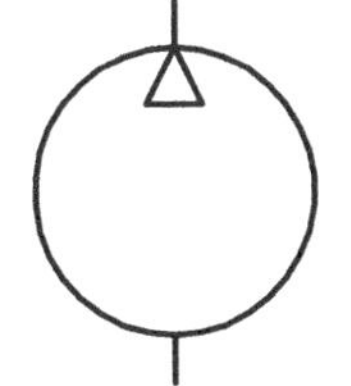

G

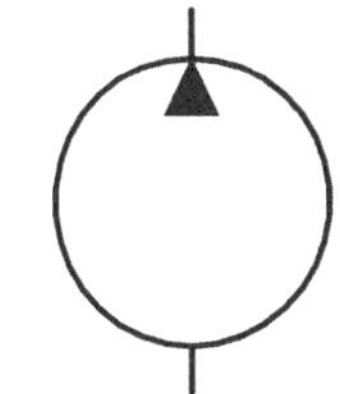

C

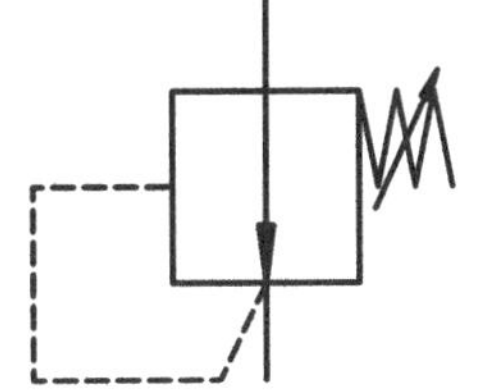

H

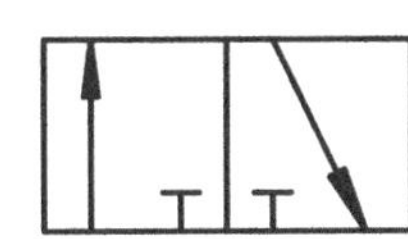

D

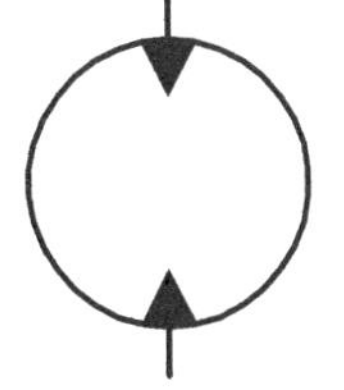

I

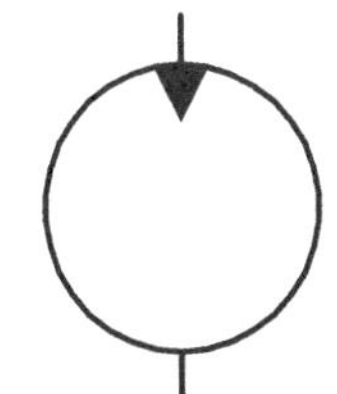

E

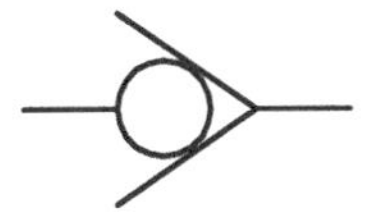

J

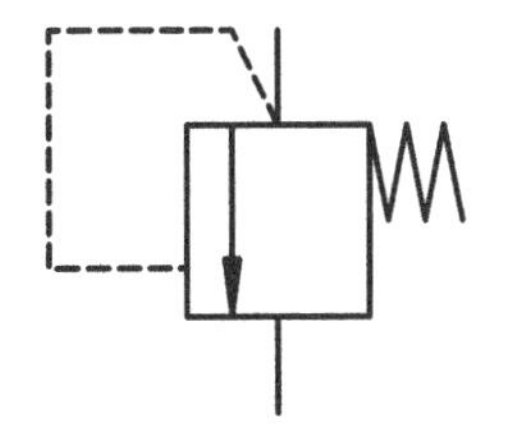

Name: ______________________________ **Date:** ____________________

ANSI SYMBOL RECOGNITION

1. Identify all of the indicated symbols in the circuit.

   a. _______ Reservoir

   b. _______ Compressor

   c. _______ Lubricator

   d. _______ Filter

   e. _______ Actuator

   f. _______ Directional control valve

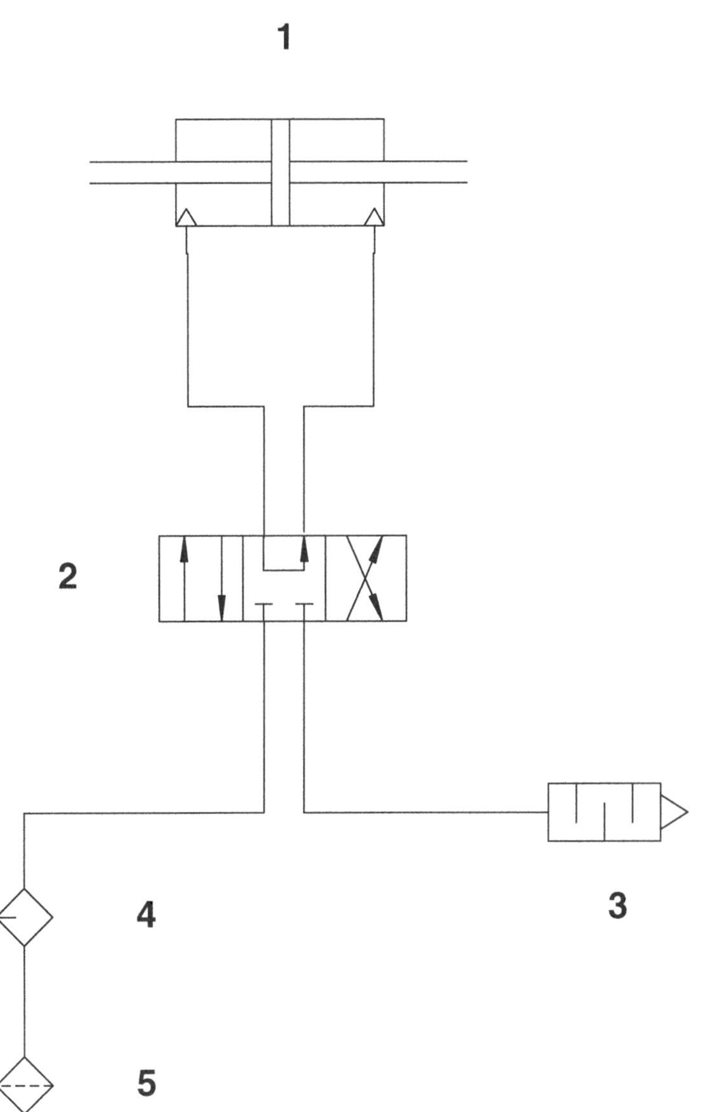

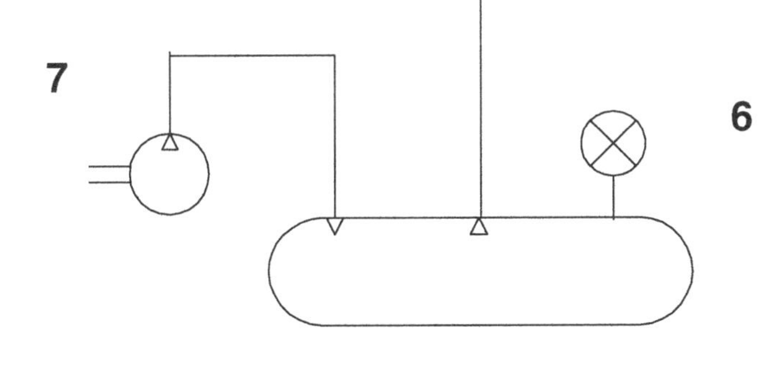

Name: ______________________ Date: ______________

MECHANICAL DRAWINGS: SCHEMATIC—PART 1

1. Name the three components that are pointed out in the drawing on page 80.

   a. ______________________

   b. ______________________

   c. ______________________

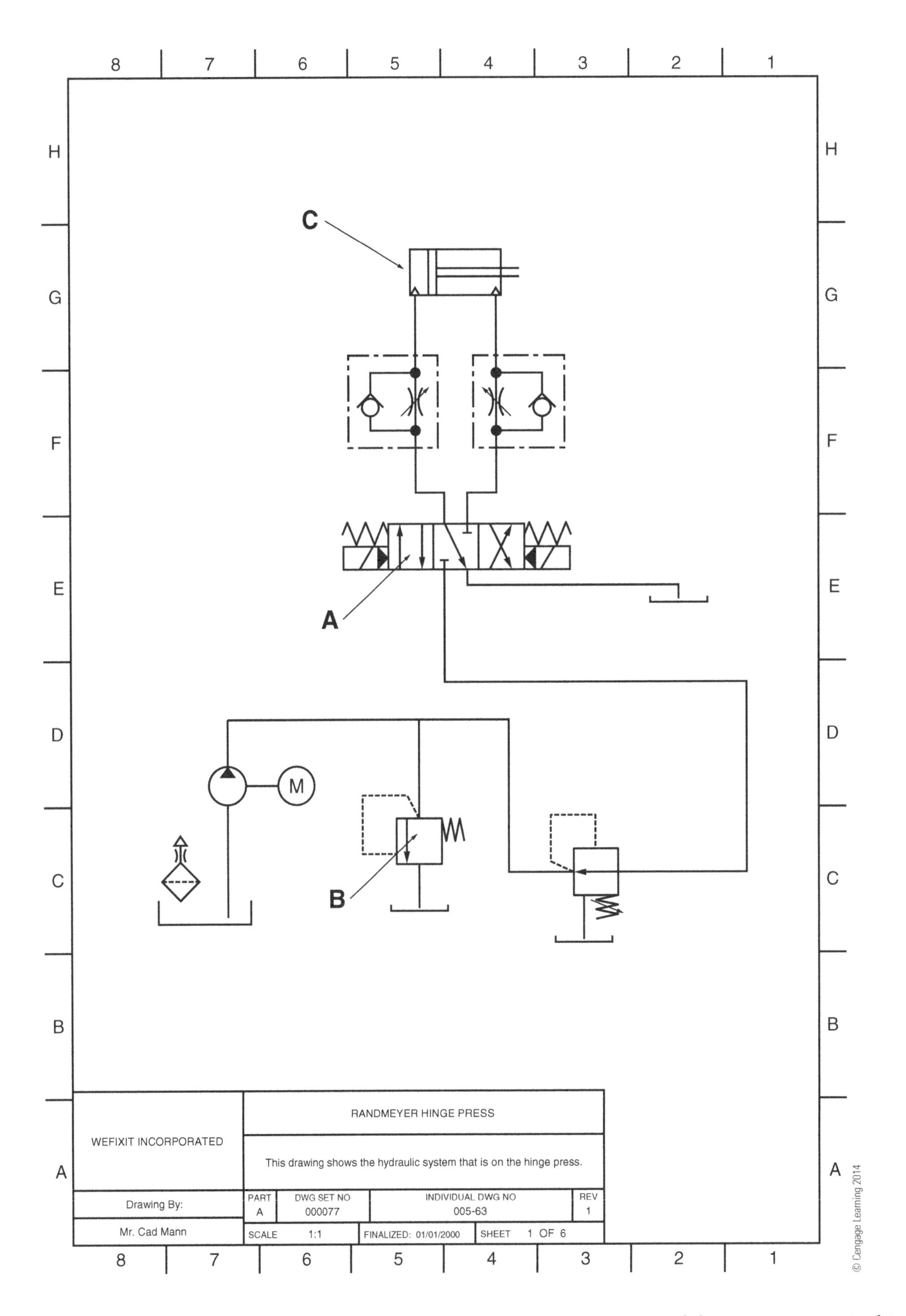

8
7
6
5
4
3
2
1
H
G
F
E
D
C
B
A
C
A
B
M
RANDMEYER HINGE PRESS
WEFIXIT INCORPORATED
This drawing shows the hydraulic system that is on the hinge press.
Drawing By:
Mr. Cad Mann
PART
A
DWG SET NO
000077
INDIVIDUAL DWG NO
005-63
REV
1
SCALE 1:1
FINALIZED: 01/01/2000
SHEET 1 OF 6

Name: ______________________ Date: ______________

MECHANICAL DRAWINGS: SCHEMATIC—PART 2

1. Connect the symbols to create a schematic that represents the pictorial diagram on page 82.

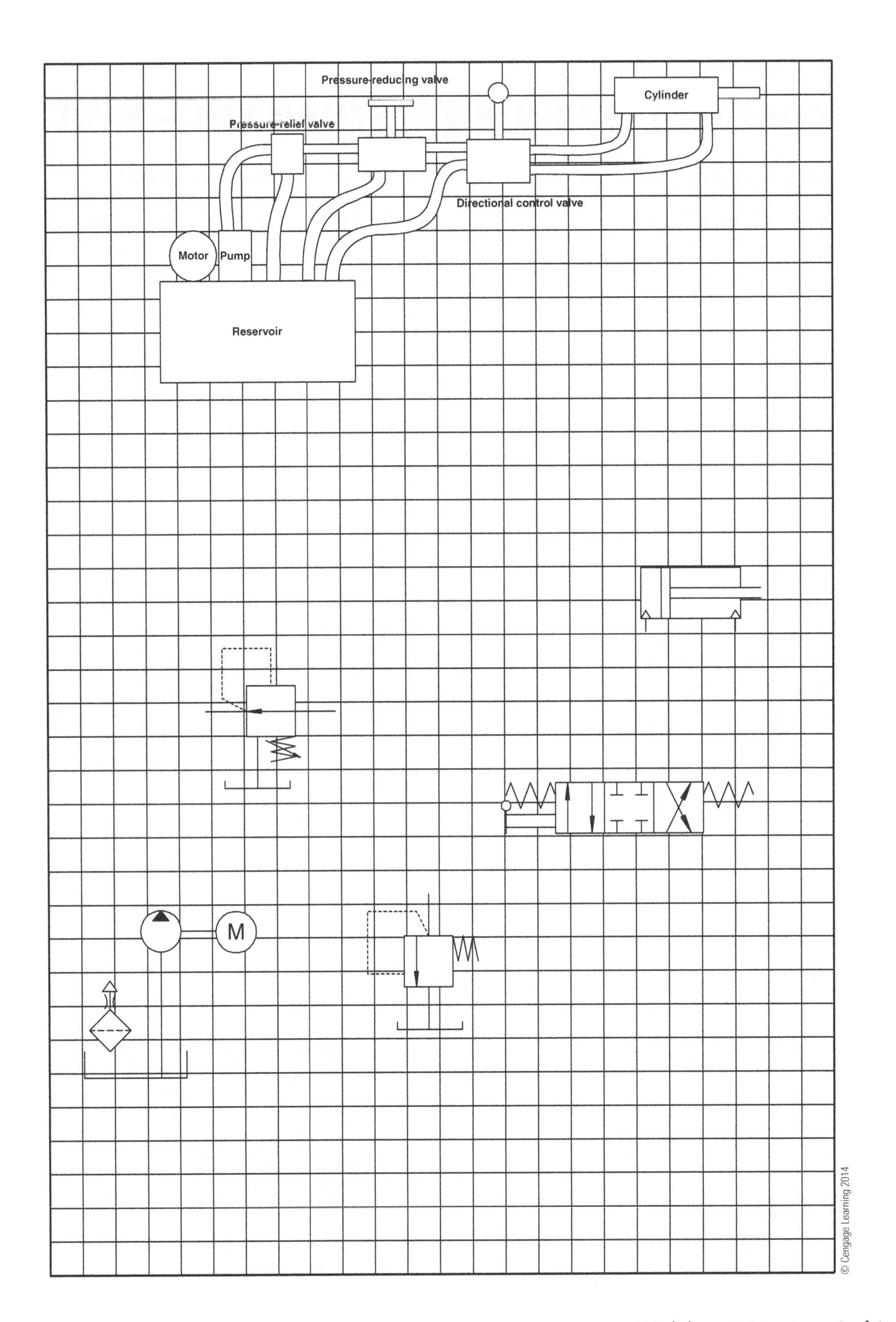
Pressure-reducing valve
Cylinder
Pressure-relief valve
Directional control valve
Motor
Pump
Reservoir
M

Name: ____________________

Date: ____________________

MECHANICAL DRAWINGS: SCHEMATIC—PART 3

1. Connect the components that are in the figure according to the schematic that has been provided on page 84.

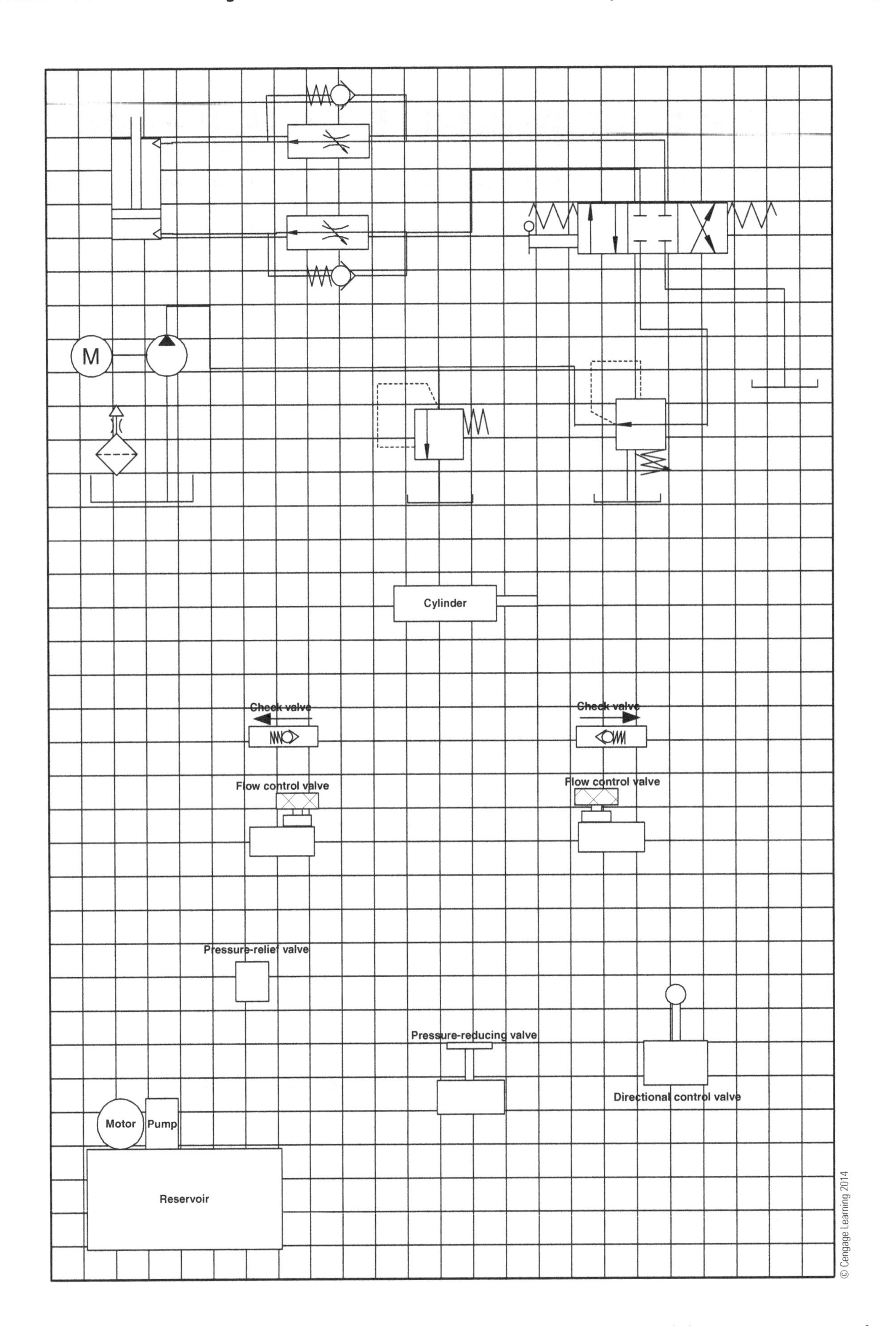

M
Cylinder
Check valve
Check valve
Flow control valve
Flow control valve
Pressure-relief valve
Pressure-reducing valve
Directional control valve
Motor
Pump
Reservoir

Name: ______________________________ Date: ____________________

MECHANICAL DRAWINGS: SCHEMATIC—PART 4

1. Draw the cylinder, control valve, pump, and pressure-reducing valve in the schematic on page 86, to accomplish the following requirements.

   a. Fluid flows from the reservoir, through the pump, and is supplied to the rest of the system.

   b. The pressure-reducing valve is used to lower the operating pressure prior to the control valve.

   c. The control valve must control the extension and retraction of the cylinder.

   d. The cylinder is connected to the ram.

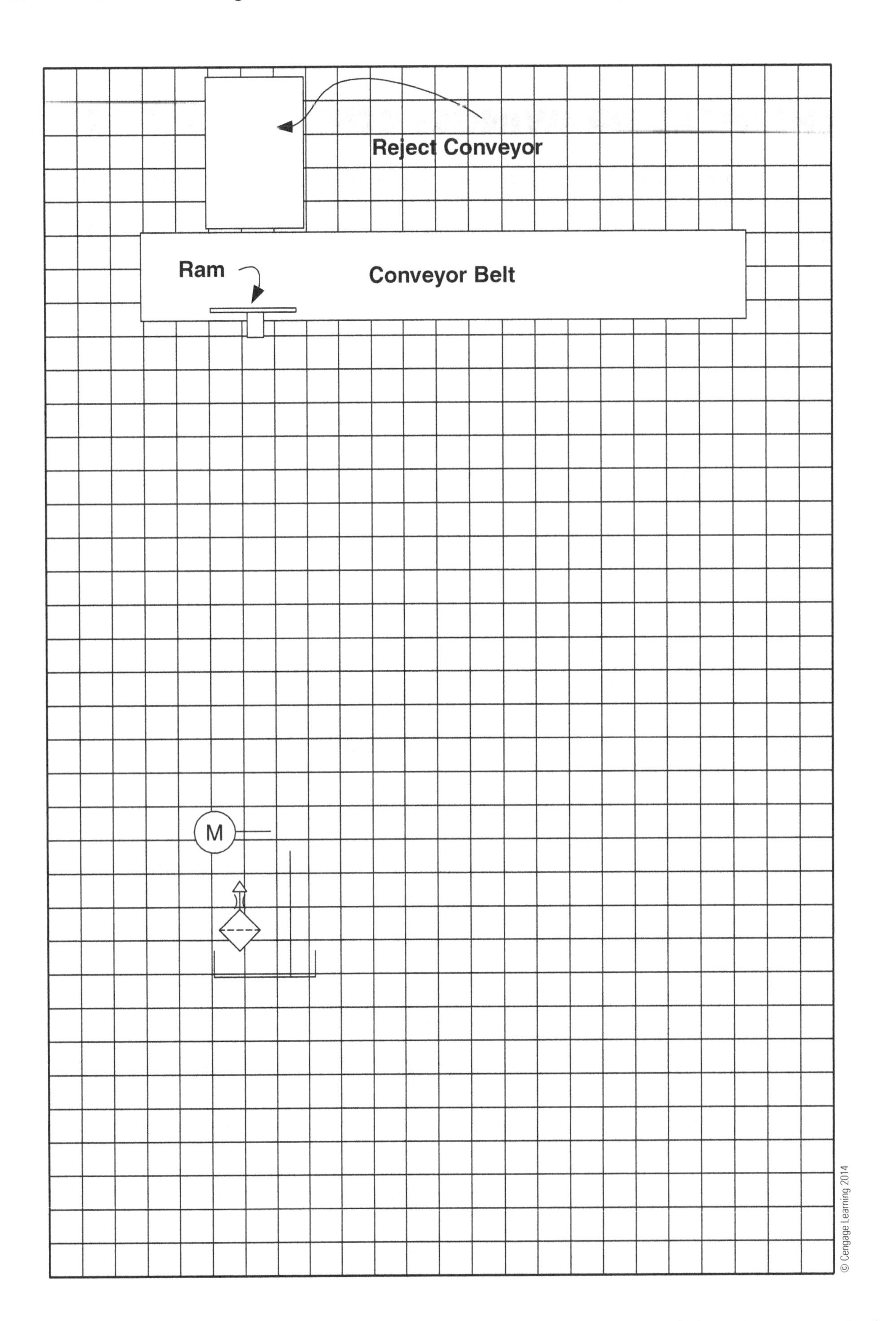
Reject Conveyor
Ram
Conveyor Belt
M
© Cengage Learning 2014

Name: ______________________________ Date: ____________________

ELECTRICAL SYMBOLS

1. Match the name to the appropriate symbol.

   _______ Chassis ground

   _______ Battery

   _______ Lowest circuit potential

   _______ DC power source

   _______ Highest circuit potential

   _______ AC power source

   _______ Ground

A 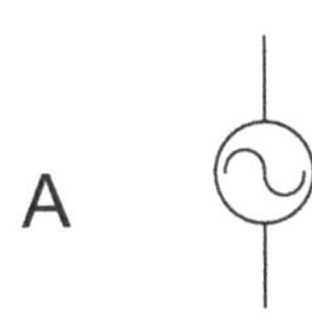

E 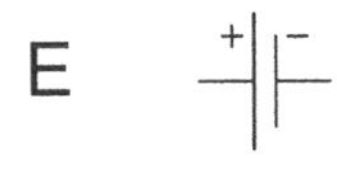

B 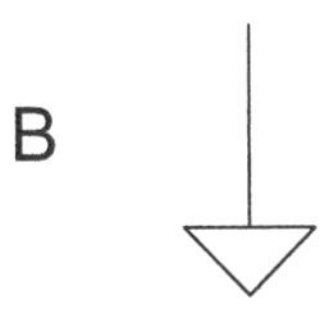

F 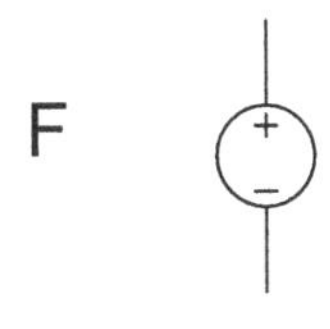

C 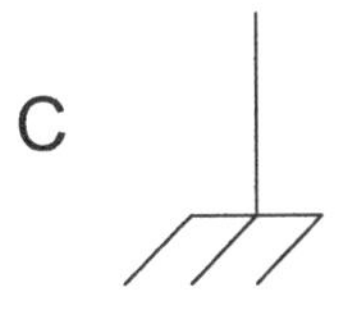

G 

D 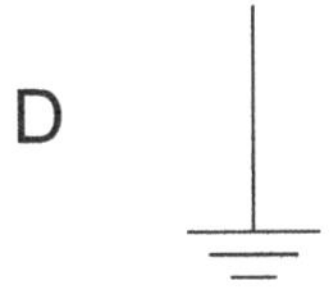

2. Match the name to the appropriate symbol (a lettered symbol may be used more than once).

   _______ Fuse

   _______ Circuit breaker

   _______ Overload

   _______ Heater

A 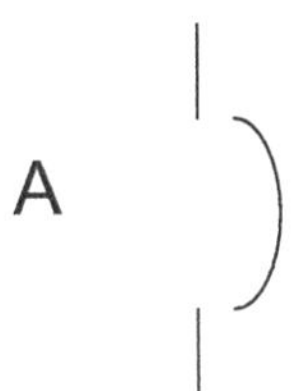

D 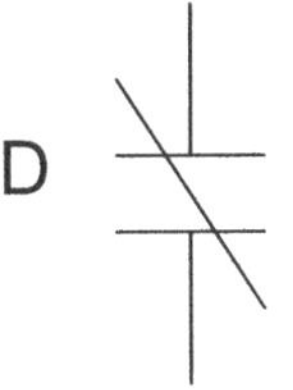

B 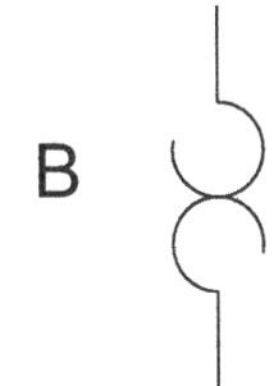

E

C 

3. Match the name to the appropriate symbol.

_______ SPST—NO

_______ SPST—NC

_______ SPDT—center off

_______ SPDT—1 NO, 1 NC

_______ DPDT—NO

_______ DPDT—NC

_______ DPDT—center off

_______ DPDT—1 NO–1 NC

A

B

C

D

E

F

G

H

4. Match the name to the appropriate symbol.

_______ Pushbutton—NO

_______ Pushbutton—NC

_______ Pushbutton—1 NO, 1 NC

_______ Selector switch—4 position—break before make

_______ Selector switch—4 position—make before break

_______ Foot switch—NO

A

D

B

E

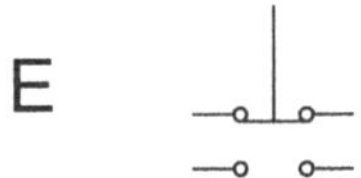

C

F

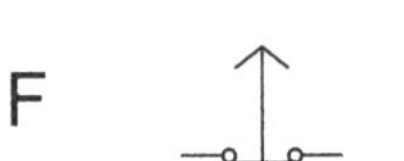

5. Match the name to the appropriate symbol.

_______ Limit switch—NO

_______ Limit switch—NC

_______ Limit switch—NO held closed

_______ Limit switch—NC held open

_______ Proximity switch—NO

_______ Proximity switch—NC

_______ Flow switch—NO

_______ Flow switch—NC

_______ Liquid level switch—NO

_______ Liquid level switch—NC

_______ Pressure switch—NO

_______ Pressure switch—NC

_______ Temperature switch—NO

_______ Temperature switch—NC

A

H 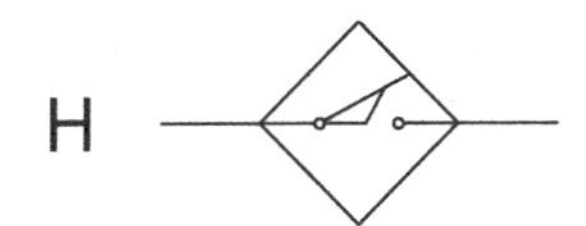

B

I 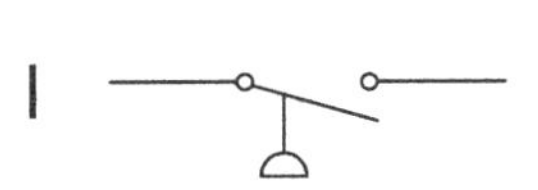

C

J 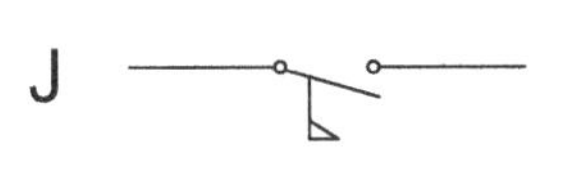

D

K 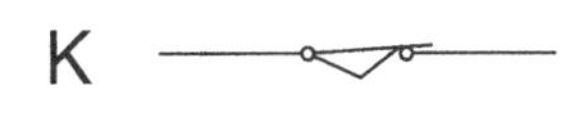

E

L 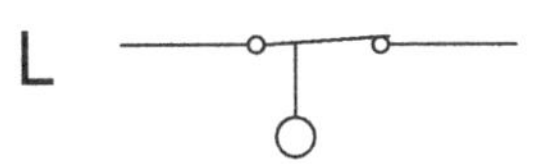

F

M 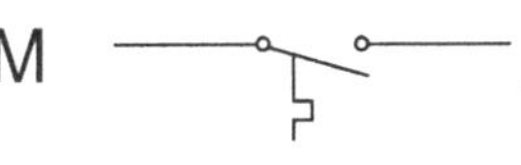

G

N 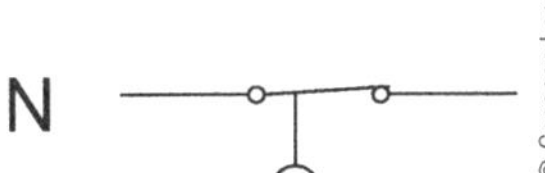

6. Match the name to the appropriate symbol.

_______ Contact—NO

_______ Contact—NC

_______ Relay coil

_______ Time-delay relay coil

_______ Time delay—NOTC

_______ Time delay—NCTO

_______ Time delay—NOTO

_______ Time delay—NCTC

_______ DC motor

_______ AC motor

_______ DC generator

_______ AC generator (alternator)

_______ DC motor-generator set

_______ AC motor-generator set

_______ Series field

_______ Shunt field

A

G 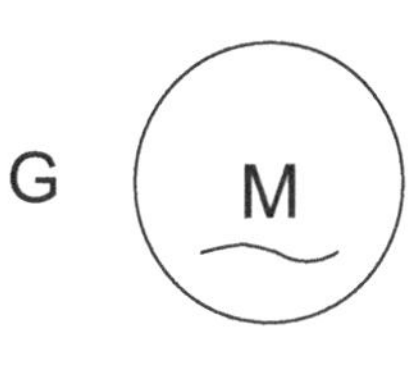

M

CR

B 

H

N 

C

I

O

MG

D

J 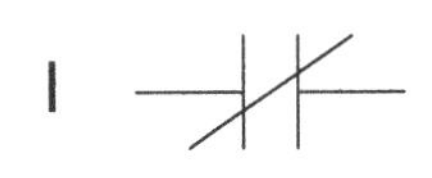

P

E

G

K

F 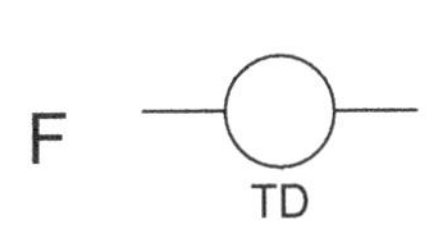

L

M

7. Match the name to the appropriate symbol (lettered symbols may be used more than once).

_______ Resistor—fixed

_______ Resistor—variable

_______ Potentiometer

_______ Capacitor—fixed

_______ Capacitor—variable

A 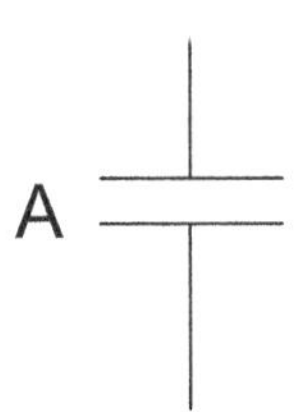

E 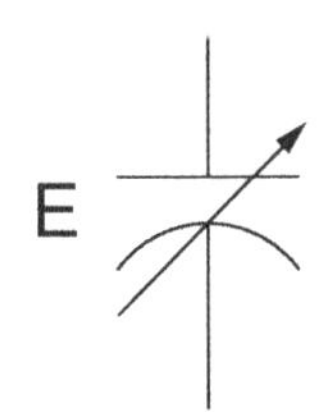

I 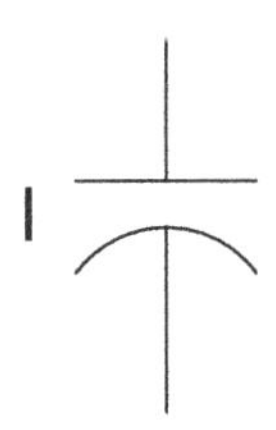

B

F 

J 

C 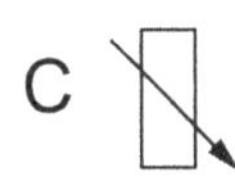

G 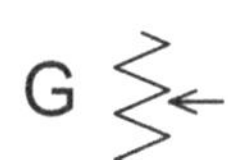

D 

H 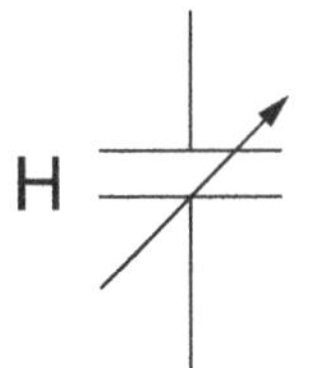

8. Match the name to the appropriate symbol (lettered symbols may be used more than once).

_______ Inductor—air core—fixed

_______ Inductor—air core—variable

_______ Inductor—iron core—fixed

_______ Inductor—iron core—variable

_______ Inductor—iron core—multiple taps

_______ Transformer—air core—fixed

_______ Transformer—air core—variable

_______ Transformer—iron core—fixed

_______ Transformer—iron core—variable

_______ Transformer

_______ Transformer—variable

_______ Current transformer

_______ Potential transformer

A 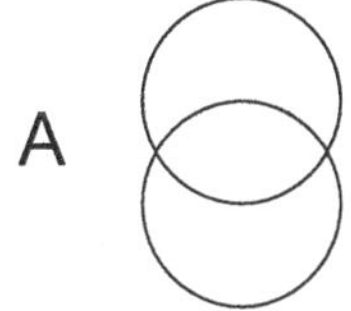

I 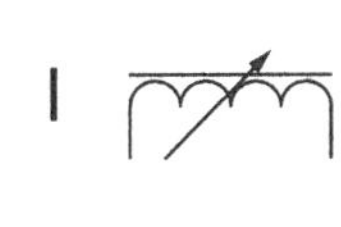

B 

J 

C

K 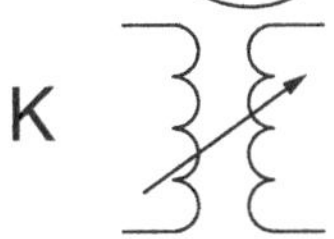

D

L 

E

M 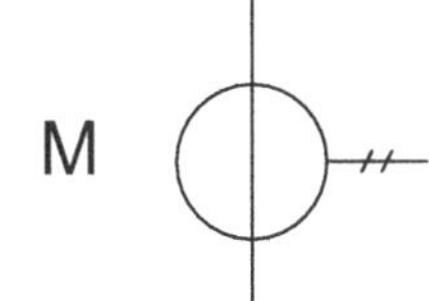

F

N 

G

O 

H

9. Match the name to the appropriate symbol.

_______ Diode

_______ Zener diode

_______ Tunnel diode

_______ Light-emiting diode (LED)

_______ Photo diode

_______ Varactor diode

_______ Bridge rectifier

_______ NPN transistor

_______ PNP transistor

_______ Darlington transistor—NPN type

_______ Darlington transistor—PNP type

_______ Unijunction transistor (UJT)

A 

G 

B 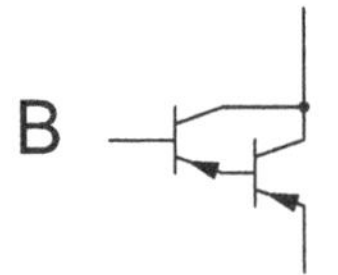

H 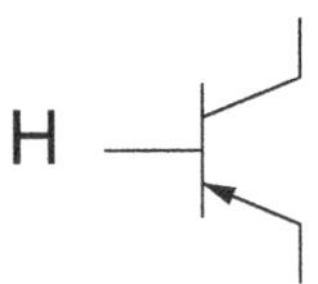

C 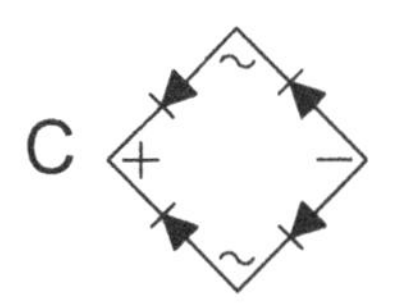

I 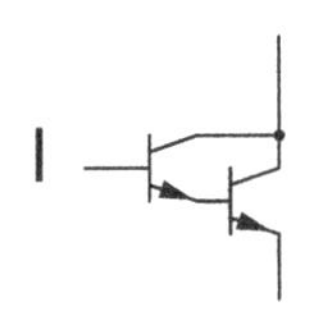

D 

J 

E 

K 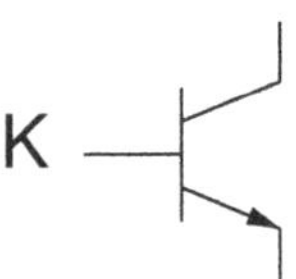

F 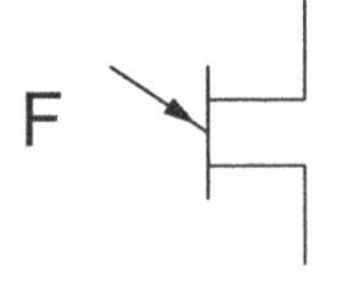

L 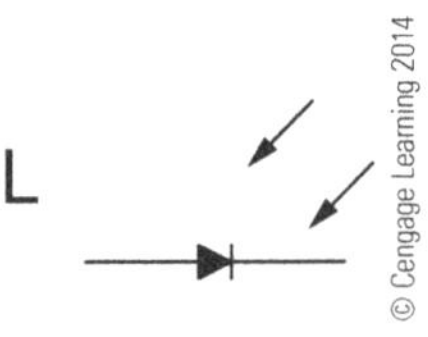

10. Match the name to the appropriate symbol.

_______ Junction field-effect transistor (JFET)—P channel

_______ Junction field-effect transistor (JFET)—N channel

_______ E-MOSFET—P channel

_______ E-MOSFET—N channel

_______ DE-MOSFET—P channel

_______ DE-MOSFET—N channel

_______ Diac

_______ Silicon-controlled rectifier (SCR)

_______ Silicon-controlled switch (SCS)

_______ Triac

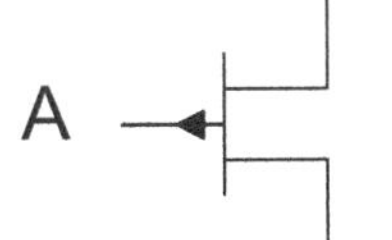

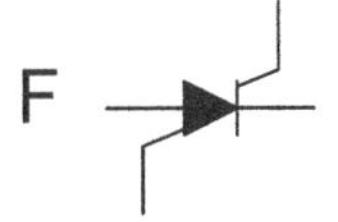

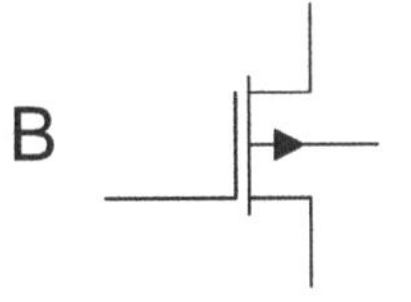

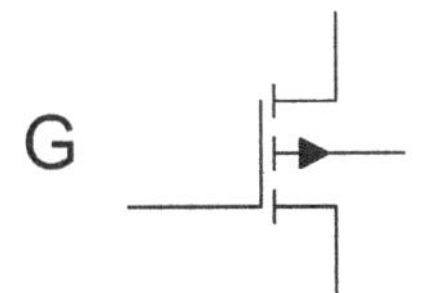

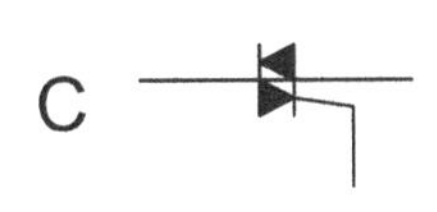

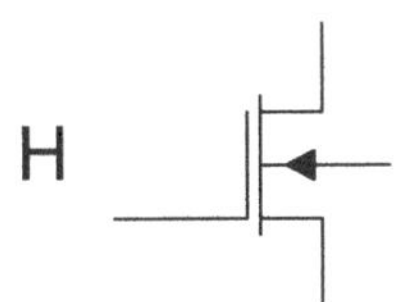

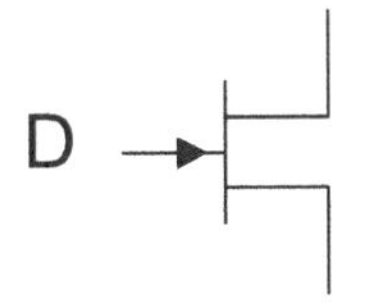

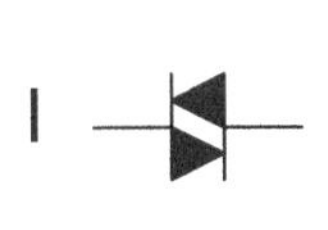

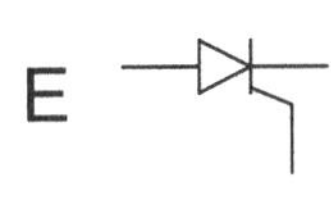

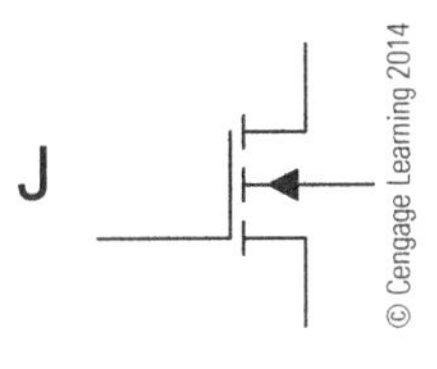

11. Match the name to the appropriate symbol.

_______ Op-amp

_______ Buffer amplifier

_______ Inverter

_______ AND gate

_______ NAND gate

_______ OR gate

_______ NOR gate

_______ XOR gate

_______ XNOR gate

_______ D type flip-flop

_______ Latch flip-flop

_______ J–K flip-flop

_______ R–S flip-flop

A 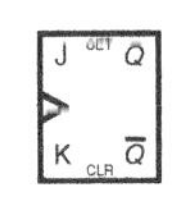

F 

K 

B

G 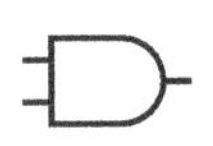

L 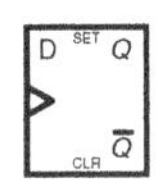

C 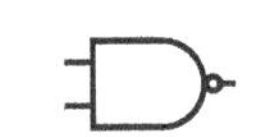

H 

M 

D 

I 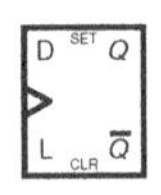

E

J

Name: ______________________ Date: ______________

SINGLE-LINE DRAWINGS

1. Given the single-line drawing, identify the following items.

   a. ______________________

   b. ______________________

   c. ______________________

   d. ______________________

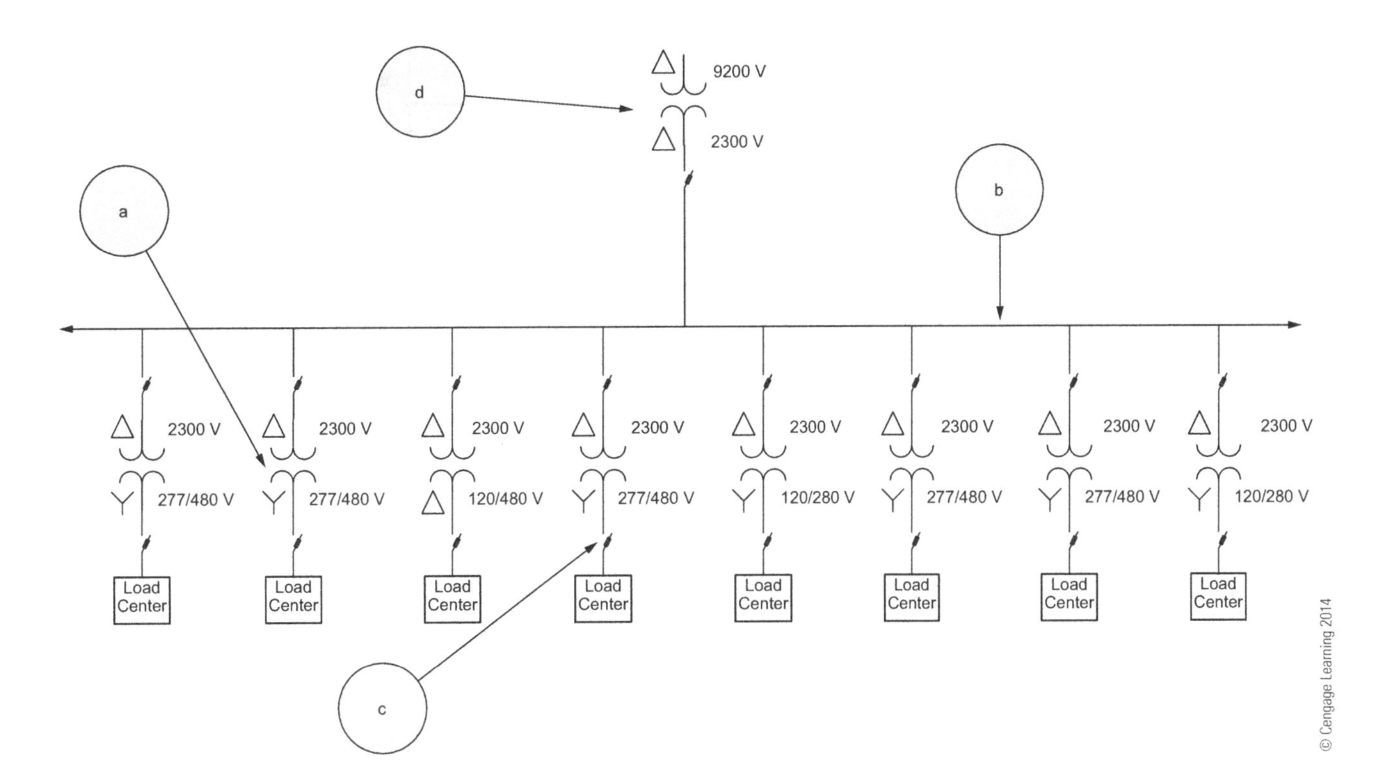

2. Convert the given pictorial diagram to a single-line drawing.

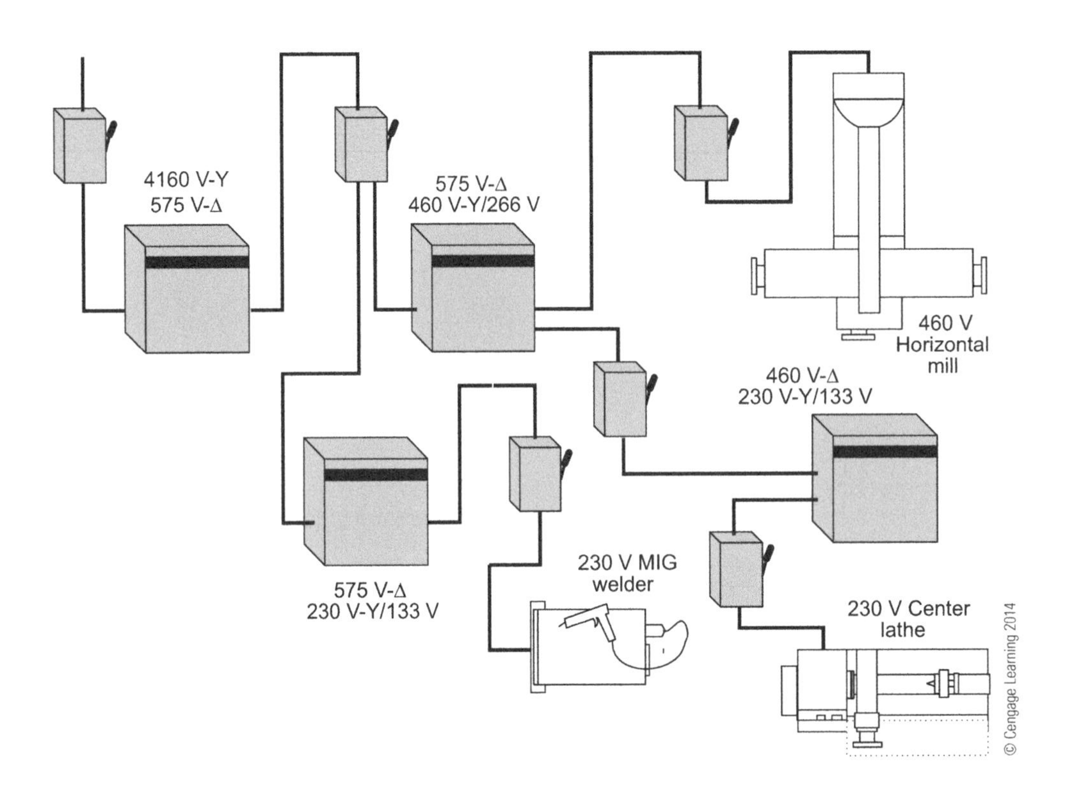

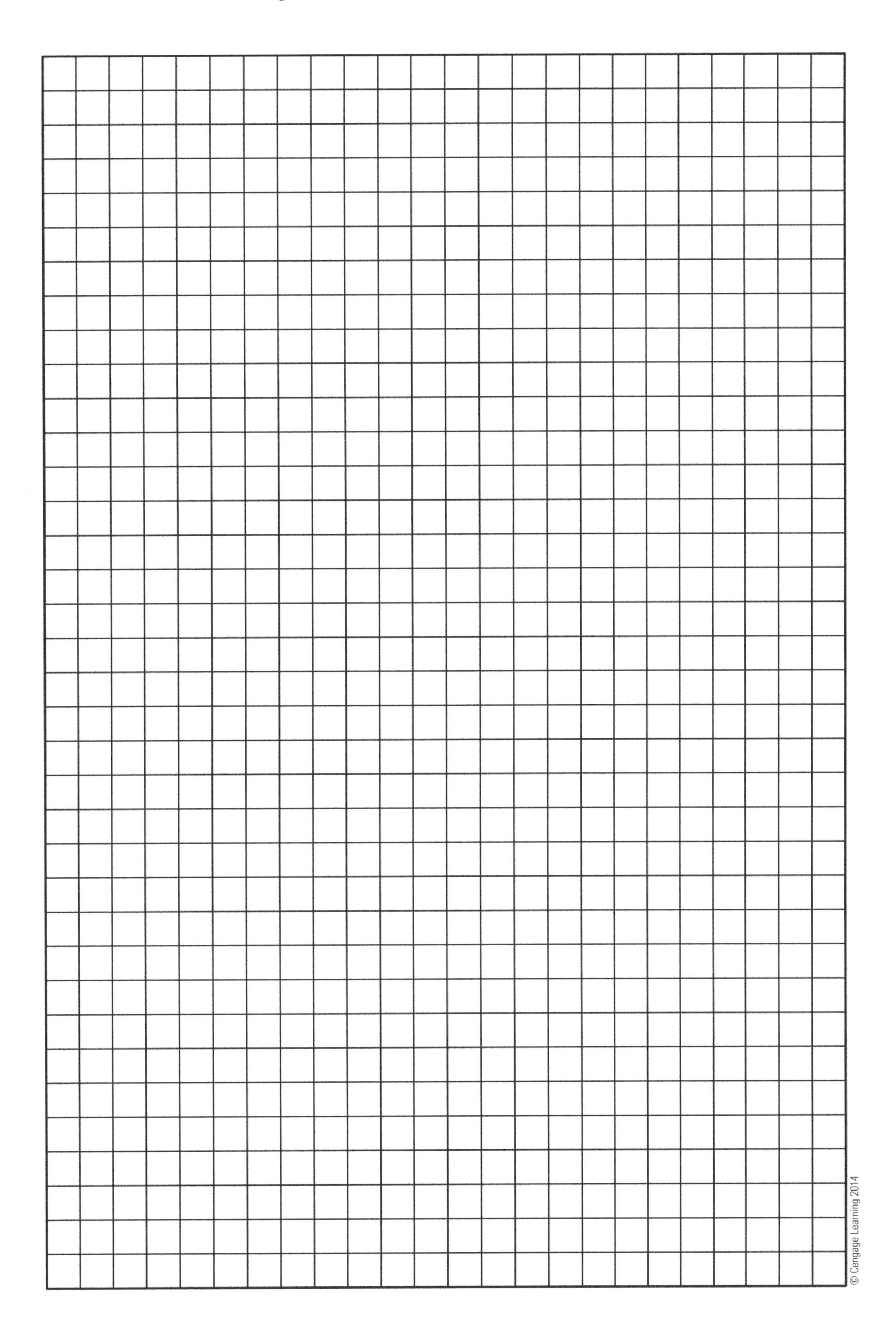

Name: ______________________ Date: ______________

PICTORIAL DIAGRAMS

1. Given the pictorial diagram, identify the following items.

   a. ______________________

   b. ______________________

   c. ______________________

   d. ______________________

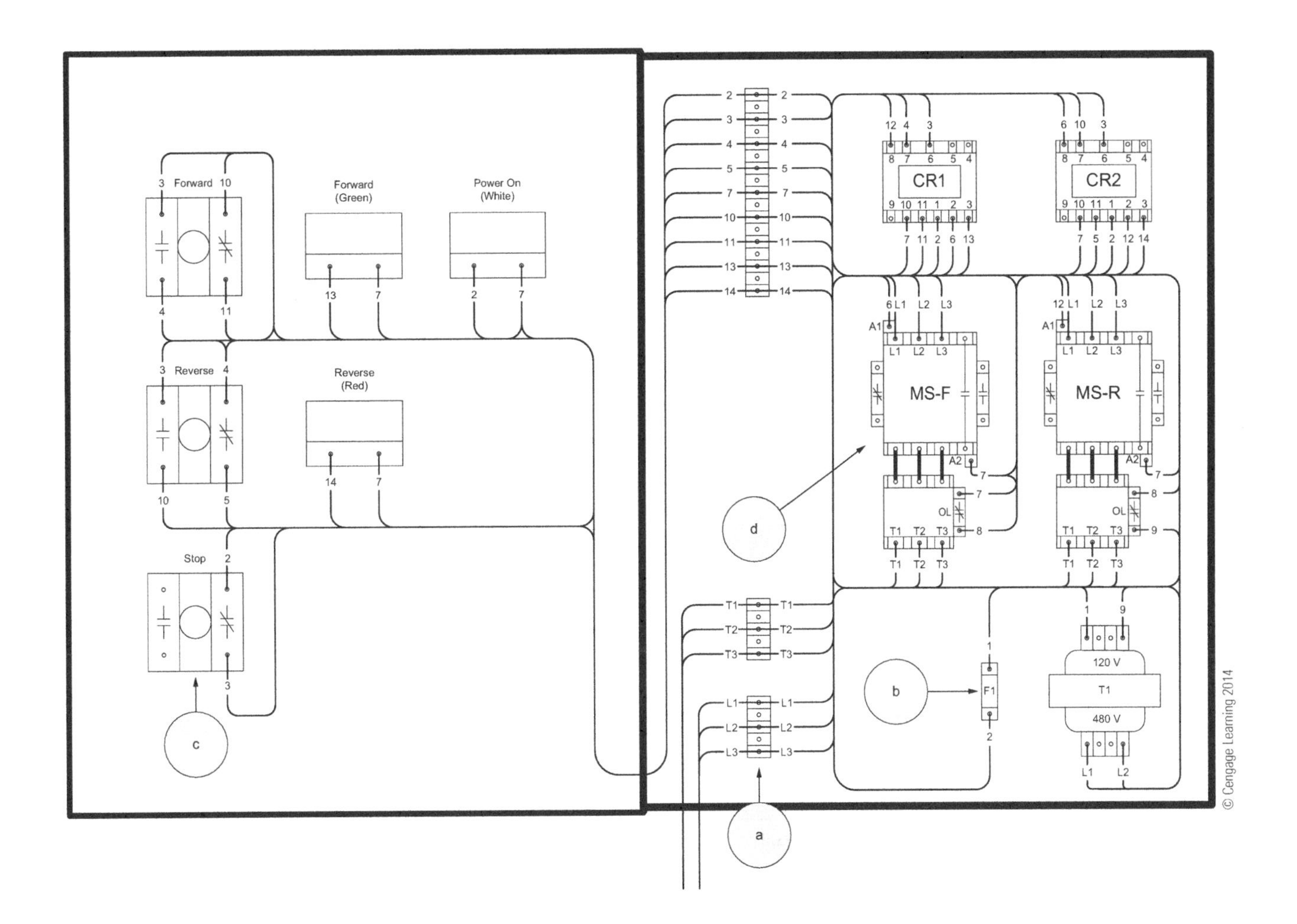

2. Convert the given schematic diagram to a pictorial diagram.

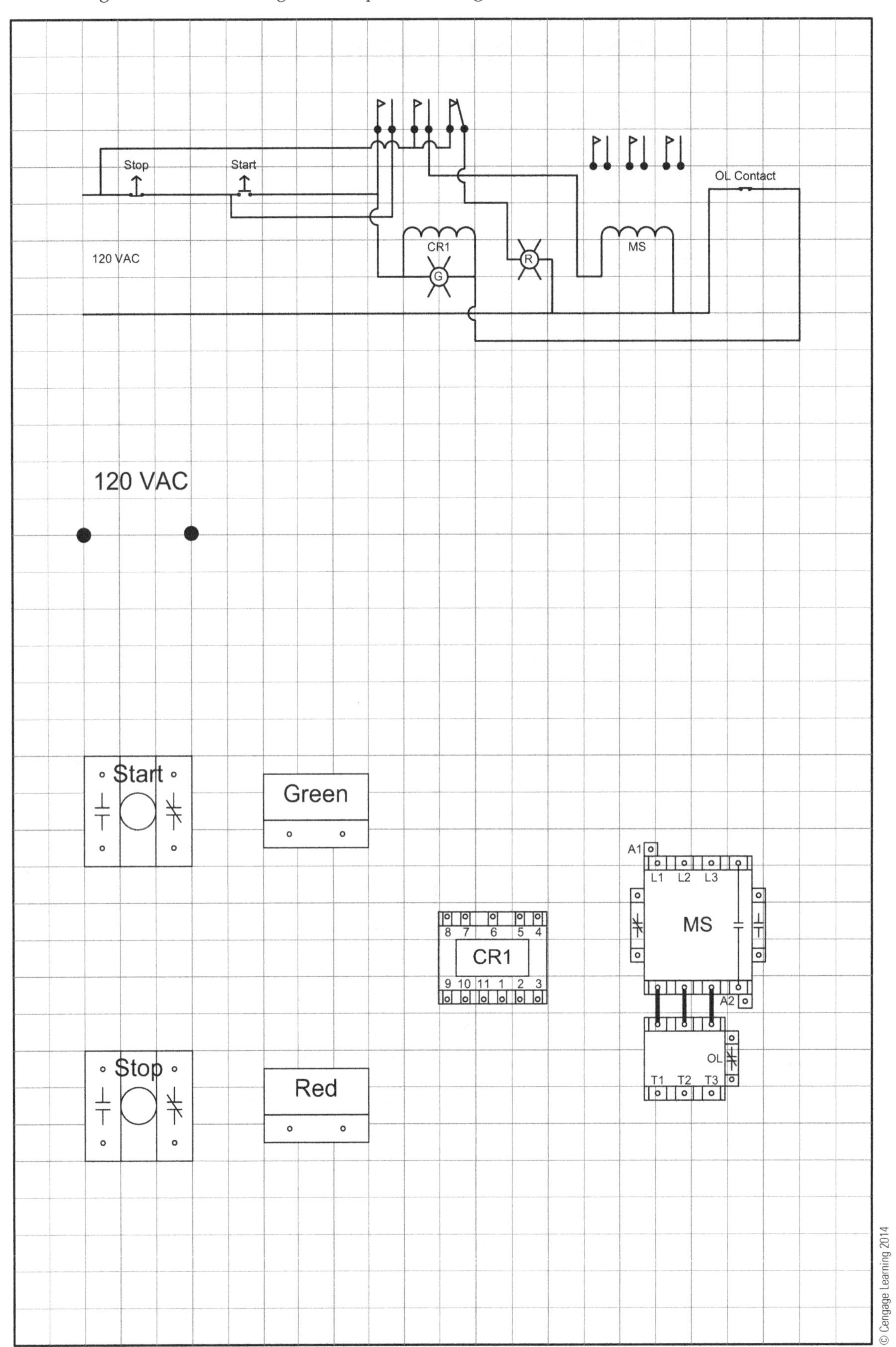

3. Convert the given ladder diagram to a pictorial diagram.

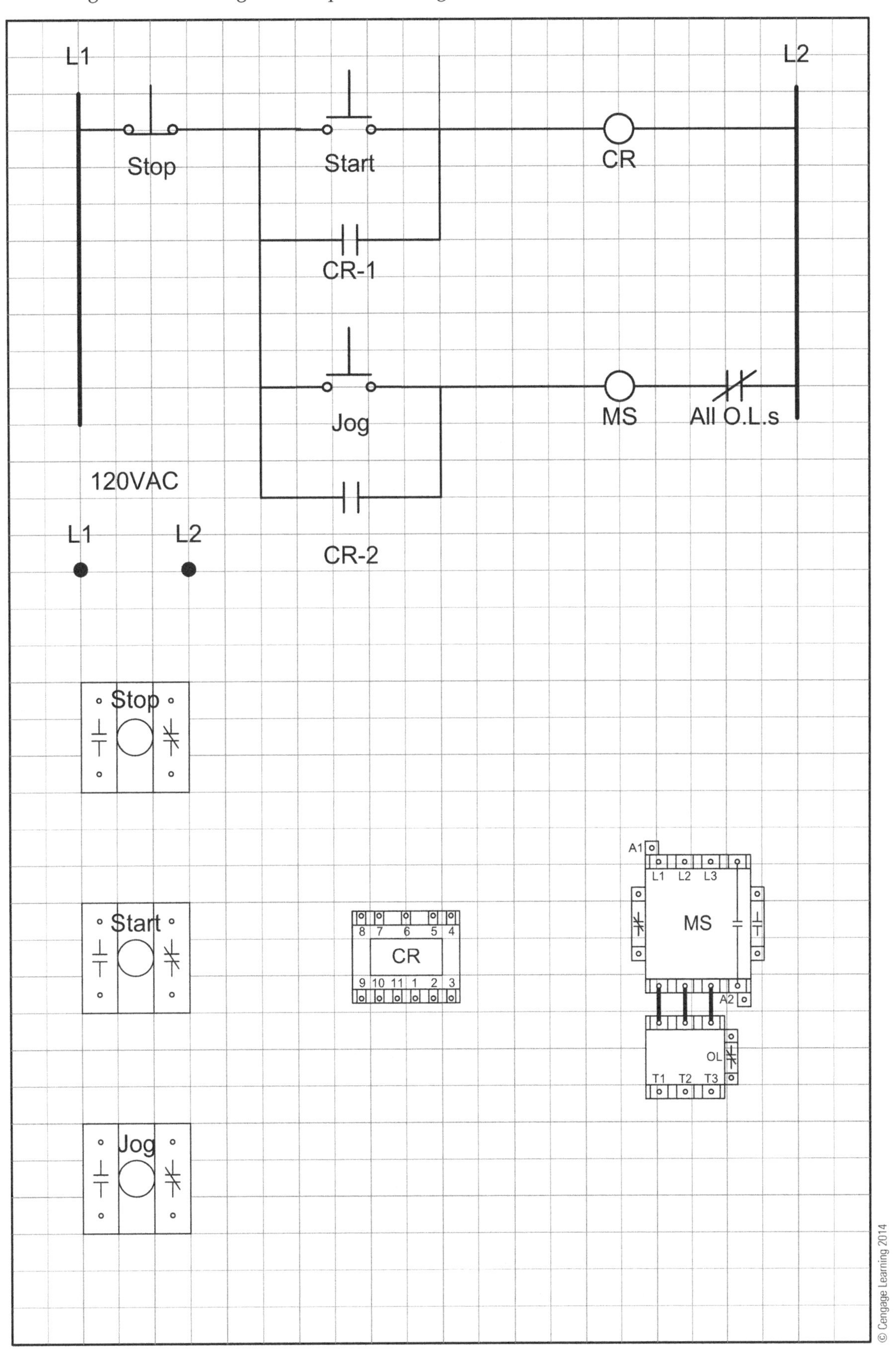

Name: ______________________________ Date: ______________________

SCHEMATIC DIAGRAMS

1. Given the schematic diagram, identify the following items.

   a. ______________________________

   b. ______________________________

   c. ______________________________

   d. ______________________________

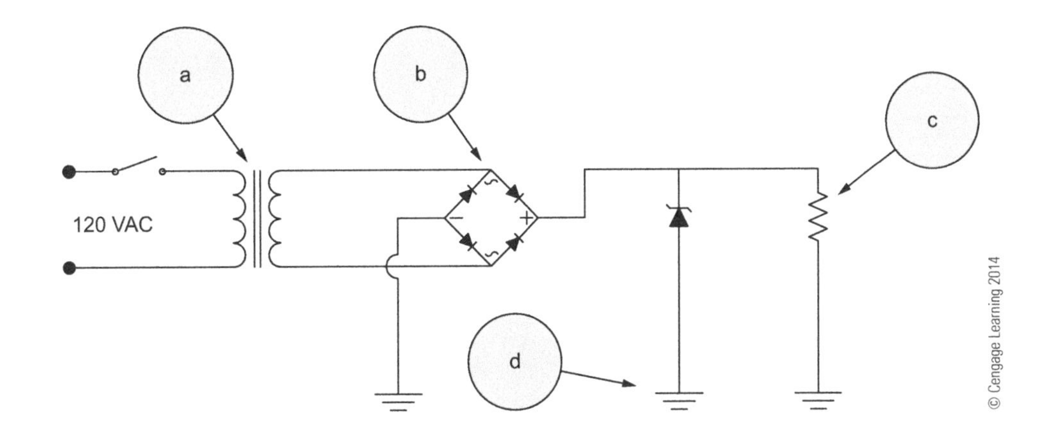

2. Convert the given pictorial diagram to a schematic diagram.

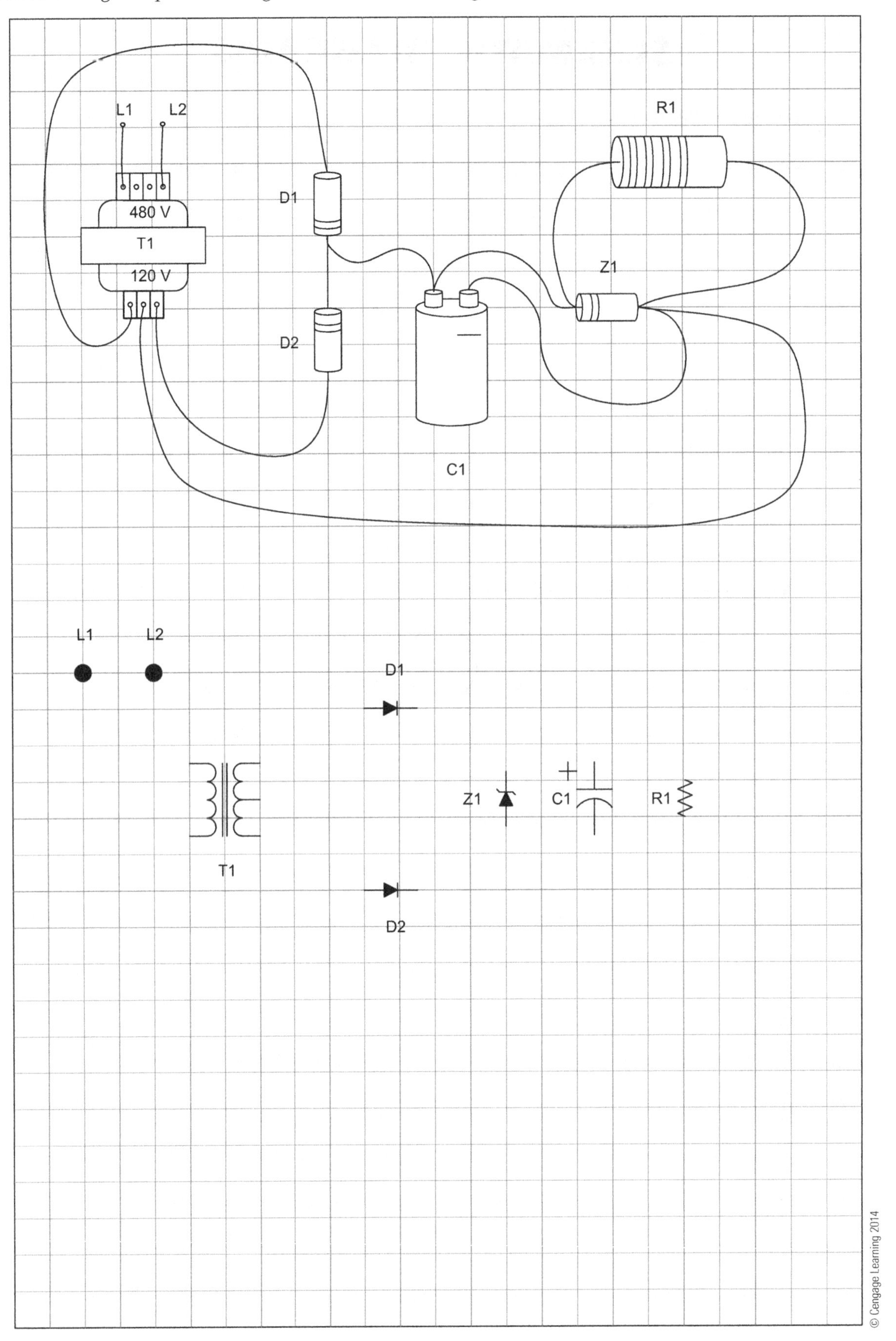

3. Convert the given ladder diagram to a schematic diagram.

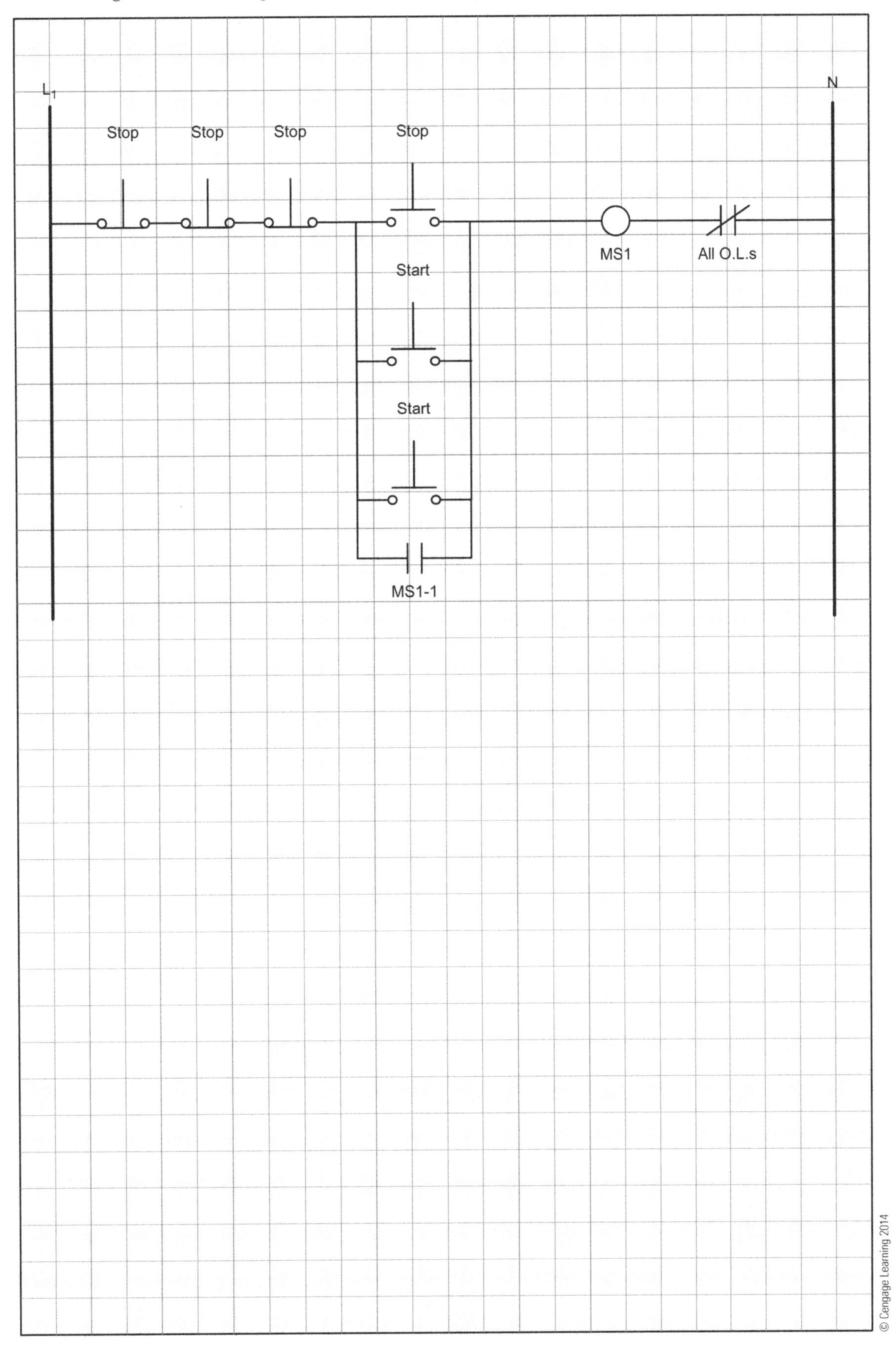

Name: ______________________ Date: ______________

LADDER DIAGRAMS

1. Given the ladder diagram, identify the following items.

   a. ______________________

   b. ______________________

   c. ______________________

   d. ______________________

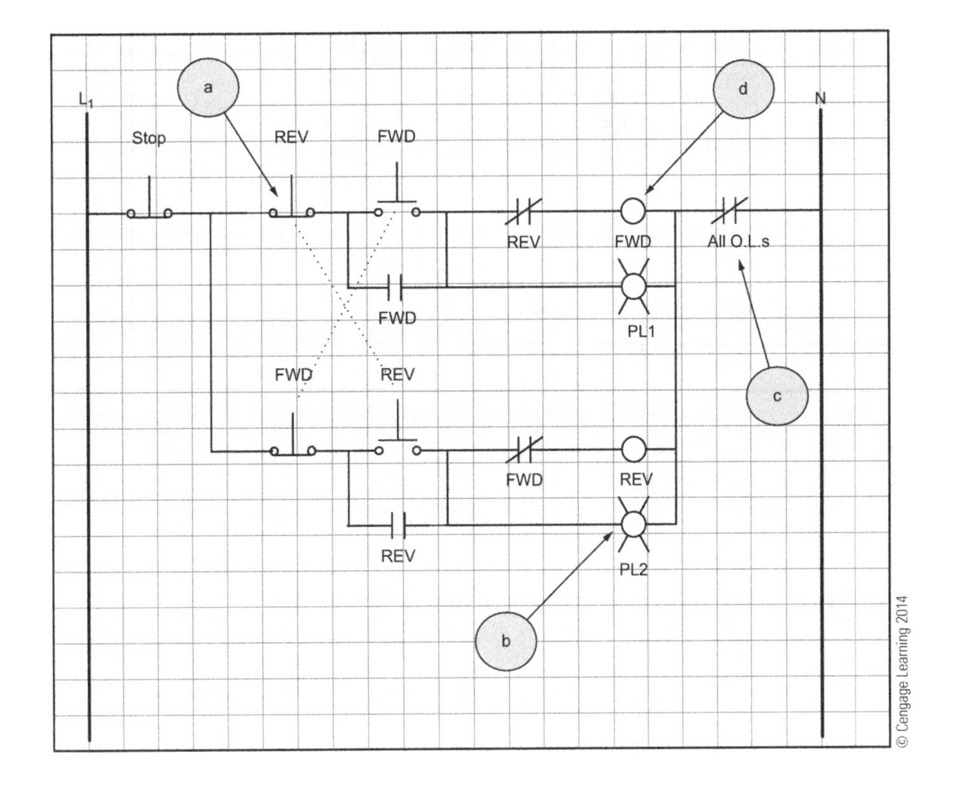

2. Convert the given ladder diagram to a pictorial diagram.

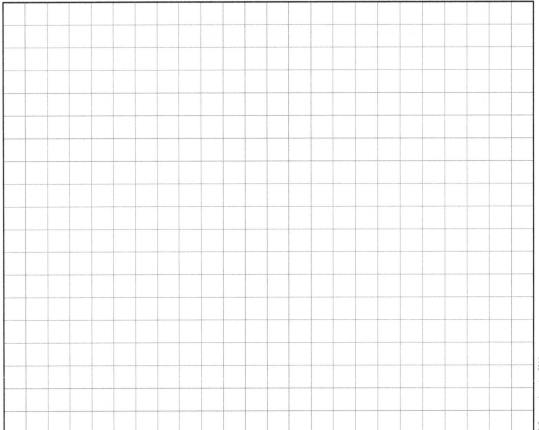

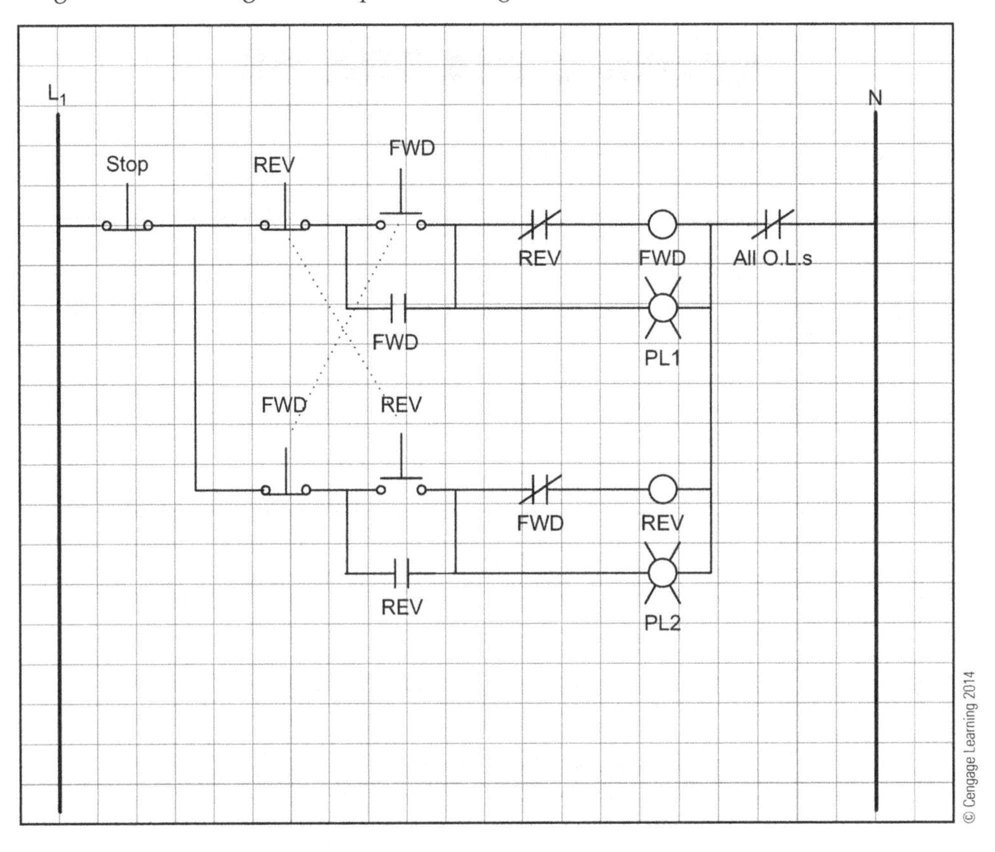

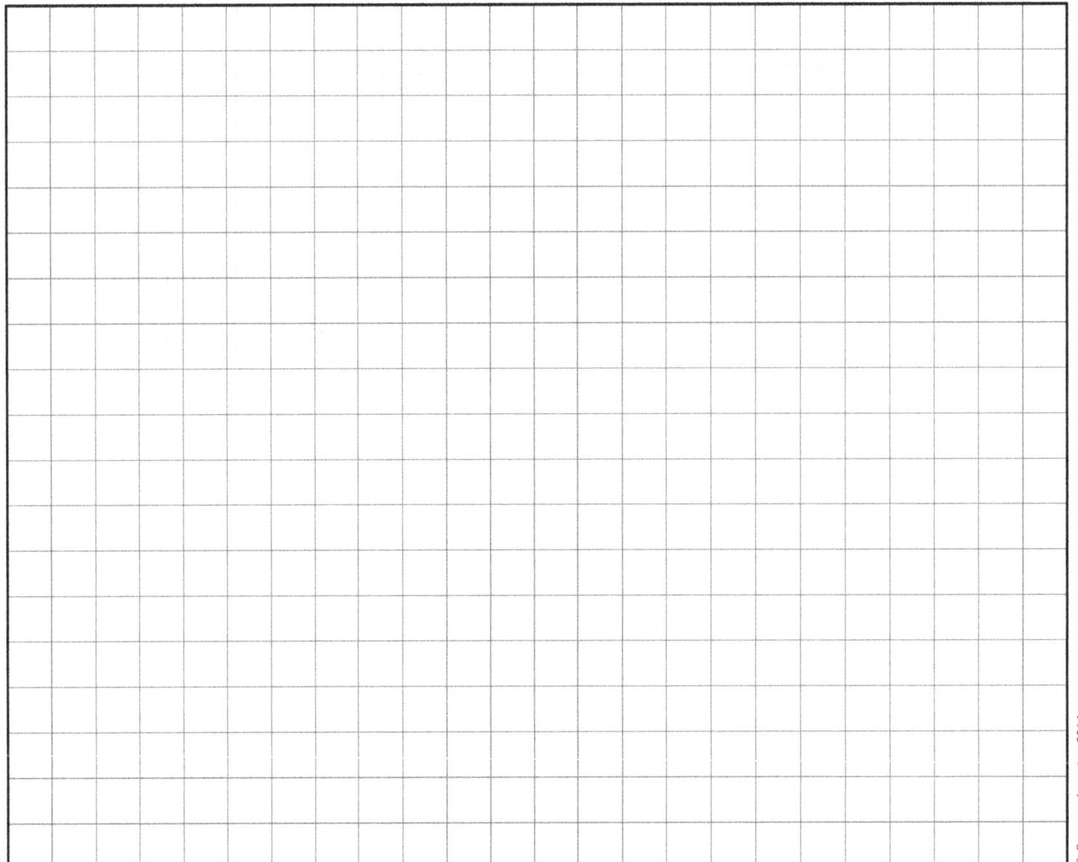

3. Convert the given ladder diagram to a schematic diagram.

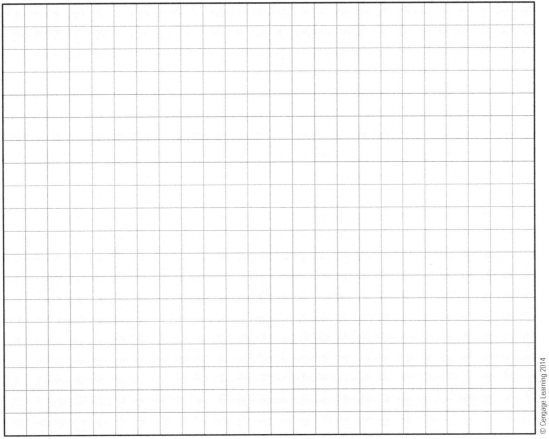

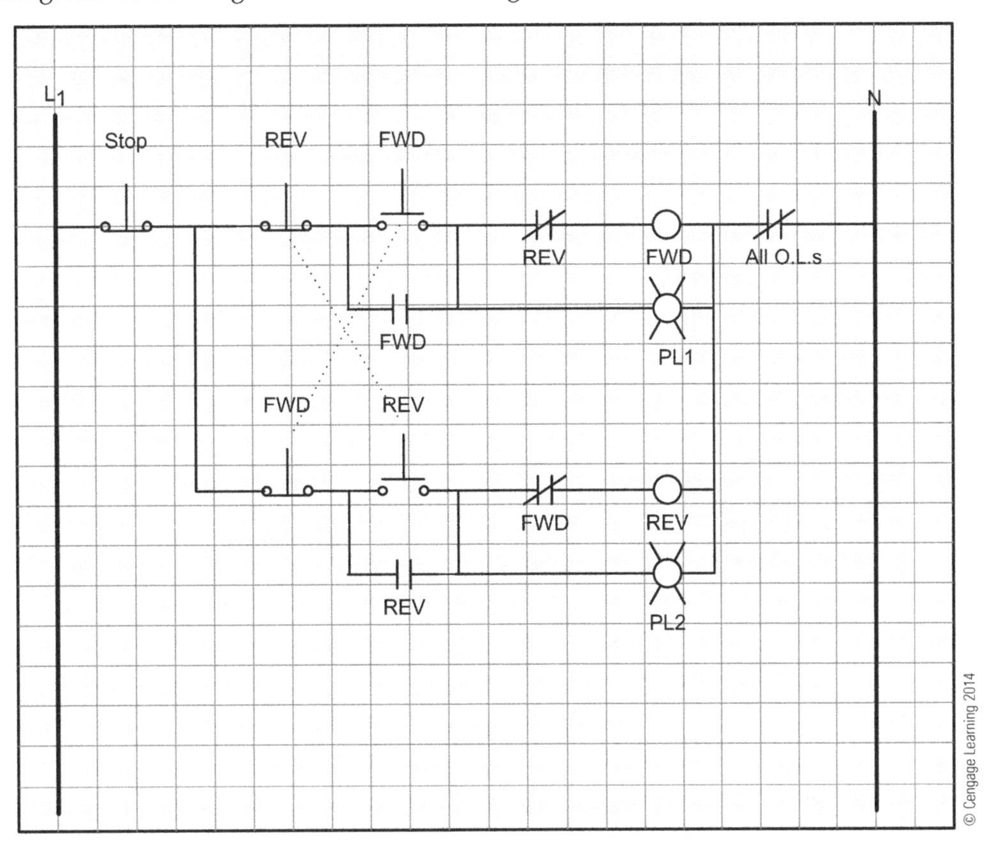

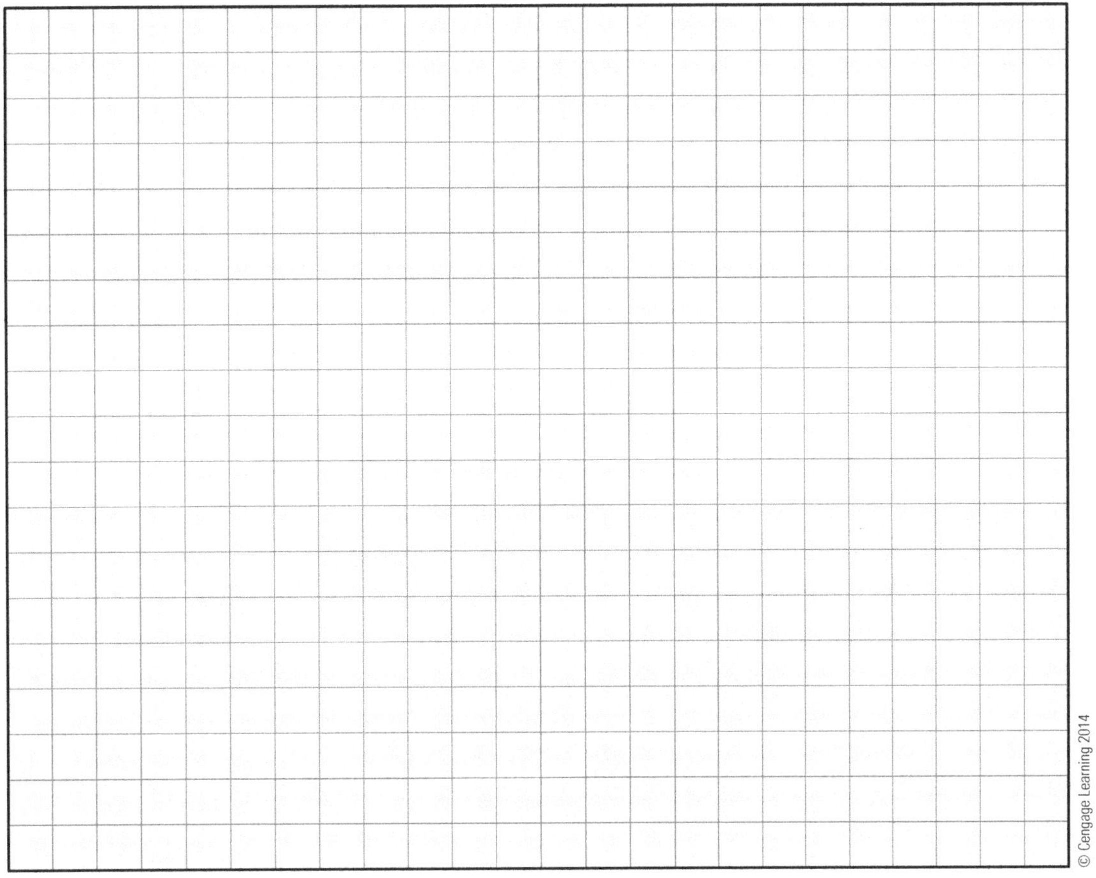

Name: ______________________ Date: ______________

WELDING SYMBOLS—PART 1

TYPE OF WELD

1. Match each weld to the correct ANSI symbol.

1. Plug or slot weld
2. Flare bevel
3. Vee butt weld
4. Flare vee butt weld
5. Bevel butt weld
6. U butt weld
7. Back weld
8. Square butt weld
9. Fillet weld
10. J butt weld

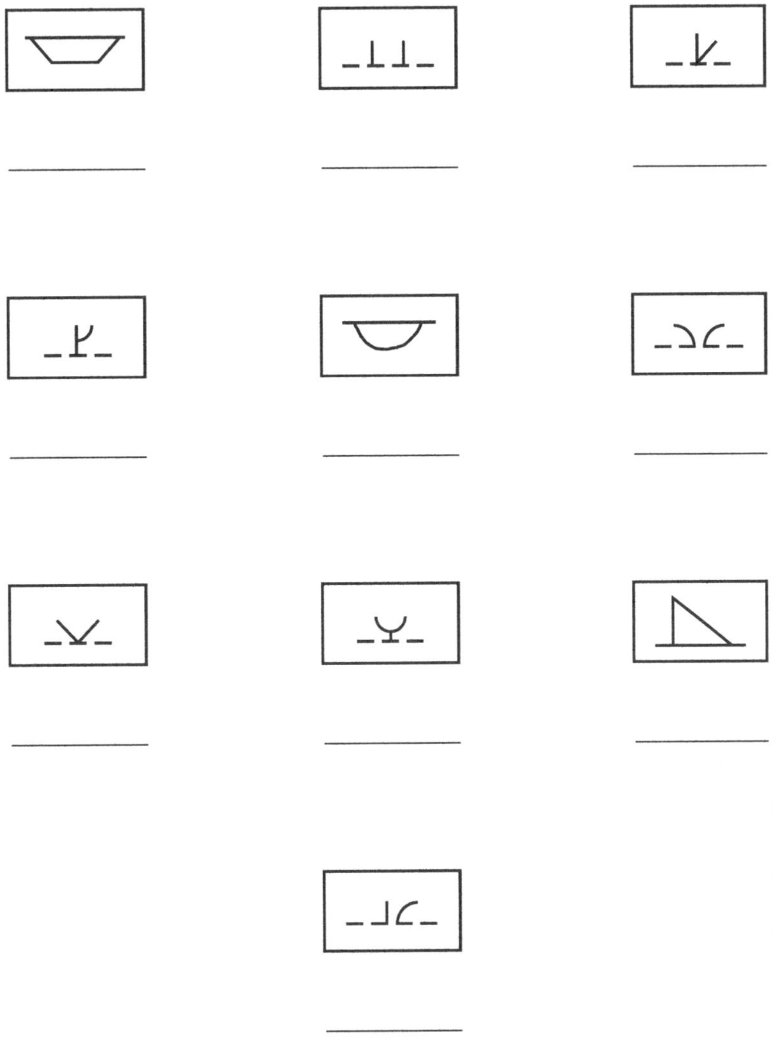

______ ______ ______

______ ______ ______

______ ______ ______

______

Name: ______________________ Date: ______________

WELDING SYMBOLS—PART 2

WELD, POSITION, METHOD, AND FINISH

1. Match each ANSI symbol to the correct definition by drawing a line from the symbol to the definition.

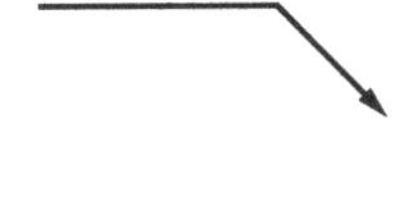

Projection

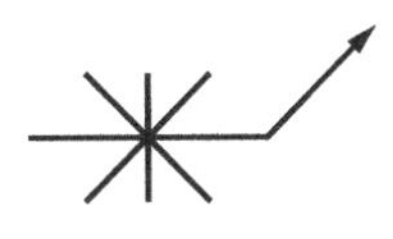

Concave

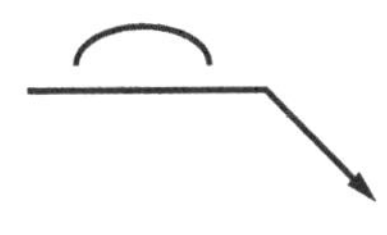

Flush

Weld position

Field weld

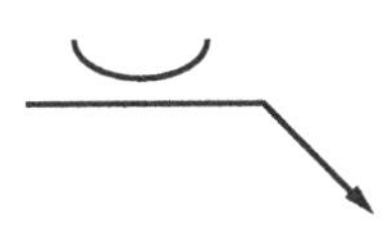

Spot

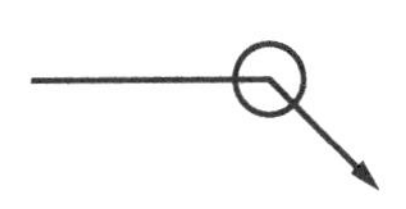

Convex

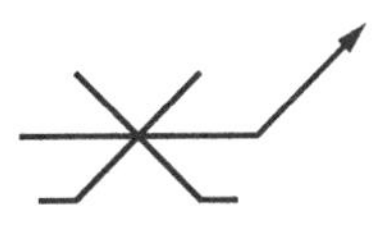

Seam

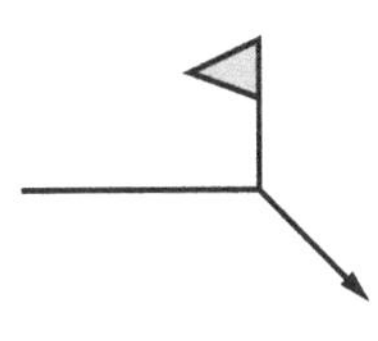

Melt through

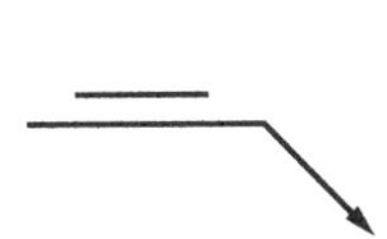

Weld all around

Name: ______________________ Date: ______________

WELDING SYMBOLS—PART 3

PRINT INTERPRETATION

1. Identify what each indicated item in the symbol represents.

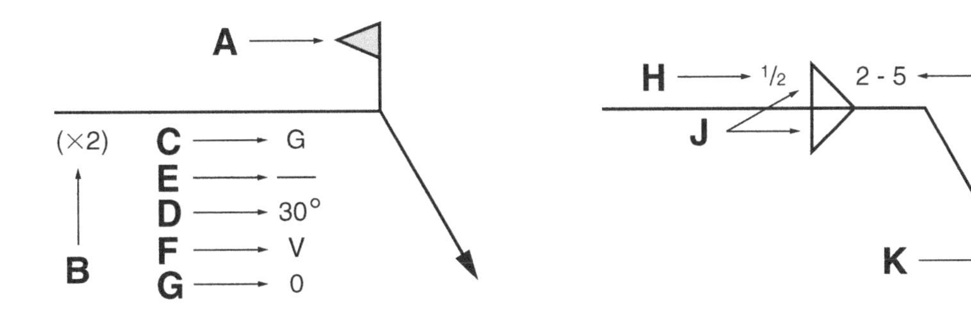

a. ______________________

b. ______________________

c. ______________________

d. ______________________

e. ______________________

f. ______________________

g. ______________________

h. ______________________

i. ______________________

j. ______________________

k ______________________

2. Draw the ANSI symbol that is used to identify when a weld has to be made in the field.

3. Draw the ANSI symbol for a fillet weld that must be placed on both sides of the joint.

Name: ______________________   Date: ______________

WELDING DRAWINGS

1. Draw the correct symbol for each description below.
   a. A fillet weld on the opposite side
   b. Field weld, U groove, no root opening, 30° angle
   c. Fillet weld, ground weld, convex form
   d. Fillet weld, both sides
   e. Weld around the part in the field

______________ a   ______________ b   ______________ c

______________ d   ______________ e

Name: ______________________________ **Date:** ____________________

UNITS OF MEASURE—CONVERSIONS 1

Refer to Appendix G to convert the units below from U.S. customary units to SI units. Show your math.

1. 4 in^2 = __________ cm^2
2. 600 ft = __________ meters
3. 1 in. = __________ cm
4. 1 mile = __________ meters
5. 15 miles = __________ km
6. 1.5 in. = __________ mm
7. 30 yards = __________ meters
8. 83 yards = __________ km

Name: ______________________________ Date: ____________________

UNITS OF MEASURE—CONVERSIONS 2

Refer to Appendix G to convert the units below from SI units to U.S. customary units. Show your math.

1. 30 cm^2 = __________ in^2
2. 600 meters = __________ ft
3. 10 cm = __________ in.
4. 100 meters = __________ miles
5. 15 km = __________ miles
6. 25 mm = __________ in.
7. 30 meters = __________ yards
8. 83 km = __________ yards

Name: ______________________ Date: ______________

CIRCLE PROPERTIES

1. Identify the properties of the circle in the graphic below.

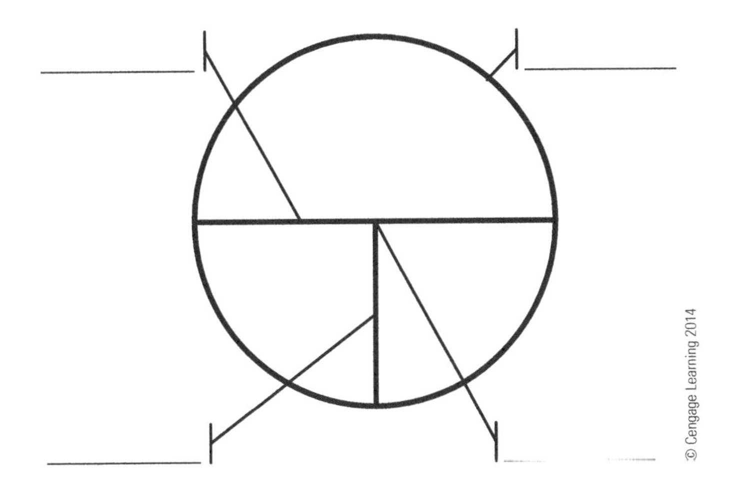

Name: ______________________ **Date:** ______________

AREA

1. What is the formula for the area of a circle when only the radius is known?

2. What is the formula for the area of a circle when only the diameter is known?

3. What is the method and formula for finding the area of a circle when only the circumference is known?

4. What is the surface area of a cylinder that has a 10 in. circumference and a height of 20 in.? (Show your work)

5. What is the area of a square where the length of one side equals 2 ft?

6. What is the area for the following shape? (Show your work)

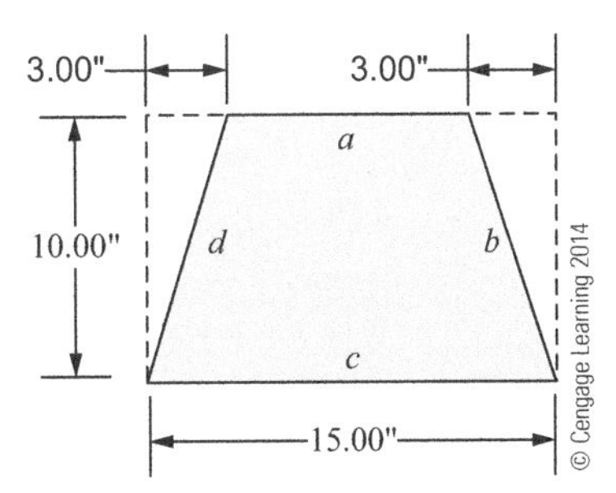

7. What is the formula for the area of a triangle?

8. What is the surface area for the following shape? (Show your work)

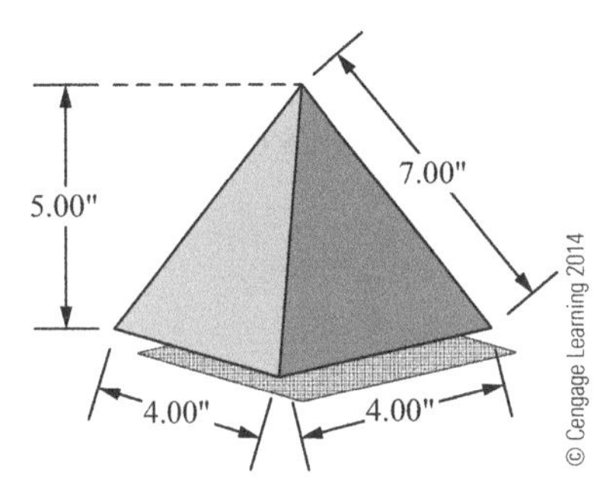

Name: ______________________ Date: ______________

VOLUME

Solve for volume for each of the following figures. (Show your work for each)

1.

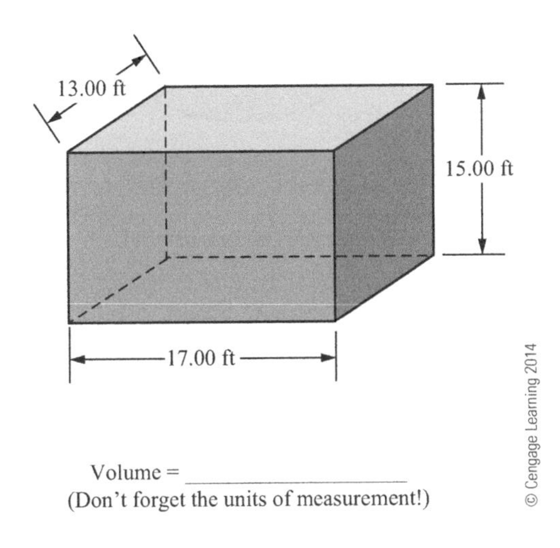

Volume = ______________
(Don't forget the units of measurement!)

2.

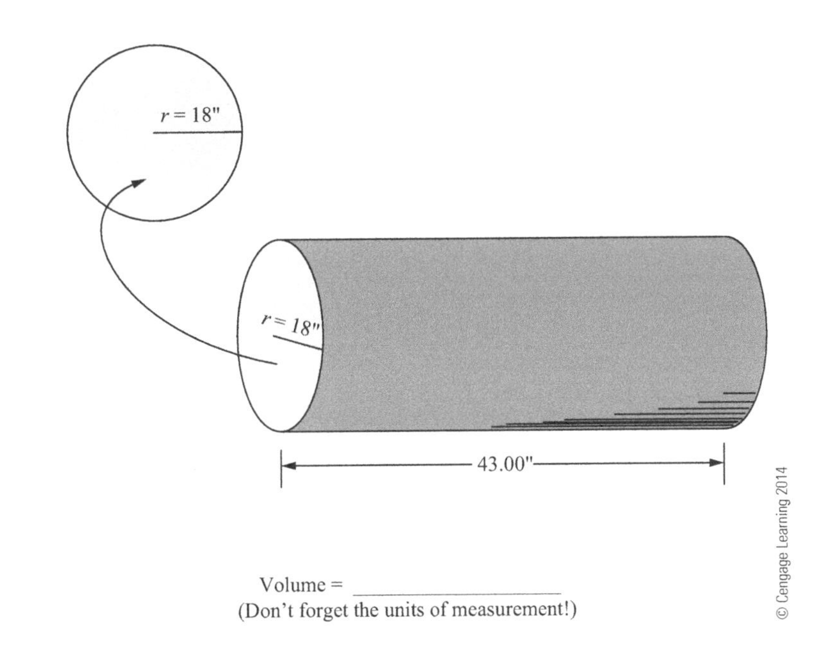

Volume = ______________
(Don't forget the units of measurement!)

3.

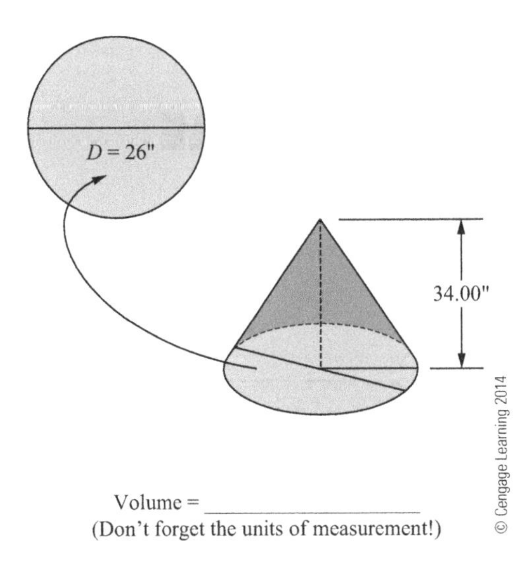

Volume = ____________________
(Don't forget the units of measurement!)

4.

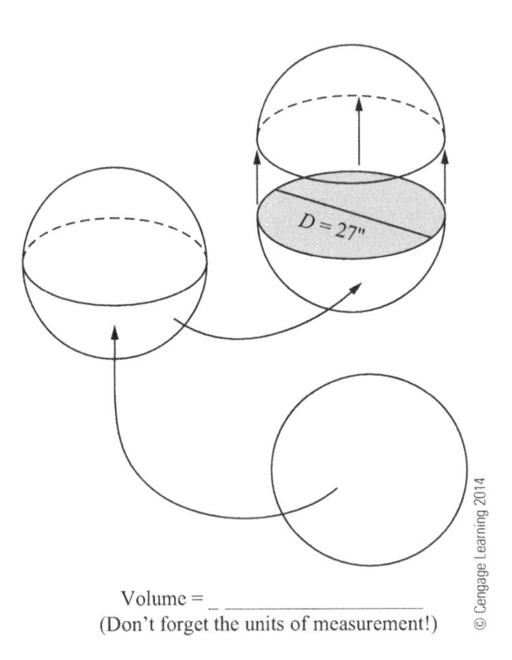

Volume = ____________________
(Don't forget the units of measurement!)

5.

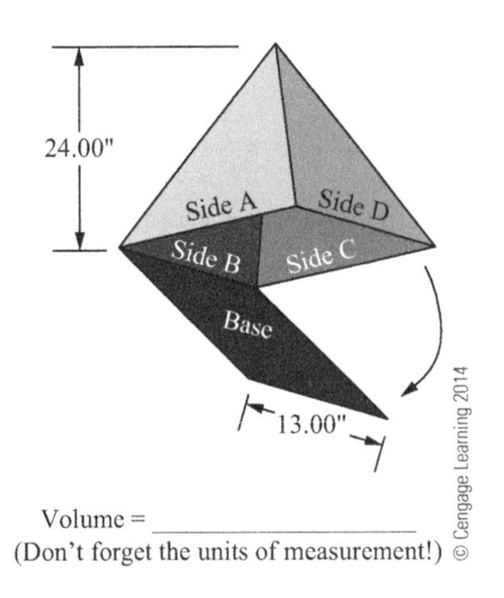

Volume = ____________________
(Don't forget the units of measurement!)

Name: ______________________________ Date: ____________________

SLING TERMINOLOGY

1. Define the following terms:
   a. Basket hitch

   b. Master link

   c. Chocker hitch

   d. Proof load

   e. Proof test

   f. CAP

   g. WLL

   h. SWL

   i. Vertical hitch

   j. Sling

Name: ______________________ Date: ______________

WEB SLING TYPES & RATINGS

1. Draw a line from each web sling to the correct type.

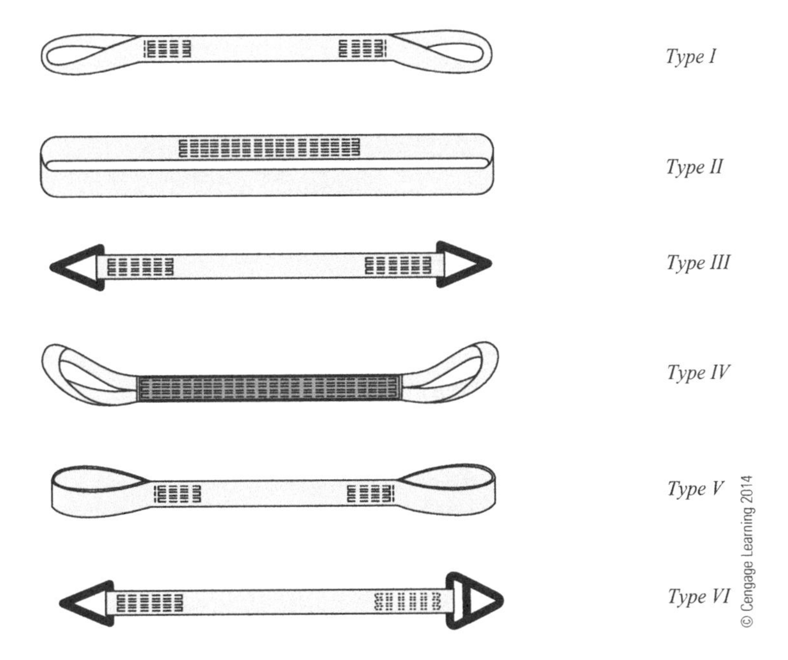

Name: ______________________ Date: ______________

ELEMENTS OF A KNOT

1. Name the following elements to a knot.

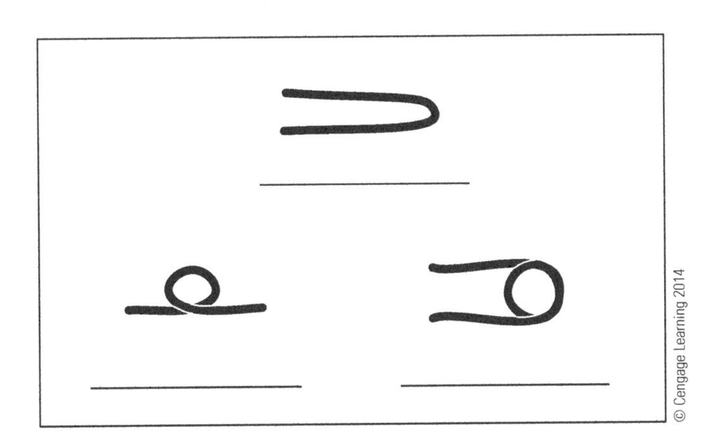

Name: ______________________ Date: ______________

IDENTIFYING KNOTS

1. Name the following knots.

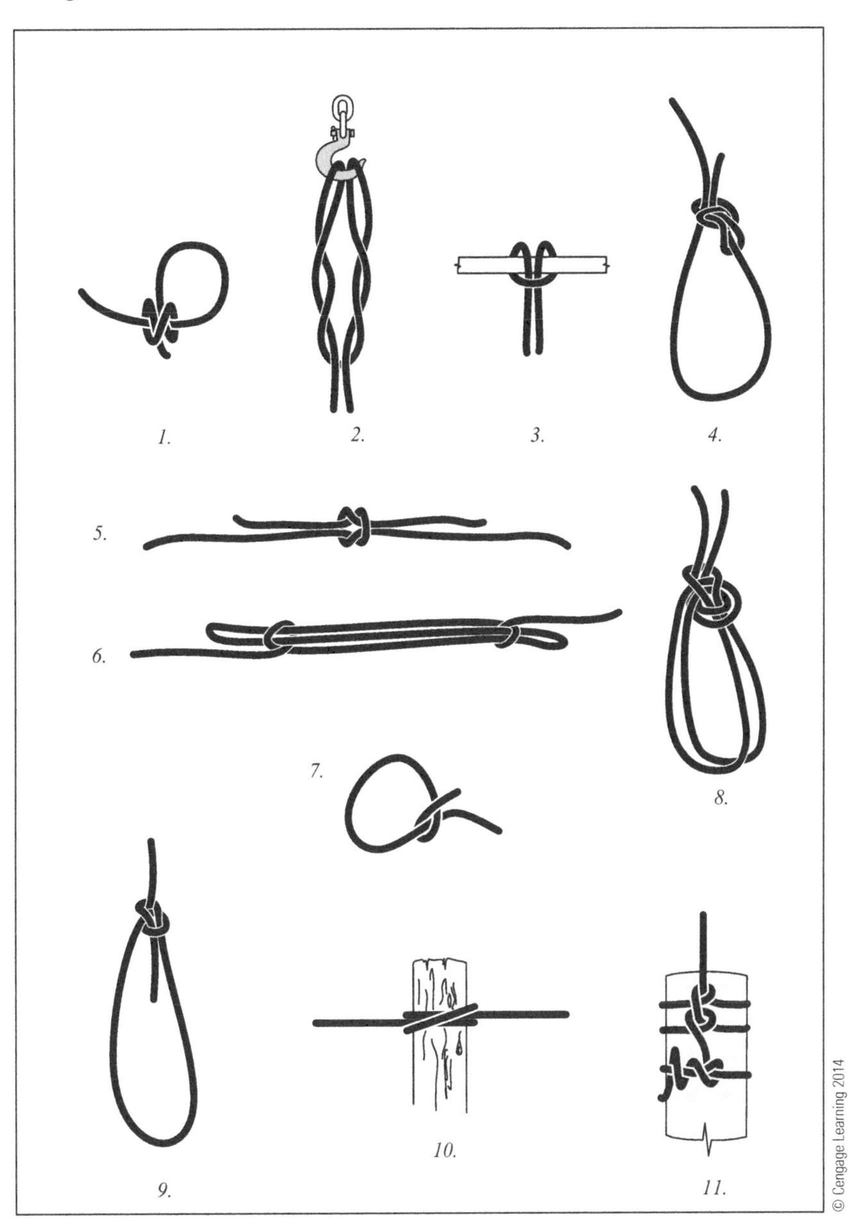

1. ______________
2. ______________
3. ______________
4. ______________
5. ______________
6. ______________
7. ______________
8. ______________
9. ______________
10. ______________
11. ______________

Name: ______________________   Date: ______________

IDENTIFYING WIRE ROPE LAY

1. Match the proper lay to the correct graphic below. (Note: There is one more choice than is needed.)
   a. Right, regular lay
   b. Right, lang lay
   c. Right, alternating lay
   d. Left, regular lay
   e. Left, lang lay
   f. Left, alternating lay

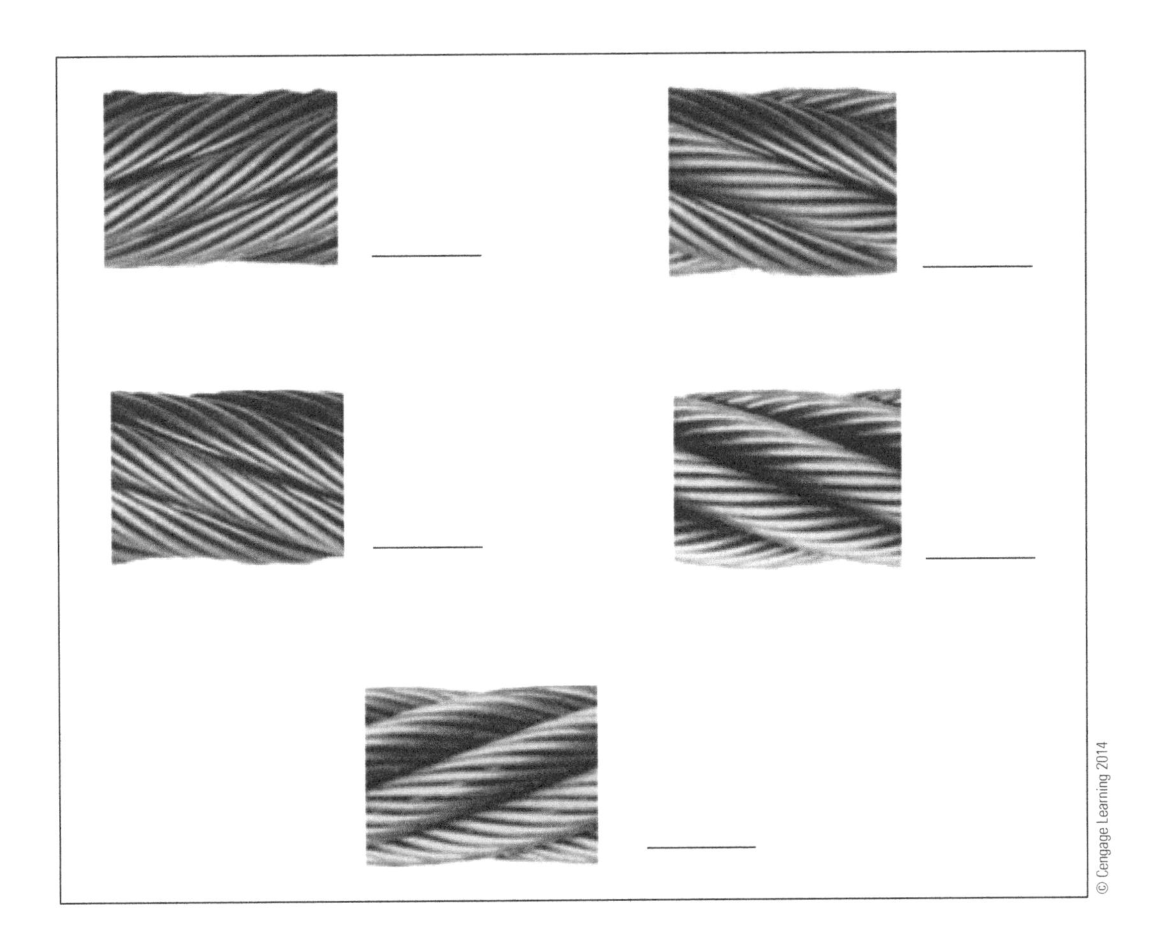

Name: ______________________ Date: ______________

LOAD PER LEG CALCULATIONS

1. Calculate the load/leg for each of the following examples. (You must show your work.)

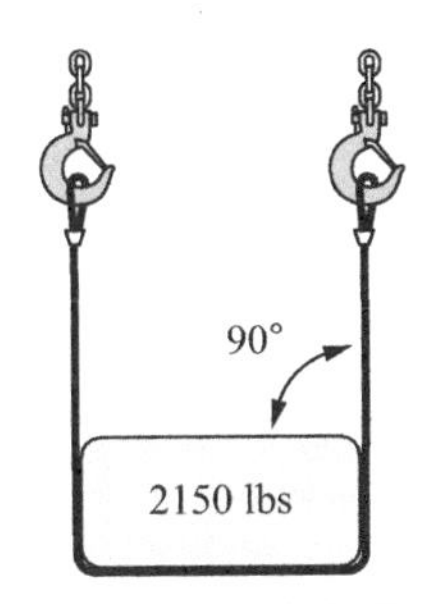

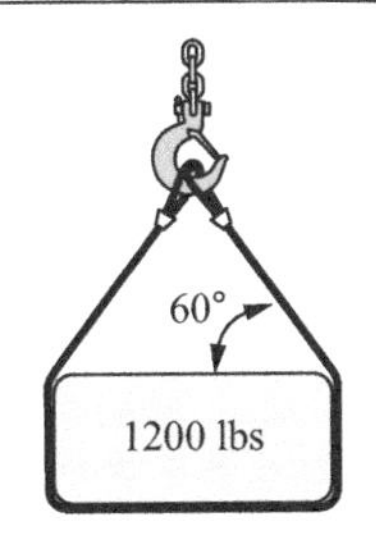

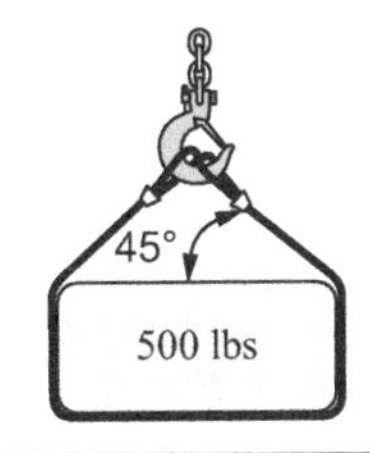

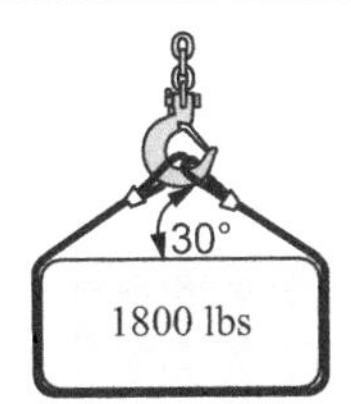

Name: ______________________ Date: ______________

SHACKLE IDENTIFICATION

1. Identify each of the shackles below.

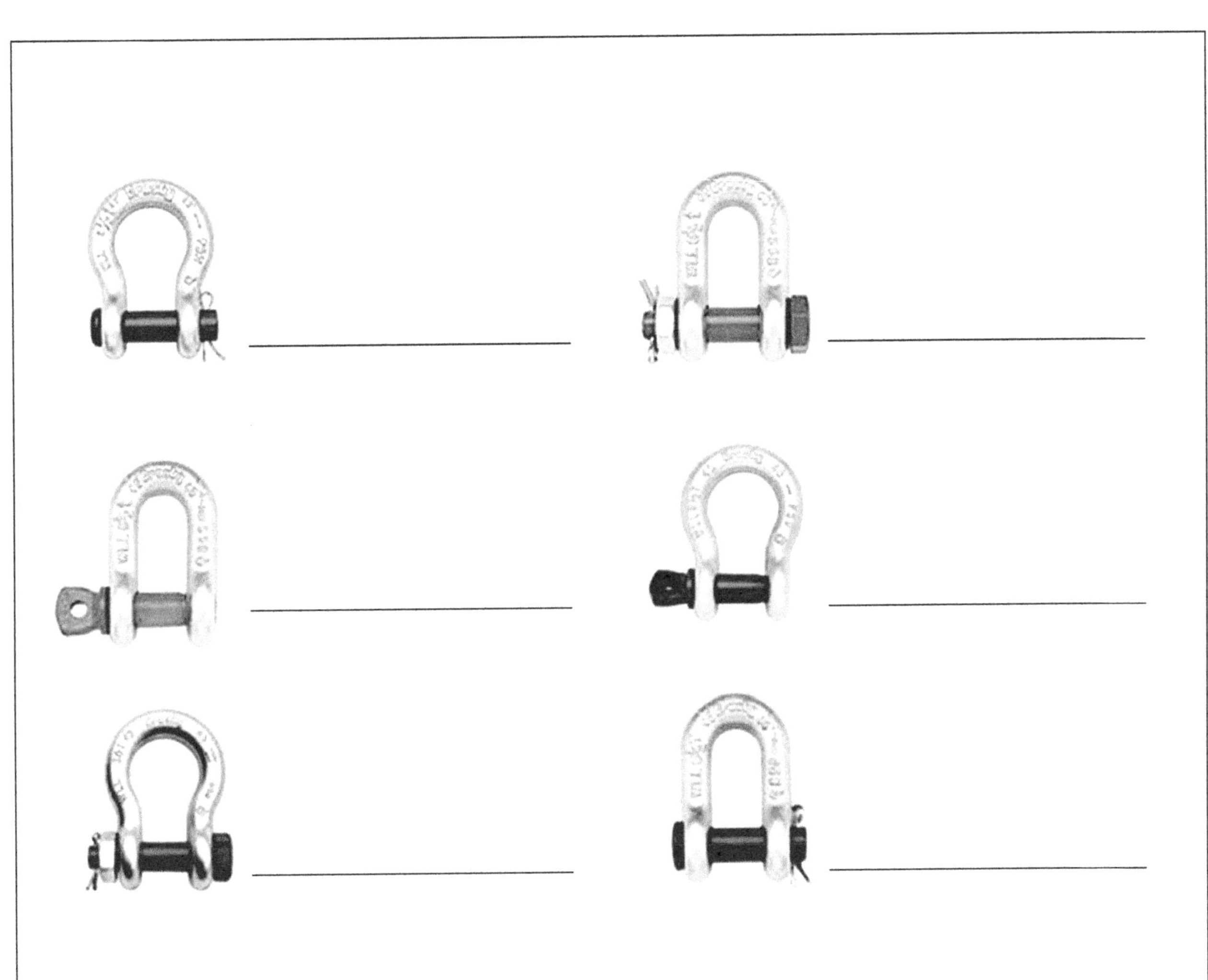

Name: ______________________ Date: ______________

CHAIN ATTACHMENT IDENTIFICATION

1. Identify each of the components below.

______________________ ______________________

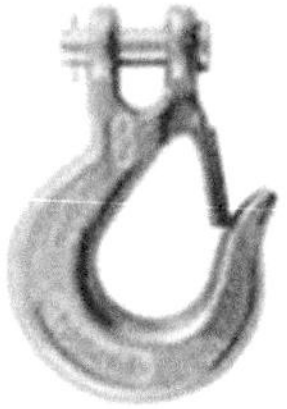

______________________ ______________________

______________________ ______________________

______________________ ______________________

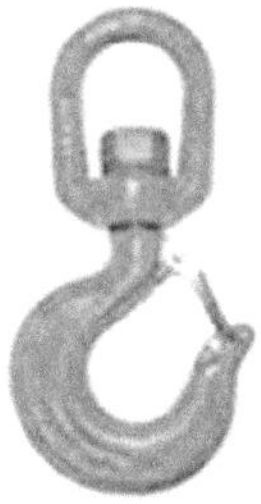

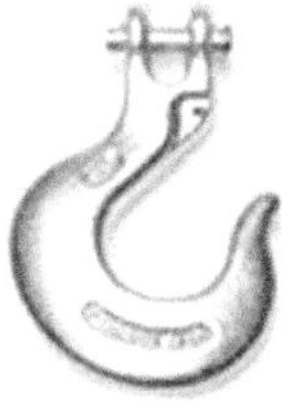

Name: ______________________________   **Date:** ____________________

FLAT BELTS

1. Match the thickness specifications for first-quality leather belting.

   a. Medium single ________ 15⁄64 to 17⁄64

   b. Heavy single ________ 21⁄64 to 23⁄64

   c. Light double ________ 3⁄16 to 7⁄32

   d. Medium double ________ 9⁄32 to 5⁄16

   e. Heavy double ________ 5⁄32 to 3⁄16

2. Calculate the belt speed of a belt if the motor is running at 1800 rpm and the pulley has a diameter of 6 in. Show your work.

3. Refer to the drawing and calculate the length of the belt that is in the drawing. Show your work.

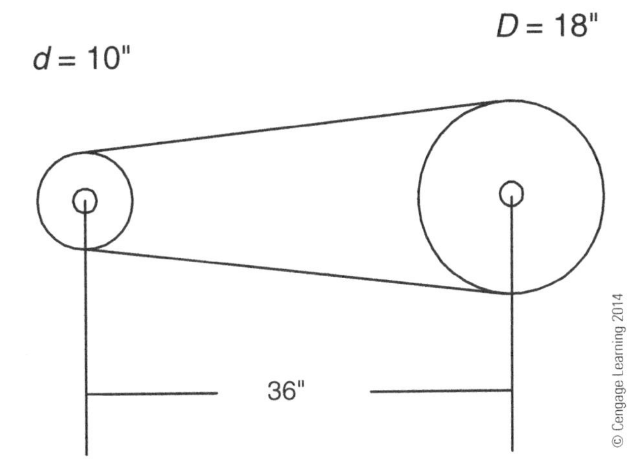

4. How would you correct this tracking problem?

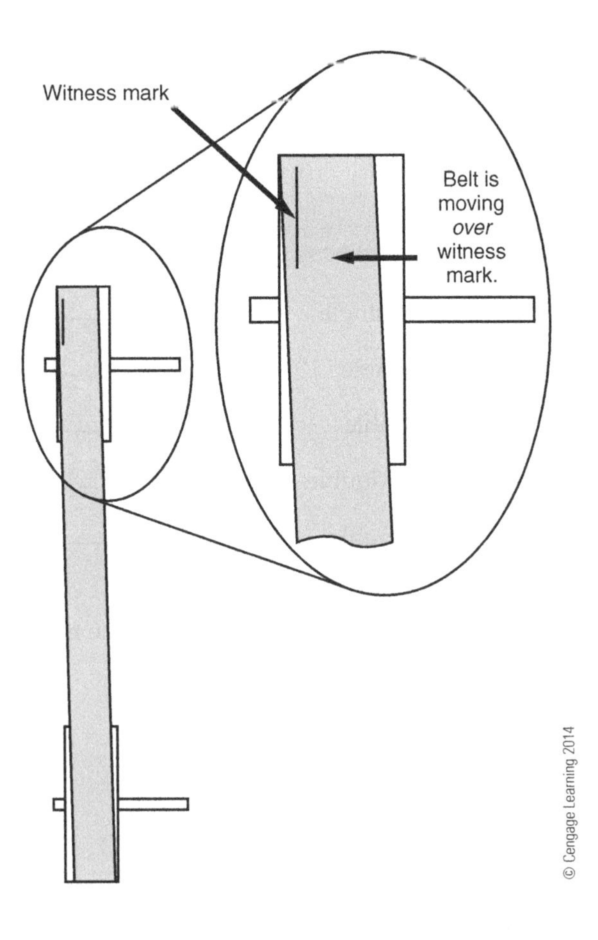

5. How would you correct this tracking problem?

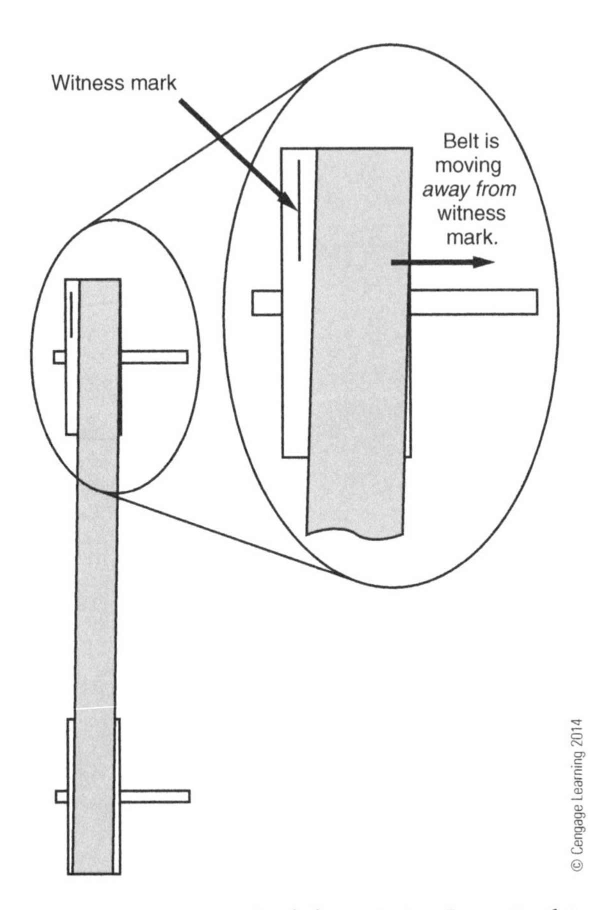

Name: ______________________ Date: ______________

V-BELTS: POWER TRANSMISSION

1. Circle the drawing below that shows the correct method for transmitting power using a V-belt.

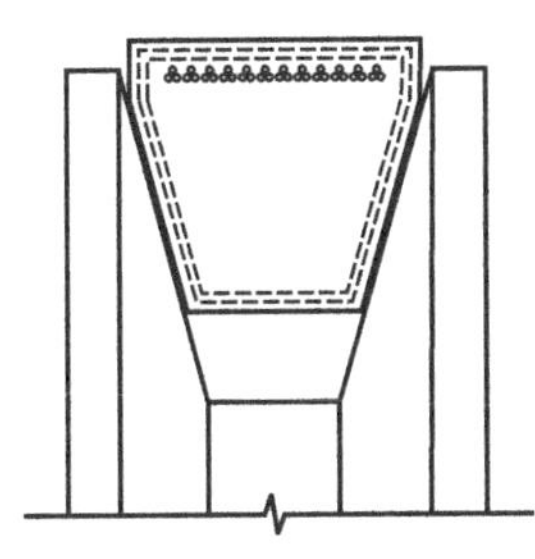

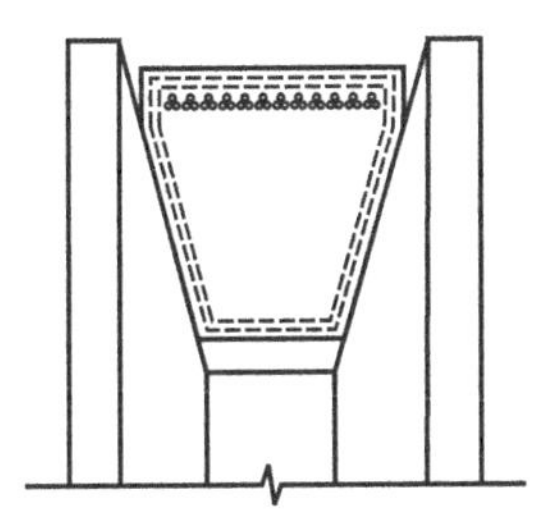

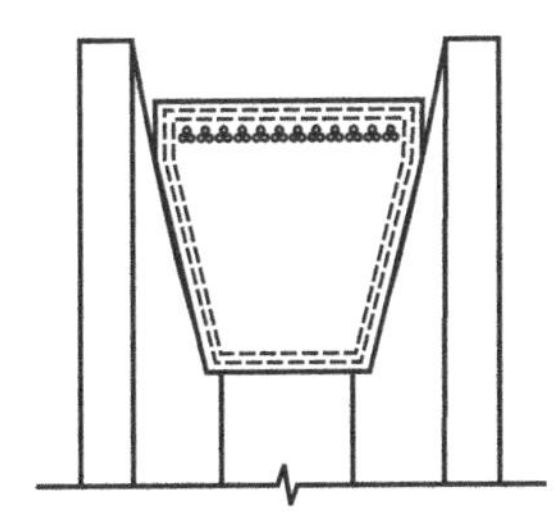

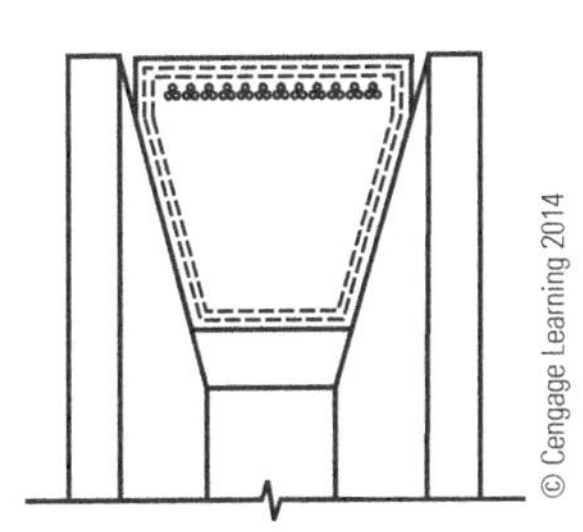

Name: ______________________ Date: ______________

V-BELTS: BELT DEFLECTION

1. What would be the proper amount of belt deflection for each of the V-belt systems shown?

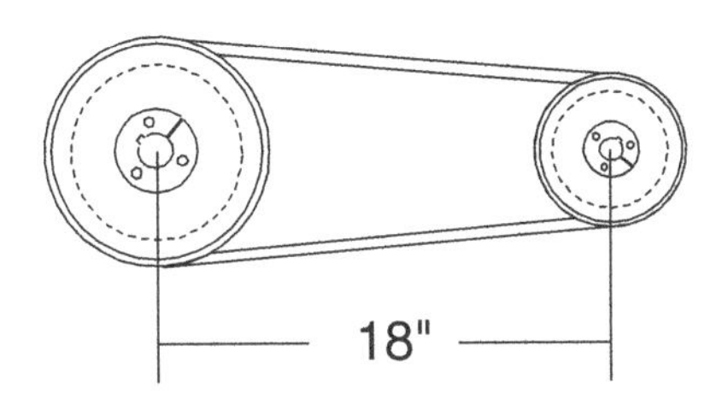

A ____________

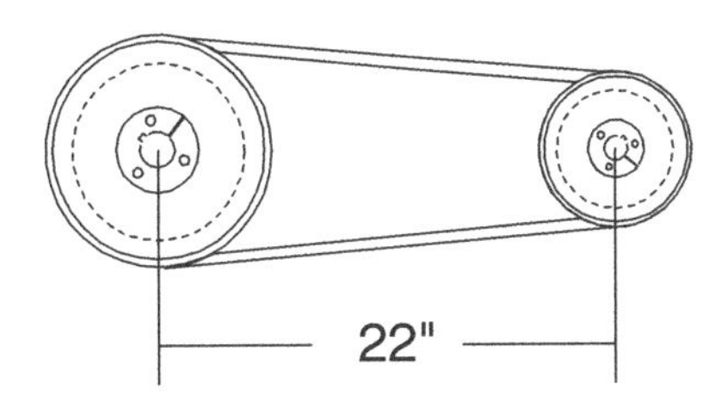

B ____________

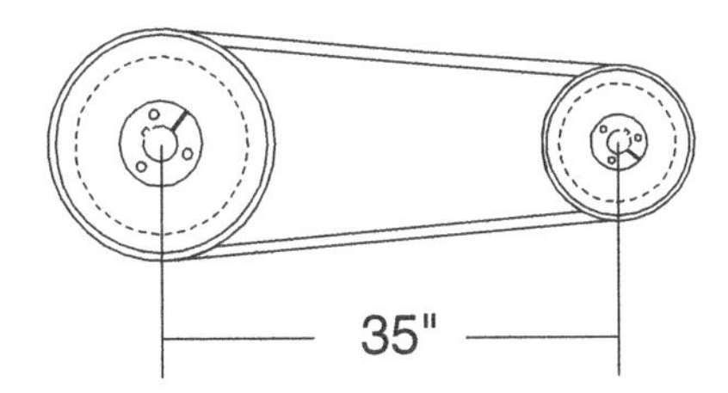

C ____________

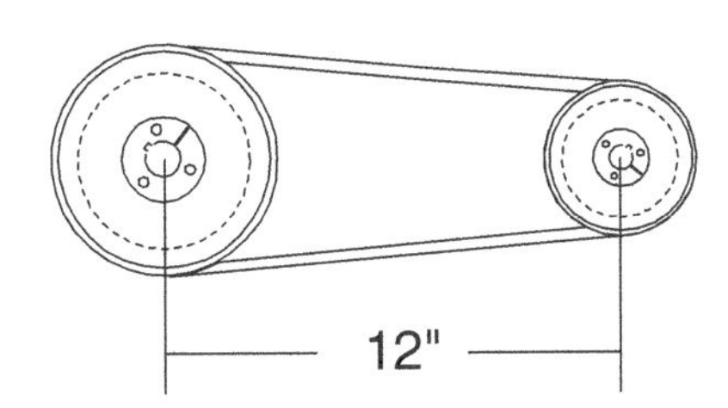

D ____________

Name: ______________________________ Date: ________________

V-BELTS: BELT SELECTION

1. Match each belt to the correct cross section.

1. A 2. B 3. C 4. D 5. E 6. 2L

7. 3L 8. 4L 9. 5L 10. 3V 11. 5V 12. 8V

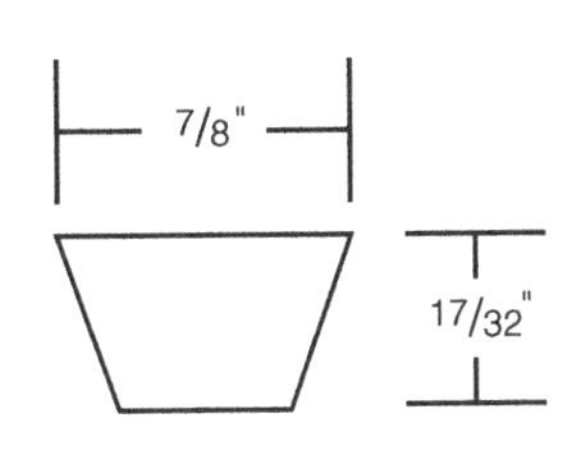

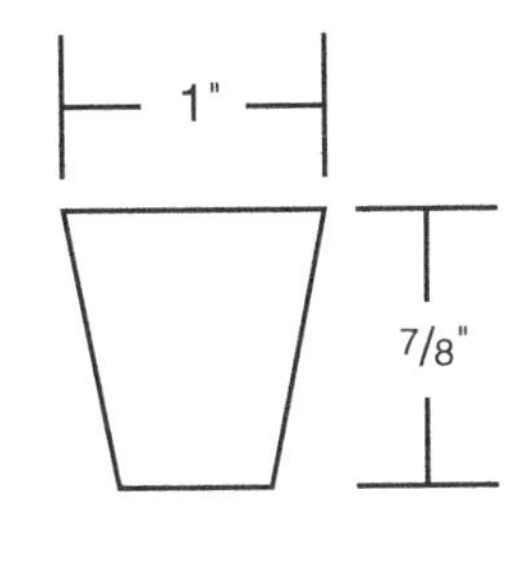

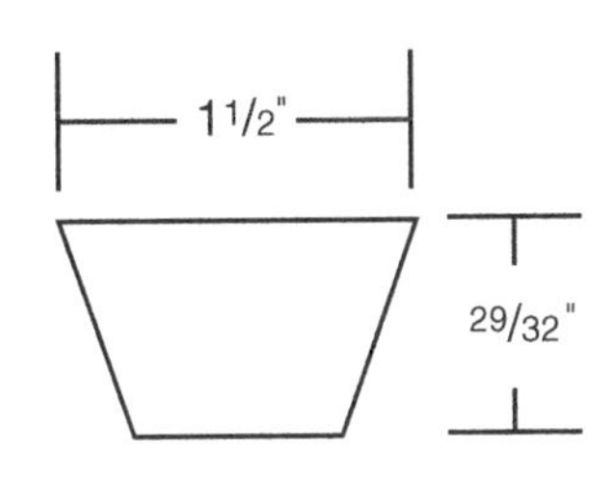

_______ _______ _______

1/2" 5/16"

21/32" 13/32"

3/8" 7/32"

_______ _______ _______

1/4" 5/32"

21/32" 3/8"

3/8" 5/16"

_______ _______ _______

1 1/4" 3/4"

5/8" 17/32"

1/2" 5/16"

_______ _______ _______

Name: ______________________　　Date: ______________

V-BELTS: DOUBLE V-BELT SELECTION

1. Match each belt to the correct cross section.

1. AA　　2. BB　　3. CC　　4. DD

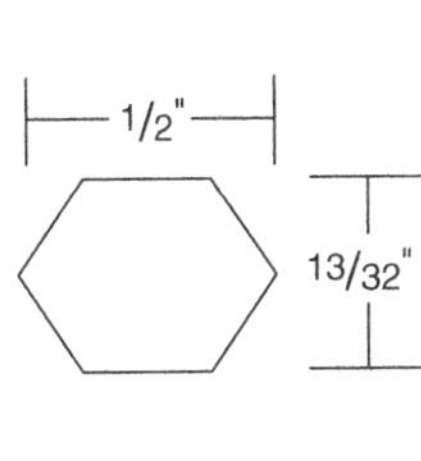

____

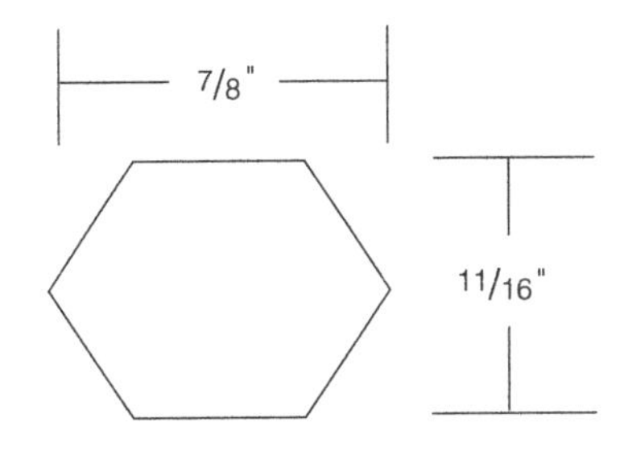

____

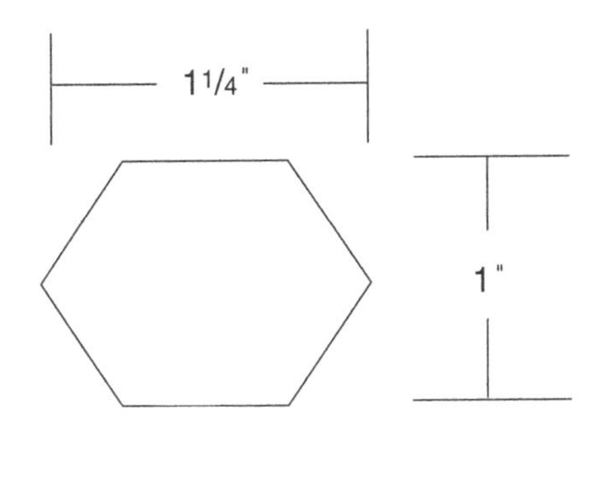

____

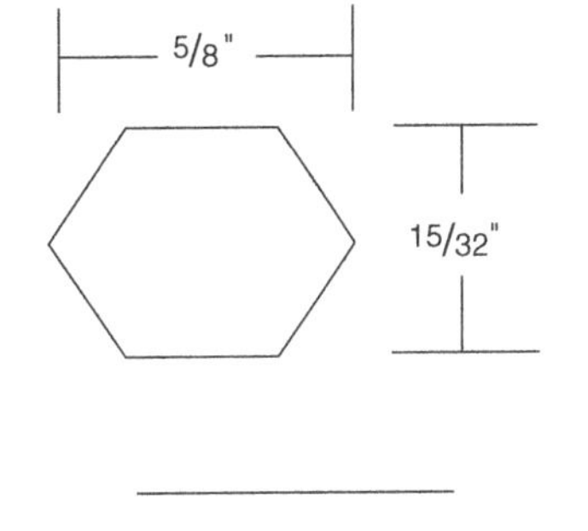

____

Name: ______________________ Date: ______________

V-BELTS: V-BELT PITCH LENGTH

1. What would be the belt pitch length of a V-belt if the distance between the shafts is 24 in., the diameter of the large pulley is 10 in., and the diameter of the small pulley is 6 in.? Both pulleys are B pulleys. Use the table on page 92 of the text to select the proper belt once you have finished calculating the needed length. Show your work.

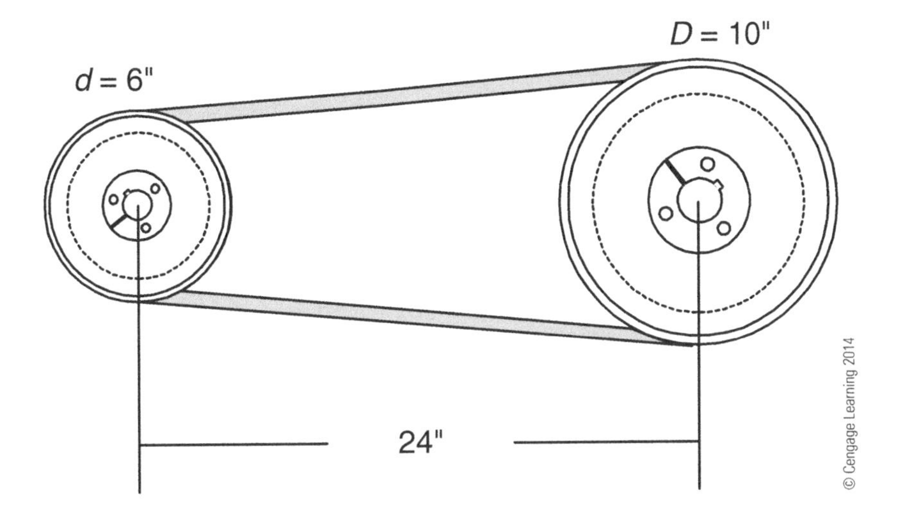

Name: ______________________ Date: ______________________

POSITIVE-DRIVE BELTS: PITCH

1. Draw the pitch line of the positive-drive belt shown in the drawing.

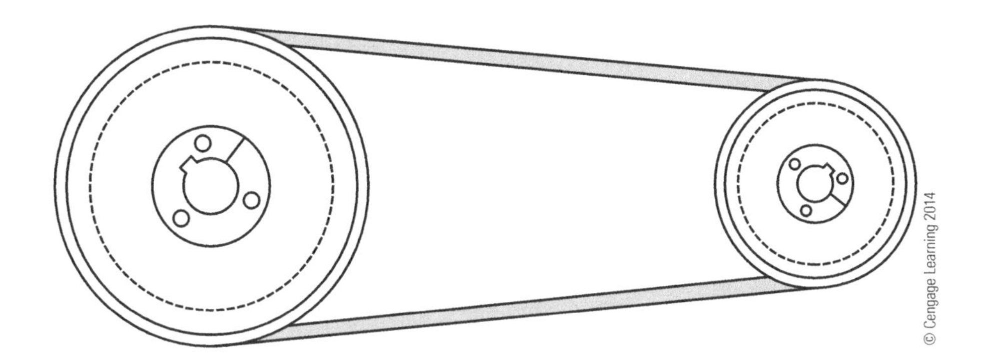

Name: ______________________________ Date: ______________

POSITIVE-DRIVE BELTS: SELECTION

1. Match the standard positive-drive belt to the appropriate pitch.

   a. Mini-extra-light (MXL)
   b. Extra-light (XL)
   c. Light (L)
   d. Heavy (H)
   e. Extra-heavy (XH)
   f. Double extra-heavy (XXH)

1. _______ Pitch—1⁄5 in.

2. _______ Pitch—1 1⁄4 in.

3. _______ Pitch—2⁄25 in.

4. _______ Pitch—1⁄2 in.

5. _______ Pitch—7⁄8 in.

6. _______ Pitch—3⁄8 in.

Name: ______________________________ Date: ____________________

ROLLER CHAIN COMPONENTS

1. Identify the following components.

A ____________________

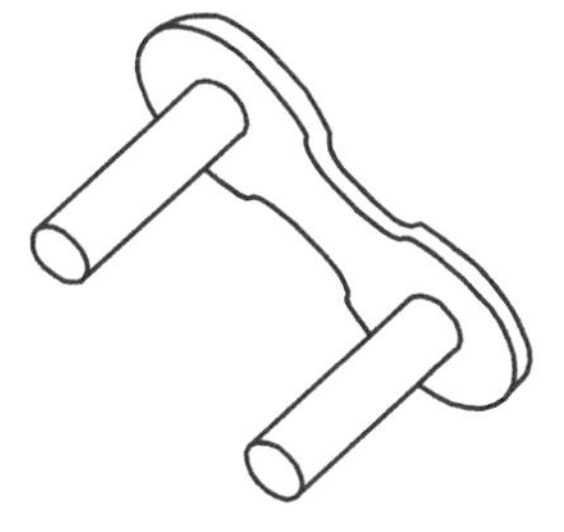

B ____________________

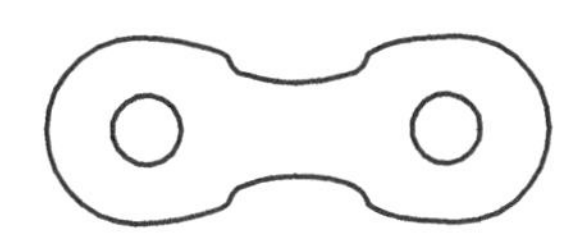

C ____________________

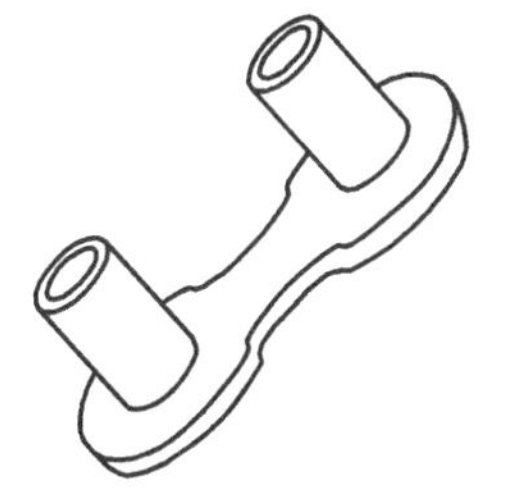

D ____________________

Name: ______________________ Date: ______________

ROLLER CHAIN IDENTIFICATION

1. Look at the identification codes and write a description for each one on the lines that have been provided. All of the numbers and letters in the identification codes must be described.

   a. 140H-2 ______________________

   b. 25 ______________________

   c. 60H ______________________

   d. 41 ______________________

   e. 35 ______________________

   f. 80H-4 ______________________

   g. 40 ______________________

   h. 120H ______________________

   i. 160H-3 ______________________

Name: ______________________ Date: ______________

SPROCKETS

1. Identify all five sprockets shown in the drawing.

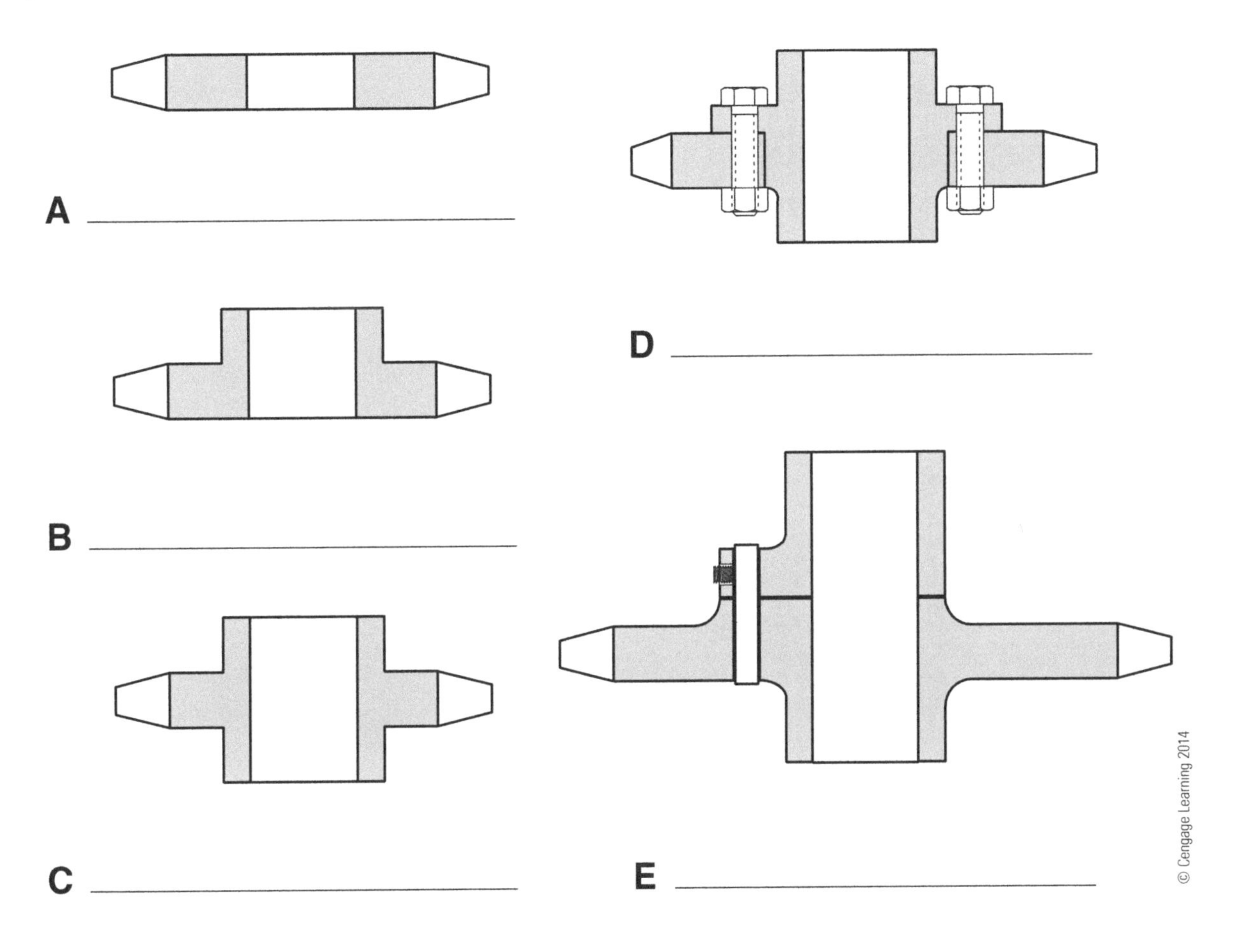

Name: ______________________________ Date: ____________________

GEARS AND GEARBOXES: PITCH CIRCLE

1. Draw the pitch circle for each gear shown in the drawing.

Name: ______________________________ Date: ______________________

GEARS AND GEARBOXES: PITCH LINE

1. On the gear set in the drawing, draw an arrow that indicates where the pitch line is.

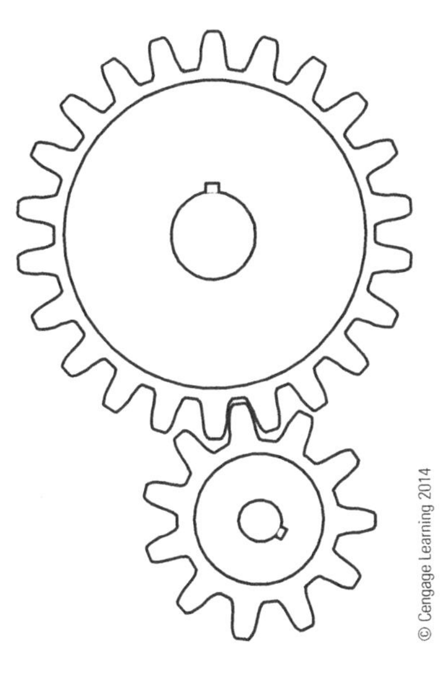

Name: ______________________ Date: ______________

GEARS AND GEARBOXES: PITCH DIAMETER

1. Show where to measure for the pitch diameter of each gear shown in the drawing.

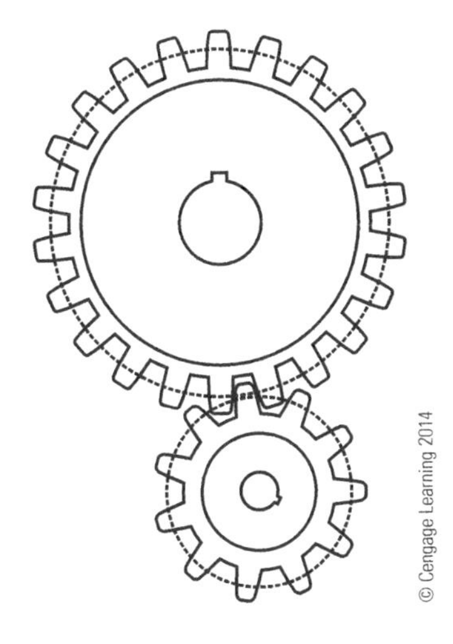

Name: ______________________ Date: ______________

GEARS AND GEARBOXES: CIRCULAR PITCH

1. Show where to measure for the circular pitch of the gear that is in the drawing.

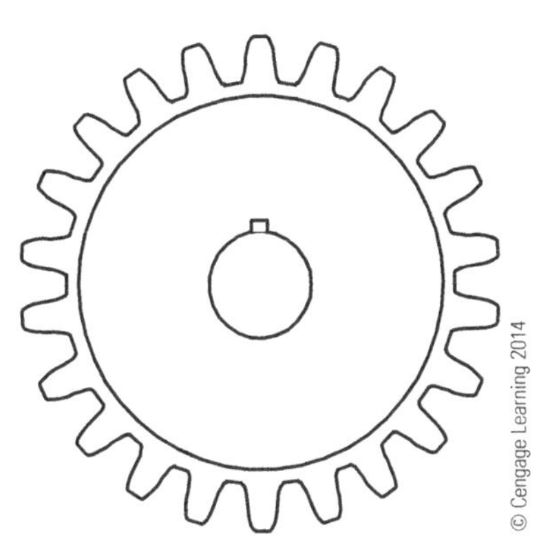

Name: ______________________________ Date: ____________________

GEARS AND GEARBOXES: GEAR RATIO

1. What is the gear ratio for each set of gears shown in the drawing?

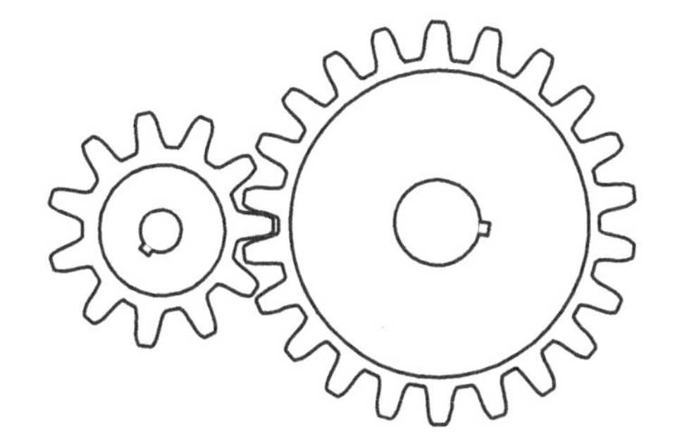

A ______________________

B ______________________

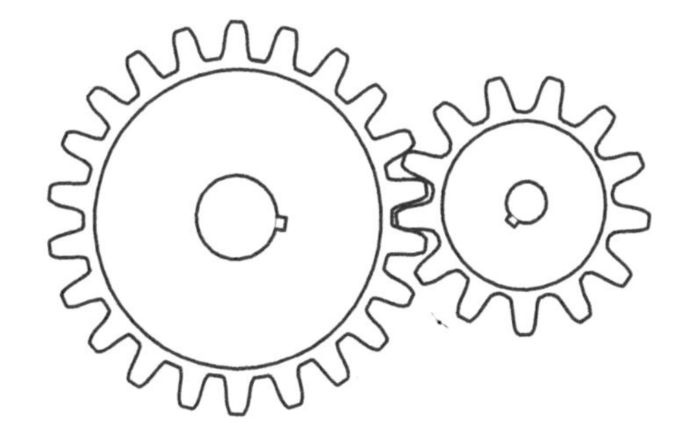

C ______________________

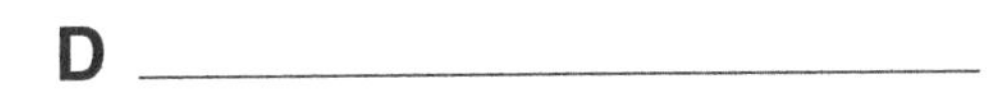

D ______________________

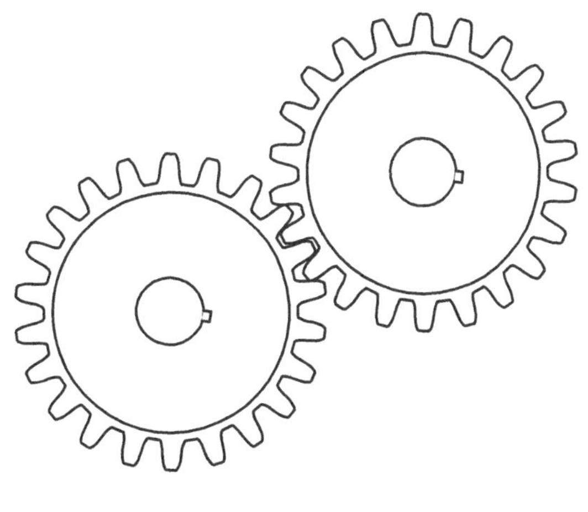

E ______________________

Name: ______________________ Date: ______________

GEARS AND GEARBOXES: GEAR TYPES

1. Match the gears to their proper names.

A

B

C

D

E

_______ Herringbone

_______ Worm

_______ Spur

_______ Bevel

_______ Helical

Name: ______________________ Date: ______________

PULLEY SPEED CALCULATIONS

1. Solve for the speed of the driven pulley. Show your work.

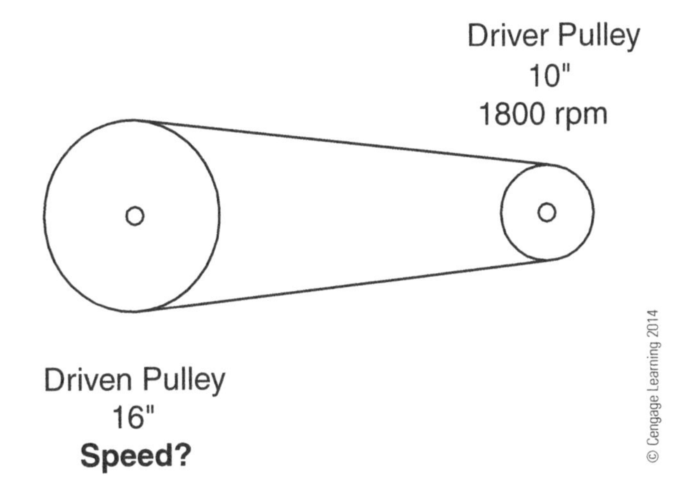

2. Solve for the speed of the driver pulley. Show your work.

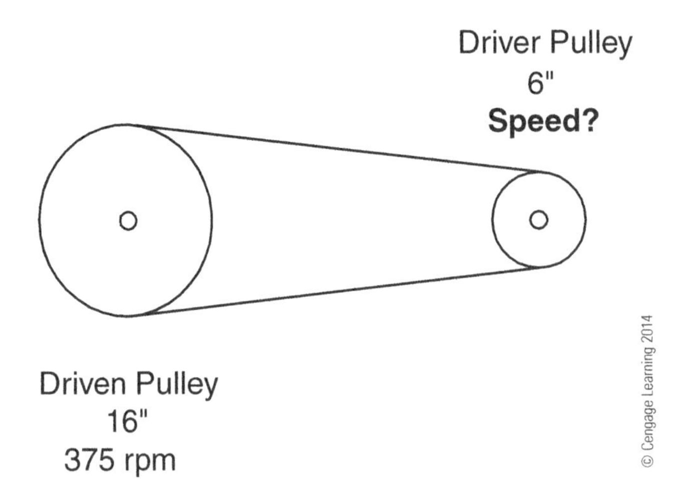

Name: ______________________________ Date: ______________________

PITCH DIAMETER CALCULATIONS

1. Solve for the pitch diameter of the driven pulley. Show your work.

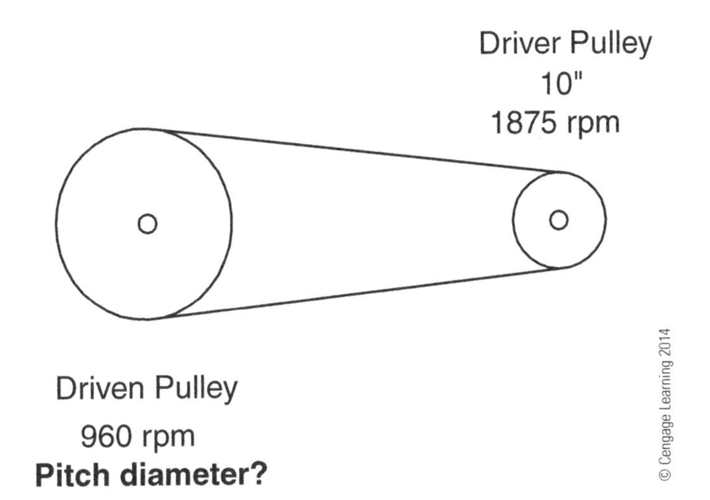

2. Solve for the pitch diameter of the driver pulley. Show your work.

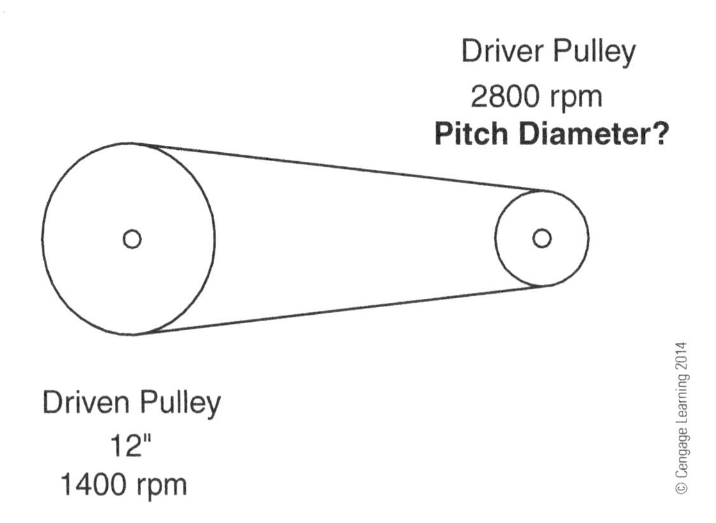

Name: ______________________ Date: ______________

SPEED CALCULATIONS: RATIO

1. Solve for the speed of the driven pulley using nothing but the ratio. Show your work.

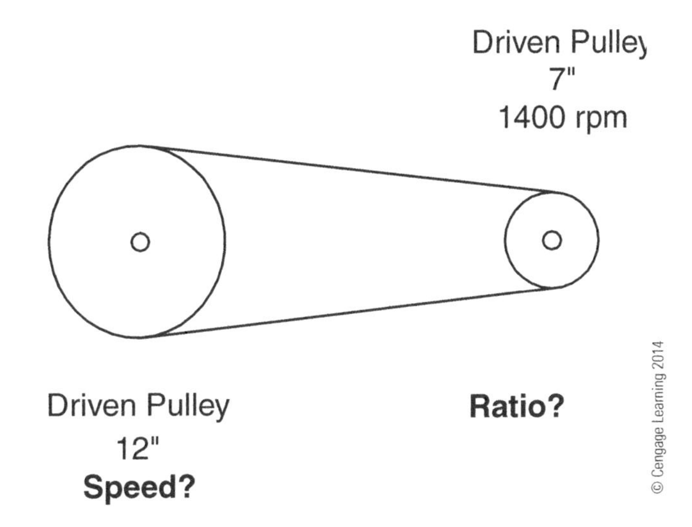

2. Solve for the pitch diameter of the driver pulley using nothing but the ratio. Show your work.

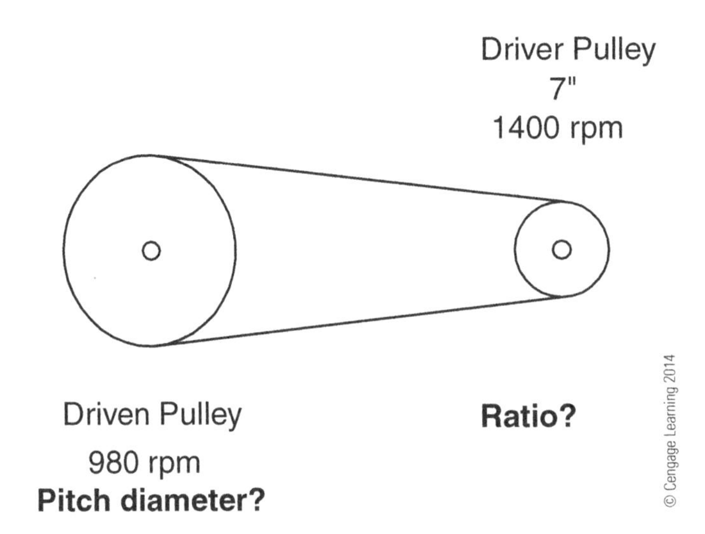

Name: ______________________ Date: ______________

SPEED CALCULATIONS: GEARS

1. Calculate for each unknown in each gear set that is in the drawing. Show your work for each problem.

A

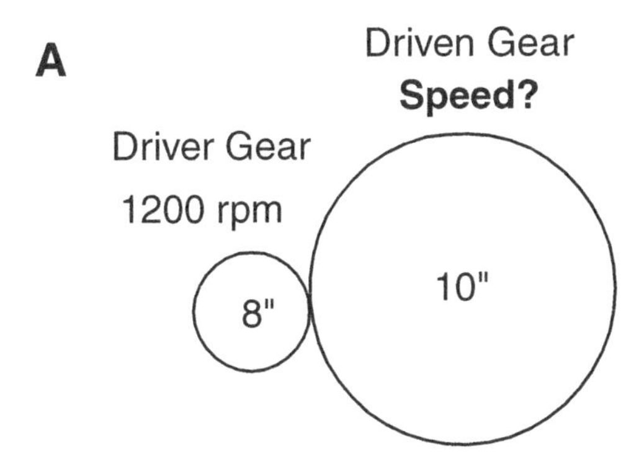

B

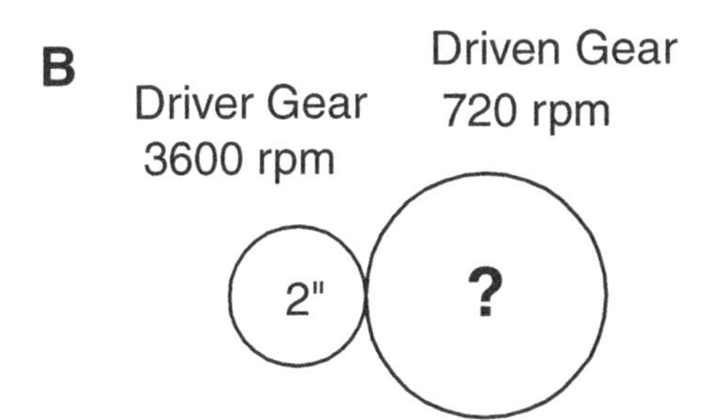

C

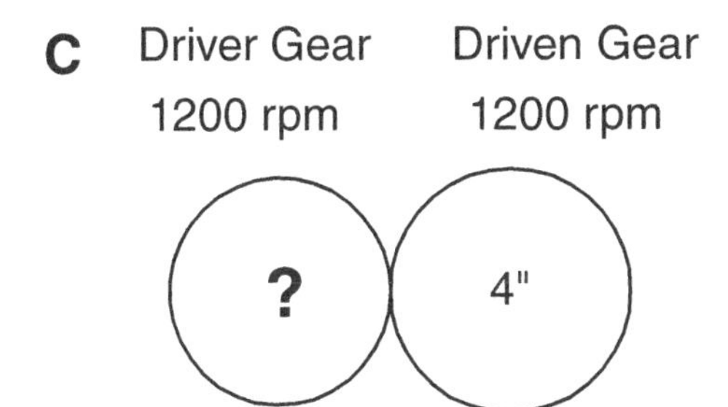

D

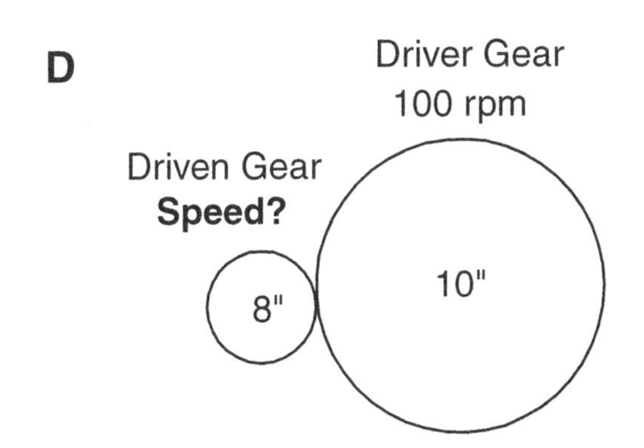

Name: ______________________________ Date: ______________________

GEAR ROTATION

1. Choose the direction of rotation for each gear that is indicated with an arrow.

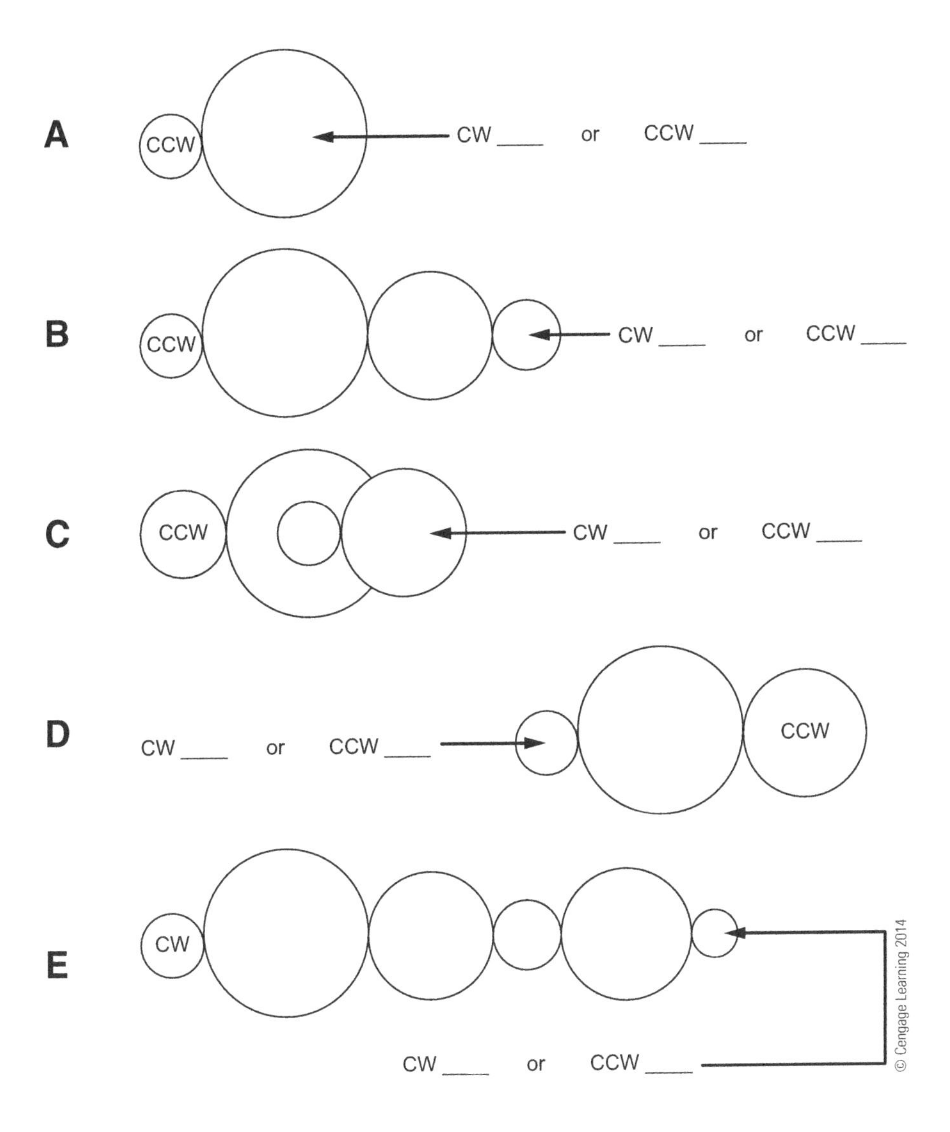

Name: ______________________ Date: ______________

BEARING LOADS

1. Match the bearing load to the correct graphic.

   ________ Combination load

   ________ Radial load

   ________ Axial load

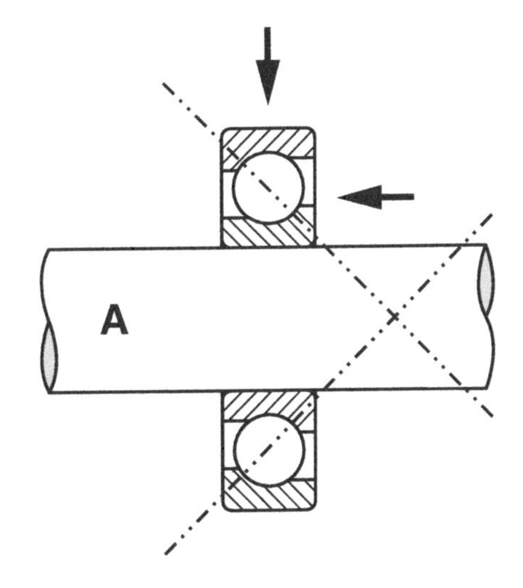

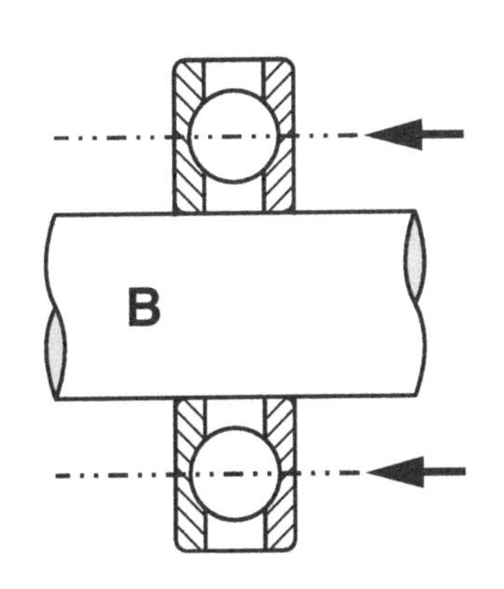

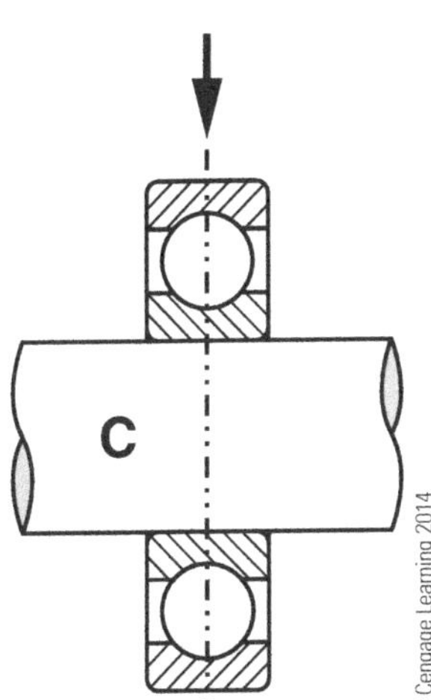

Name: ______________________________ Date: ______________

BEARING CONSTRUCTION

1. Name all of the bearing parts that are in the drawings.

   a. ______________________________

   b. ______________________________

   c. ______________________________

   d. ______________________________

   e. ______________________________

   f. ______________________________

   g. ______________________________

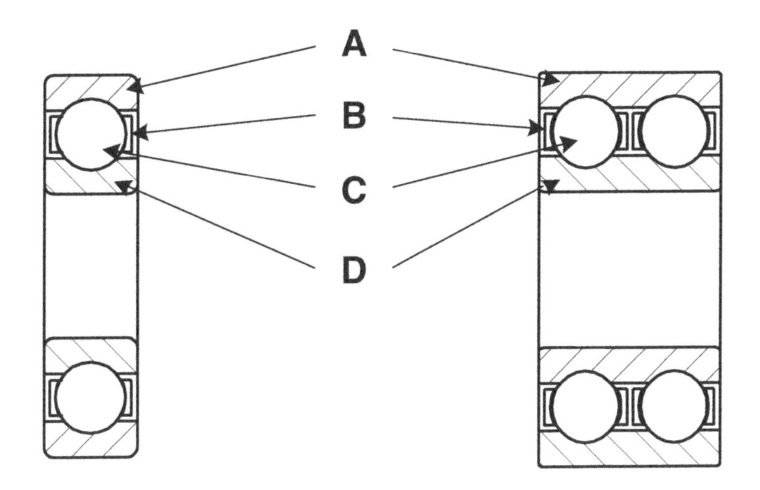

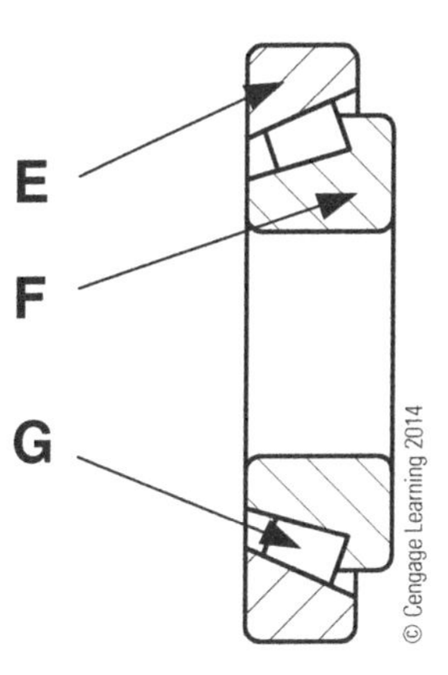

Name: ____________________ Date: ____________________

BEARING SERIES

1. Match each series to its description by drawing lines from the series to the description.

a. Series 100	1. Medium series
b. Series 200	2. Heavy series
c. Series 300	3. Extra-light series
d. Series 400	4. Light series

2. Circle the graphic that best demonstrates the four different series of bearings.

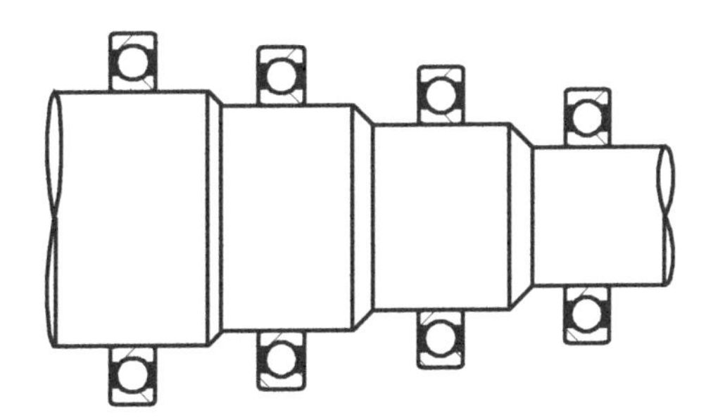

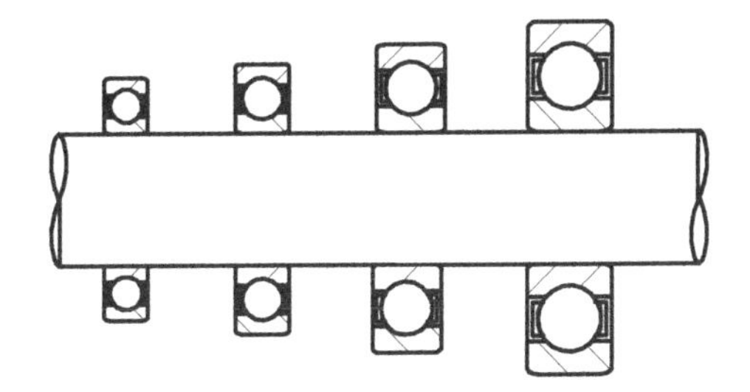

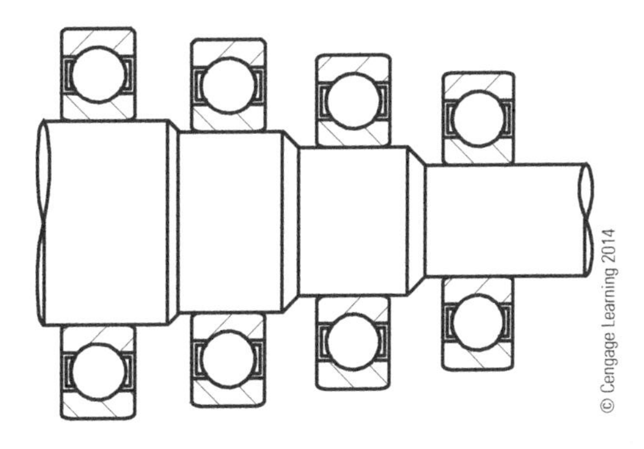

Name: ______________________________ Date: ____________________

BEARING TYPE—PART 1

MISCELLANEOUS BEARINGS

1. Match each graphic to the correct bearing type that is listed.

   _______ Deep groove ball bearing

   _______ Needle bearing

   _______ Taper-roller bearing

   _______ Ball-thrust bearing

   _______ Angular-contact ball bearing

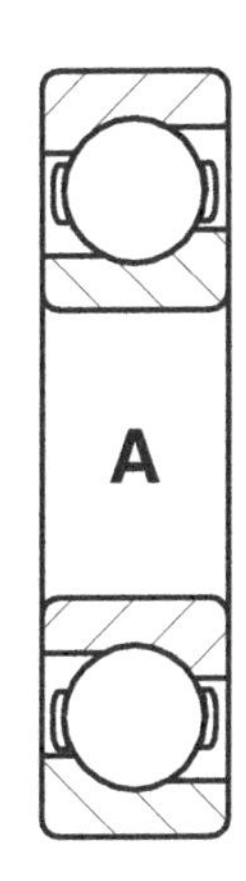

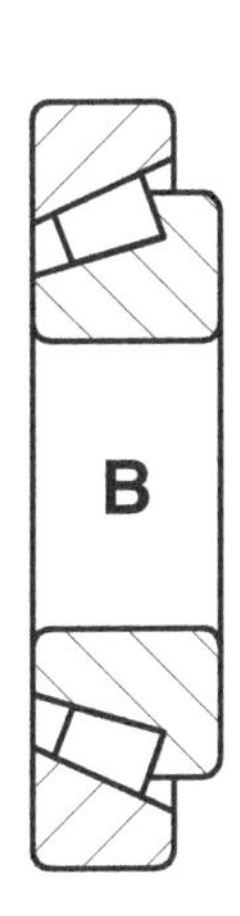

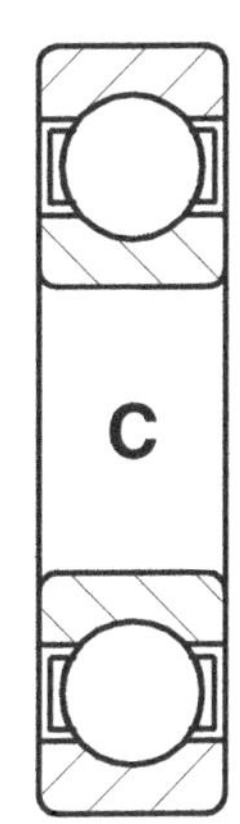

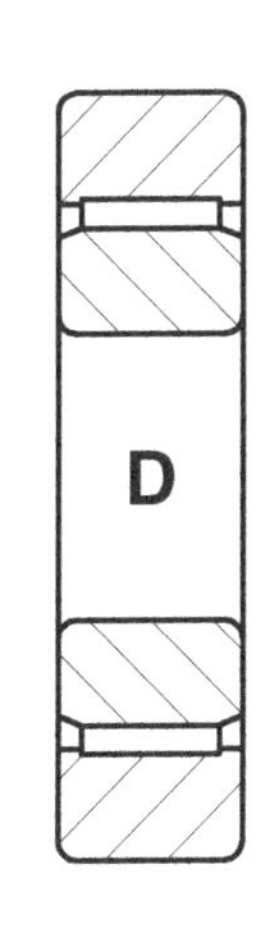

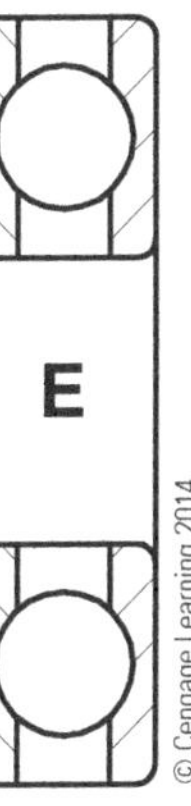

Name: ______________________________ Date: ______________________

BEARING TYPE—PART 2

BALL BEARINGS

1. Match each graphic to the correct type of ball bearing that is listed.

_______ Deep groove ball bearing

_______ Double-row deep groove ball bearing

_______ Self-aligning double-row deep groove ball bearing

_______ Angular-contact ball bearing

_______ Thrust bearing

_______ Angular-contact double-element ball bearing

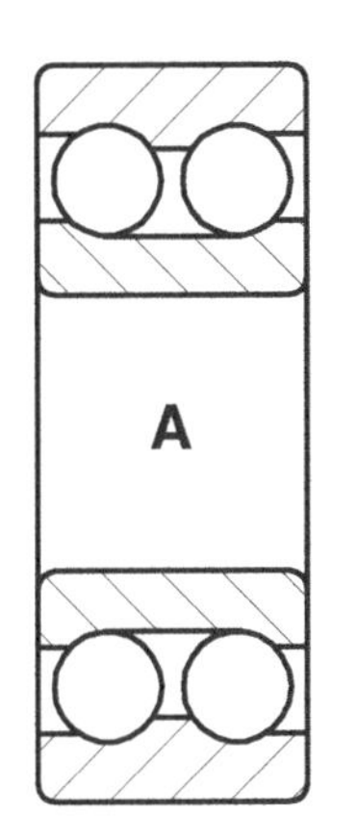

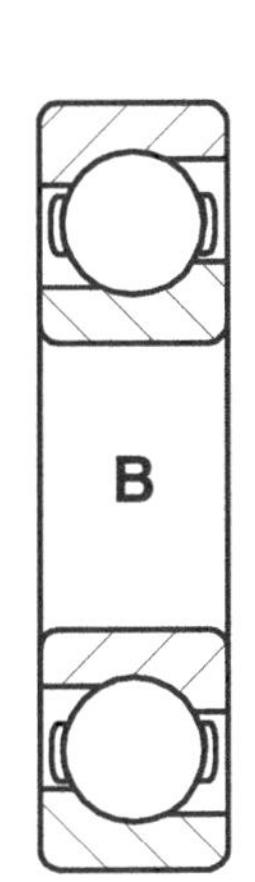

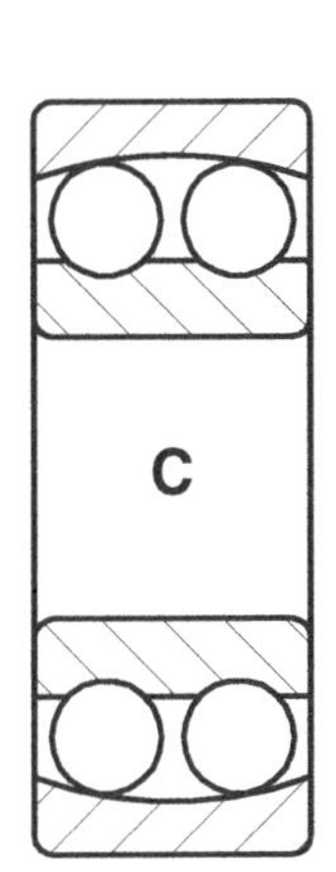

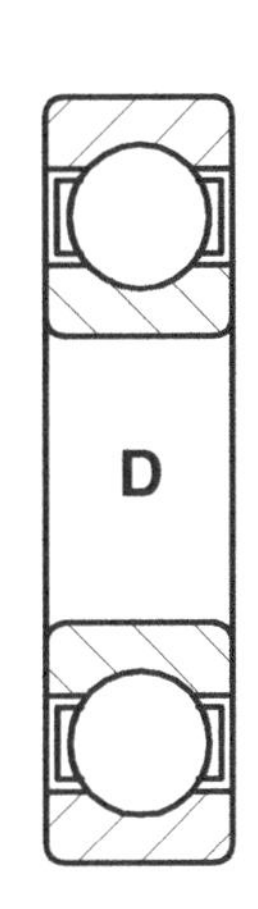

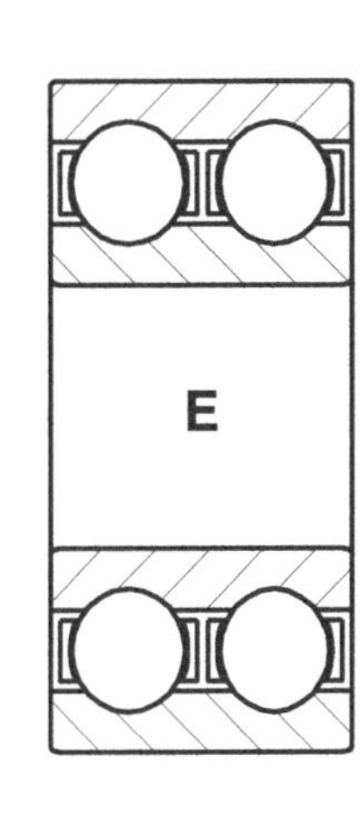

Name: ______________________ Date: ______________

BEARING TYPE—PART 3

ROLLER BEARINGS

1. Match each graphic to the correct type of roller bearing that is listed.

_______ Nonseparable straight roller bearing

_______ Double-row spherical-roller bearing

_______ Self-aligning spherical-roller bearing

_______ Straight roller bearing with a separable inner ring

_______ Straight roller bearing with a separable outer ring

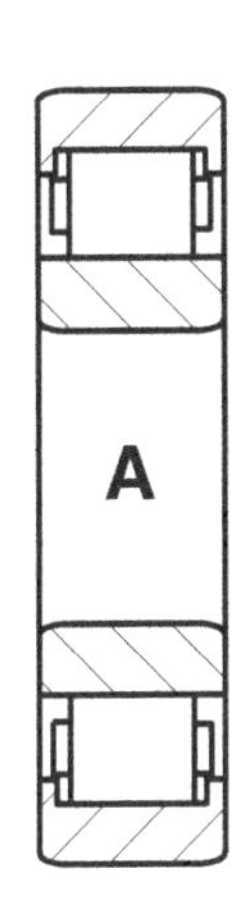

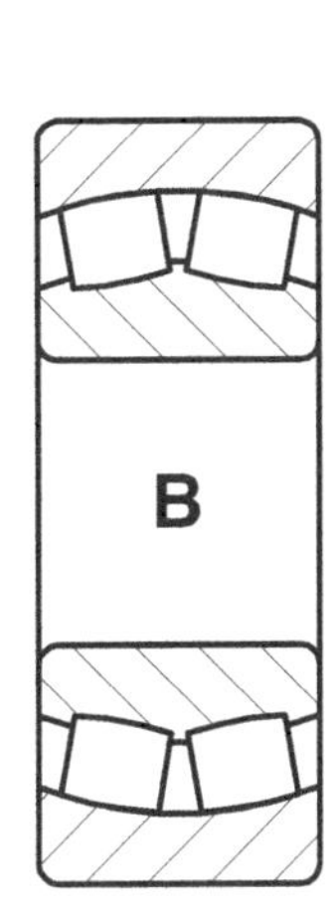

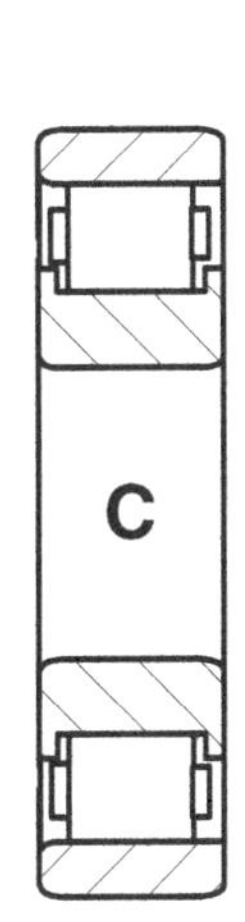

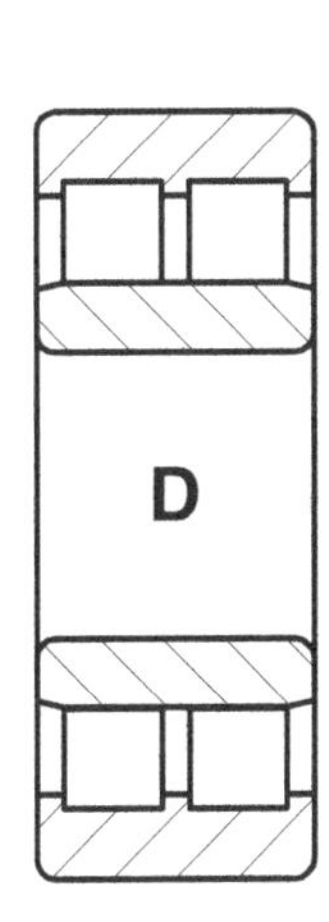

Name: ______________________________ Date: ______________

BEARING TYPE—PART 4

1. Match each graphic to the correct type of plain bearing that is listed.

_______ Split bearing

_______ Stave bearing liner

_______ Thrust washers

_______ Solid bearings

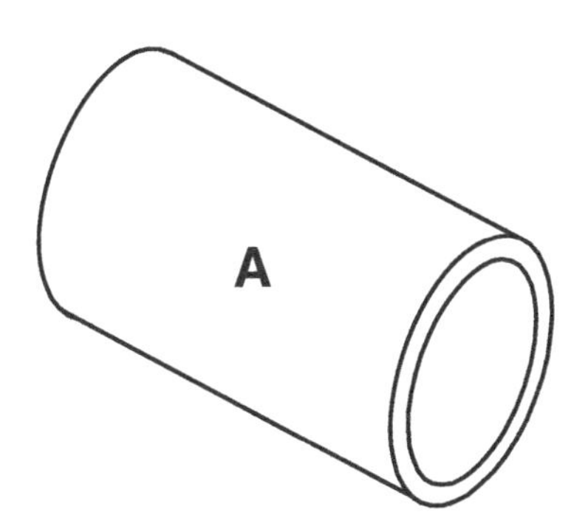

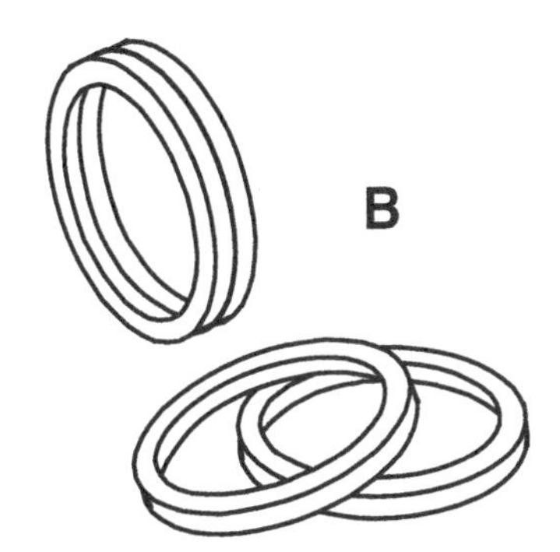

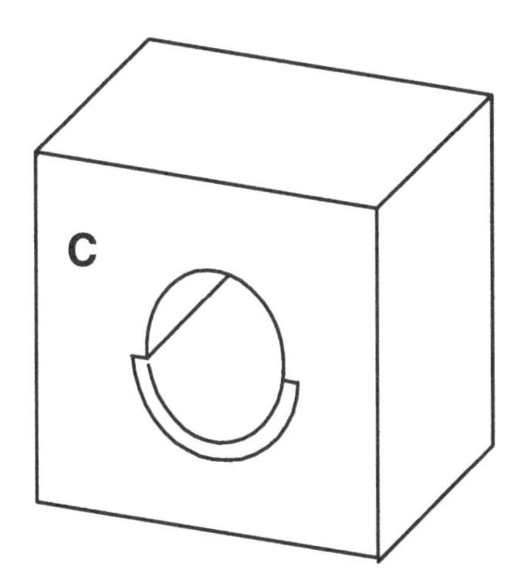

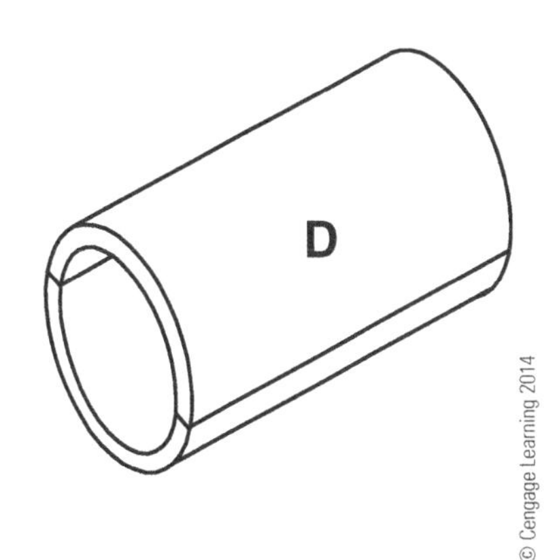

Name: ______________________________ Date: ______________

BEARING FAILURE

1. Match the cause of failure to the symptom. For each listed cause, choose the correct numbered description. You can use each cause only once.

a. Overloading or excessive thrust
b. Contamination
c. Fluting
d. High temperatures
e. Spalling
f. Improper lubrication
g. False Brinell damage
h. Pitting
i. Misalignment
j. Moisture

1. _______ This occurs when continual impacting forces (such as vibrations) are passed from one ring to the other, through the rolling elements, when there is no rotation of the shaft. Indentations are formed on the outer races from the impacts. This causes the rolling elements, as they begin to rotate, to create heat as they encounter these evenly spaced indentations.

2. _______ This is indicated by more wear on one side of the bearing than on the other side. Also, opposing sides may show signs of wear. Another indication is uneven wear on the rolling element.

3. _______ The discoloring of the raceways and the rolling elements indicates this. The metal is usually darkened with a bluish-purple coloring where the overheating occurred. It is not uncommon for the bearing to become deformed as a result of excessive internal temperatures.

4. _______ No lubrication will cause friction and overheating within the bearing. Overlubrication can place internal pressure on the bearing because the rolling elements will have to move the excessive amount of lubrication within the bearing as well as the load.

5. _______ This is an indication of electrical current flow through the bearing. The problem occurs where the contact is made between the outer race and the rolling elements and between the rolling elements and the inner race. The presence of thin lines etched into the races makes it easy to recognize this condition.

6. _______ Rusting surfaces are an indication of this cause of failure. Oxidation occurs when moisture is present and lubrication is lacking. This is commonly referred to as fretting corrosion. If a bearing has a suitable amount of lubrication, oxidation should not occur even when the bearing is in a moist environment.

7. _______ This occurs mostly when welding currents pass through the bearing.

8. _______ This occurs any time a foreign particle enters the bearing. This usually occurs when the bearing is operating in a dirty environment. The damage is usually in the form of deformity. This will cause damage to the races and rolling elements as well.

9. _______ This is indicated by the flaking away of metal pieces due to metal fatigue. This occurs when the rolling element and the bearing race begin to flex because an excess load is being applied to them. This flexing (distortion) is momentary and repetitive. As the metal begins to fatigue, microscopic fractures begin to appear. This causes the metal to begin flaking.

10. _______ Thrust damage is indicated by marks on the shoulder or upper portions of the inner and outer races. There will also be anywhere from a slight discoloration to heavy galling.

Name: ______________________________ Date: ______________________

PULLEY AND SPROCKET ALIGNMENT

1. What type of misalignment is this?

____

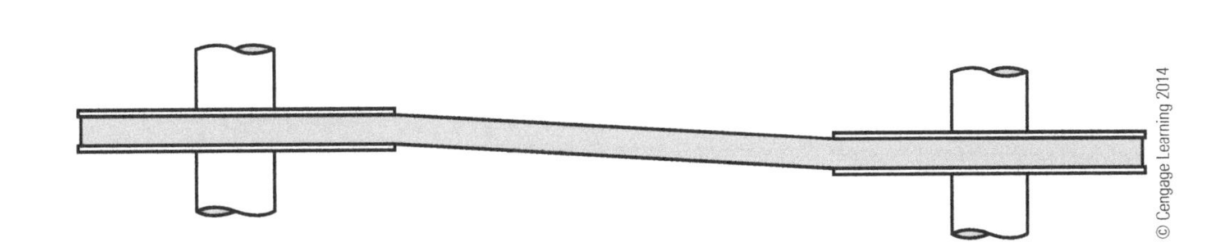

2. Describe what should be done to correct the misalignment in question 1.

____

____

____

3. What type of misalignment is this?

____

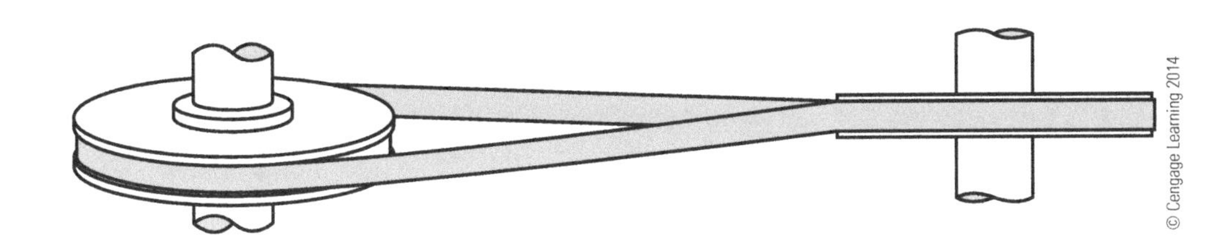

4. What should be done to correct the misalignment in question 3?

____

____

____

5. What type of misalignment is this?

____

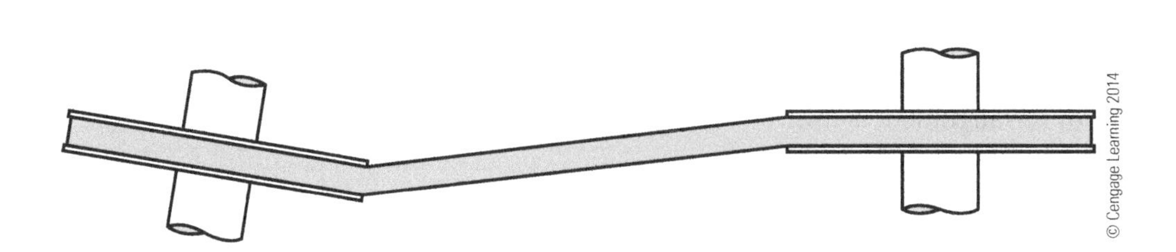

6. What should be done to correct the misalignment in question 5?

____

____

____

Name: ______________________ Date: ______________

COUPLING ALIGNMENT: COUPLING FLANGE

1. Name the lettered parts of the flange.

   a. ______________________

   b. ______________________

   c. ______________________

   d. ______________________

   e. ______________________

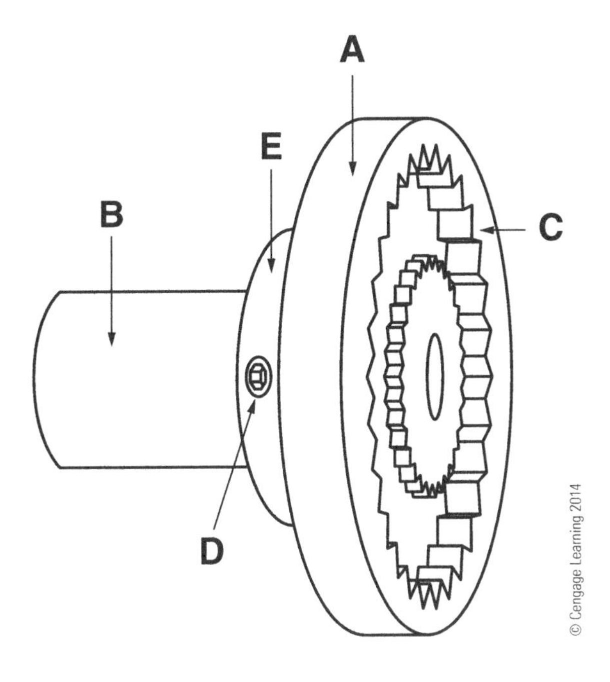

Name: ______________________ Date: ______________

COUPLING ALIGNMENT: SHIMS

1. Identify which of the following pictures shows a shim.

   ______

A

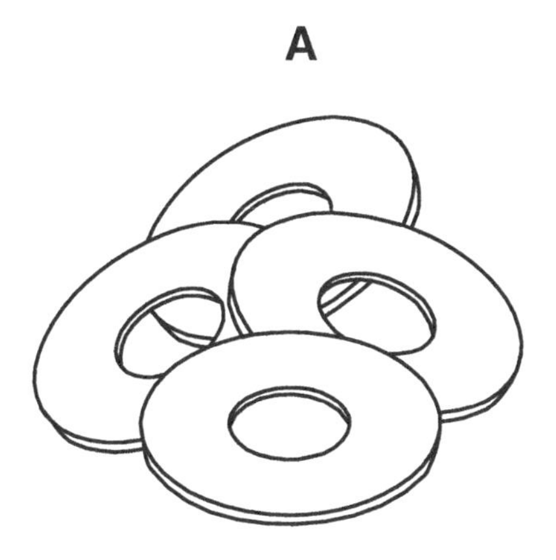

B

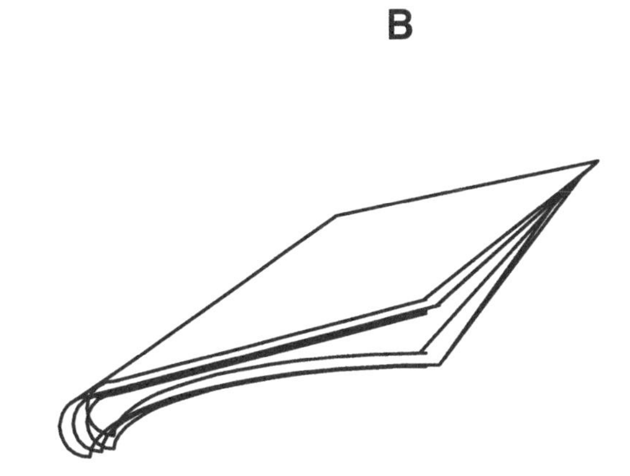

C

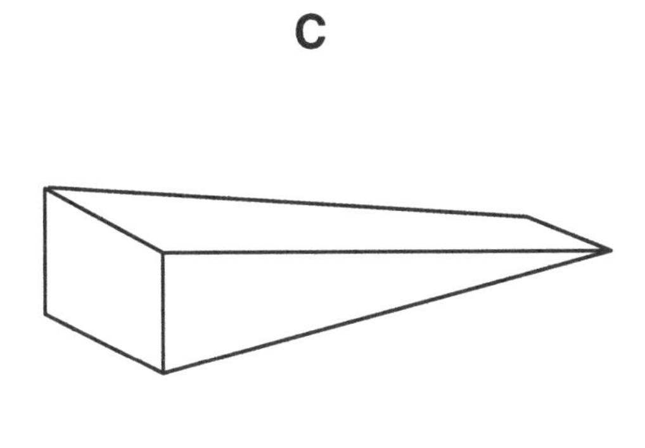

D

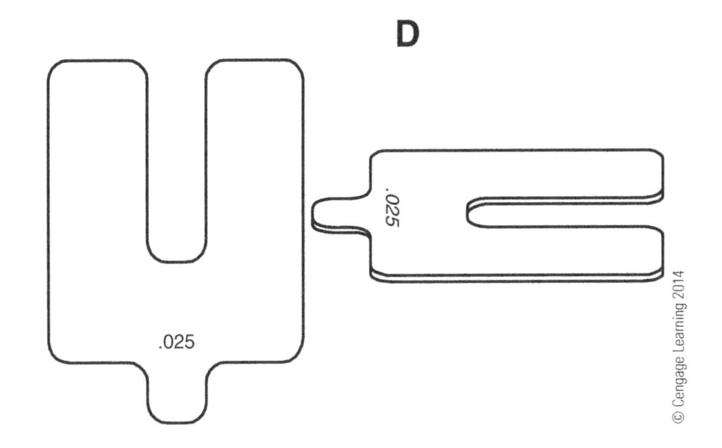

Name: ______________________ Date: ______________

COUPLING ALIGNMENT: SOFT FOOT

1. Match the correct graphic to each type of soft foot.

   1. _______ Angular

   2. _______ Parallel

   3. _______ Springing

   4. _______ Induced

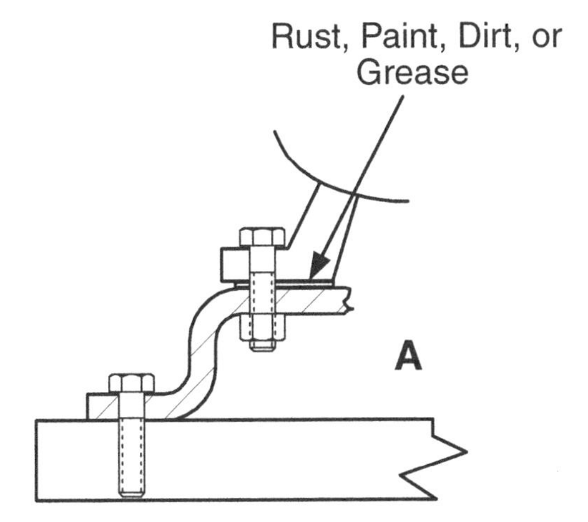

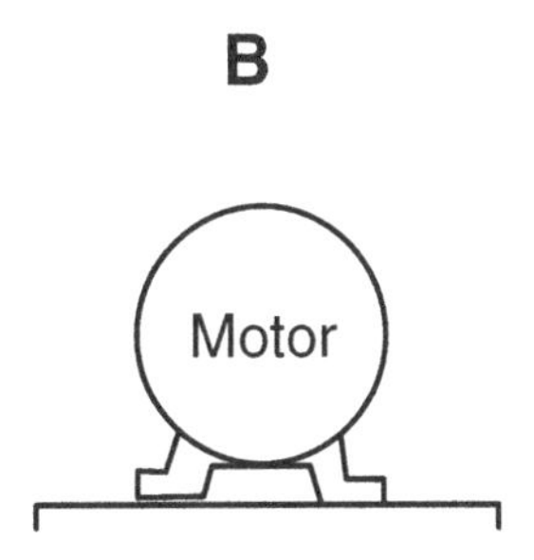

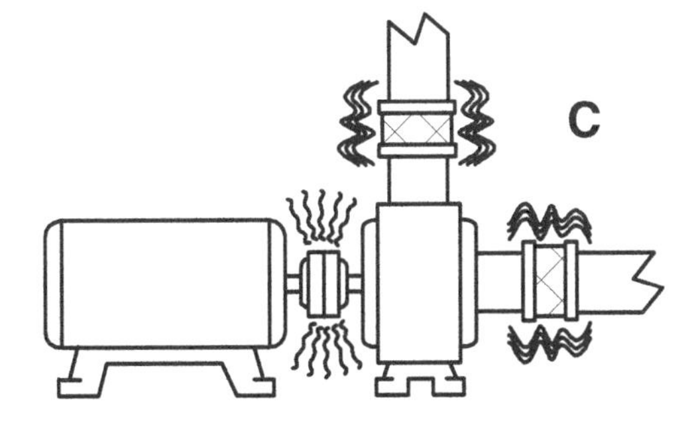

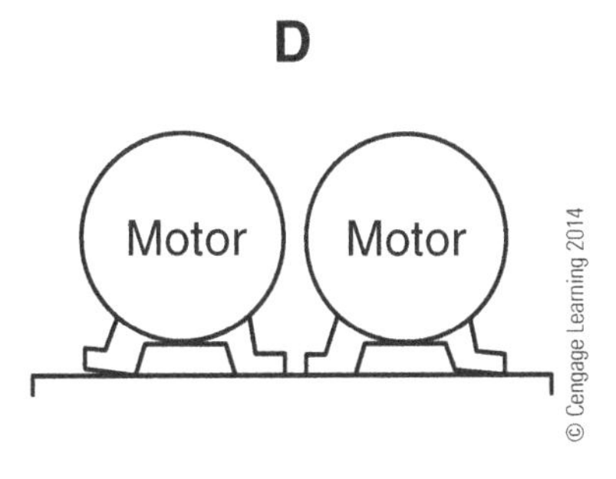

Name: ______________________________ Date: ______________________

COUPLING ALIGNMENT: HORIZONTAL/VERTICAL

1. Draw, in the space provided below, a motor and a pump from the horizontal view.

2. Draw, in the space provided below, a motor and a pump from the vertical view.

Name: ______________________ Date: ______________

COUPLING ALIGNMENT: MISALIGNMENTS

1. Match the misalignment to its graphical representation.

   1. ______ No misalignment
   2. ______ Angular misalignment
   3. ______ Offset and angular misalignment
   4. ______ Offset misalignment

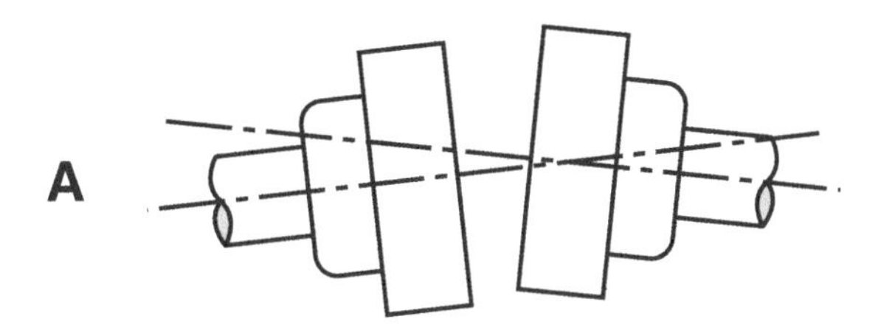

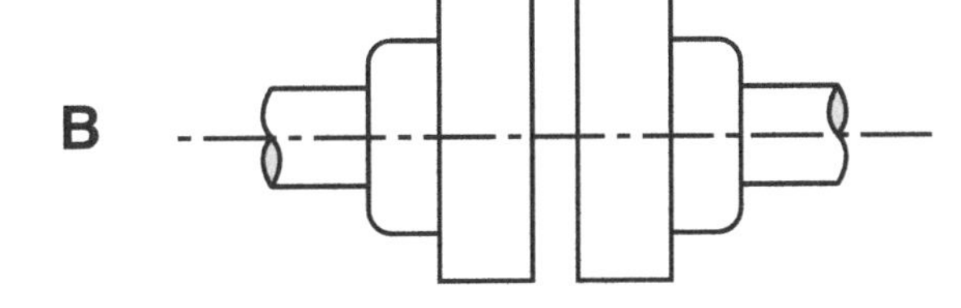

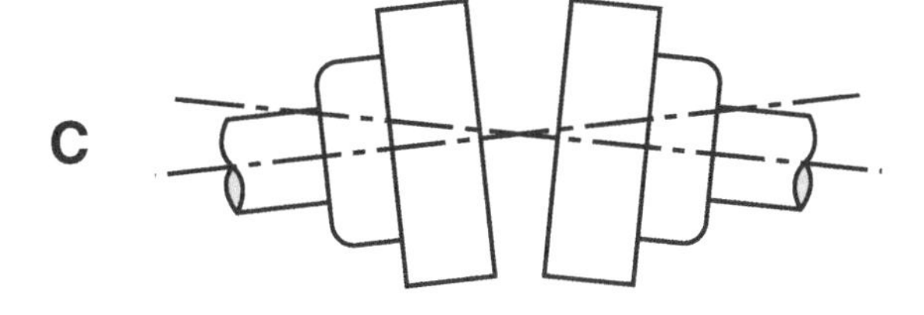

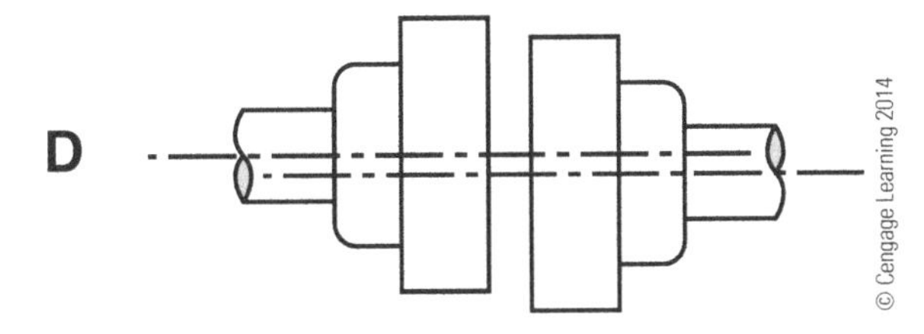

Name: ______________________________ Date: ______________________

COUPLING ALIGNMENT: TOTAL INDICATOR READING

1. What is the total indicator reading (TIR) for each of the drawings?

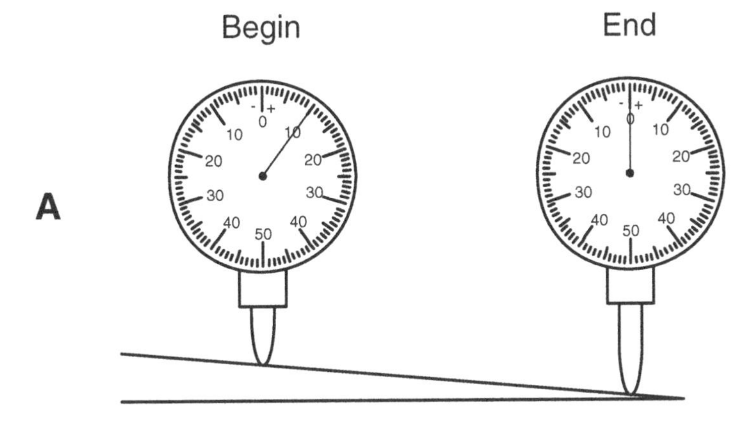

TIR = ____________

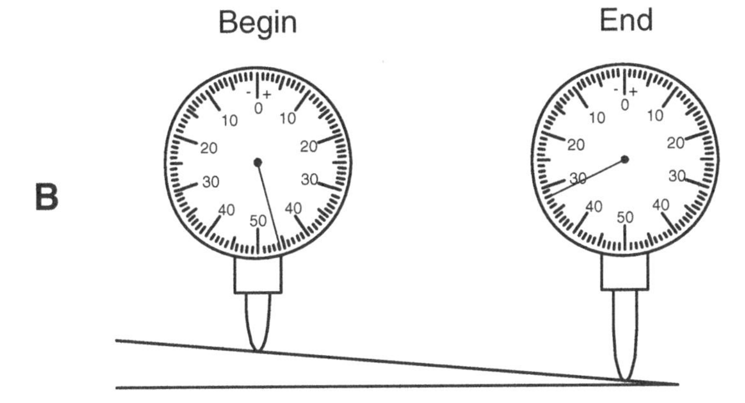

TIR = ____________

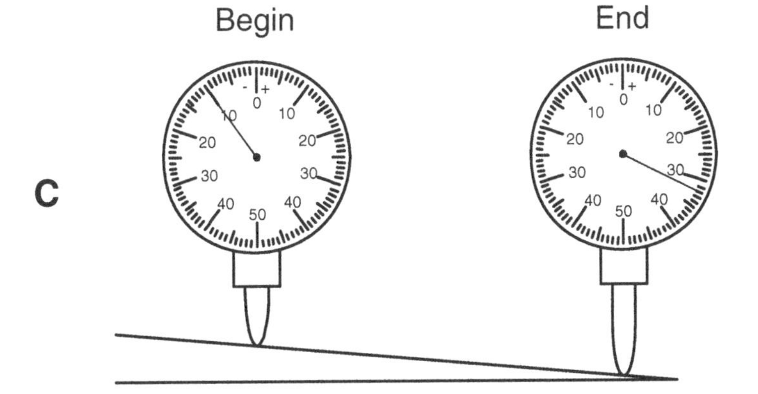

TIR = ____________

Name: ______________________________ Date: ______________

DIAL INDICATOR READINGS

1. Indicate on the drawing where to take the dial indicator readings when checking for an offset misalignment in the horizontal position.

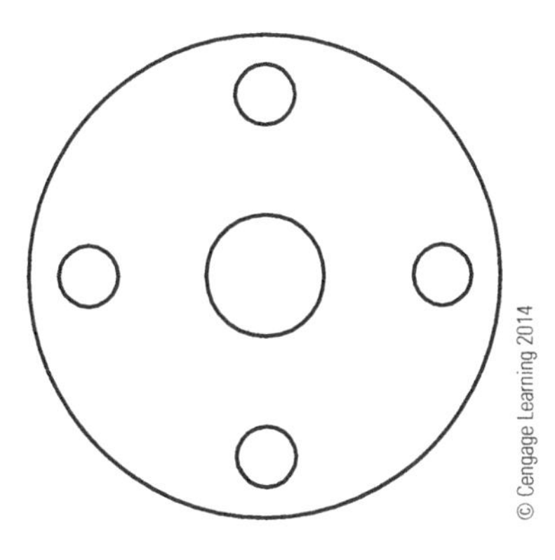

2. Indicate on the drawing where to take the dial indicator readings when checking for an angular misalignment in the vertical position.

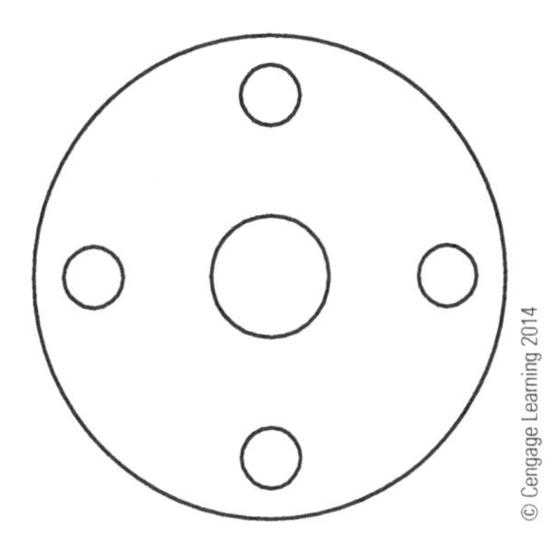

3. Indicate on the drawing where to take the dial indicator readings when checking for an offset misalignment in the vertical position.

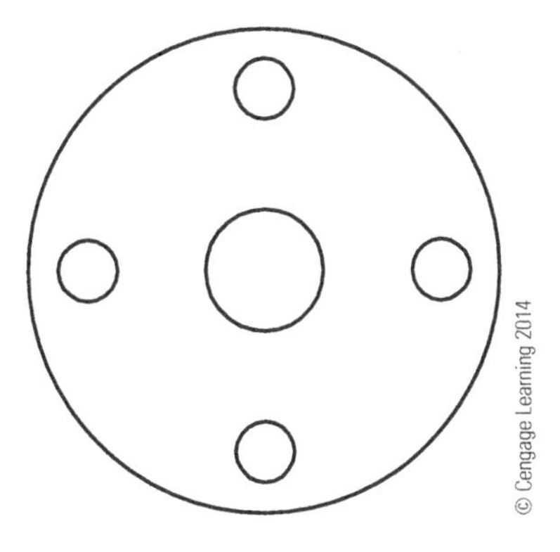

4. Indicate on the drawing where to take the dial indicator readings when checking for an angular misalignment in the horizontal position.

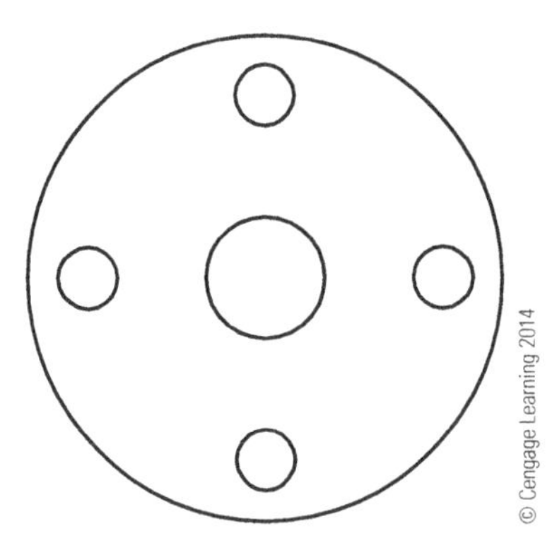

5. Indicate on the drawing where the dial indicator needle would be pointing if the dial indicator were zeroed prior to moving and the plunger moved 0.028 in.

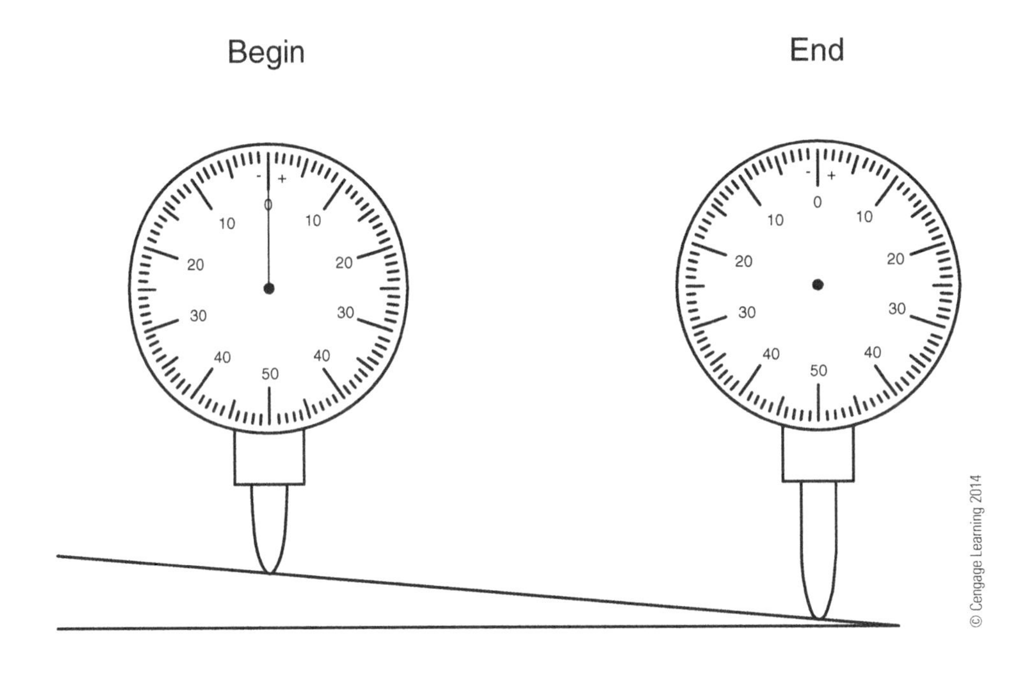

Name: ______________________ Date: ______________

COUPLING SHAFT ALIGNMENT METHODS

1. Match each figure to the correct method of alignment.

   1. _______ Feeler gauge method

   2. _______ Dial indicator method

   3. _______ Reverse dial indicator method

   4. _______ Laser alignment method

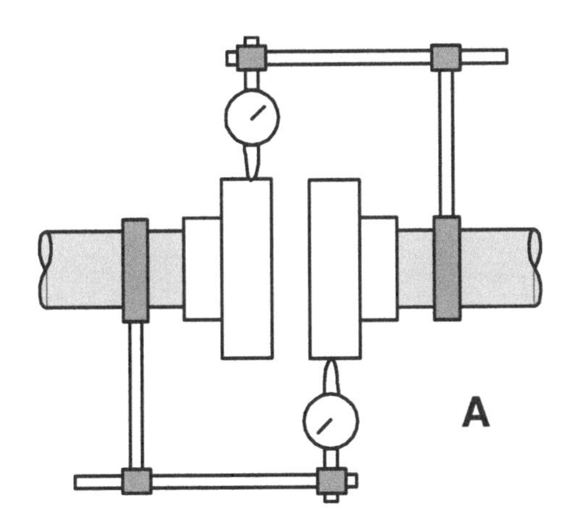

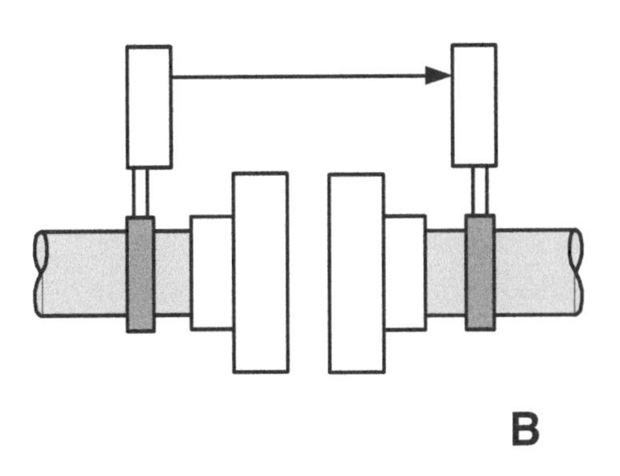

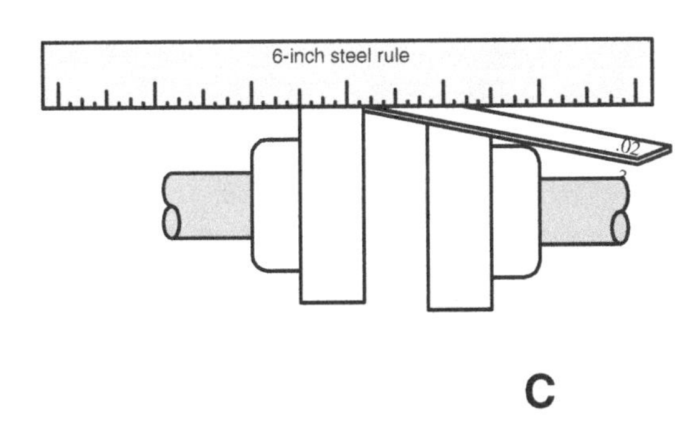

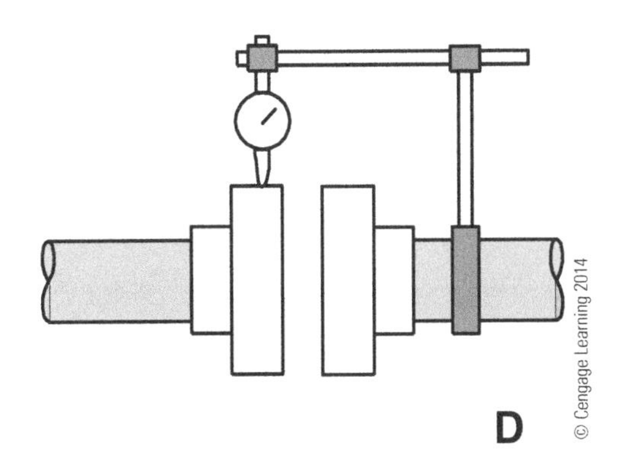

Name: ______________________ Date: ______________

COUPLING SHAFT ALIGNMENT

1. Number the following procedures in the proper chronological order.

a. _______ Check for any vertical offset misalignment.

b. _______ Slide the coupling flanges toward each other until each is at the end of the shaft on which it is mounted and lock them down using the setscrew that is present on the hub.

c. _______ Tighten the motor and the gearbox to the baseplate to the required torque specifications.

d. _______ Loosely thread the bolts that will hold the motor and gearbox down into the baseplate. Do not tighten them down yet.

e. _______ Check for vertical angular misalignment.

f. _______ Repeat the previous steps until all of the alignments are within the specified tolerance of the coupling.

g. _______ Next, check for horizontal angular misalignment.

h. _______ If horizontal offset misalignment is found to exist, then adjust the jackscrews that should be adjusted to eliminate the offset misalignment.

i. _______ Set the motor and gearbox on the baseplate that they are to be mounted upon, making sure that the coupling flanges are on the shafts. Do not tighten the flanges down yet.

j. _______ If any vertical offset misalignment is found, place shims under the correct mounting feet of the motor and/or gearbox to correct the misalignment.

k. _______ If horizontal angular misalignment is found to exist, then adjust the jackscrews that should be adjusted as to eliminate the angular misalignment.

l. _______ If any vertical angular misalignment is found, place shims under the correct mounting feet of the motor and/or gearbox to correct the misalignment.

m. _______ Check for horizontal offset misalignment.

Name: ______________________ Date: ______________

SHIM CALCULATION: DIAL INDICATOR METHOD—PART 1

1. Calculate the thickness of the shim stock that will be needed to correct a vertical angular misalignment. Here are the readings:

   12 o'clock position: 0.000 in.

   6 o'clock position: 0.056 in.

   Diameter traveled by plunger: 2¾ in.

   Distance between feet: 7 in.

Name: ______________________________ Date: ____________________

SHIM CALCULATION: DIAL INDICATOR METHOD—PART 2

1. Calculate the thickness of the shim stock that will be needed to correct a vertical offset misalignment. Here are the readings:

   12 o'clock position: 0.000 in.

   6 o'clock position: 0.177 in.

   Rod sag: 0.009 in.

Name: ______________________________ Date: ______________

LUBRICATION

1. Match the word to its definition.

   a. Hydrocarbon

   b. Semisolid

   c. Viscosity

   d. Lubricant

   _______ A substance that reduces friction by providing a smooth surface of film over parts that move against each other

   _______ Any compound containing mostly hydrogen and carbon

   _______ A gel-like substance that has the characteristics of both a solid and a liquid

   _______ The internal friction of a lubricant, caused by the molecular attraction

2. Match the appropriate method of lubrication to its drawing.

   1. _______ Submersion—ring

   2. _______ Wick

   3. _______ Submersion—splash

   4. _______ Submersion—chain

   5. _______ Drip

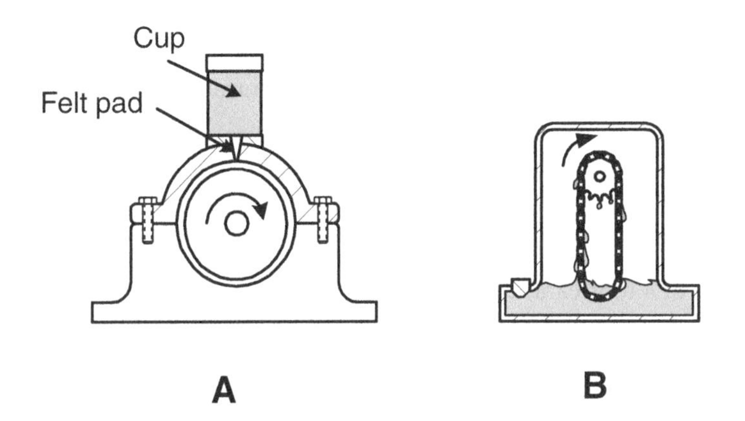

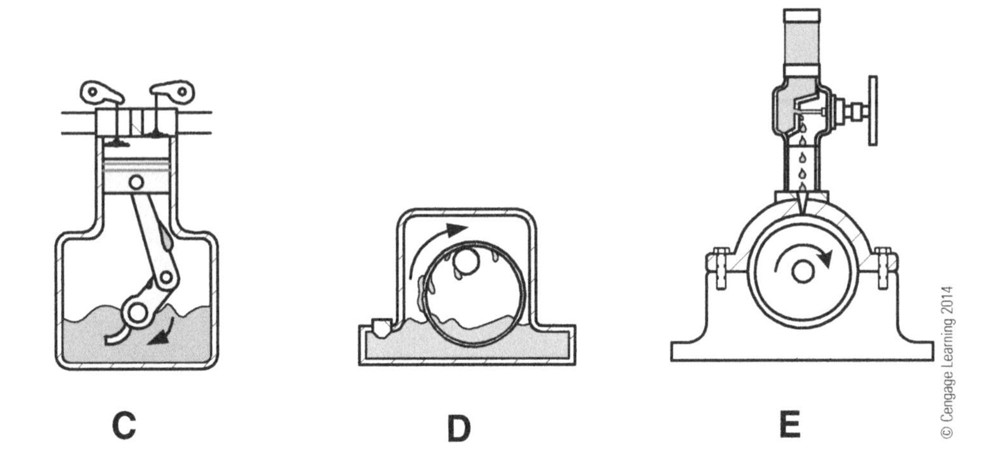

Name: ______________________ Date: ______________

PACKING MATERIAL

1. List the various types of fibers that are used for packing in a packing seal.

    ____

    ____

    ____

    ____

    ____

2. Of the fibers listed in question 1, which is the most dangerous?

    ____

Name: ______________________________ Date: ____________________

TYPES OF PACKING MATERIAL

1. Match each description to the correct type of packing material.

   a. Twisted fiber packing

   b. Interlocking packing

   c. Square-braided packing

   d. Braid-over-braid packing

1. _______ This packing material can be made from cotton, plastic, or leather and is manufactured to have a square-shaped cross section. It is not uncommon for this packing to have metal wires within the strands of packing material. This is to add strength and give the packing better shape-holding characteristics. Because this type of packing is usually used in heavier applications, it is impregnated with oil or grease. This packing is generally stronger than the twisted type of packing because the fibers are braided.

2. _______ This type of packing is the strongest of all the packing types. This type has interlocking fibers that are resistant to fraying or unraveling. This type of packing is manufactured to have a square cross section. It is impregnated with a lubricant as well. This type of packing is used for heavy-duty applications where it would not be desirable to use the other types of packing.

3. _______ This type of packing has a circular cross section. The center is sometimes composed of lead wires, which, as mentioned before, give strength to the packing and give the packing better shape-holding characteristics. The lead wires are covered with a braided jacket, which in turn is covered with a braided jacket. It is for this reason that this packing is sometimes referred to as a jacket-over-jacket. As with the others, this type of packing is also impregnated with lubricants.

4. _______ This type of packing is the most widely used type of packing. It is usually made with twisted strands of cotton that have been lubricated with mineral oil and graphite. Because this packing is simply twisted and not interlocked, it is not as strong as the other types of packing.

Name: ______________________ Date: ______________

STUFFING BOX SEAL—PART 1

1. Label the parts of the stuffing box seal in the drawing.

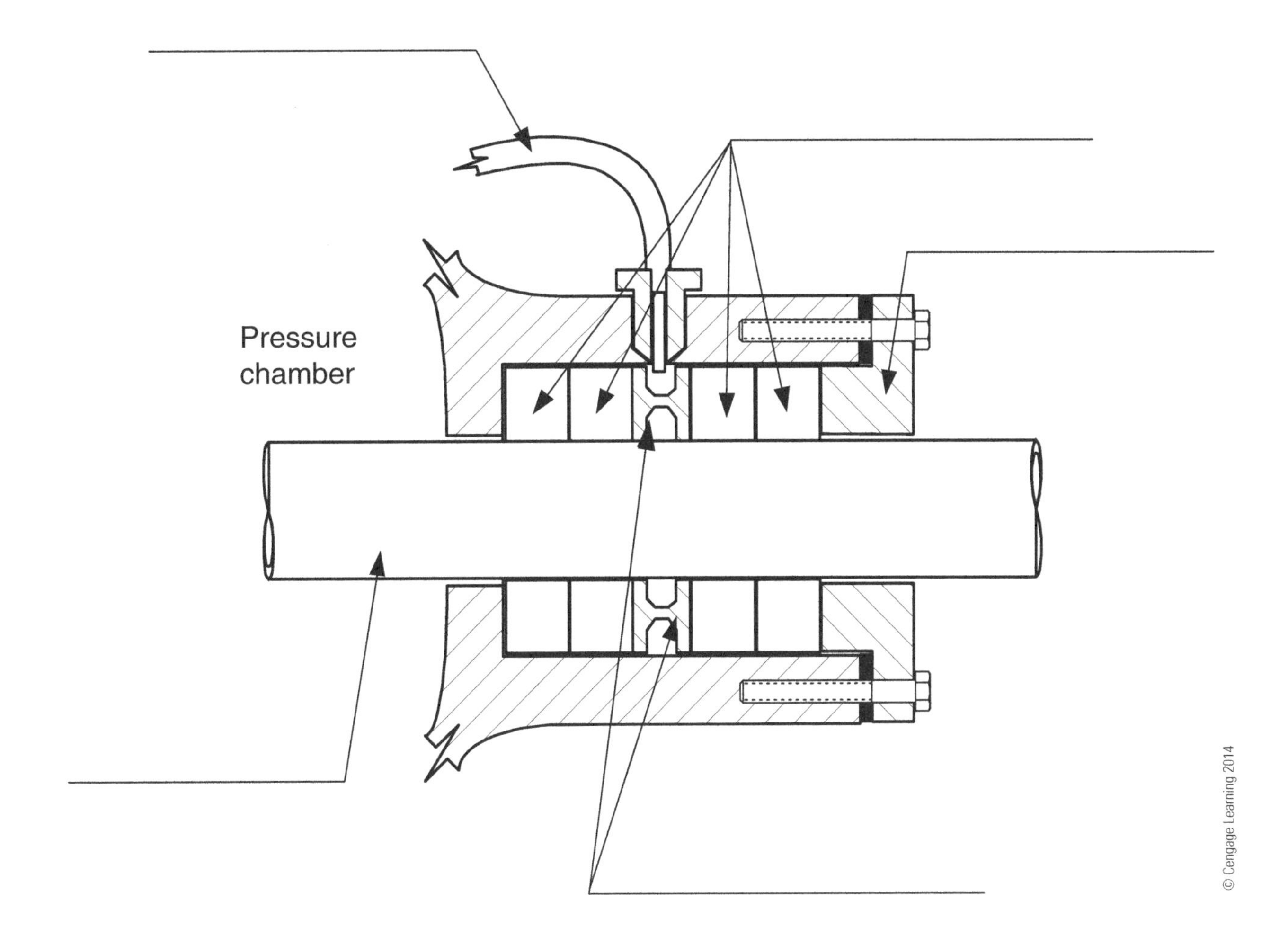

Name: ______________________________ Date: ____________________

STUFFING BOX SEAL—PART 2

1. Number in proper chronological order the following procedures for installing packing into a seal.

   a. _______ Wrap the packing material around the shaft at least five times and precut all of the rings without scarring the shaft.

   b. _______ Fill the vessel or cavity with the fluid to be used during operation. Monitor the seal. If there is an excessive amount of leakage, slowly tighten the follower until the desired amount of leakage is present. If there is no leakage at all within a few minutes of filling the vessel or cavity, then loosen up the follower slightly until the desired amount of leakage is present.

   c. _______ Install the follower. If everything was done correctly, the follower should barely enter the stuffing box.

   d. _______ Set the first ring on the shaft and gently tamp the ring into the stuffing box. Make sure that the ring goes into the stuffing box squarely, to ensure that the packing does not bind or roll.

   e. _______ Repeat the process for the next two rings, making sure to firmly tamp each ring into position against the preceding ring.

   f. _______ Install the final two rings, making sure to firmly tamp each ring into position.

   g. _______ Allow the unit to run for a few hours, and then check the leakage. The follower may need to be slightly adjusted after the unit has been run.

   h. _______ Remove the follower from the stuffing box and remove all of the old packing material.

   i. _______ Separate all of the rings.

   j. _______ Place the lantern ring (if one is used) on the shaft and gently slide it into position, making sure not to allow the front edge to pass the lubrication port.

   k. _______ Make sure to clean all surfaces (including the follower) before trying to install any new packing material.

   l. _______ Tighten the stuffing box, alternating from side to side, until the follower has compressed the packing material about ⅓ of the way into the stuffing box. This allows the packing and lantern ring to occupy about ⅔ of the stuffing box. Remember, when tightening, it is important to keep an eye on the lantern ring through the lubrication port. Do not let it pass the port.

Name: ______________________ Date: ______________

MECHANICAL SEALS

1. Identify the three types of mechanical seals in the drawings.

A ______________________

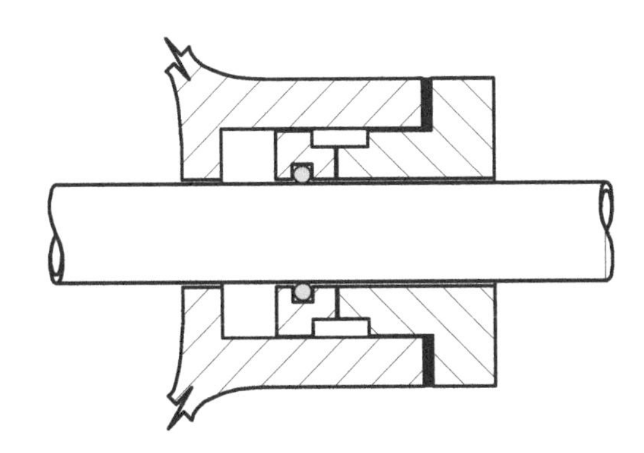

B ______________________

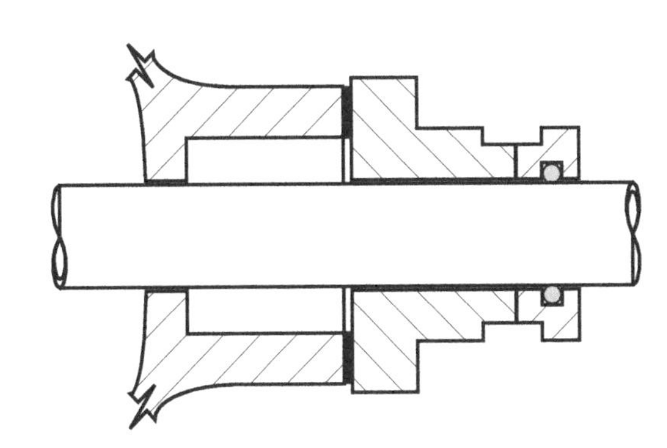

C ______________________

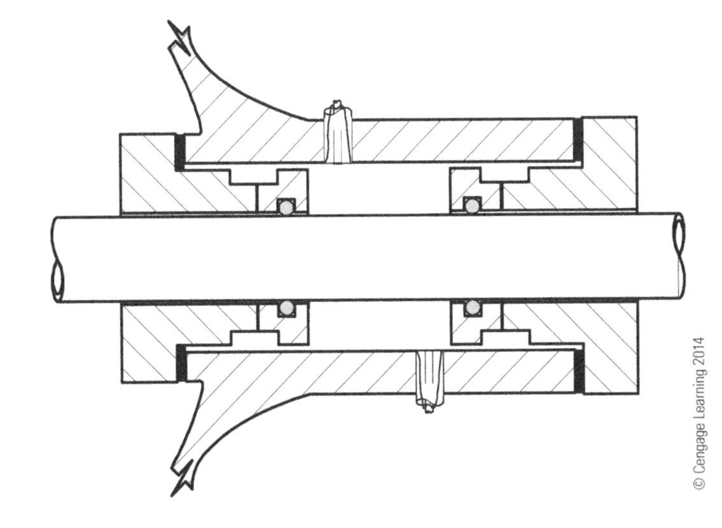

Name: ______________________ Date: ______________

RADIAL LIP SEALS

1. Identify each of the single-element radial lip seals that are in the drawing as either cased (A) or bonded (B).
   a. Cased radial lip seal, single-element
   b. Bonded radial lip seal, single-element

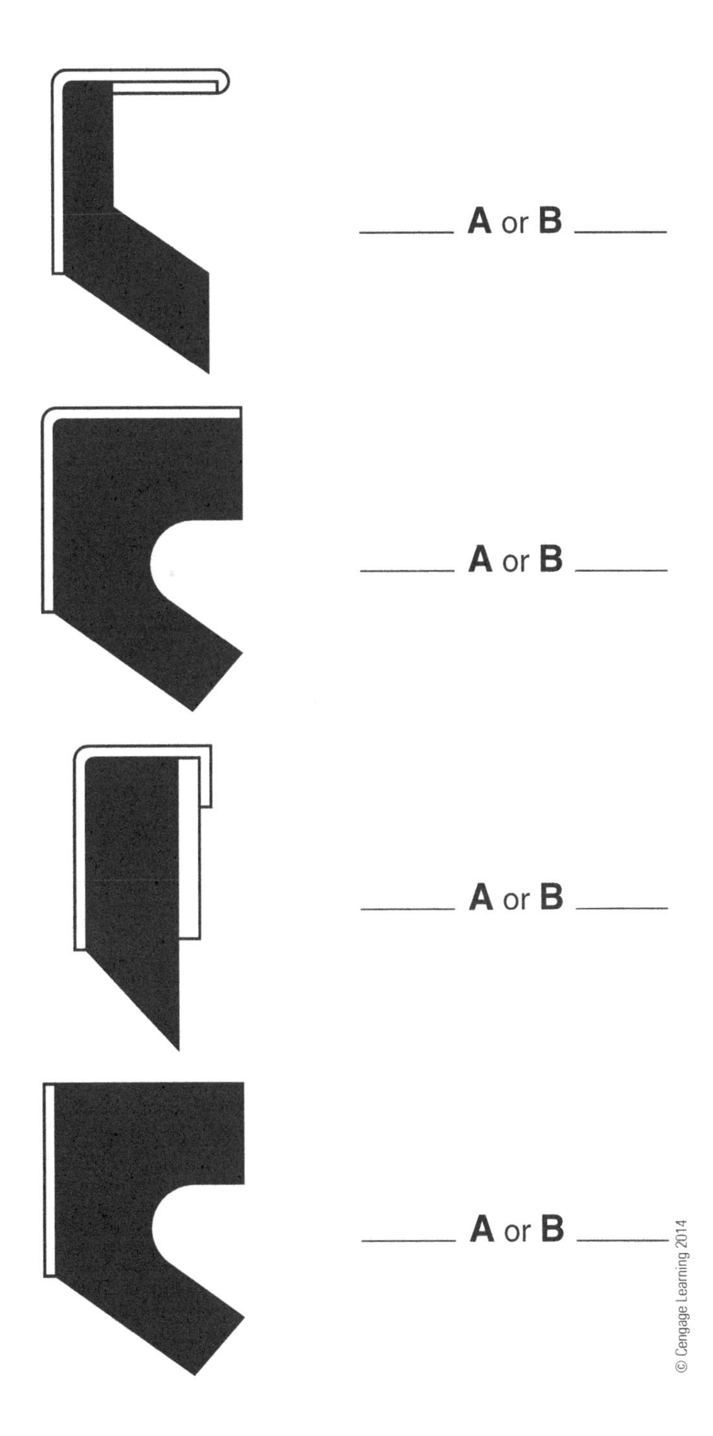

Name: ______________________________ Date: ____________________

PUMP TYPES

1. Match each graphic to its name.

   1. _______ Vane pump

   2. _______ Gear pump

   3. _______ Piston pump

A

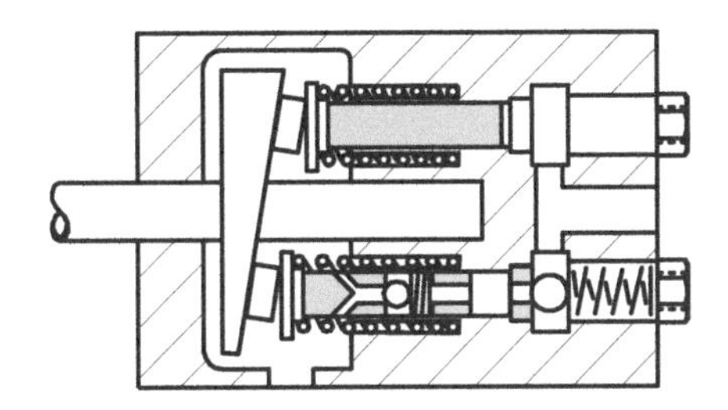

B

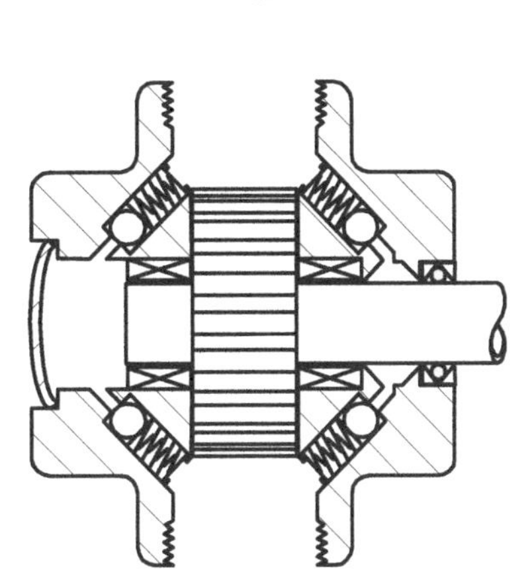

C

Name: ______________________________ Date: ____________________

PISTON PUMPS

1. Match each graphic to its name.

   1. _______ Radial piston pump

   2. _______ Straight-axis piston pump

   3. _______ Bent-axis piston pump

A

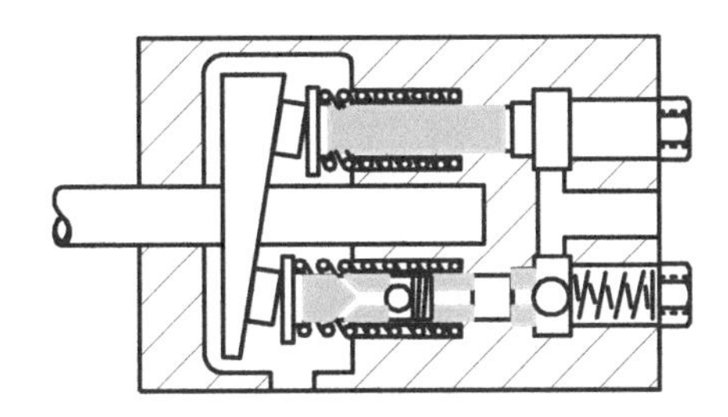

B

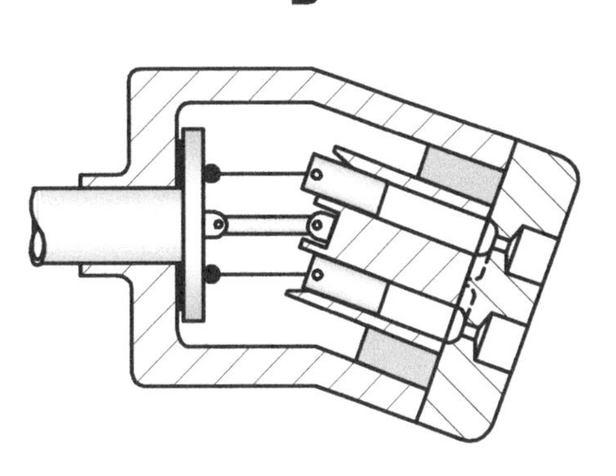

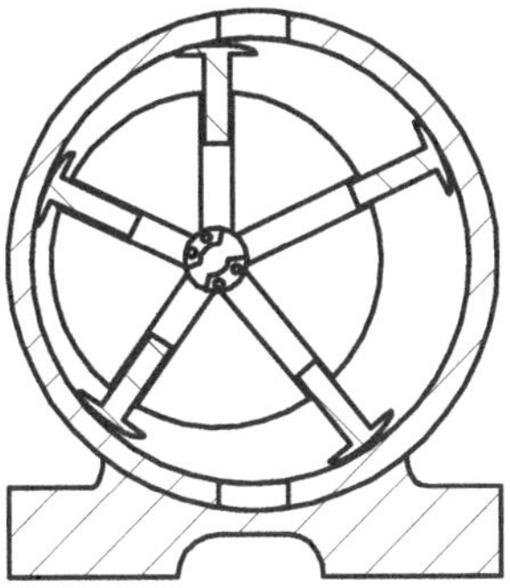

C

Name: ______________________ Date: ______________

VOLUMETRIC EFFICIENCY

1. What is the volumetric efficiency of a pump that is rated at 10 cu. in. per revolution when it is supplying only 8.765 cu. in.? Show your work.

2. What is the volumetric efficiency of a pump that is rated at 30 cu. in. per revolution when it is supplying only 27.5 cu. in.? Show your work.

Name: ______________________________ Date: ____________________

DELIVERY CAPABILITY OF A PUMP

1. What is the delivery capability of a pump that has a displacement of 25 cu. in. per revolution and a motor speed of 1800 rpm?

2. What is the delivery capability of a pump that has a displacement of 15 cu. in. per revolution and a motor speed of 3600 rpm?

Name: ______________________________ Date: ______________

POWER CALCULATIONS

1. How much power is needed to move 10,000 lb 2.5 ft in 3 minutes? Show your work.

2. How much power is needed to move 4000 lb 22 ft in 1 minute? Show your work.

Name: ______________________________ Date: ______________

HORSEPOWER CALCULATIONS

1. How much horsepower is required to move 10,000 lb 2.5 ft in 3 minutes? Show all of your work.

2. How much horsepower is required to move 100 tons 20 ft in 2 minutes? Show all of your work.

Name: ______________________ Date: ______________

UNITS OF MEASUREMENT

1. Complete the following conversion chart for units of pressure measurement.

Bars: Gauge Pressure	Bars: Absolute Pressure	psig	psia
0	1	0	14.5
1	2	______	______
2	3	29.0	______
3	4	______	58.0
4	5	______	______
5	6	______	87.0
6	7	______	______
7	8	______	______
8	9	116.0	______

Name: ______________________________ Date: ______________

STATIC HEAD PRESSURE

1. What should the gauge pressure read for the container of water in this figure?

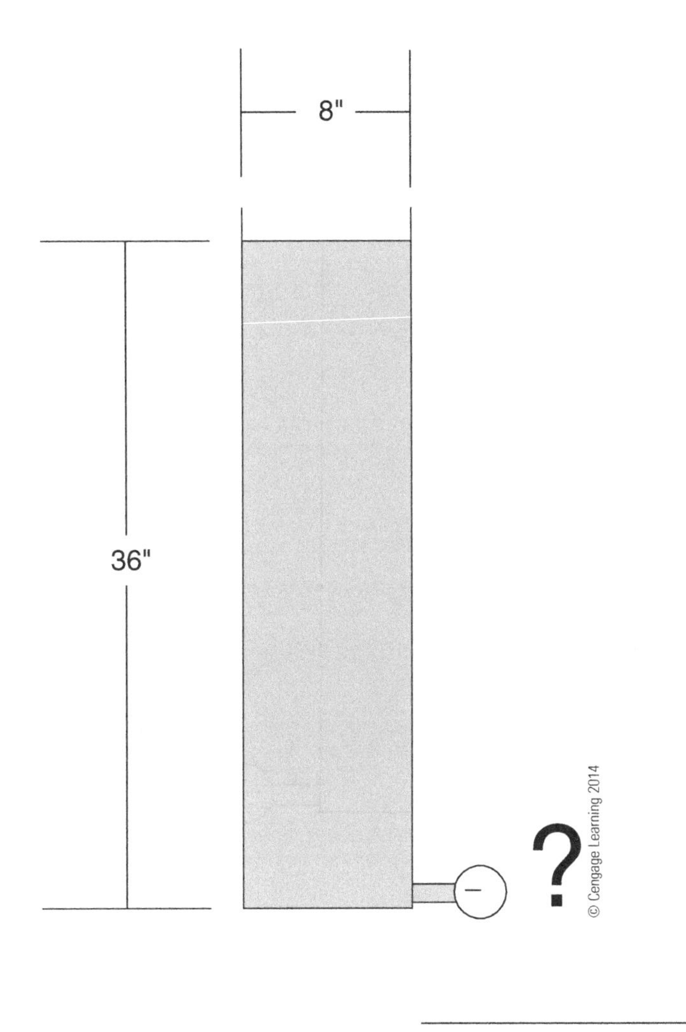

____

2. What should the gauge pressure read for the container of hydraulic oil in this figure?

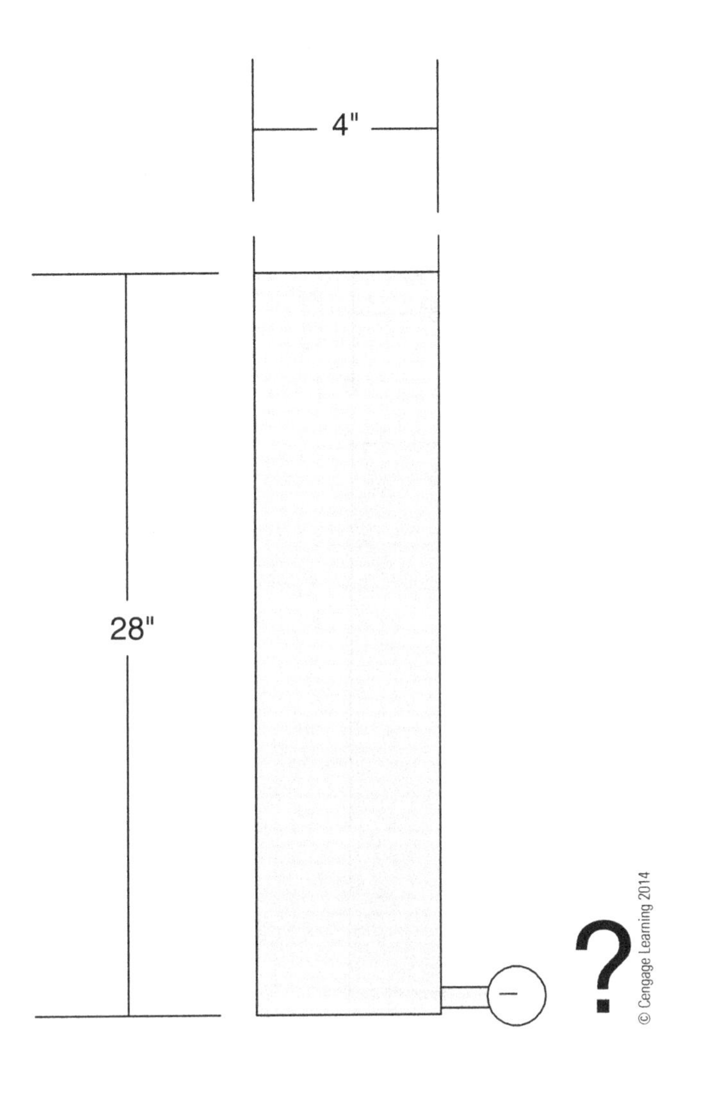

____

Name: ______________________________ Date: ______________________

HYDRAULIC PRESSURE

1. Give the pressure that would be present at each place indicated with a letter, throughout the hydraulic system, when a total of 20 lb of force is placed on cylinder A.

   a. ______________________________

   b. ______________________________

   c. ______________________________

   d. ______________________________

   e. ______________________________

   f. ______________________________

   g. ______________________________

   h. ______________________________

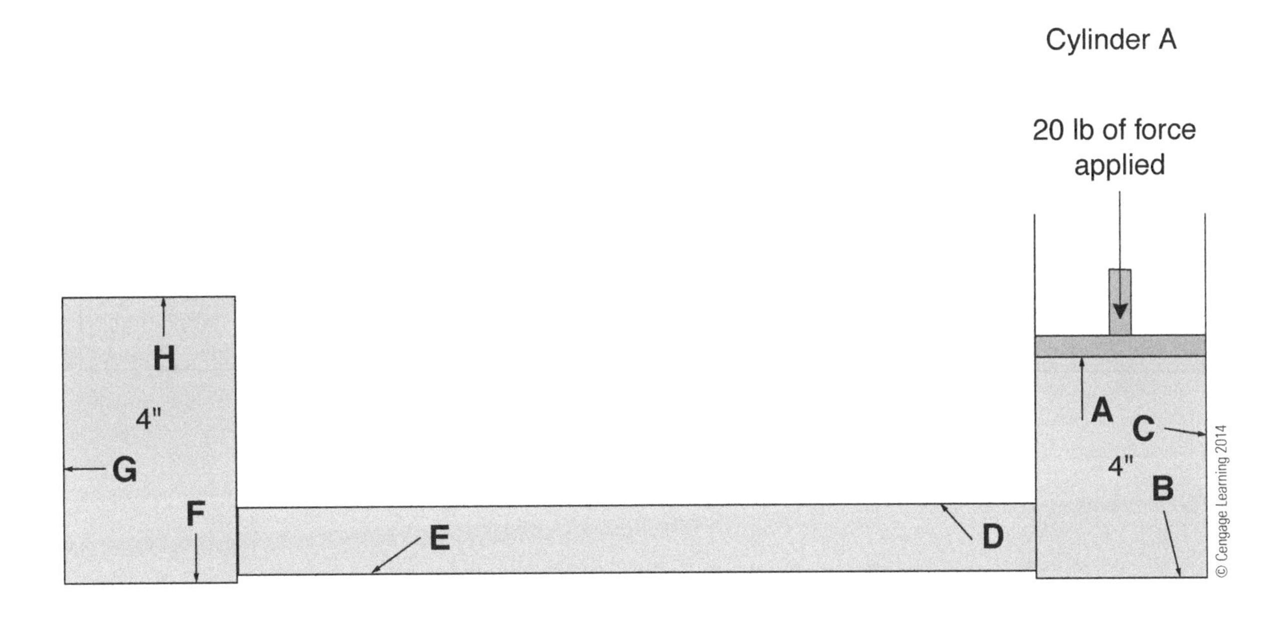

2. Give the pressure that would be present at each place which is indicated with a letter, throughout the hydraulic system, when a total of 20 pounds of force is placed on cylinder A.

a. ______________________________

b. ______________________________

c. ______________________________

d. ______________________________

e. ______________________________

f. ______________________________

g. ______________________________

h. ______________________________

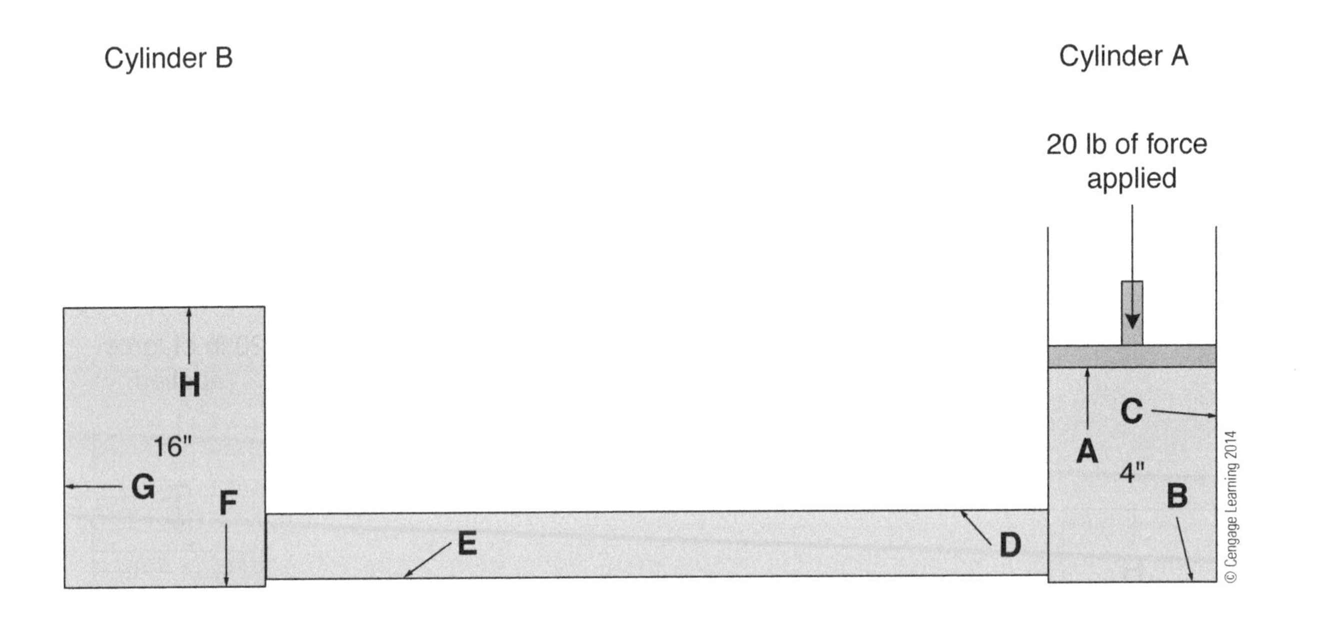

Name: ______________________________ Date: ____________________

FORCE, PRESSURE, AND AREA

1. What is the area of a piston when the force is 55 lb and the pressure is 22 psi?

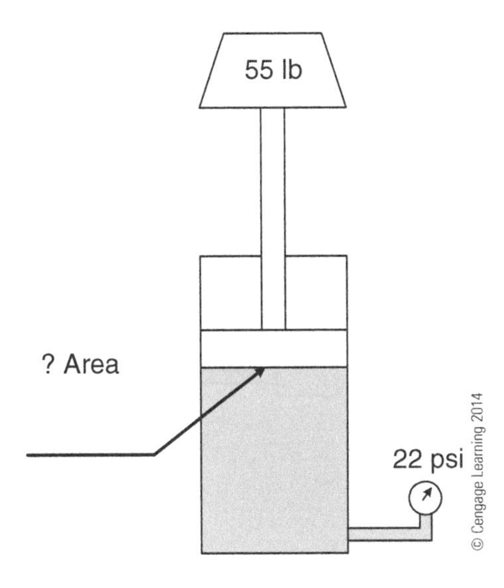

2. What is the force applied to cylinder A in the figure if the piston surface area is 6 in.^{2} and 60 psi is present in the system?

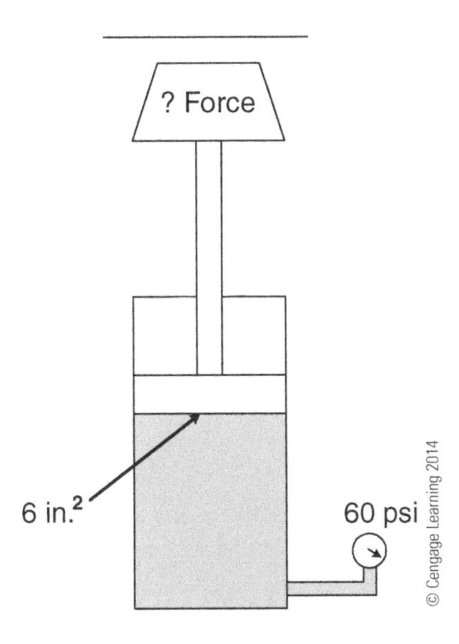

3. What is the pressure in the system shown in the figure if 1200 lb of force is applied to a piston with a surface area of 1.5 in.^{2}?

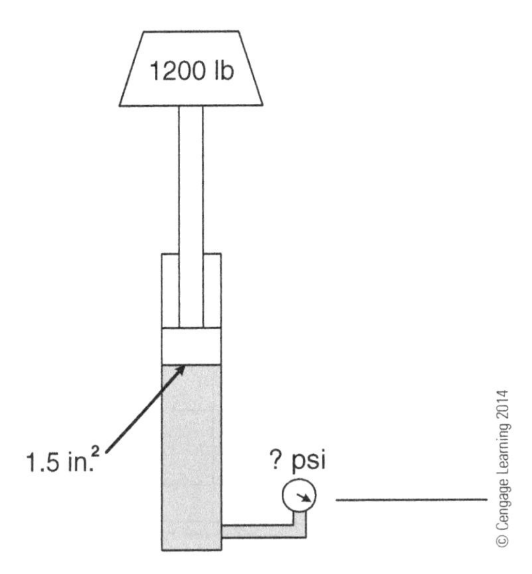

Name: ______________________ Date: ______________

FLUID CONDITIONERS

1. Match each fluid conditioner to its definition.

   a. Dryer

   b. Lubricator

   c. Filter

   d. Regulator

   1. _______ This removes contaminants from the air before the air reaches pneumatic components such as valves and actuators such as a cylinder.

   2. _______ This is used to maintain constant reduced pressures in specified locations of a pneumatic system.

   3. _______ This is used to remove friction in the components in a pneumatic system.

   4. _______ This is used to keep the components from becoming corroded because of moisture.

Name: ______________________________ Date: ______________

VACUUM

1. Select the drawing that illustrates the best vacuum.

   ______

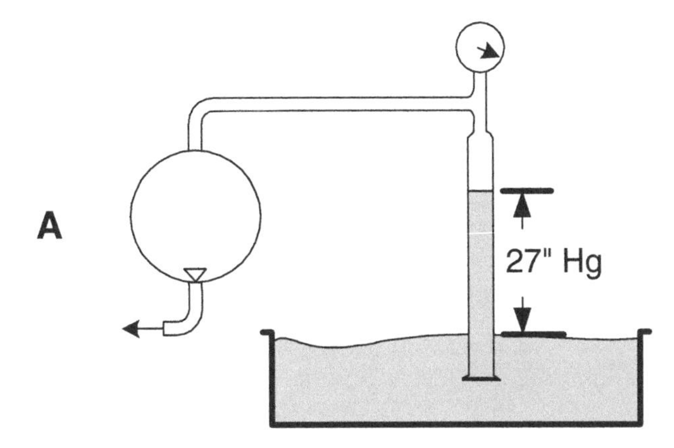

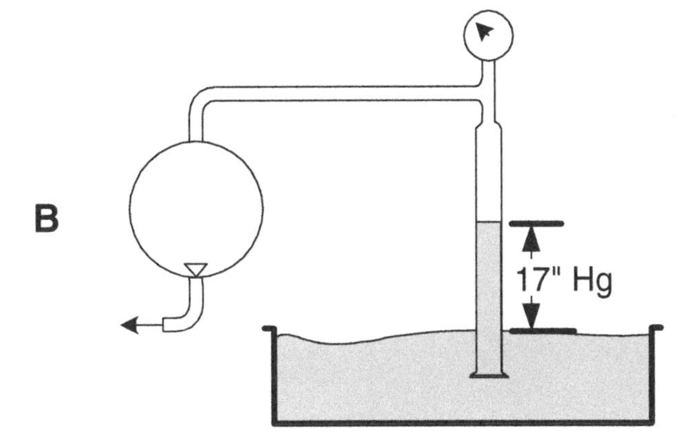

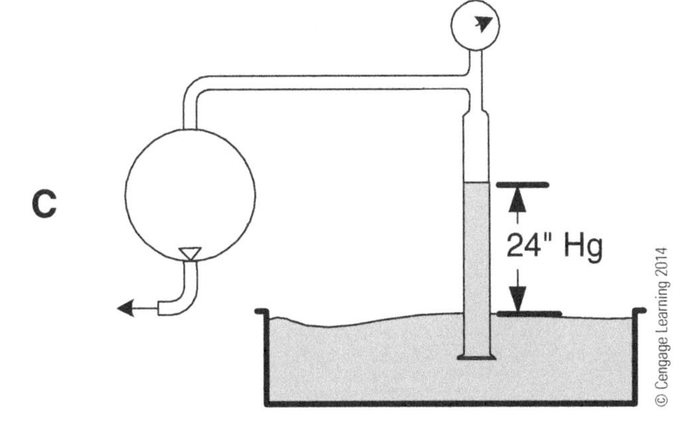

Name: ______________________________ Date: ______________________

DIRECTIONAL CONTROL VALVES

1. Indicate the type of each valve that is in the drawing.

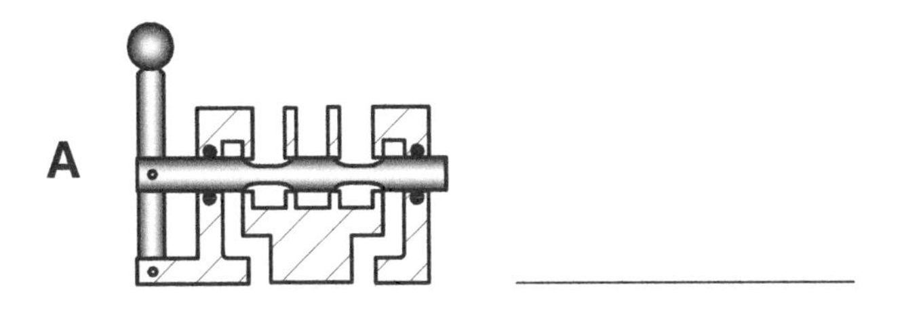

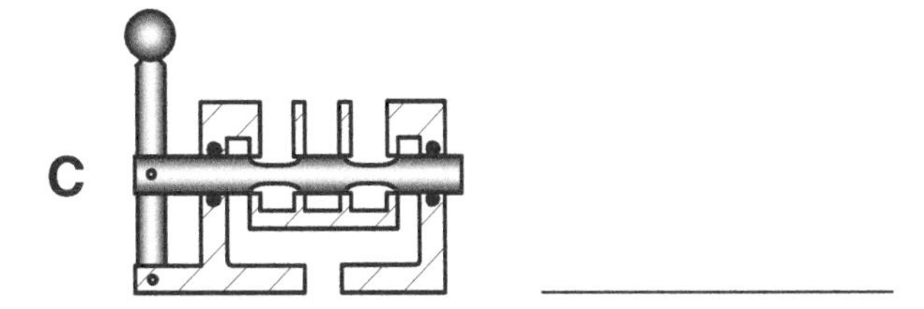

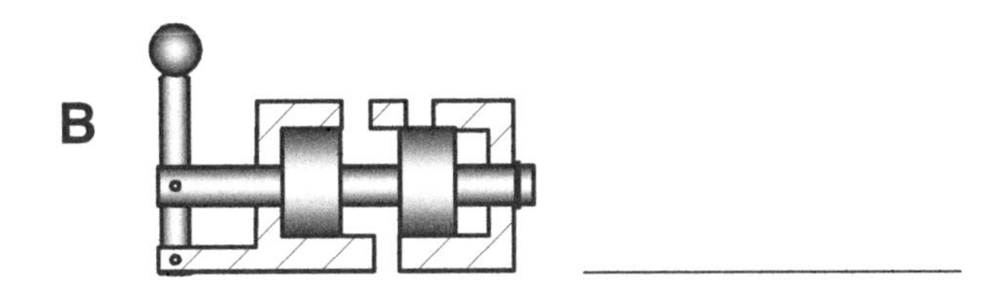

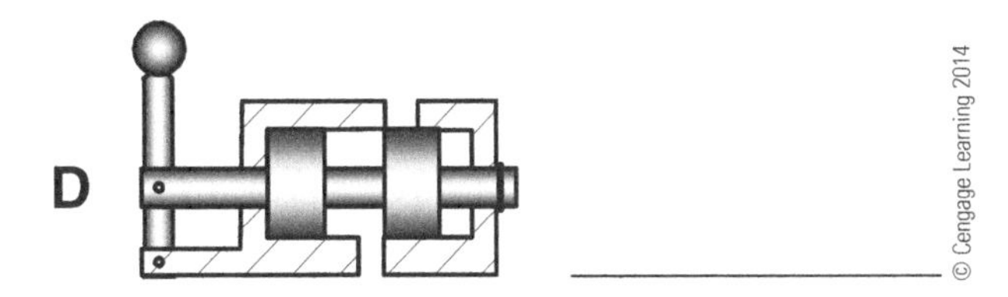

2. Label each port for each valve shown as it would be used in a hydraulic system.

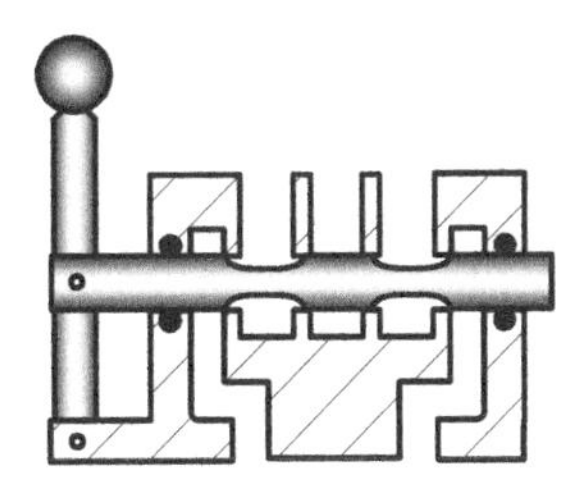

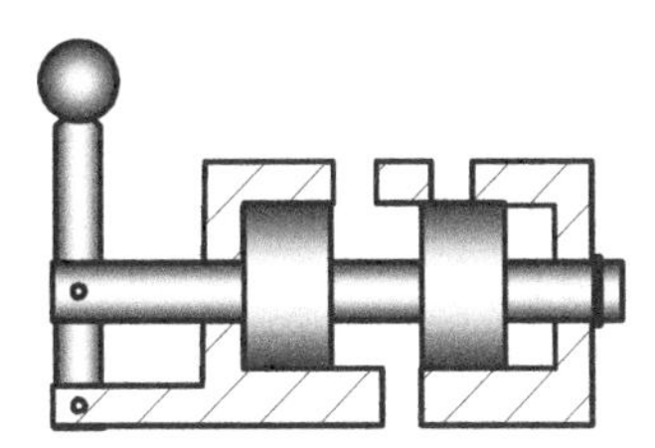

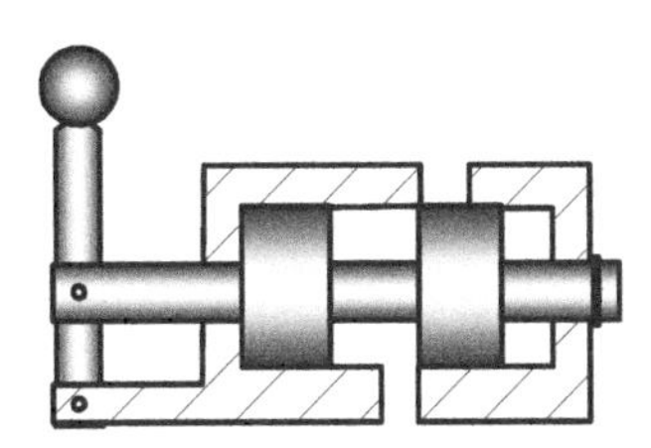

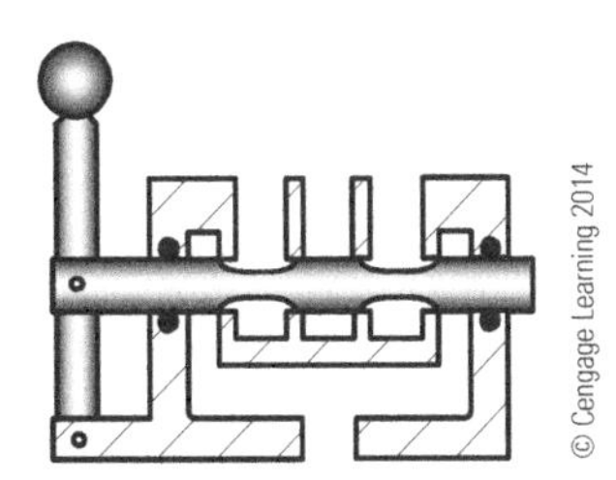

3. Label each port for each valve shown as it would be used in a pneumatic system.

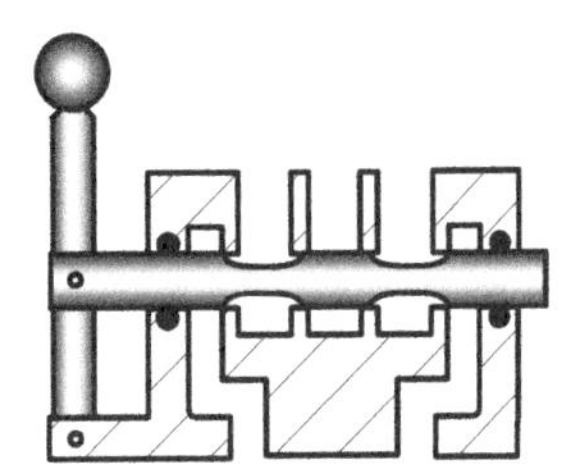

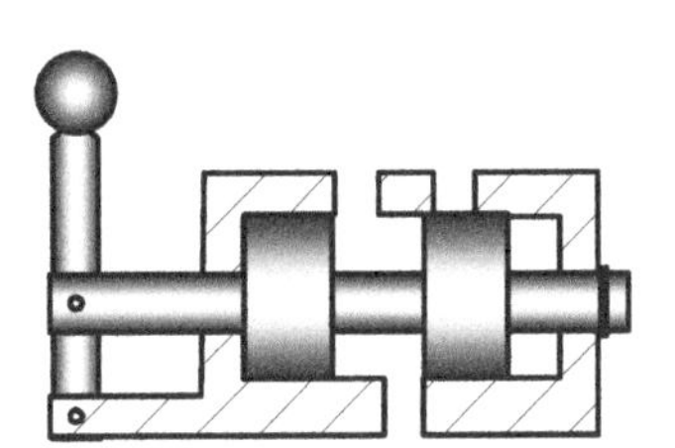

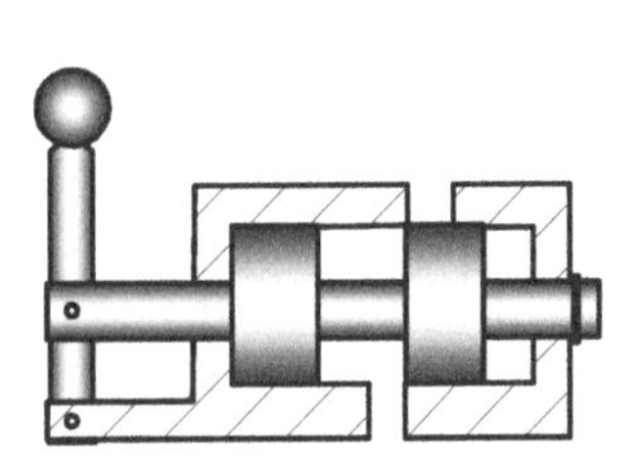

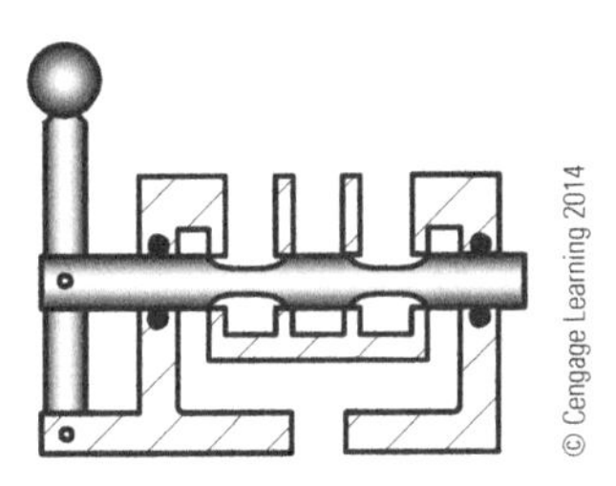

4. Draw the ANSI symbol for each valve shown in the drawing.

A

C

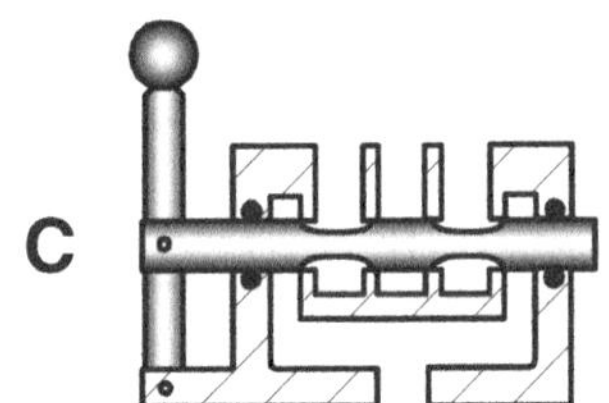

B

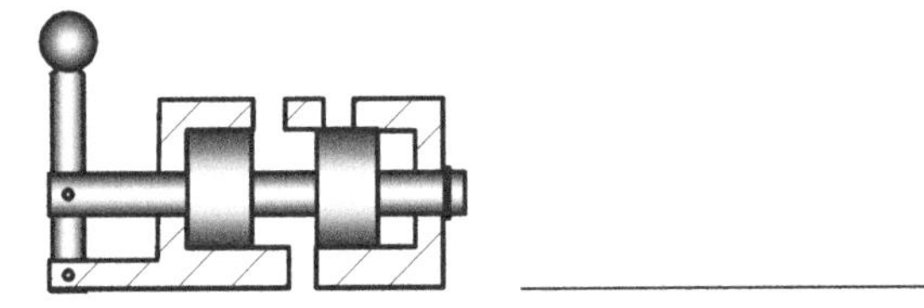

D

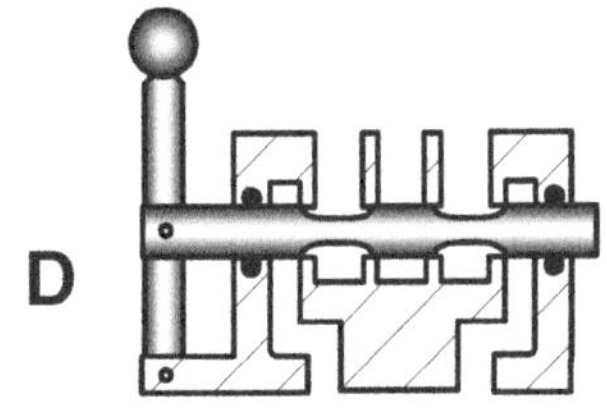

© Cengage Learning 2014

Name: ______________________________ Date: ______________

VALVES

1. From the list, choose the proper use for each numbered valve.

   a. Directional control

   b. Flow control

   c. Pressure relief

   d. Pressure control

   1. ______ Poppet valve
   2. ______ Gate valve
   3. ______ Pressure-relief valve
   4. ______ Pressure-reducing valve
   5. ______ Gate valve
   6. ______ Spool valve
   7. ______ Ball valve
   8. ______ Check valve
   9. ______ Pressure-regulating valve
   10. ______ Needle valve

Name: ______________________ Date: ______________

ANSI SYMBOL PLACEMENT

1. Add a pressure-relief valve to the circuit between the pump and the check valve.

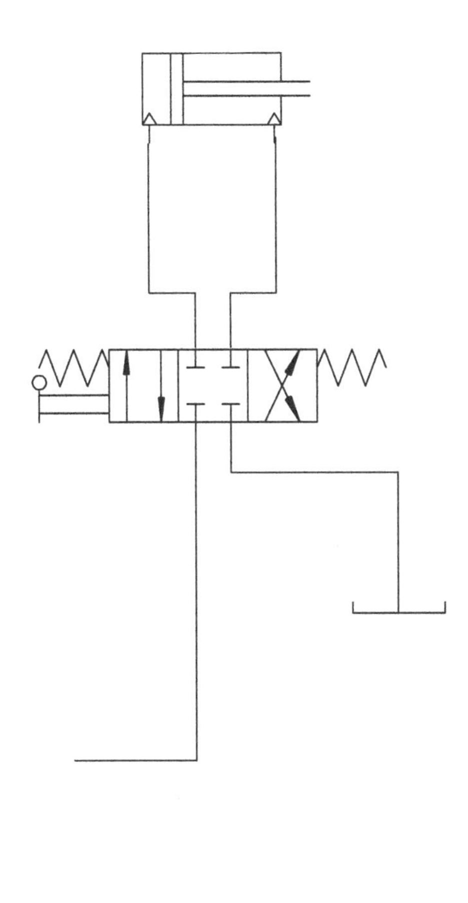

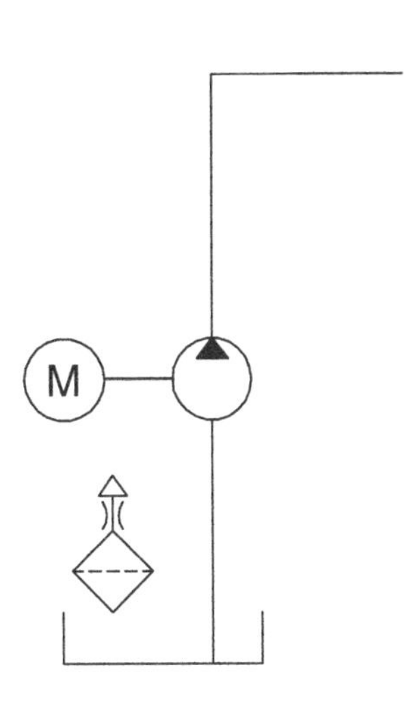

2. Add a lever-operated, spring-return, four-way, two-position directional control valve to the hydraulic system shown in the drawing.

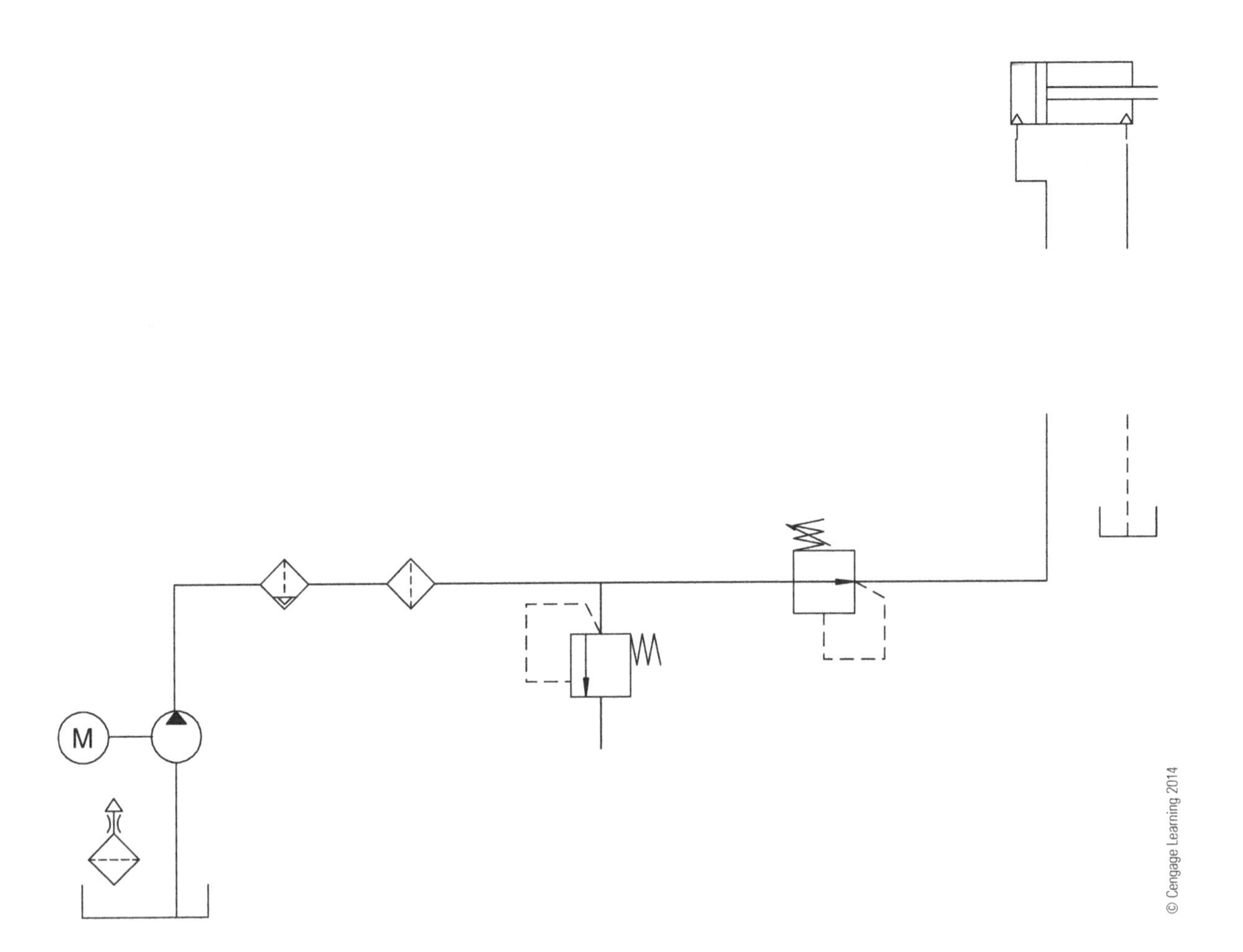

Name: ______________________ Date: ______________

CYLINDERS

1. Complete the drawing below to achieve the following:
   a. A directional control valve is used to extend the cylinder.
   b. The same directional control valve is used to retract the same cylinder.
   c. When the directional control valve is released, it centers itself, and the cylinder will stop where it is.

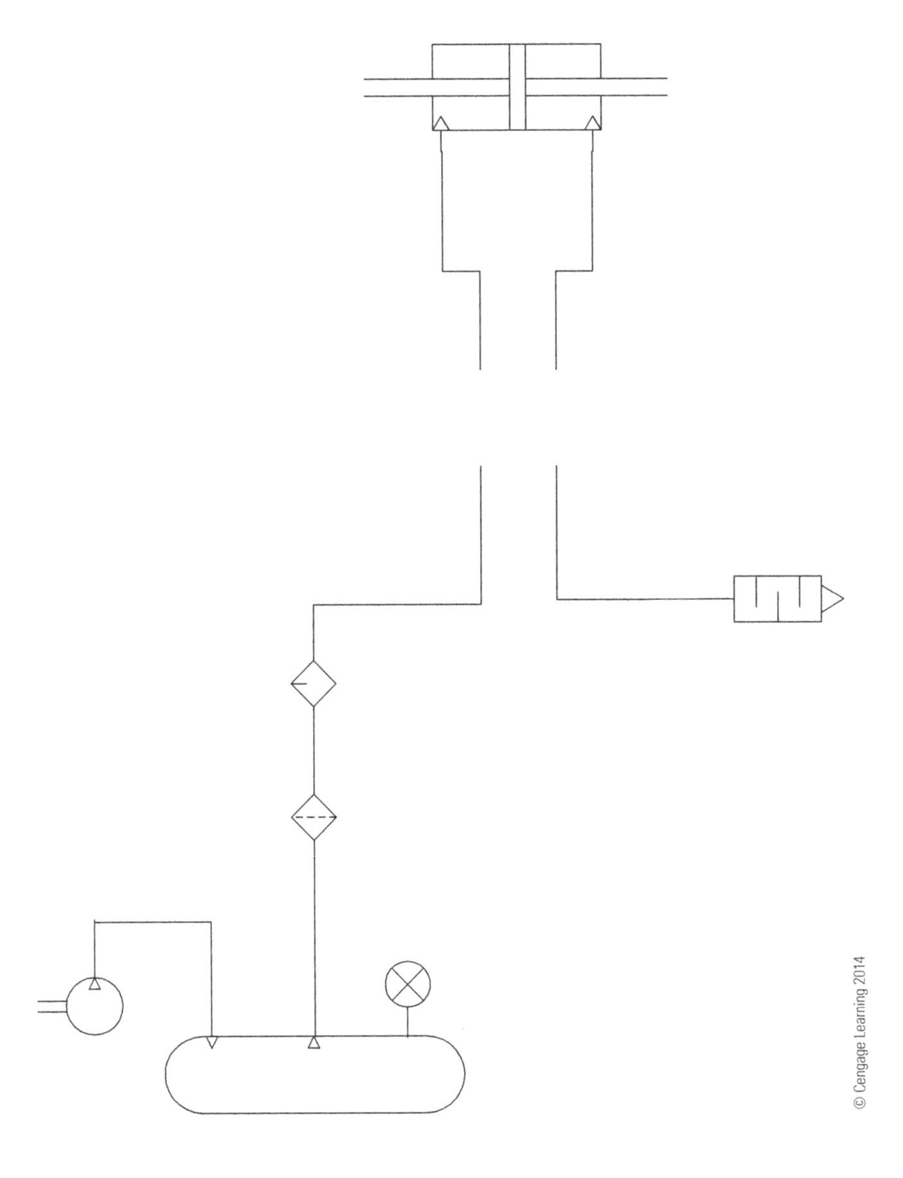

2. What type of cylinder is used in question 1?

____

3. Redraw the entire drawing, modifying the circuit to do the following:

   a. A directional control valve is used to extend the cylinder only.

   b. When the directional control valve is released, the cylinder retracts automatically.

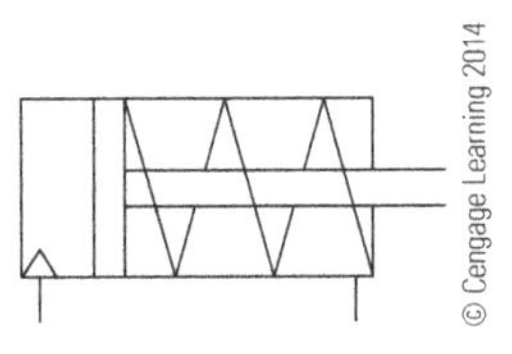

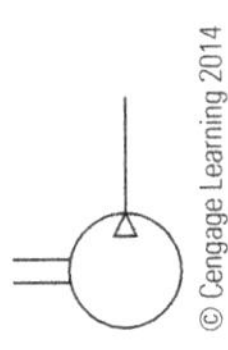

4. What type of cylinder is used in question 3?

____

Name: ______________________ Date: ______________

CIRCUIT INTERPRETATION

1. Connect the following circuit according to the schematic.

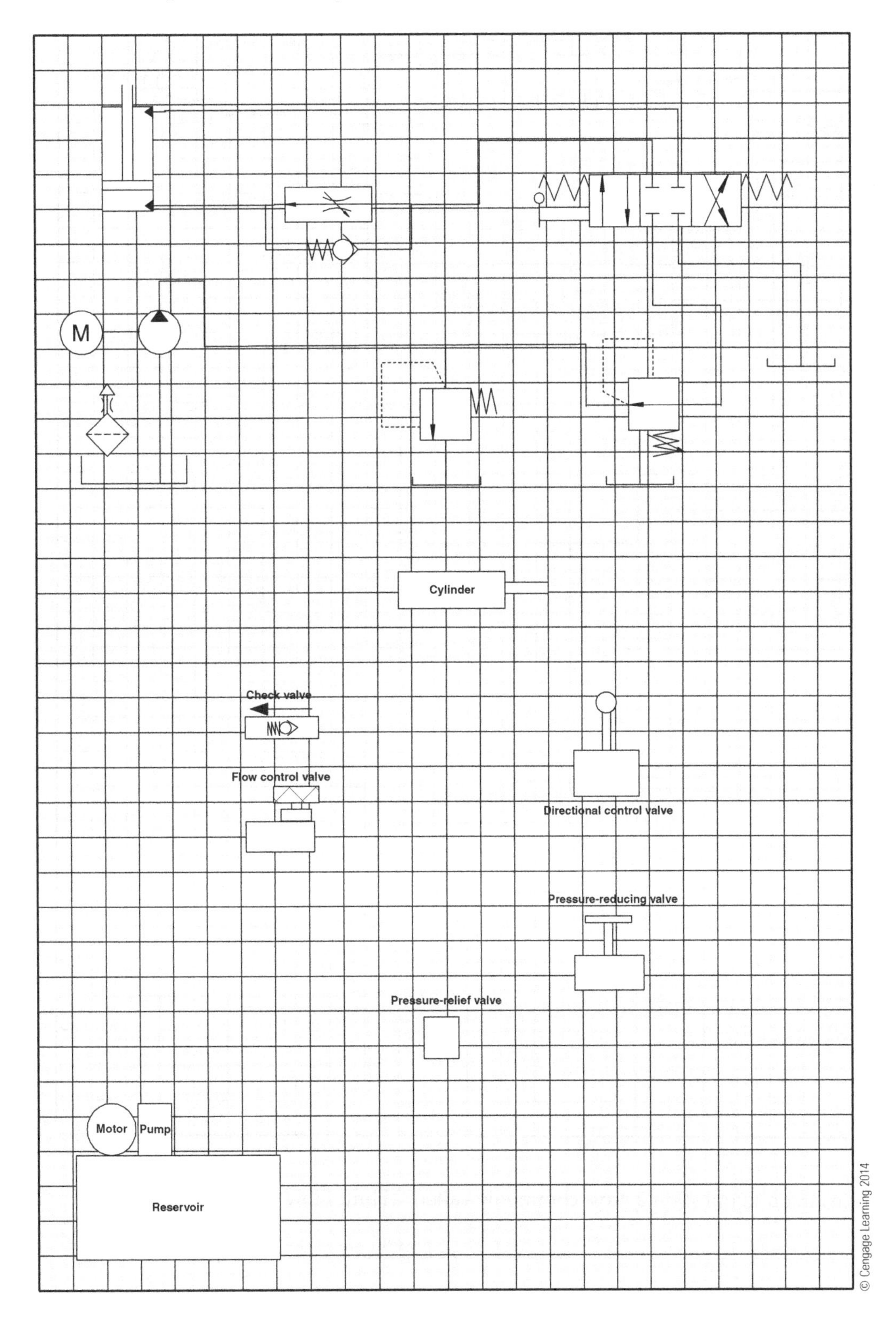

2. Draw a schematic for the circuit that is connected in the drawing.

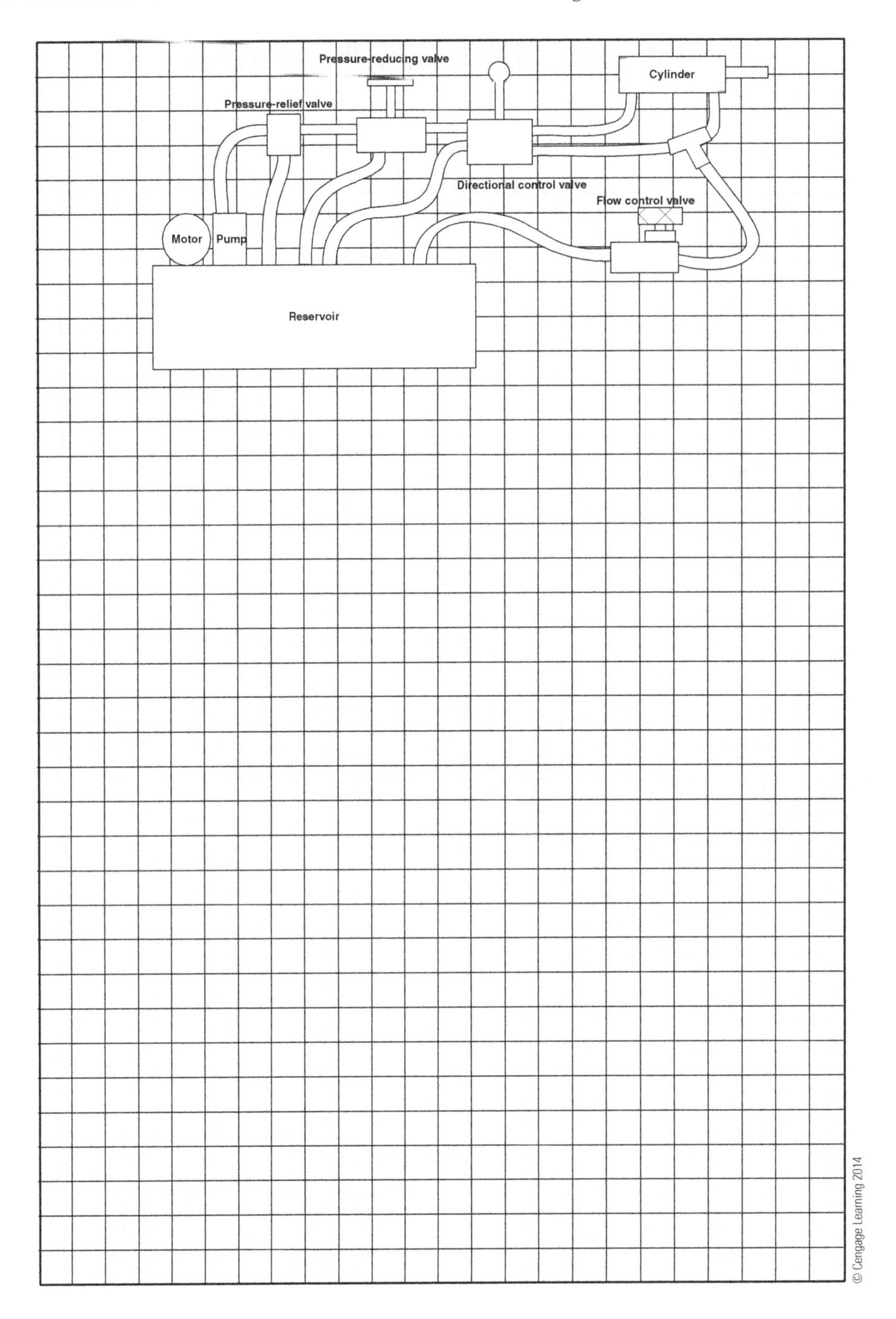

3. What is the circuit in question 2 most commonly called? (Hint: Flow Control Circuitry)

___

Name: ______________________________ Date: ______________________

PIPING TOOLS

1. Name the piping tool that should be used for each application listed.

   a. ____________________ Turn a polished chrome pipe into a fixture.

   b. ____________________ Cut a pipe. This cutting tool will leave a very smooth but sharp edge.

   c. ____________________ Check to see whether the piping system has been properly installed horizontally, vertically, or at 45°.

   d. ____________________ Manually cut threads onto a pipe.

   e. ____________________ Weld a piping system or mark a pipe for a square cut.

   f. ____________________ Cut a pipe. This cutting tool will leave a very rough edge.

   g. ____________________ Measure a length of pipe.

   h. ____________________ Find a vertical reference point by using this tool.

   i. ____________________ Turn a steel pipe into a coupling.

   j. ____________________ Remove the burrs that are on the inside diameter of the pipe after cutting.

Name: ______________________________ Date: ____________________

THREAD LENGTH

1. Complete the following gas piping thread length table.

Nominal Pipe Size	Thread Length
1/2	________
3/4	________
1	________
1 1/4	________
1 1/2	________
2	________
2 1/2	________
3	________
4	________

Name: ______________________ Date: ______________

MOISTURE COLLECTION (GAS PIPING)

1. In the space below, draw a drip that would be used to collect moisture within a gas piping system.

Name: ______________________________ Date: ______________________

GAS PIPING SUPPORT

1. On the drawing below, indicate where each support should be to meet the minimal standards for this gas piping system.

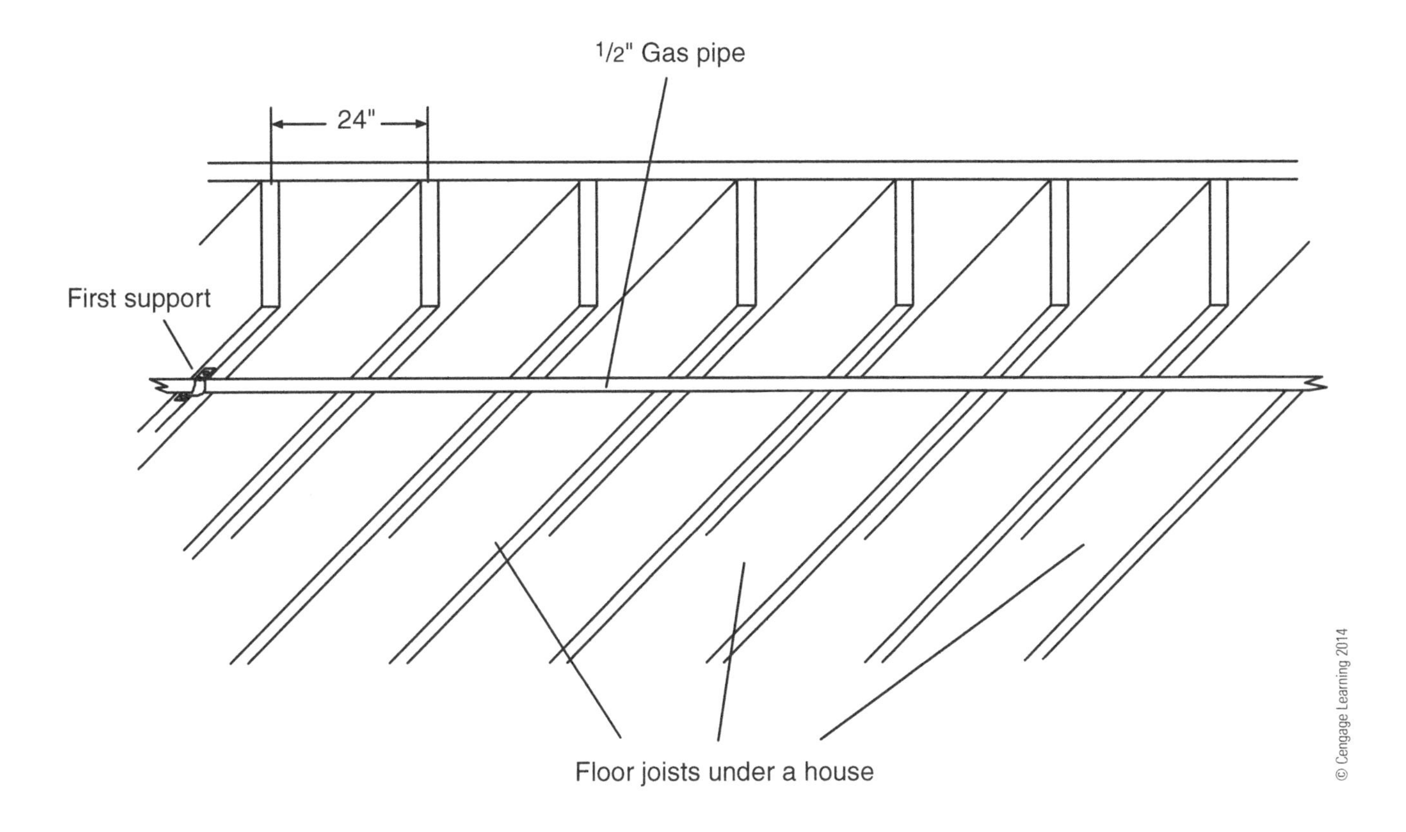

Name: ____________________ Date: ____________________

WASTE DISPOSAL

1. Identify which section of pipe in the waste disposal system that is shown in the drawing is considered to be the drainpipe.

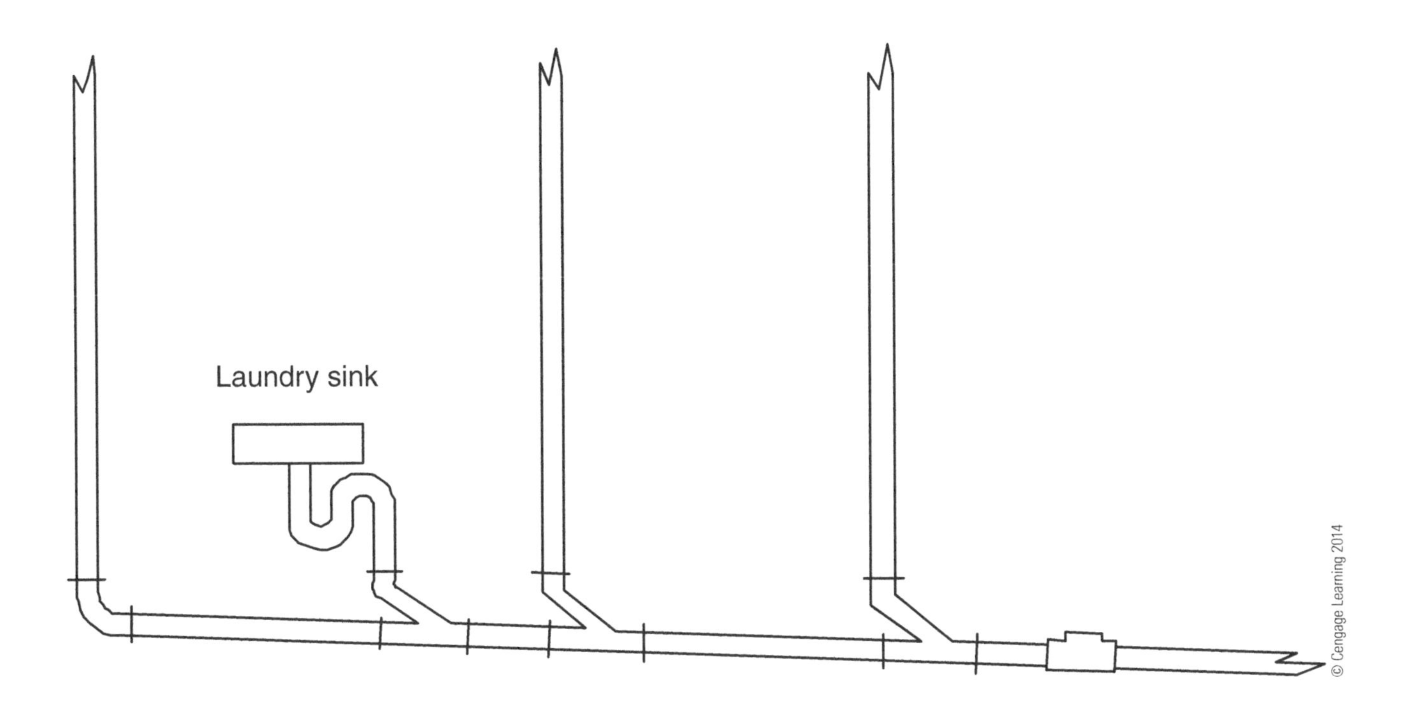

Name: ____________________ Date: ____________________

DRAINPIPE ANGLE

1. Using a straightedge, draw in the drainpipe at the correct angle for the drainage system that is in the drawing. Do not forget to connect all of the risers to the drainpipe.

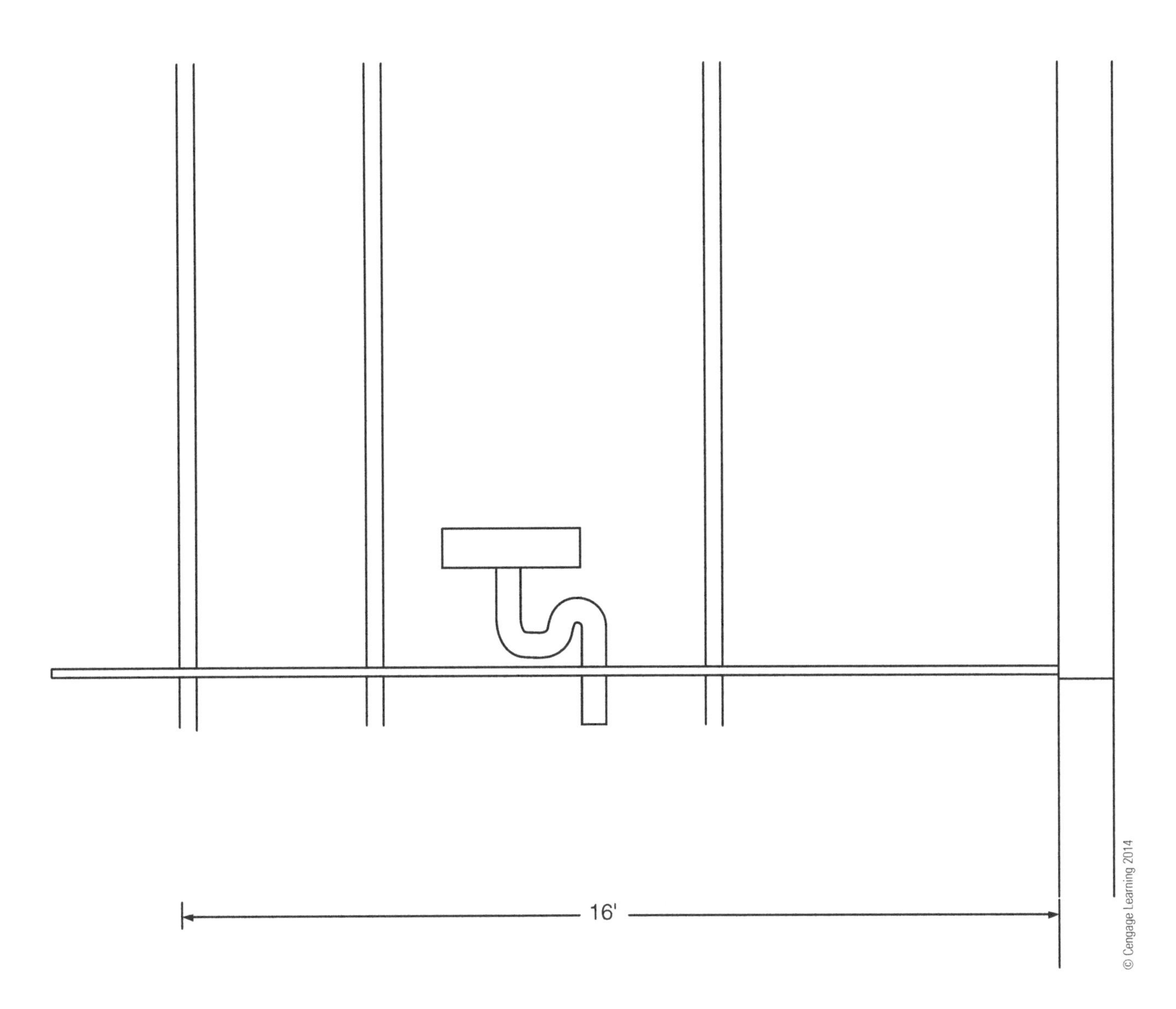

Name: ______________________ Date: ______________

DRAINPIPE SUPPORT

1. In the figure below, draw each support that should be placed under the drainpipe to meet the minimal standards for this waste disposal piping system.

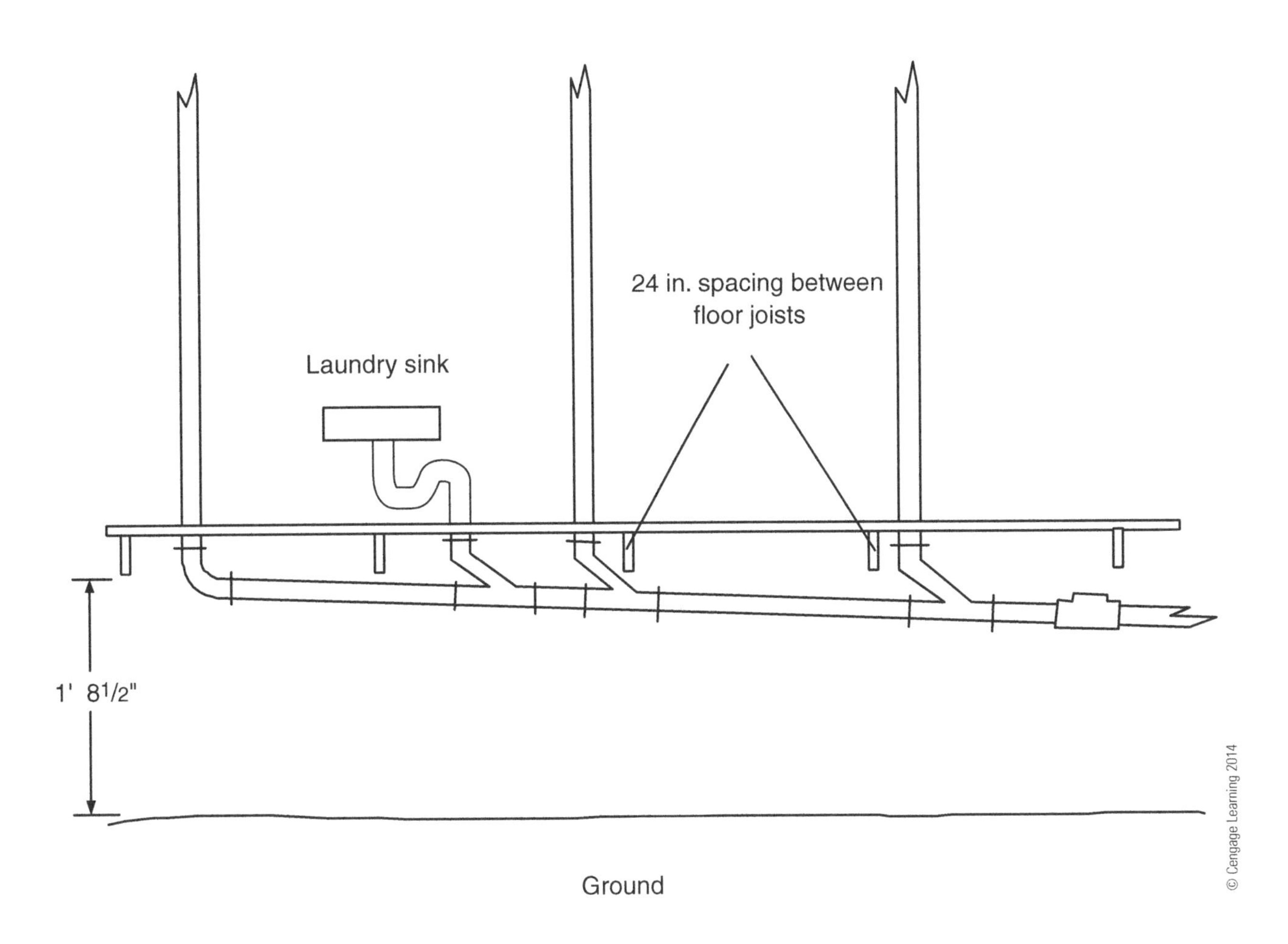

Name: ______________________________ Date: ______________

PLASTIC PIPING

1. Match each acronym to the correct type of piping.

   a. ABS ______ Polyvinyl chloride

   b. PE ______ Chlorinated polyvinyl chloride

   c. PVC ______ Polyethylene

   d. CPVC ______ Acrylonitrile butadiene styrene

   e. PEX ______ Cross-linked polyethylene

2. Name two types of piping that are used primarily for waste and disposal systems.

   ____

   ____

3. Name two types of piping that are used for water supply systems.

   ____

   ____

4. Of the five types of piping, which one is considered to be state of the art?

   ____

5. What type of piping is the plastic pipe slowly replacing?

   ____

Name: ______________________________ Date: ______________________

WATER SUPPLY SYSTEMS: COPPER PIPING

1. Each listed item is either an advantage or a disadvantage of copper piping. Place each item in the appropriate column.

   Cost Rigidity Formability Oxidation properties

   Fatigue capabilities Expansion Contraction Strength

Advantages	**Disadvantages**
____________	____________
____________	____________
____________	____________
____________	____________
____________	____________
____________	____________
____________	____________
____________	____________

Name: ______________________ Date: ______________

CAST IRON

1. Properly label the items that are in the drawing.

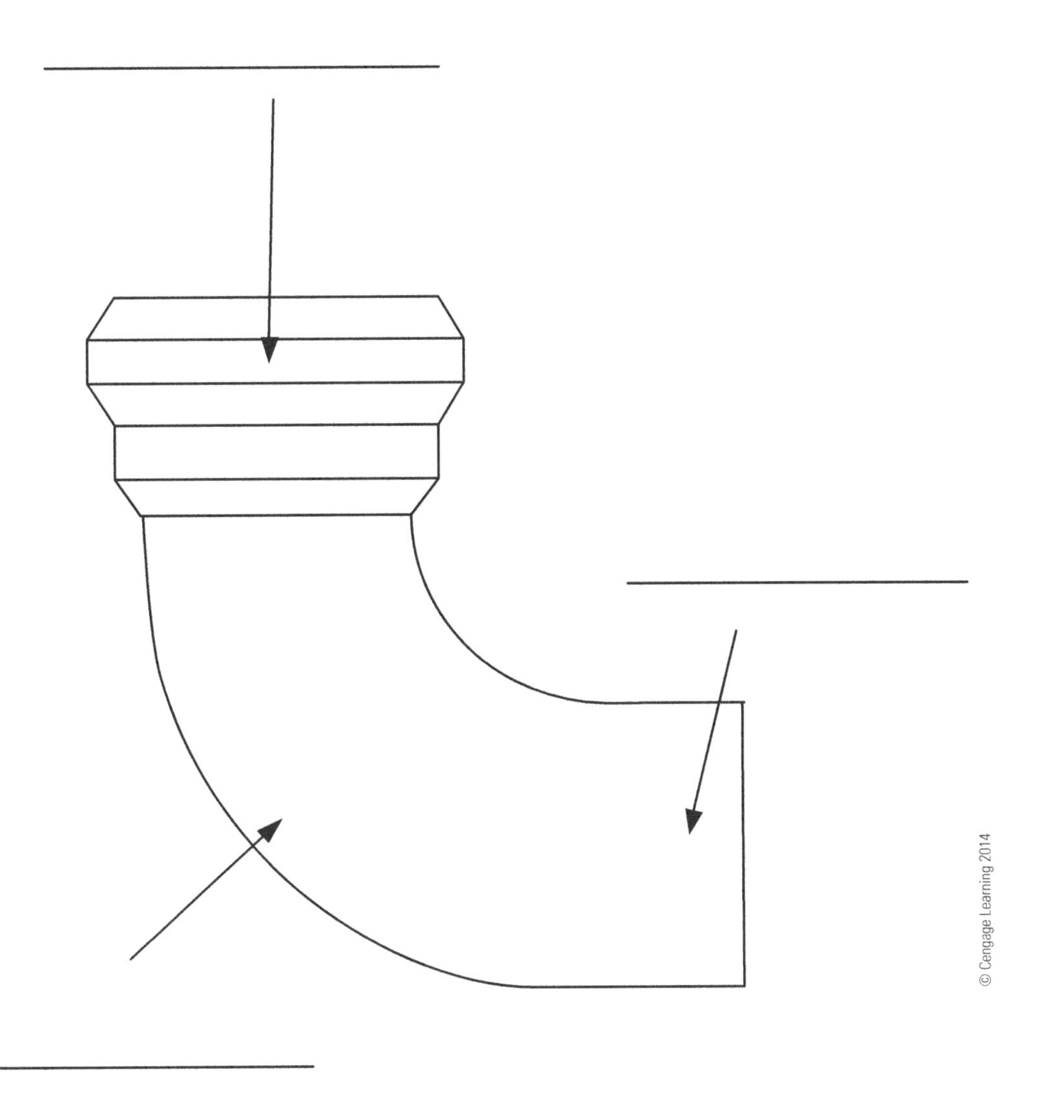

Name: ______________________ Date: ______________

WATER SUPPLY SYSTEMS: FITTING SPECIFICATIONS

1. Decipher the given specification completely.

   8½", 45°, ell., cu., swt., std.

Name: ______________________________ **Date:** ____________________

FITTING TYPES: BRANCHES

1. Match each of the fitting types with the correct graphic.

1. ________ 45° angle elbow

2. ________ Plug

3. ________ Tee

4. ________ 90° angle elbow

5. ________ Cap

6. ________ YT

A

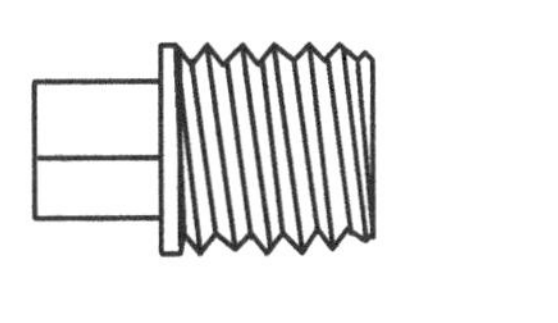

B

C

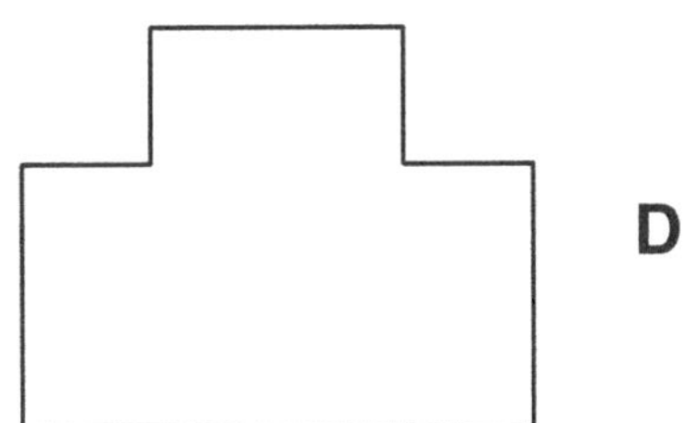

D

E

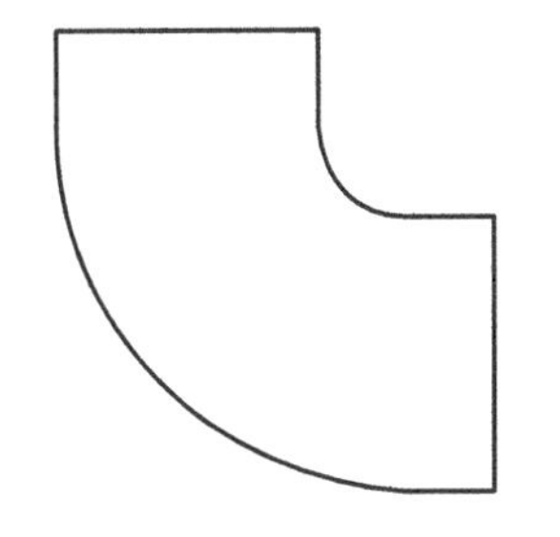

F

Name: ______________________________ Date: ____________________

FITTING APPLICATIONS

1. In the spaces provided, list all of the fittings that would be needed to accomplish the application. Also state how many of each type of fitting are needed.

   a. Two pieces of pipe need to be connected together. This fitting should be used when neither of the two pipes can be turned but they still need to be connected.

   ____

   ____

   b. A piping system is running parallel to the wall. An 8" I-beam protrudes out past the surface of the wall. Two offsets must be made to get around the I-beam.

   ____

   ____

   c. A drip must be made to collect the moisture that may be present in a pneumatic system.

   ____

   ____

   d. Two pieces of pipe are to be connected together, and there are no extenuating circumstances.

   ____

   ____

   e. A piece of pipe is to run horizontal for 5 ft and then turn vertically to the ceiling. As the pipe reaches the ceiling, the pipe has to branch off to the right (horizontal run) and continue across the ceiling.

   ____

   ____

Name: ______________________ Date: ______________

PIPING SKETCHES

1. Make a parts list including all of the components that will be needed to completely assemble the piping system shown in the figure.

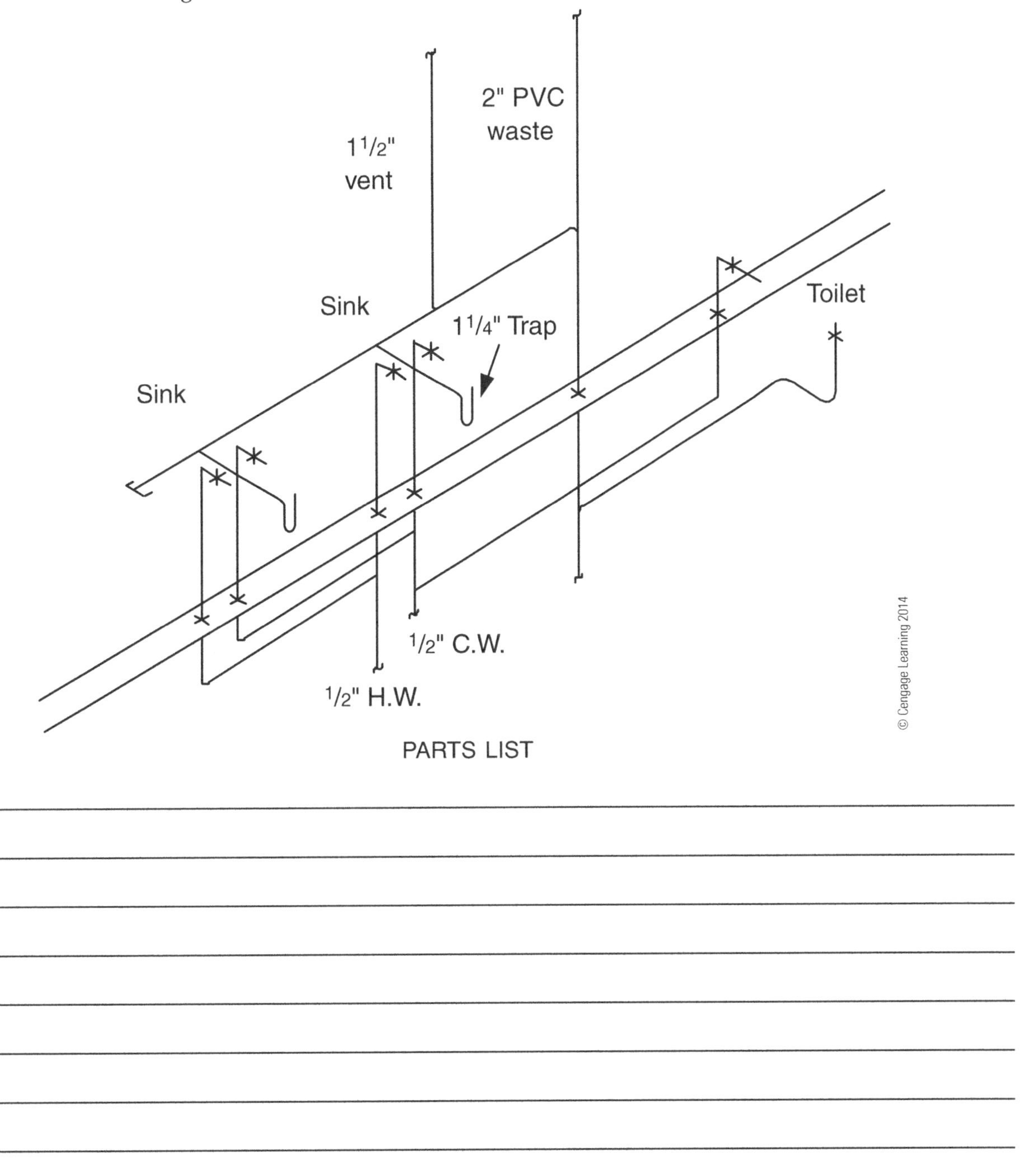

____

____

____

____

____

____

____

____

____

____

____

Name: ______________________ Date: ______________

FITTING ALLOWANCE

1. What is the end-to-end measurement of the pipe indicated with the letter A? Show your work.

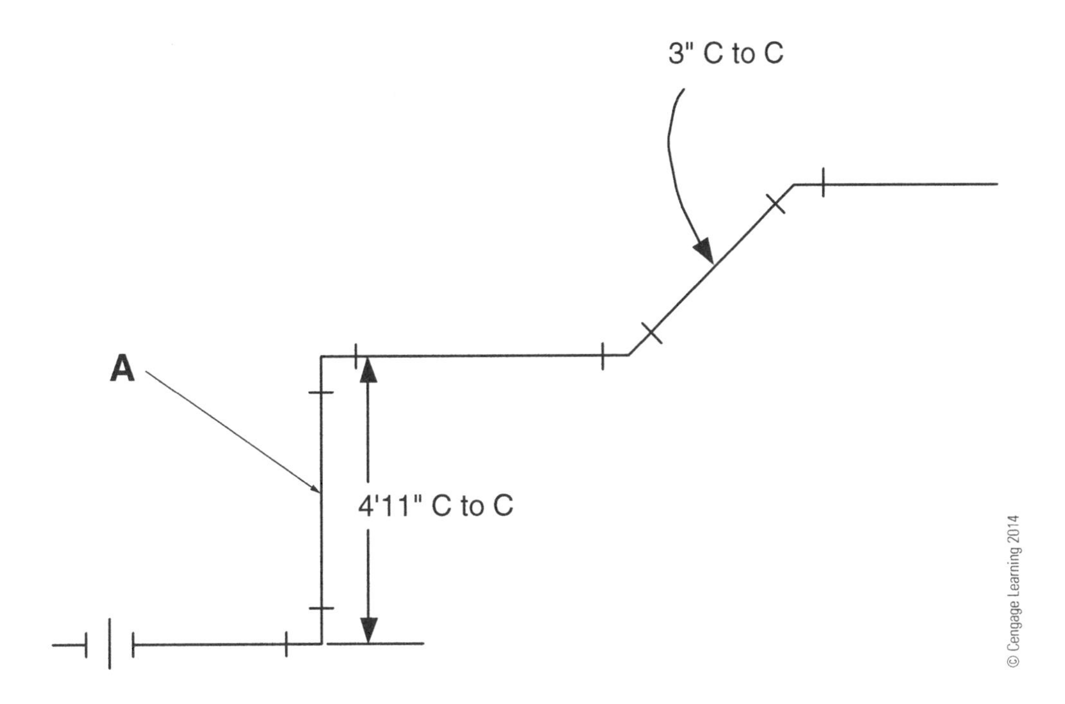
Nominal pipe size of 3/4" is being used.
Fitting allowance for 45 angle elbows = 1".
Fitting allowance for 90 angle elbows = 1 3/8".
3" C to C
A
4'11" C to C
© Cengage Learning 2014

2. What is the end-to-end measurement of the pipe indicated with the letter B? Show your work.

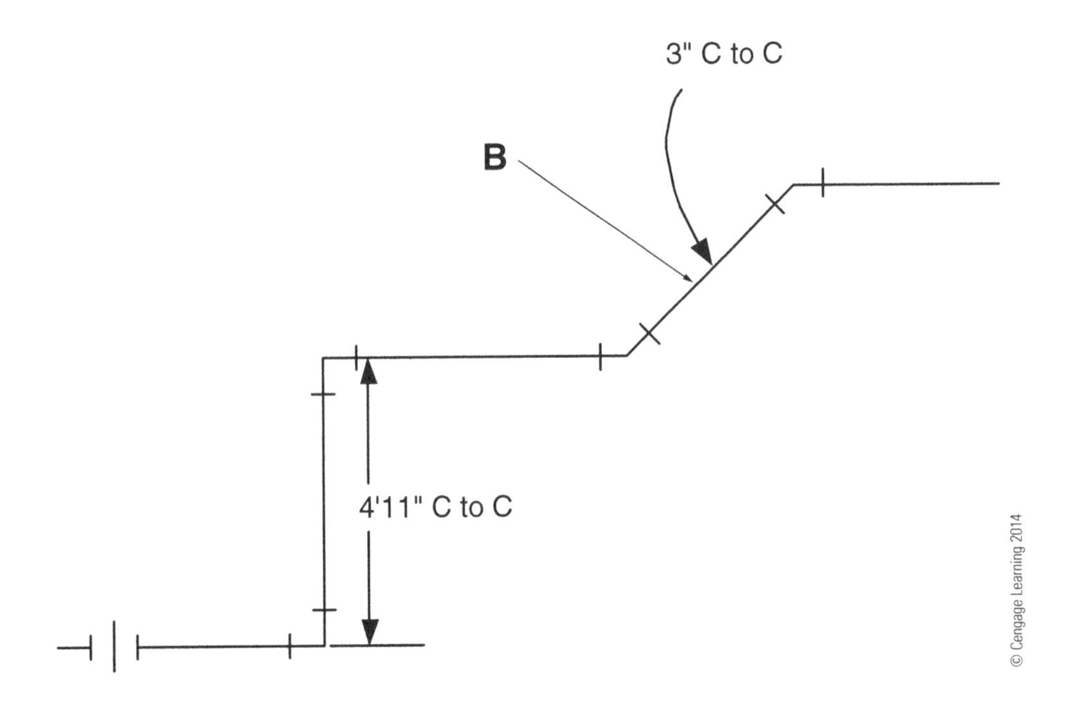
Nominal pipe size of 3/4" is being used.
Fitting allowance for 45 angle elbows = 1".
Fitting allowance for 90 angle elbows = 1 3/8".
3" C to C
B
4'11" C to C
© Cengage Learning 2014

Name: ____________________ Date: ____________

PIPE CONNECTION METHODS

1. Identify the correct pipe connection method for each piping system that is listed. You may use a pipe connection method more than once. Keep in mind that some piping systems may use more than one method.

a. Sweating/soldering
b. Threading
c. Caulking
d. Gluing
e. Welding
f. Barbed fittings/hose clamps
g. Crimping
h. Flaring

1. _______ Copper piping systems
2. _______ Polyethylene
3. _______ Cast iron
4. _______ Polyvinyl chloride
5. _______ Black iron
6. _______ Acrylonitrile butadiene styrene
7. _______ Brass
8. _______ Cross-linked polyethylene
9. _______ Stainless steel
10. _______ Chlorinated polyvinyl chloride
11. _______ Copper tubing systems

Name: ______________________________ Date: ______________

ATOMIC STRUCTURE

1. Identify the parts of the copper atom shown.

_______ Electron

_______ Proton

_______ Neutron

_______ Valence electron

_______ Nucleus

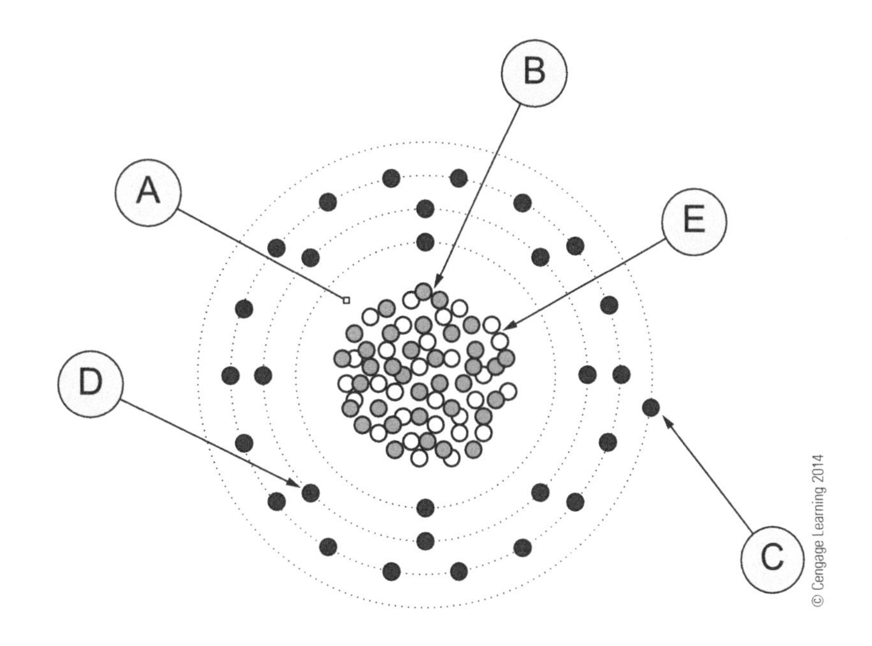

Name: ______________________________ Date: ____________________

RESISTOR COLOR CODE—PART 1

1. Match the value to the correct resistor.

______ 33 kΩ ±10%

______ 270 Ω ±5%

______ 7.5 MΩ ±10%

______ 56 Ω ±1%

______ 9100 Ω ±20%

______ 0.18 MΩ ±10%

______ 8.2 kΩ ±1%

______ 2.0 MΩ ±5%

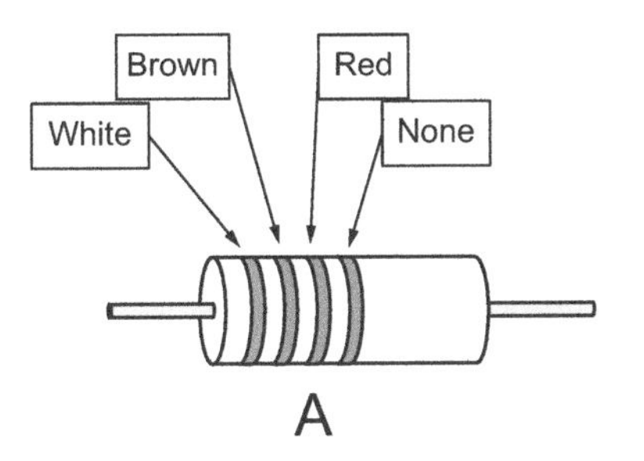

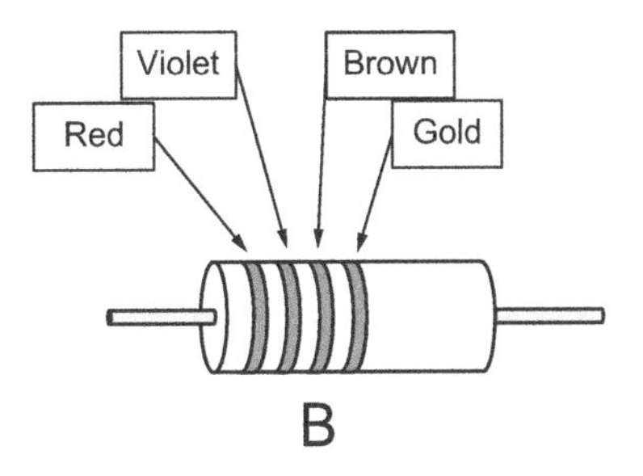

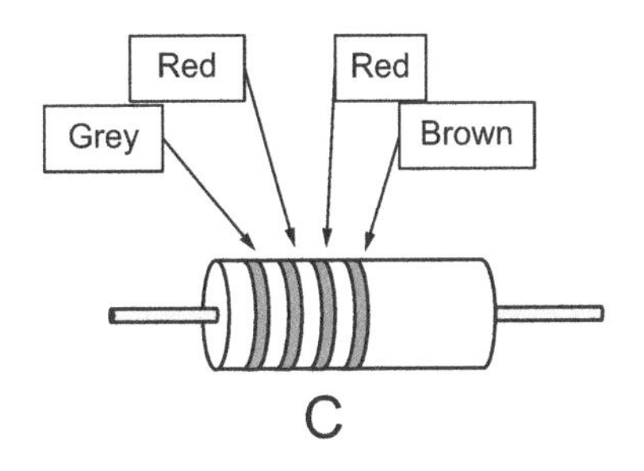

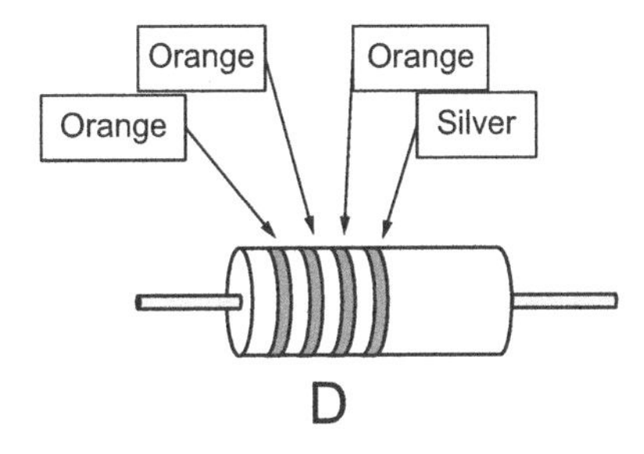

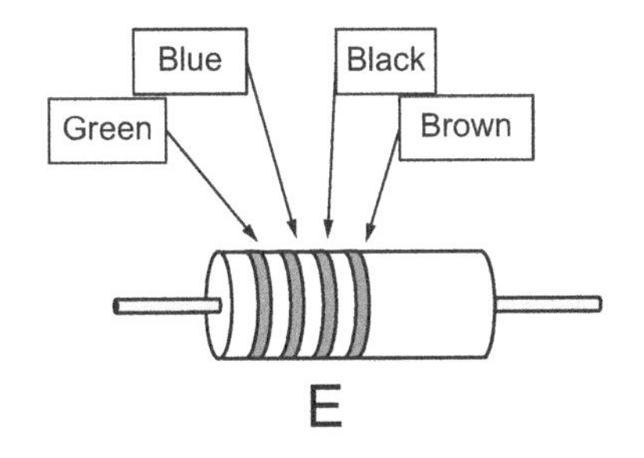

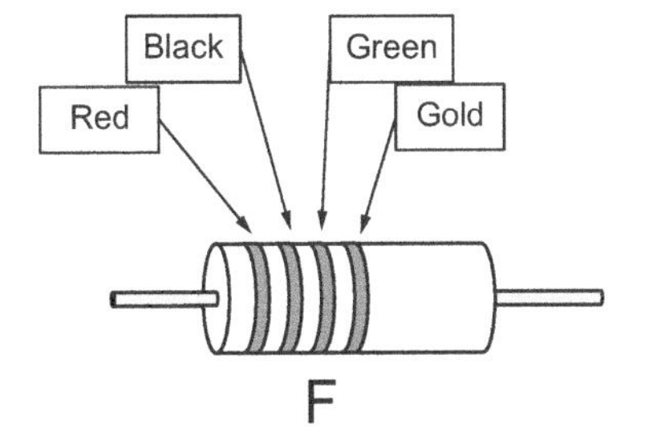

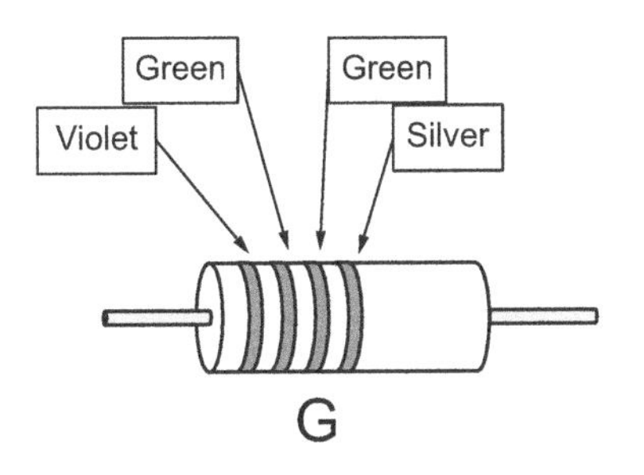

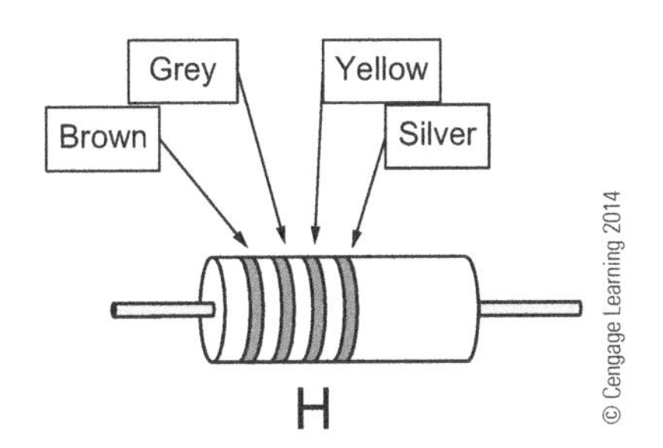

Name: ______________________________ Date: ____________________

RESISTOR COLOR CODE—PART 2

1. Match the value to the correct resistor.

_______ 77.7 kΩ ±0.5%

_______ 4.02 kΩ ±1%

_______ 1000 Ω ±0.1%

_______ 9.53 Ω ±0.5%

_______ 22.6 Ω ±0.5%

_______ 511 MΩ ±0.1%

_______ 82.5 MΩ ±1%

_______ 397 Ω ±0.1%

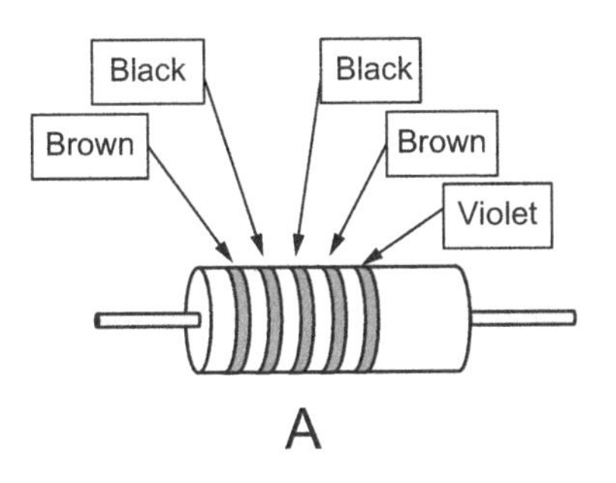

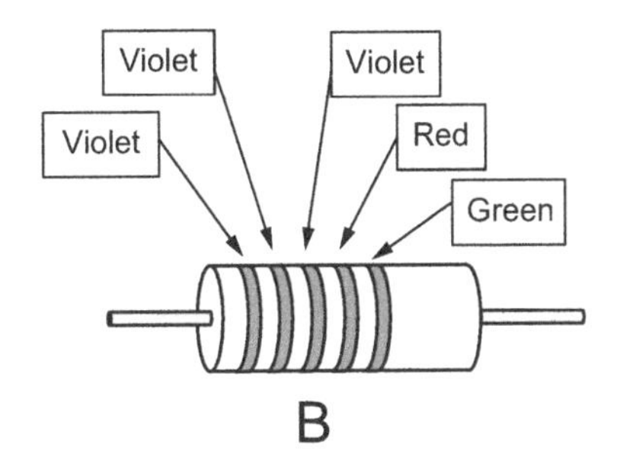

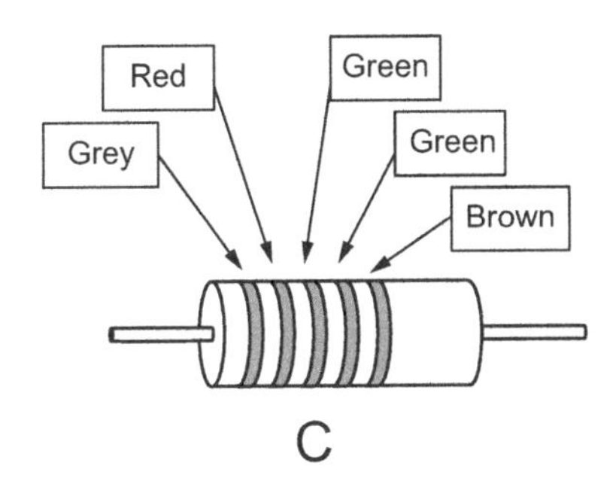

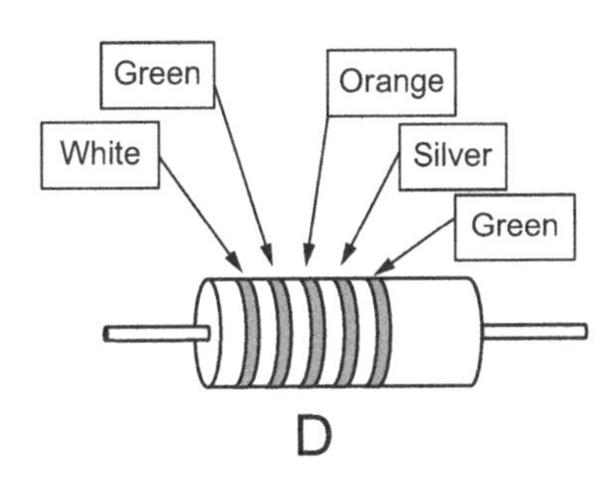

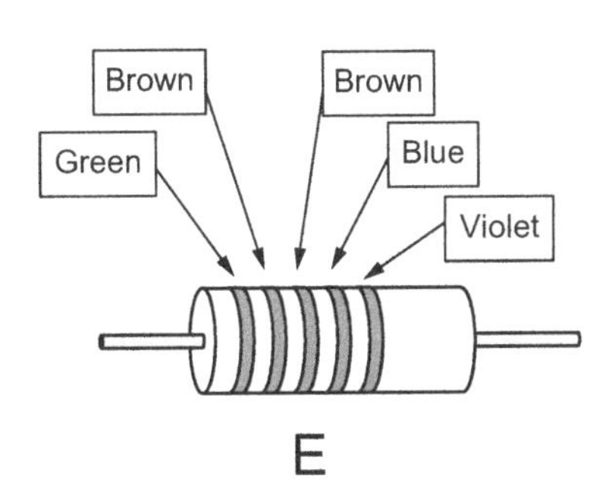

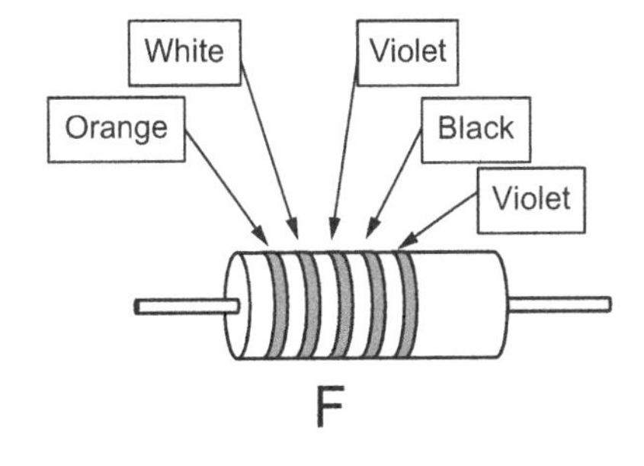

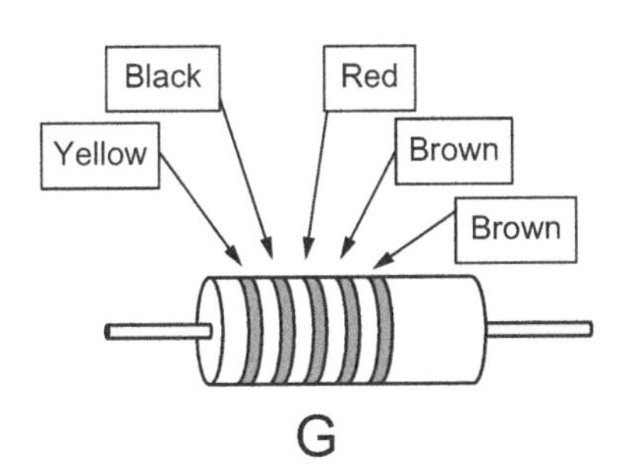

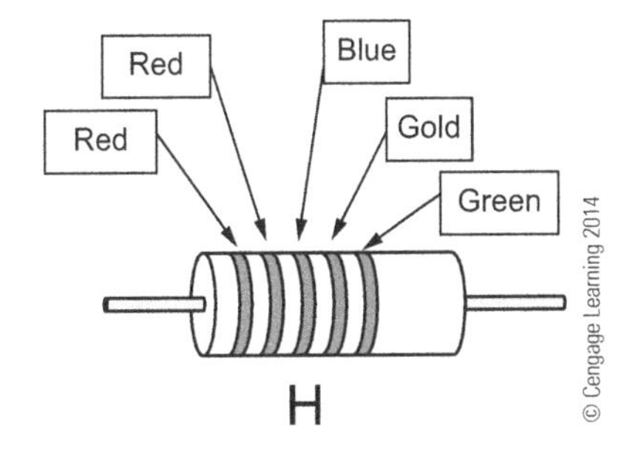

Name: ______________________________ Date: ______________

OHM'S LAW: FINDING CURRENT

1. Determine the amount of current flow.

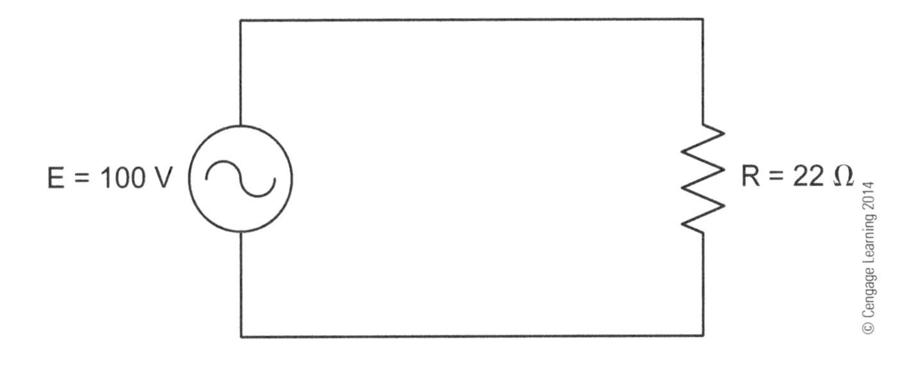

2. Determine the amount of current flow.

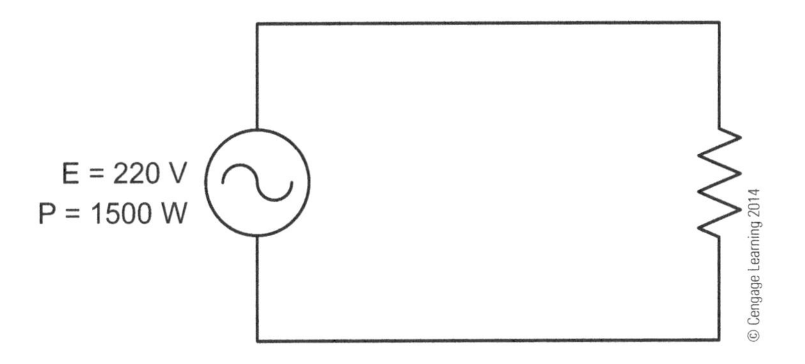

3. Determine the amount of current flow.

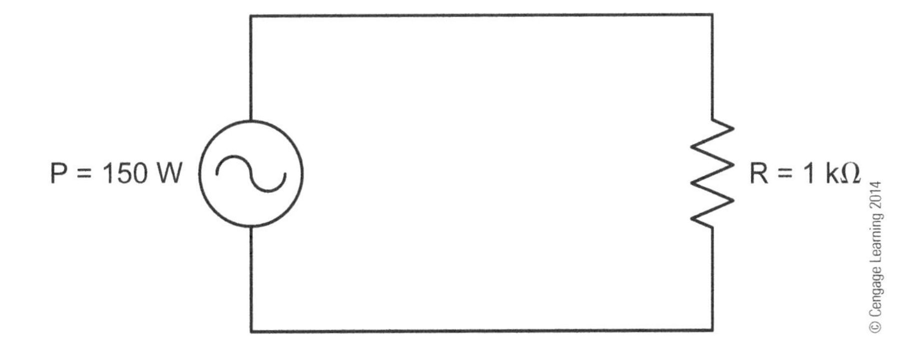

4. Determine the amount of current flow.

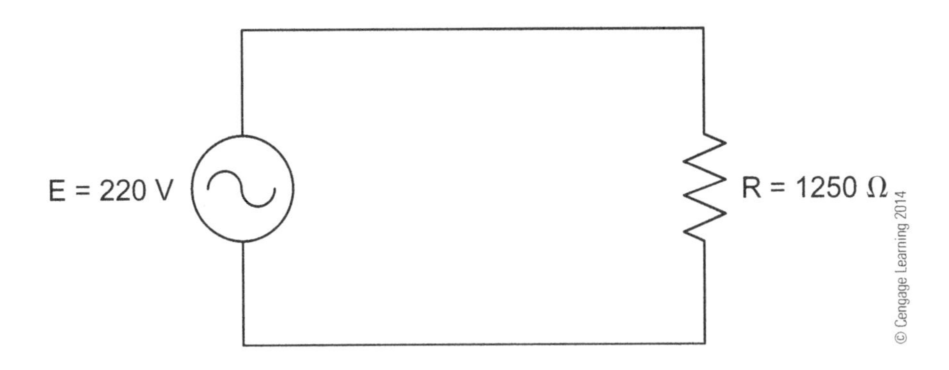

5. Determine the amount of current flow.

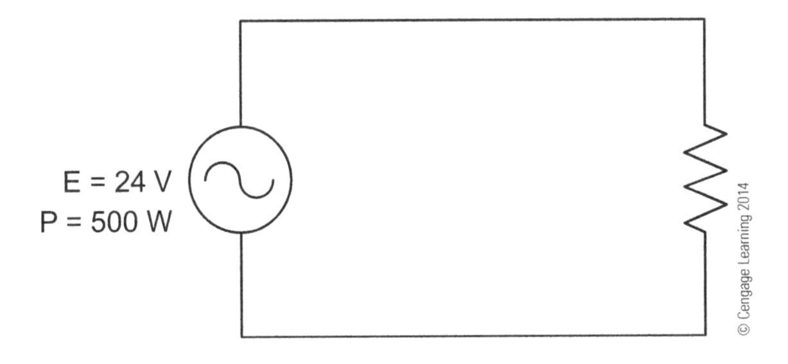

6. Determine the amount of current flow.

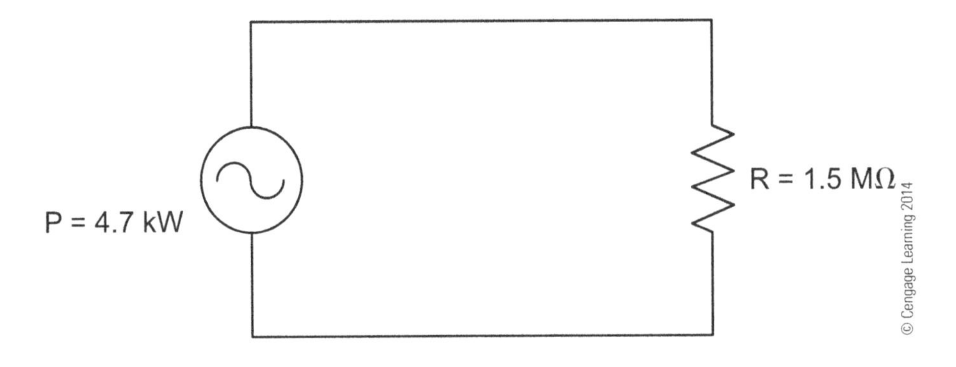

Name: ____________________ Date: ____________________

OHM'S LAW: FINDING VOLTAGE

1. Determine the amount of applied voltage.

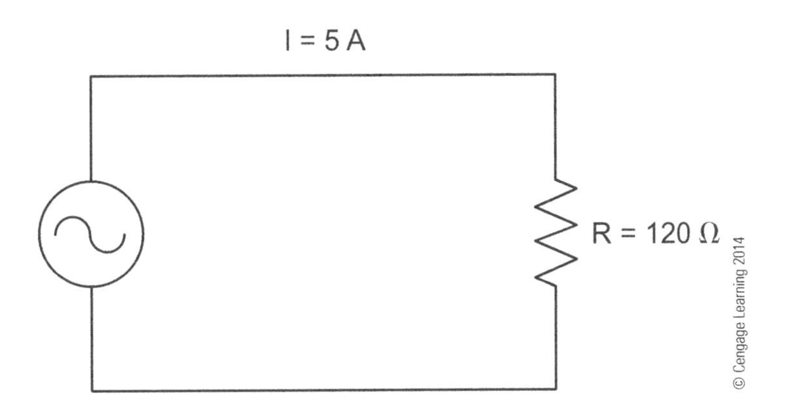

2. Determine the amount of applied voltage.

3. Determine the amount of applied voltage.

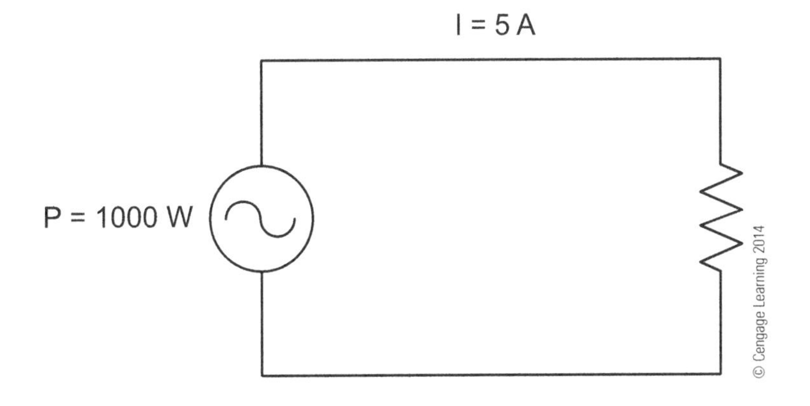

4. Determine the amount of applied voltage.

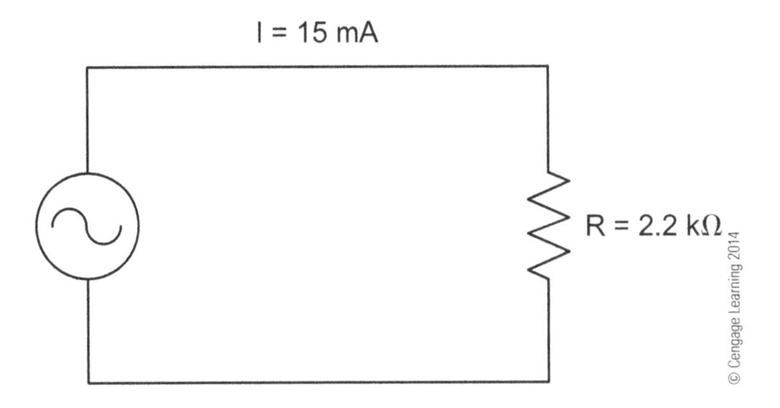

5. Determine the amount of applied voltage.

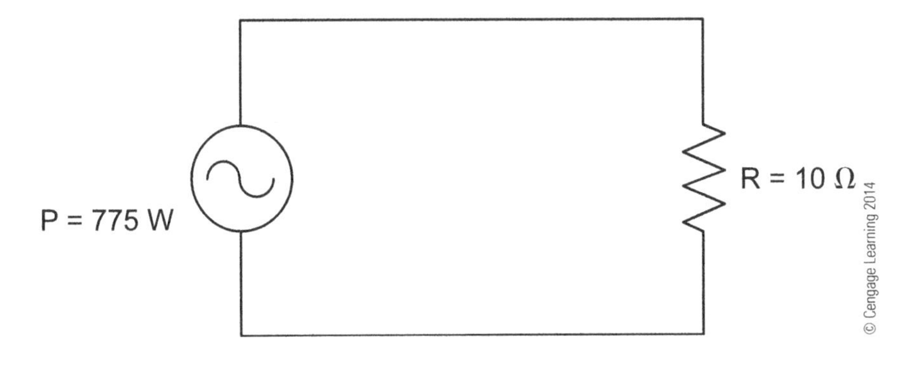

6. Determine the amount of applied voltage.

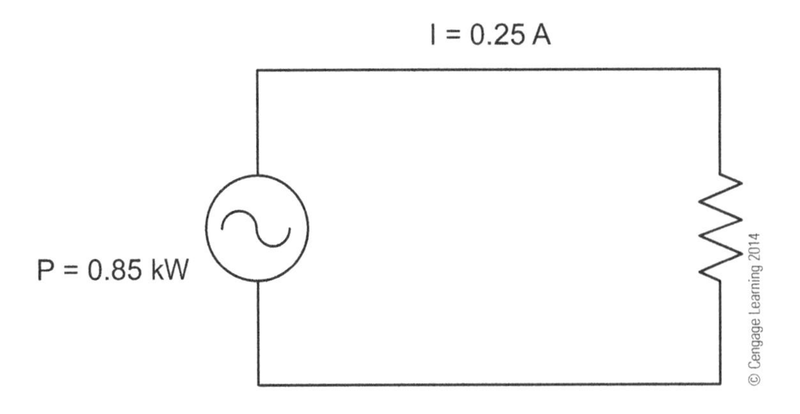

Name: ______________________________ Date: ______________

OHM'S LAW: FINDING RESISTANCE

1. Determine the amount of resistance.

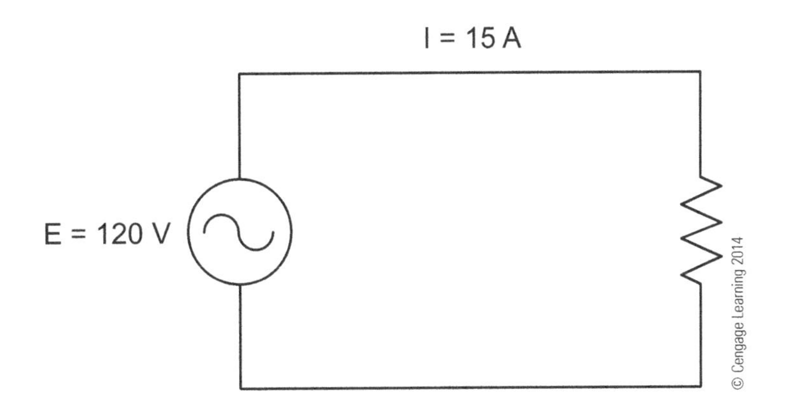

2. Determine the amount of resistance.

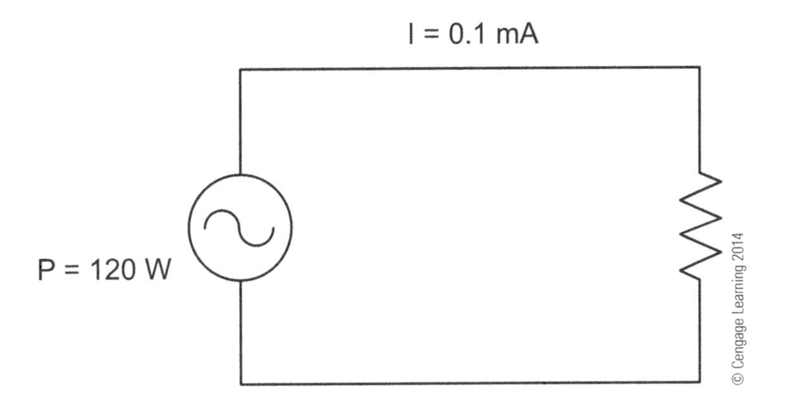

3. Determine the amount of resistance.

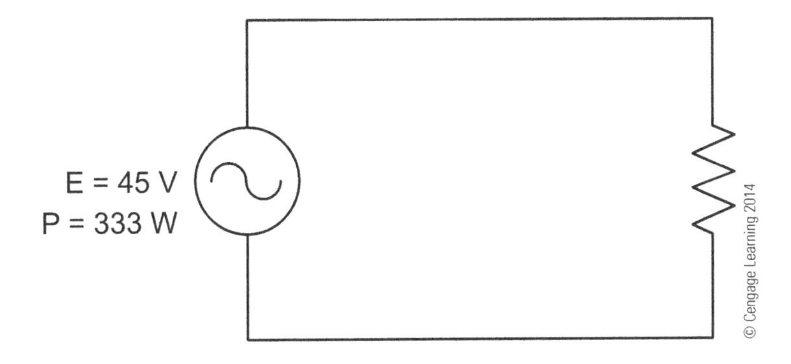

4. Determine the amount of resistance.

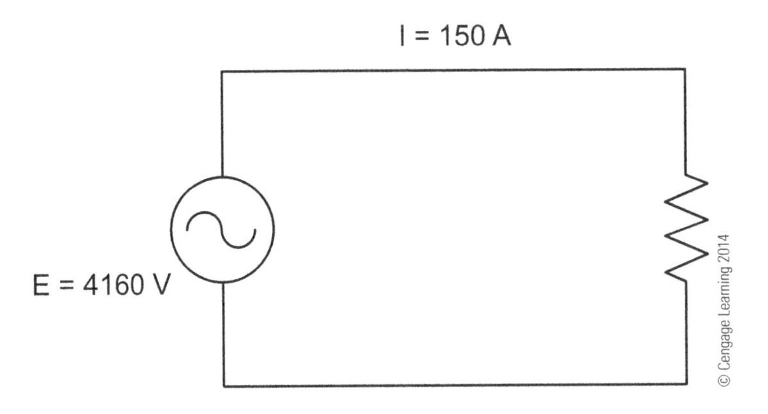

5. Determine the amount of resistance.

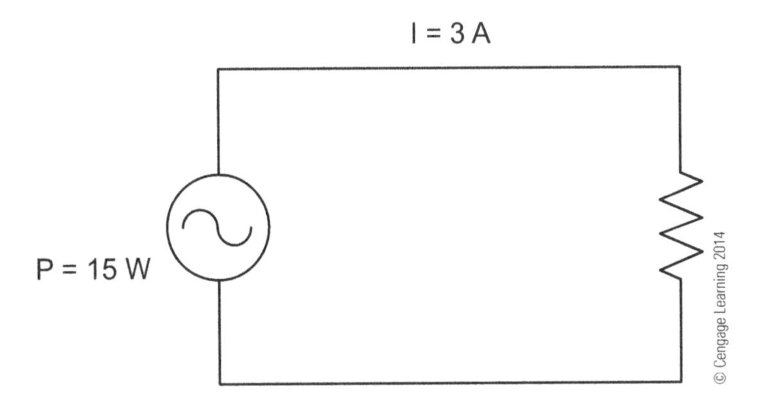

6. Determine the amount of resistance.

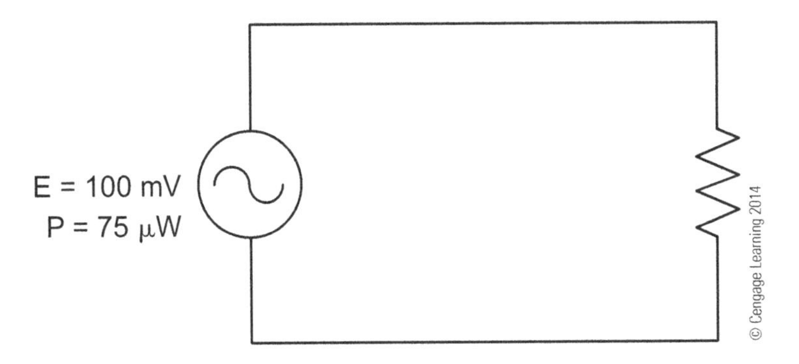

Name: ______________________________ Date: ______________

POWER LAW: FINDING POWER

1. Determine the amount of power.

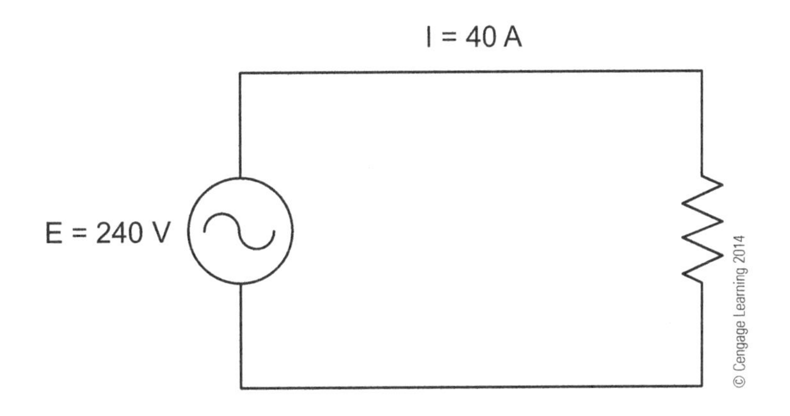

2. Determine the amount of power.

3. Determine the amount of power.

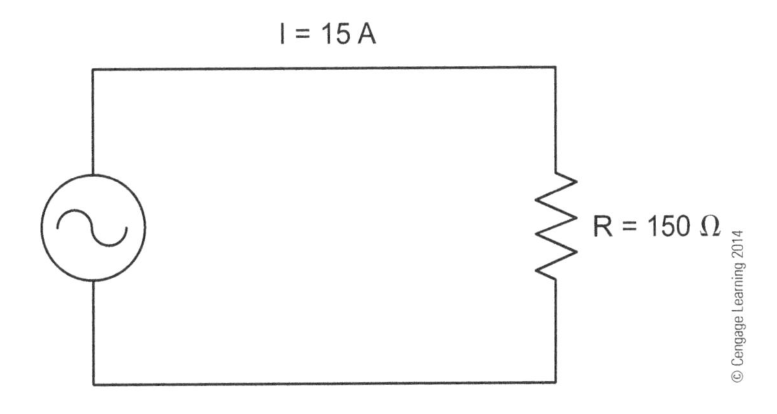

4. Determine the amount of power.

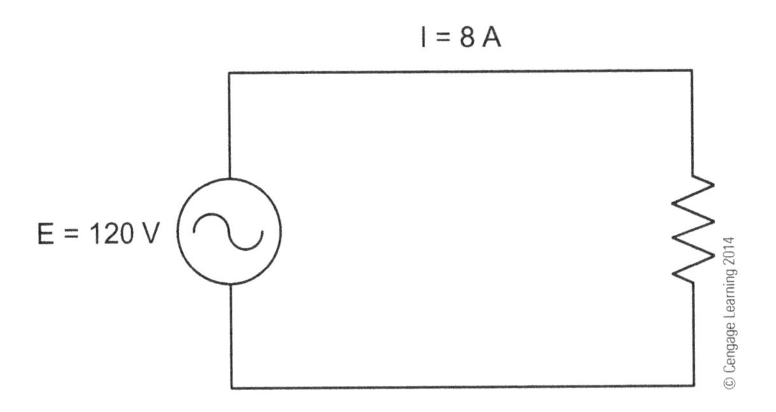

5. Determine the amount of power.

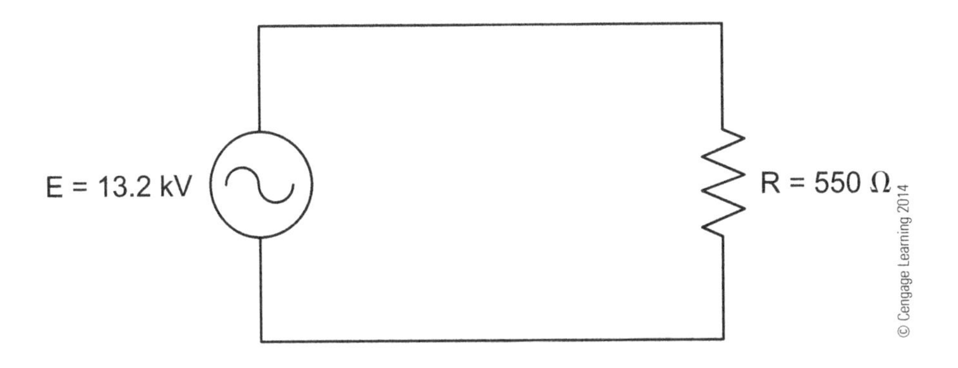

6. Determine the amount of power.

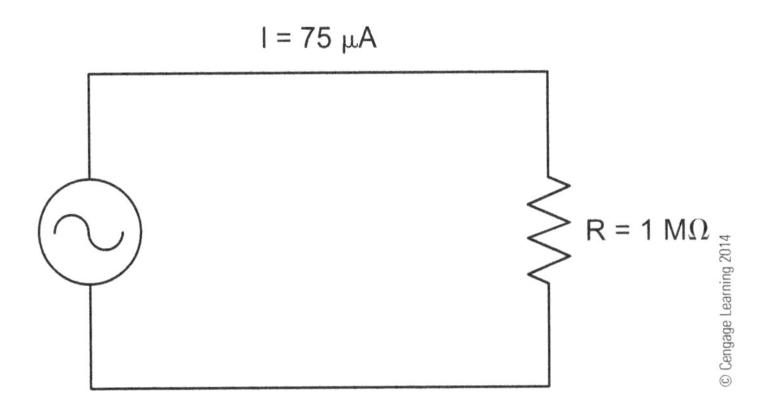

Name: ______________________ Date: ______________

APPLYING OHM'S LAW—PART 1

1. Determine the amount of current flow.

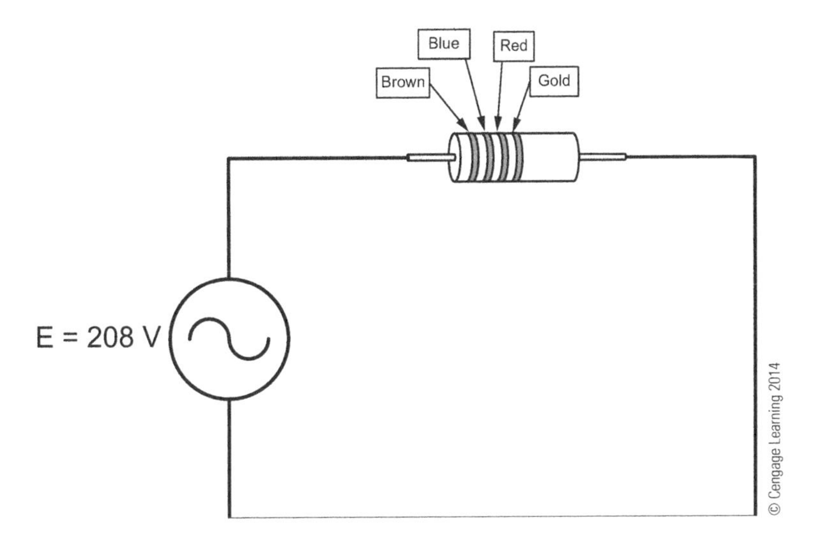

2. Determine the amount of current flow.

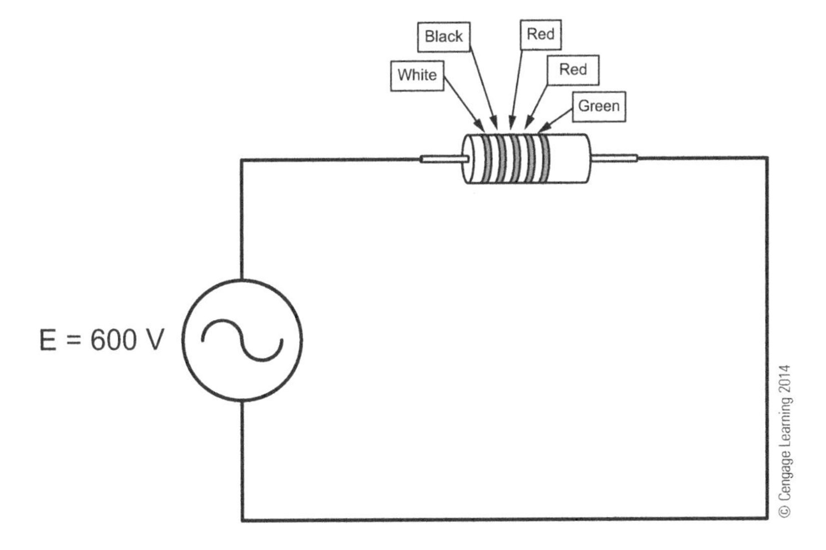

Name: ____________________ Date: ____________________

APPLYING OHM'S LAW—PART 2

1. Determine the amount of applied voltage.

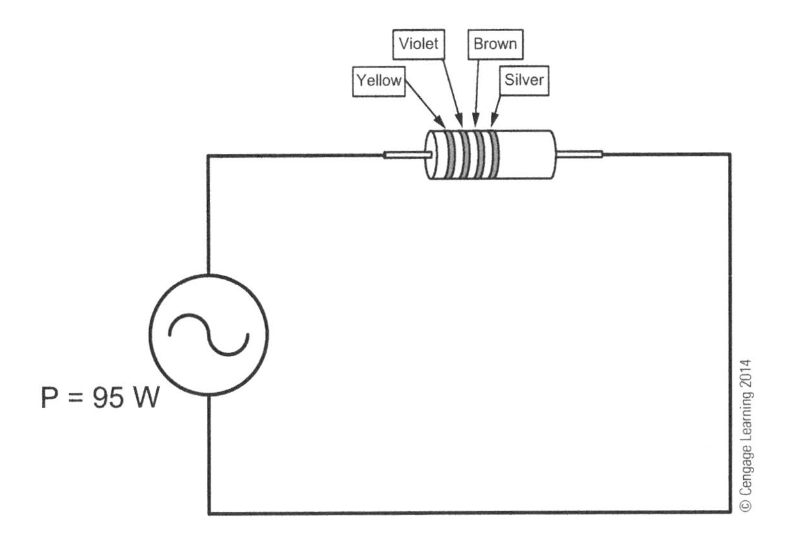

2. Determine the amount of applied voltage.

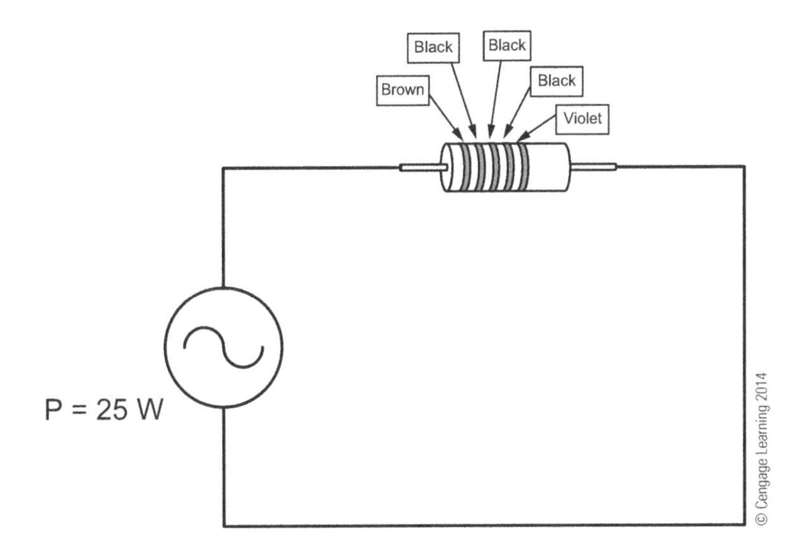

Name: ______________________ Date: ______________

APPLYING OHM'S LAW—PART 3

1. Determine color code for the amount of resistance. The tolerance cannot be determined; therefore, do not indicate a color code for the tolerance band.

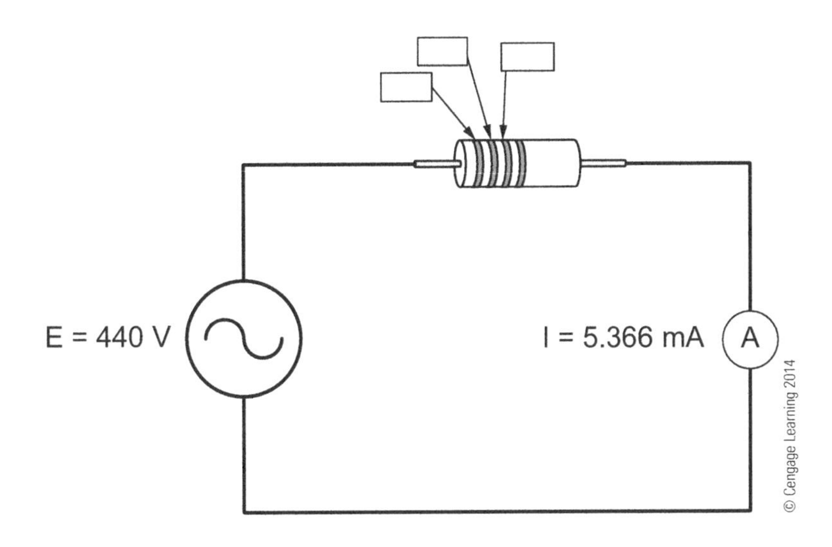

2. Determine color code for the amount of resistance. The tolerance cannot be determined; therefore, do not indicate a color code for the tolerance band.

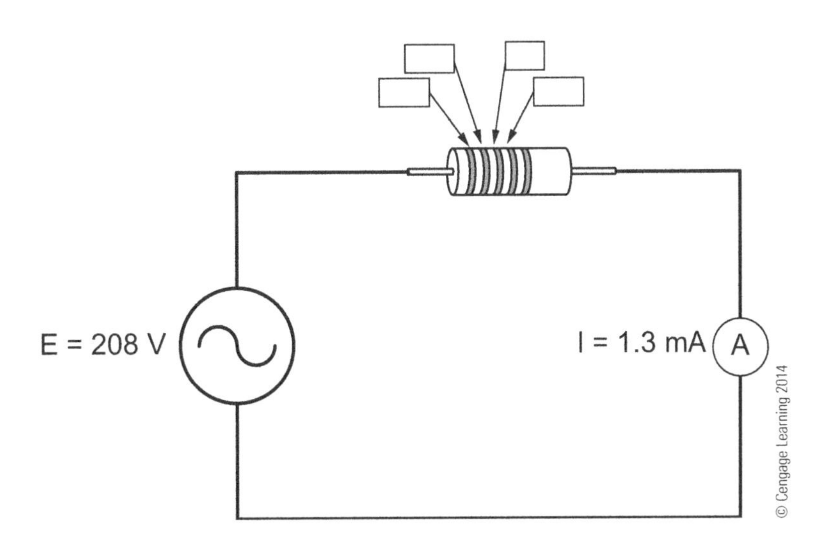

Name: ______________________ Date: ______________

APPLYING POWER LAW

1. Determine the amount of power.

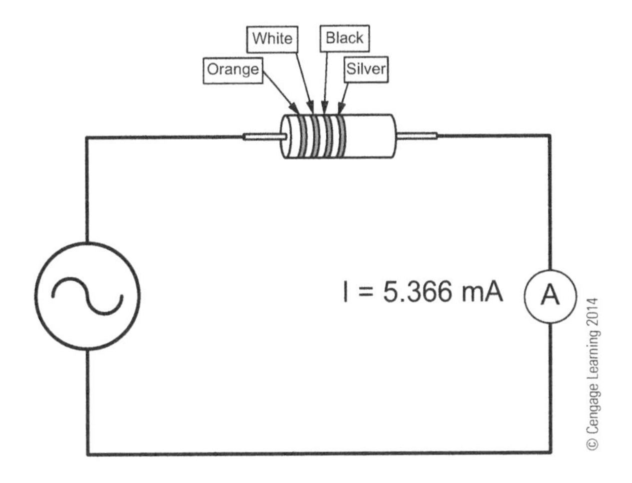

2. Determine the amount of power.

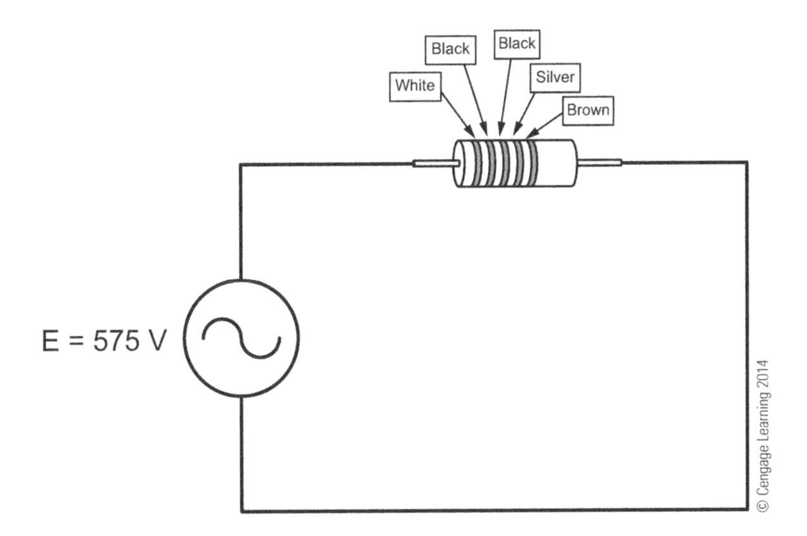

Name: ______________________________ Date: ______________________

DIGITAL MULTIMETER: MEASURING CURRENT

1. Complete the wiring of the circuit. Include the meter connections that would allow you to measure the current flowing through the light bulb. Also, indicate the position of the selector switch for measuring the current.

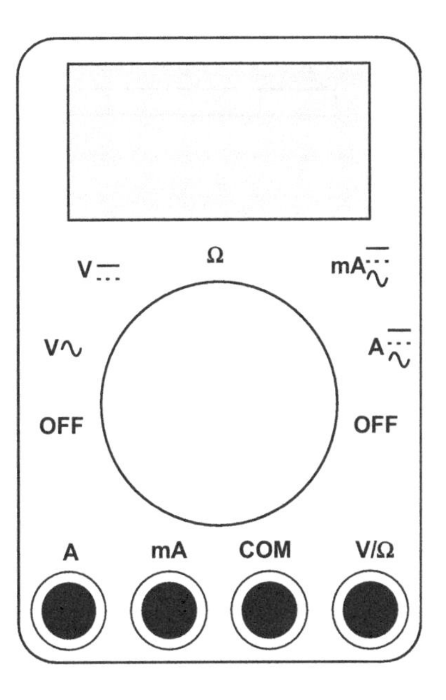

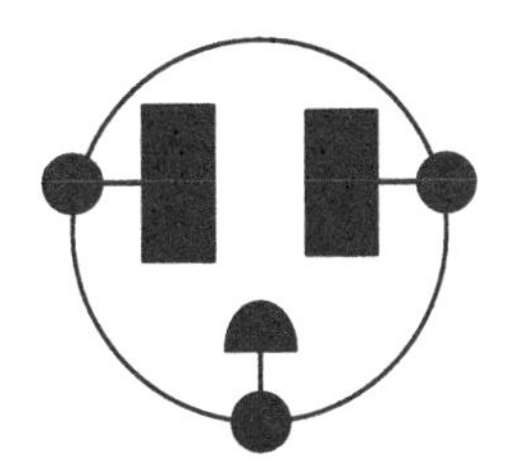

2. Complete the wiring of the circuit. Include the meter connections that would allow you to measure the current flowing through the single-phase motor. Also, indicate the position of the selector switch for measuring the current.

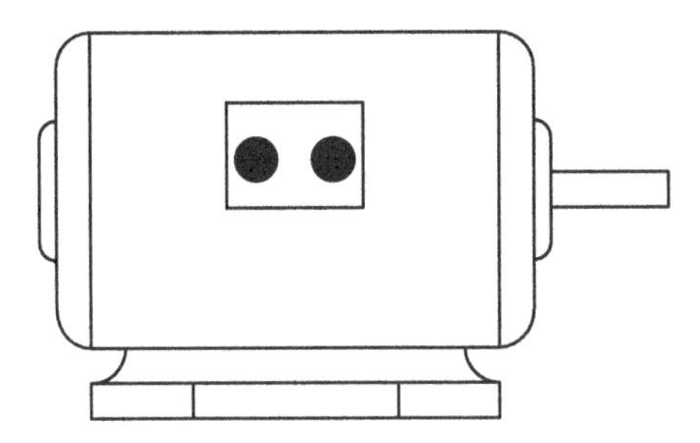

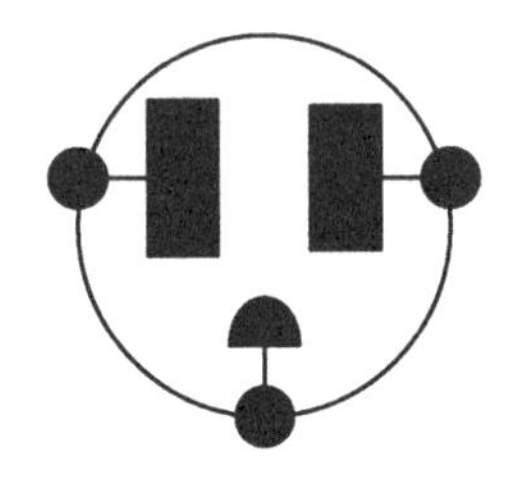

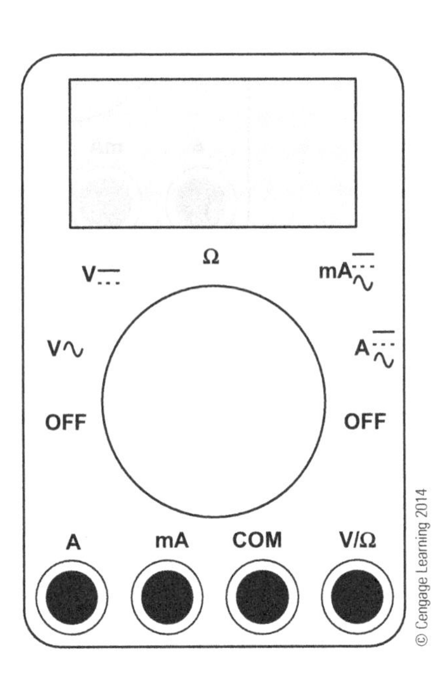

3. Complete the wiring of the circuit. Include the meter connections that would allow you to measure the current flowing through the motor starter coil. Also, indicate the position of the selector switch for measuring the current.

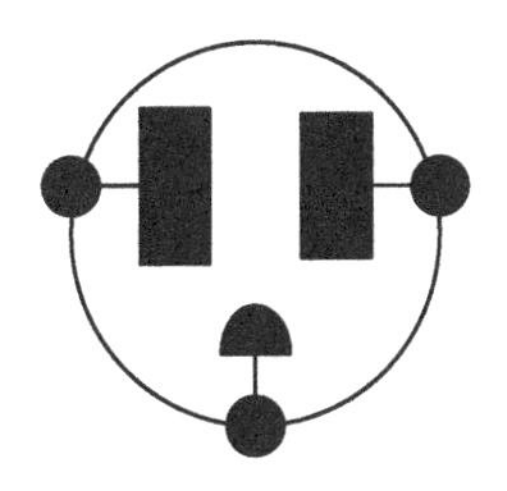

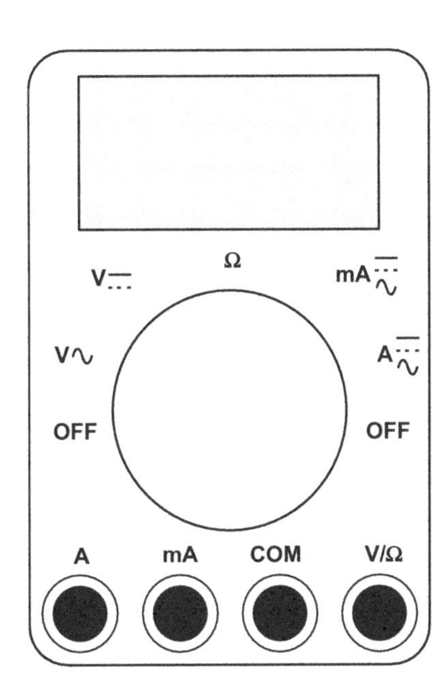

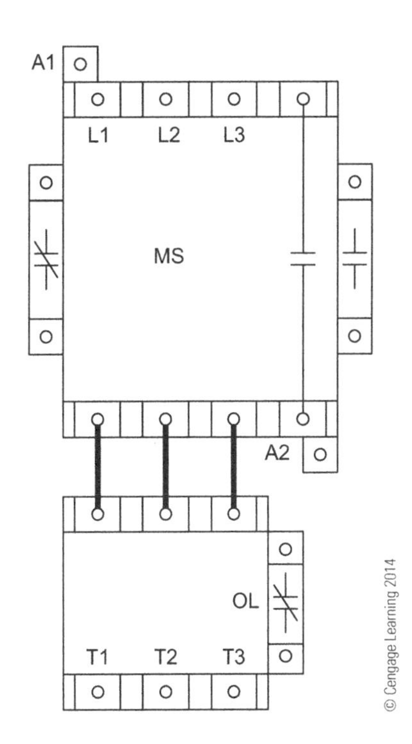

Name: ______________________ Date: ______________

DIGITAL MULTIMETER: MEASURING VOLTAGE

1. Complete the wiring of the circuit. Include the meter connections that would allow you to measure the voltage applied to the light bulb. Also, indicate the position of the selector switch for measuring the voltage.

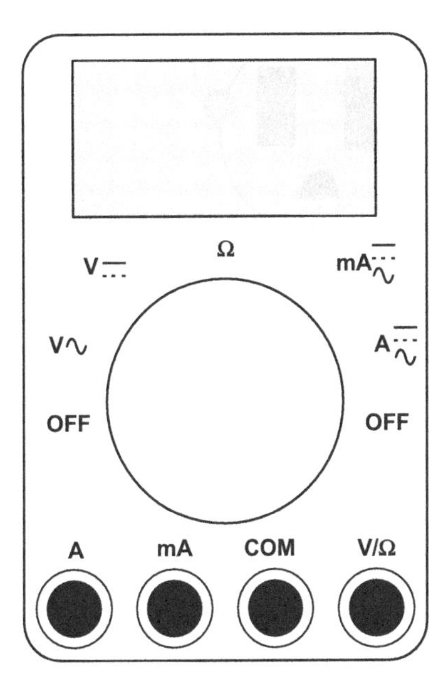

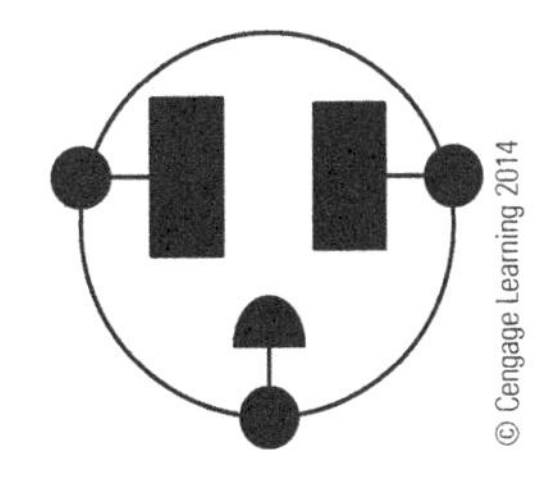

2. Complete the wiring of the circuit. Include the meter connections that would allow you to measure the voltage applied to the single-phase motor. Also, indicate the position of the selector switch for measuring the voltage.

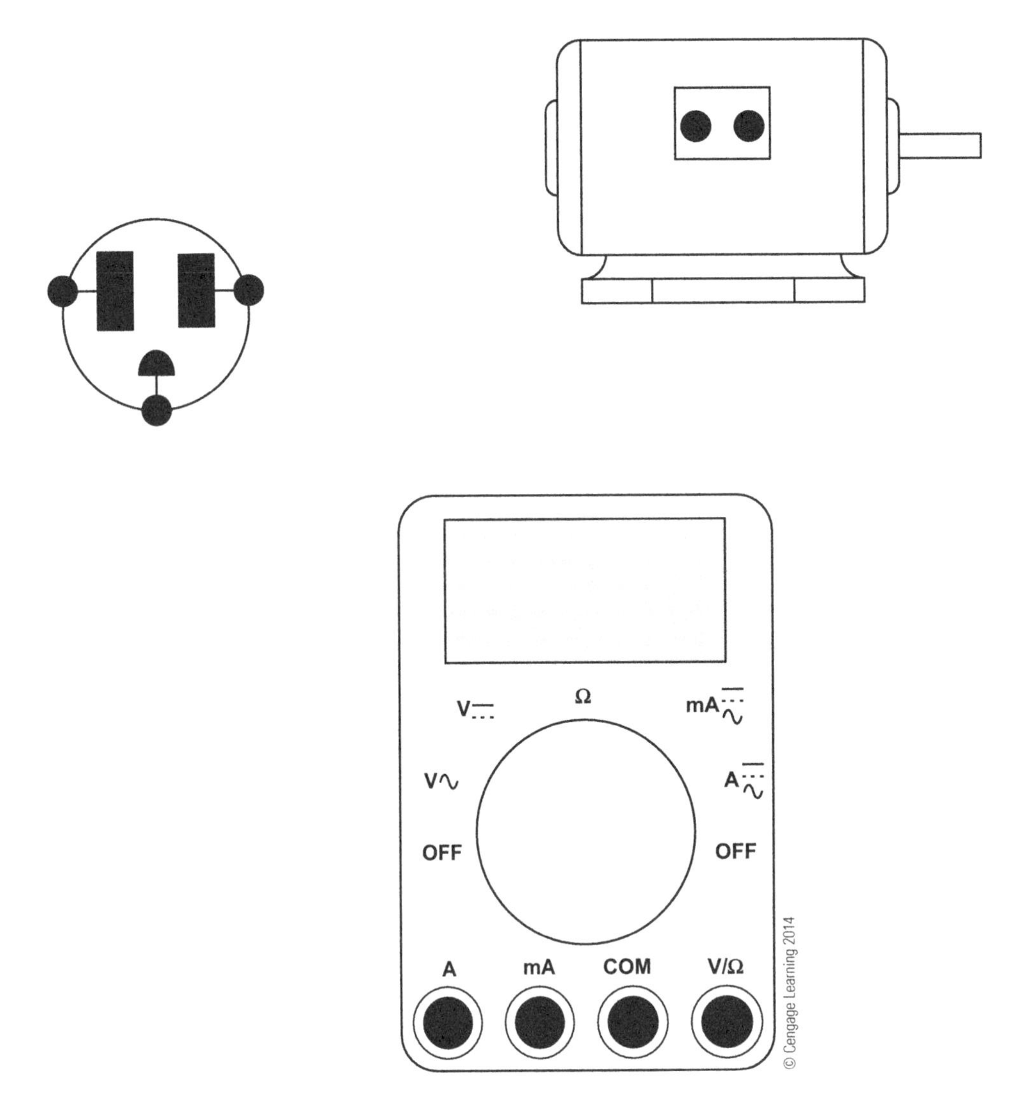

3. Complete the wiring of the circuit. Include the meter connections that would allow you to measure the voltage applied to the motor starter coil. Also, indicate the position of the selector switch for measuring the voltage.

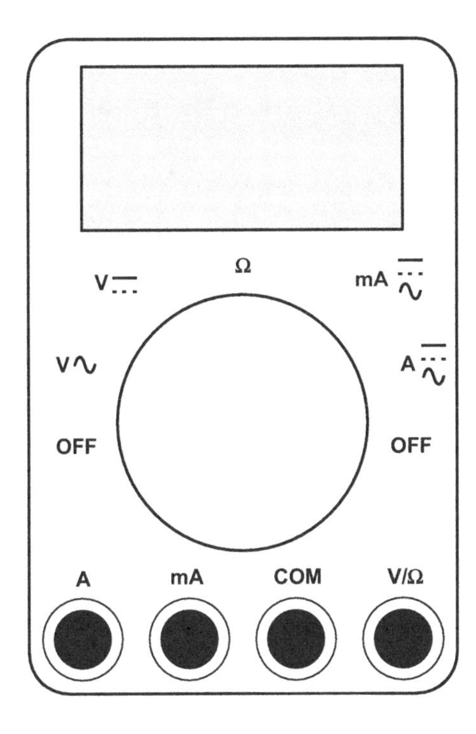

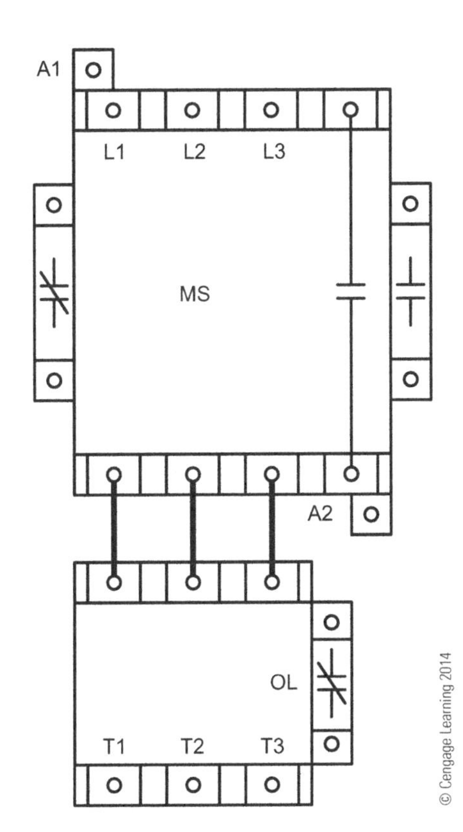

Name: ____________________ Date: ____________________

DIGITAL MULTIMETER: MEASURING RESISTANCE

1. Complete the wiring of the circuit. Include the meter connections that would allow you to measure the resistance of the light bulb. Also, indicate the position of the selector switch for measuring the resistance.

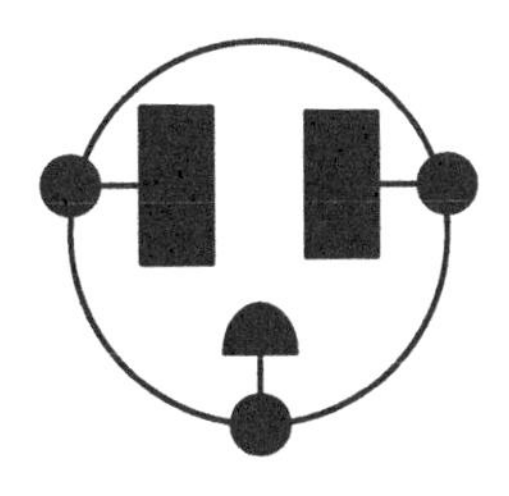

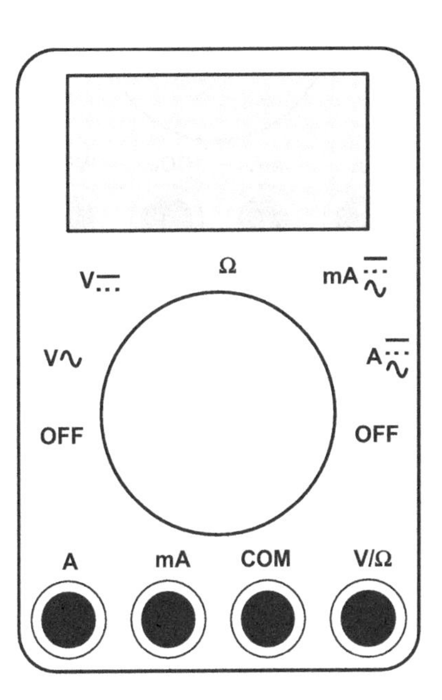

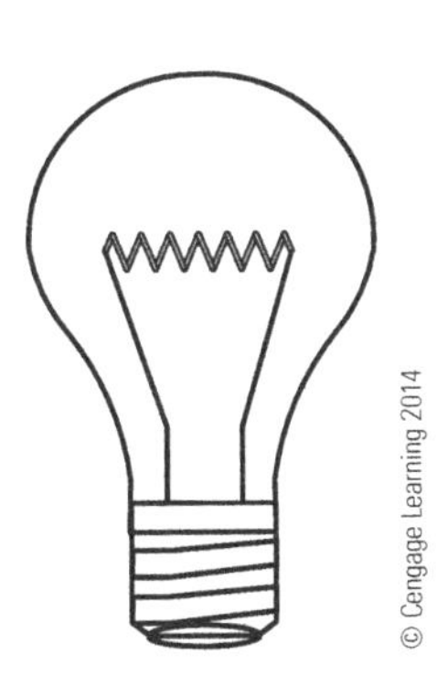

2. Complete the wiring of the circuit. Include the meter connections that would allow you to measure the resistance of the single-phase motor. Also, indicate the position of the selector switch for measuring the resistance.

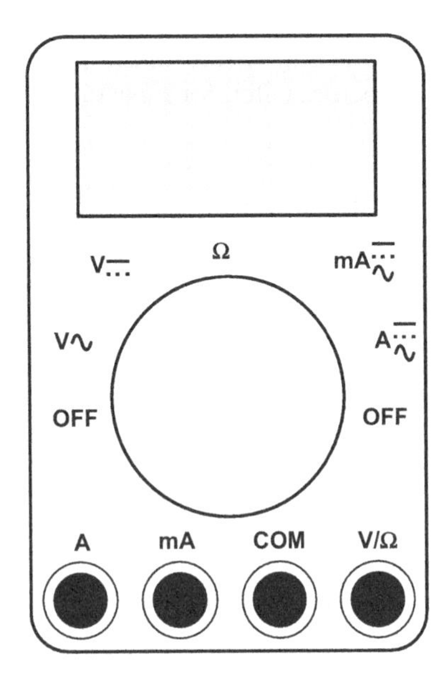

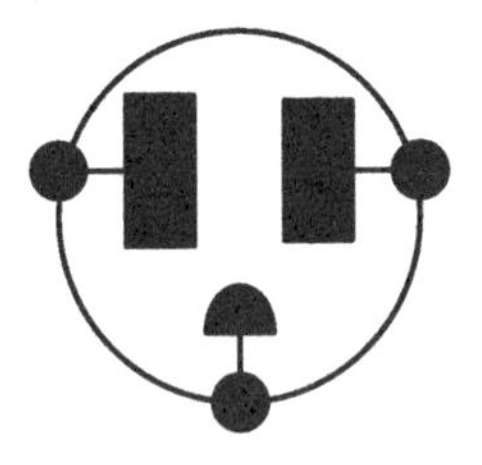

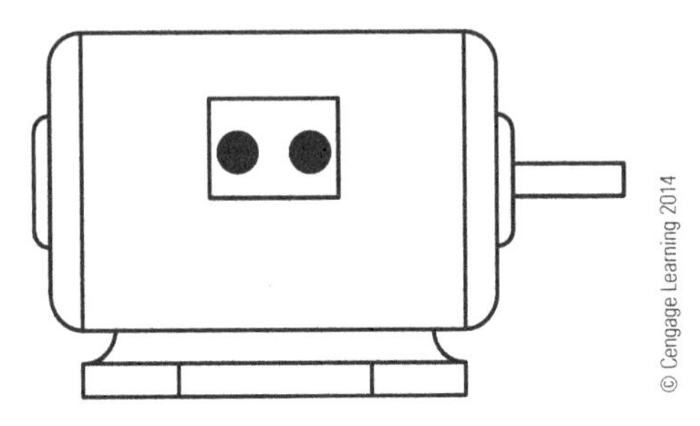

3. Complete the wiring of the circuit. Include the meter connections that would allow you to measure the resistance of the motor starter coil. Also, indicate the position of the selector switch for measuring the resistance.

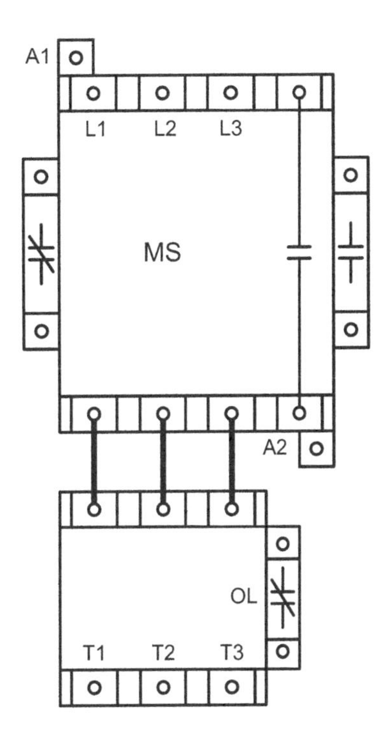

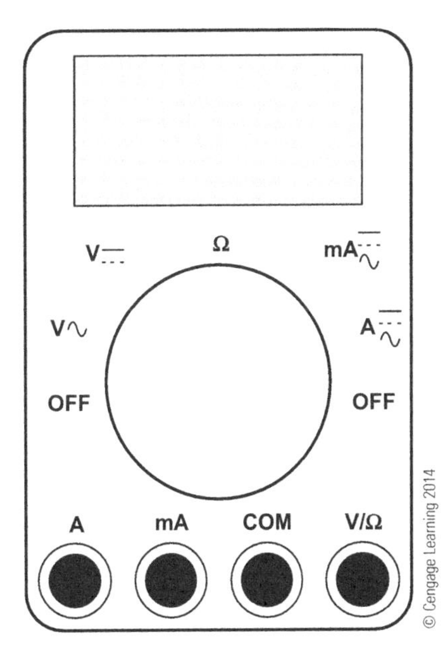

Name: ______________________________ Date: ______________

OSCILLOSCOPE: MEASURING VOLTAGE

1. Given the circuit, determine the amount of voltage indicated on channel 1 and on channel 2 of the oscilloscope.

   a. Channel 1 voltage ______________

   b. Channel 2 voltage ______________

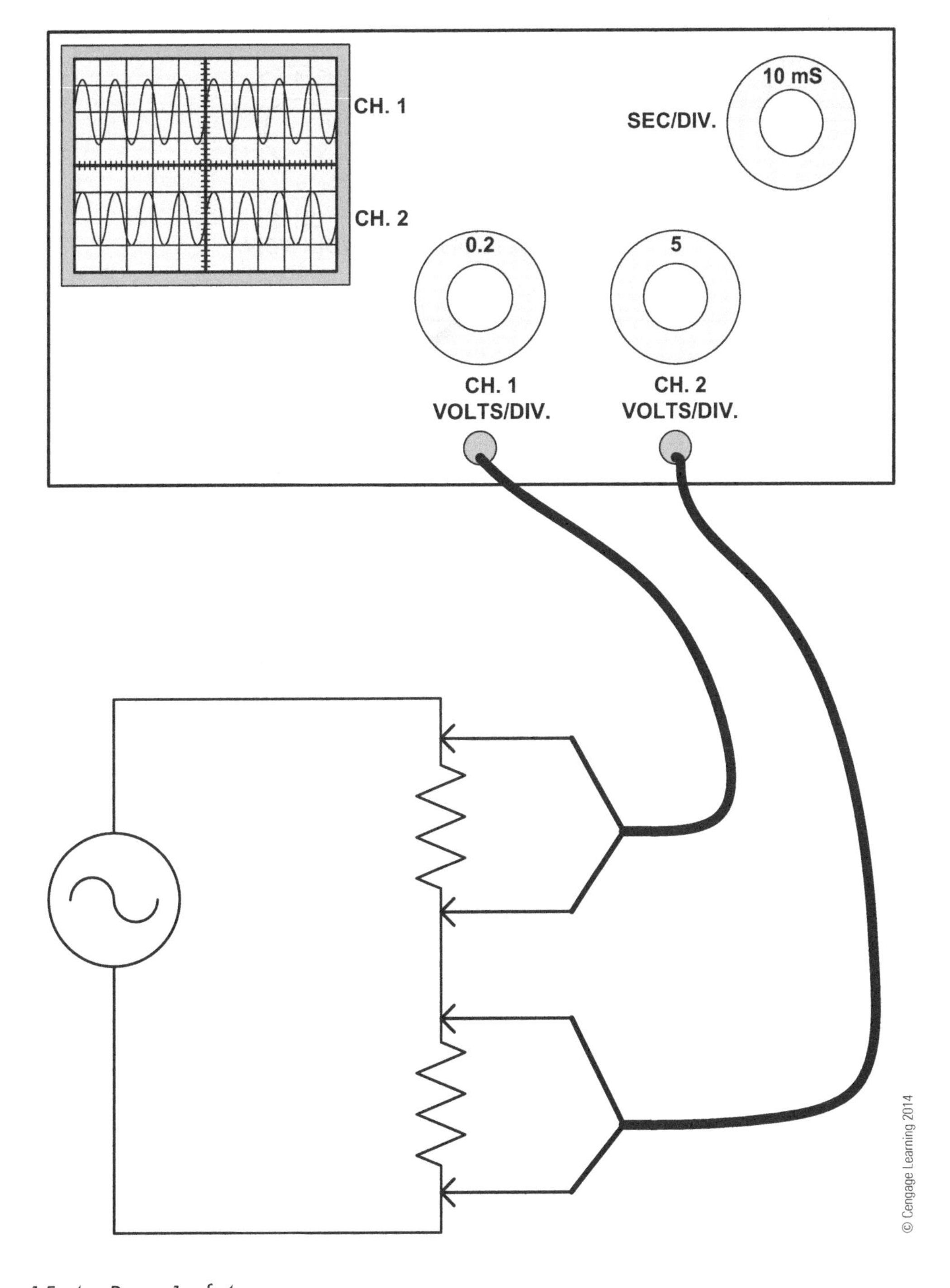

2. Given the circuit, determine the amount of voltage indicated on channel 1 and on channel 2 of the oscilloscope.

   a. Channel 1 voltage ____________________

   b. Channel 2 voltage ____________________

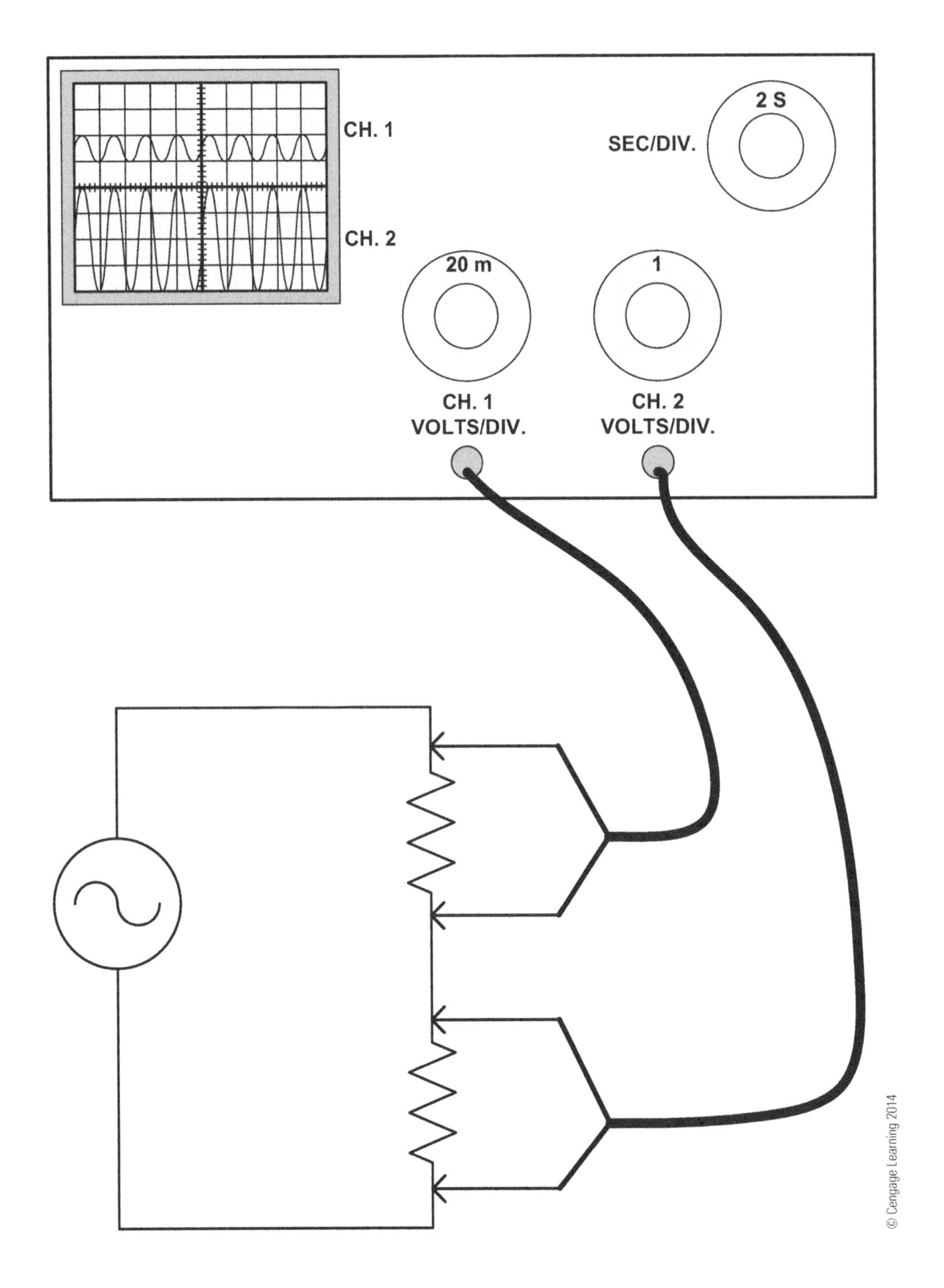

3. Given the circuit, determine the amount of voltage indicated on channel 1 and on channel 2 of the oscilloscope.

   a. Channel 1 voltage ____________________

   b. Channel 2 voltage ____________________

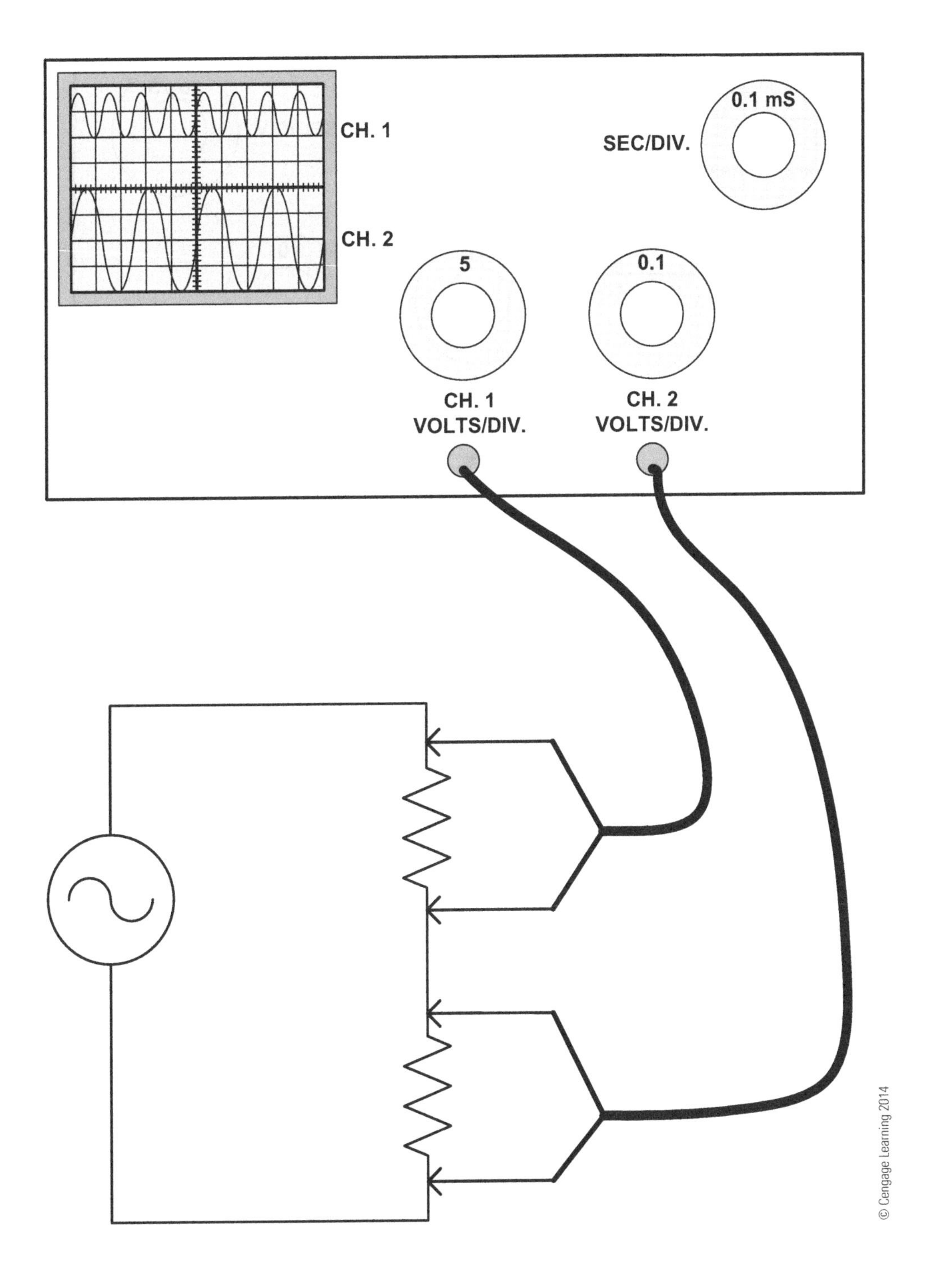

4. Given the circuit, determine the amount of voltage indicated on channel 1 and on channel 2 of the oscilloscope.

   a. Channel 1 voltage ______________________

   b. Channel 2 voltage ______________________

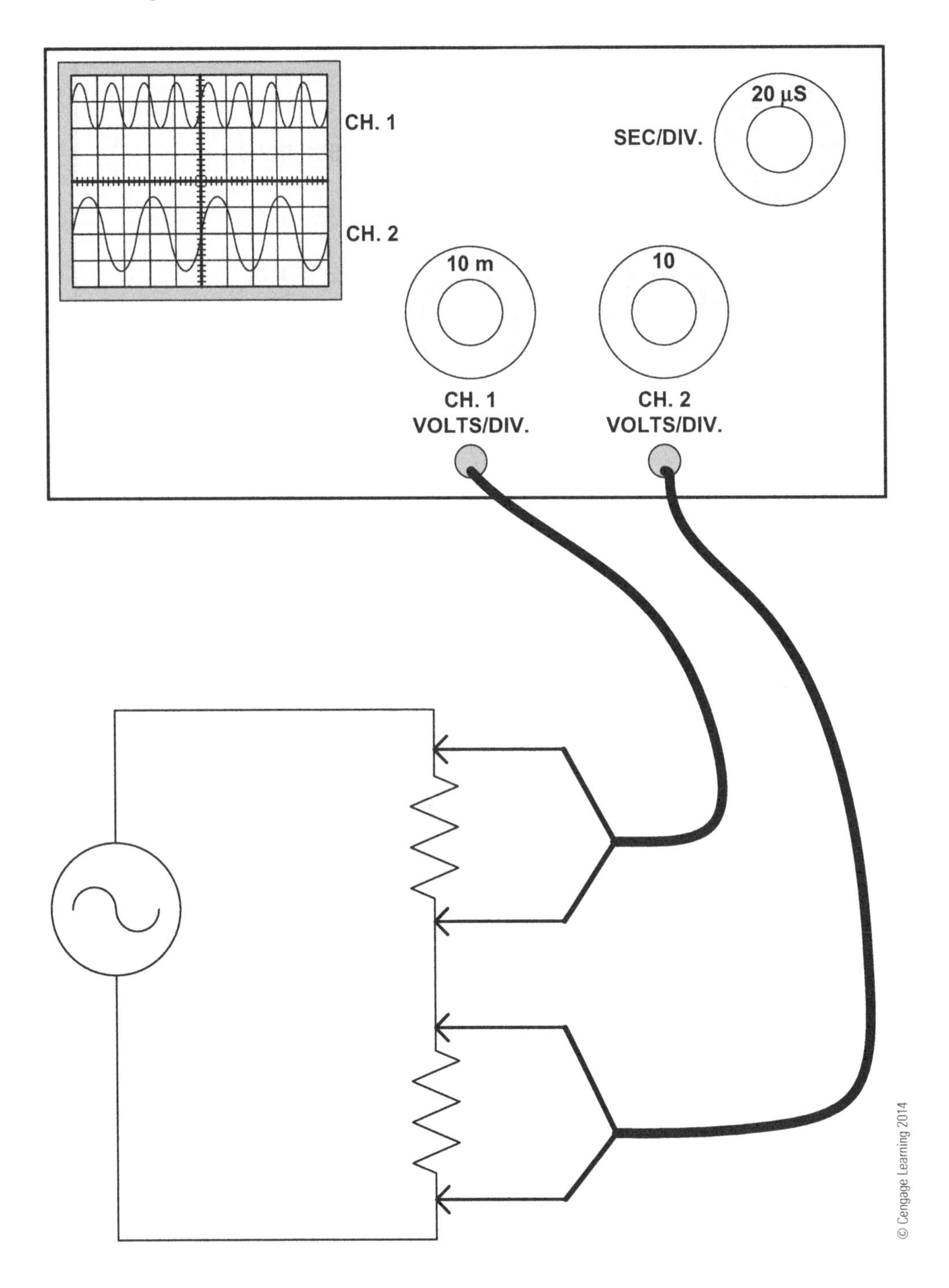

Name: ______________________ Date: ______________

OSCILLOSCOPE: MEASURING FREQUENCY

1. Given the circuit, determine the frequency of the waveform displayed on channel 1 and on channel 2 of the oscilloscope.
   a. Channel 1 frequency ______________
   b. Channel 2 frequency ______________

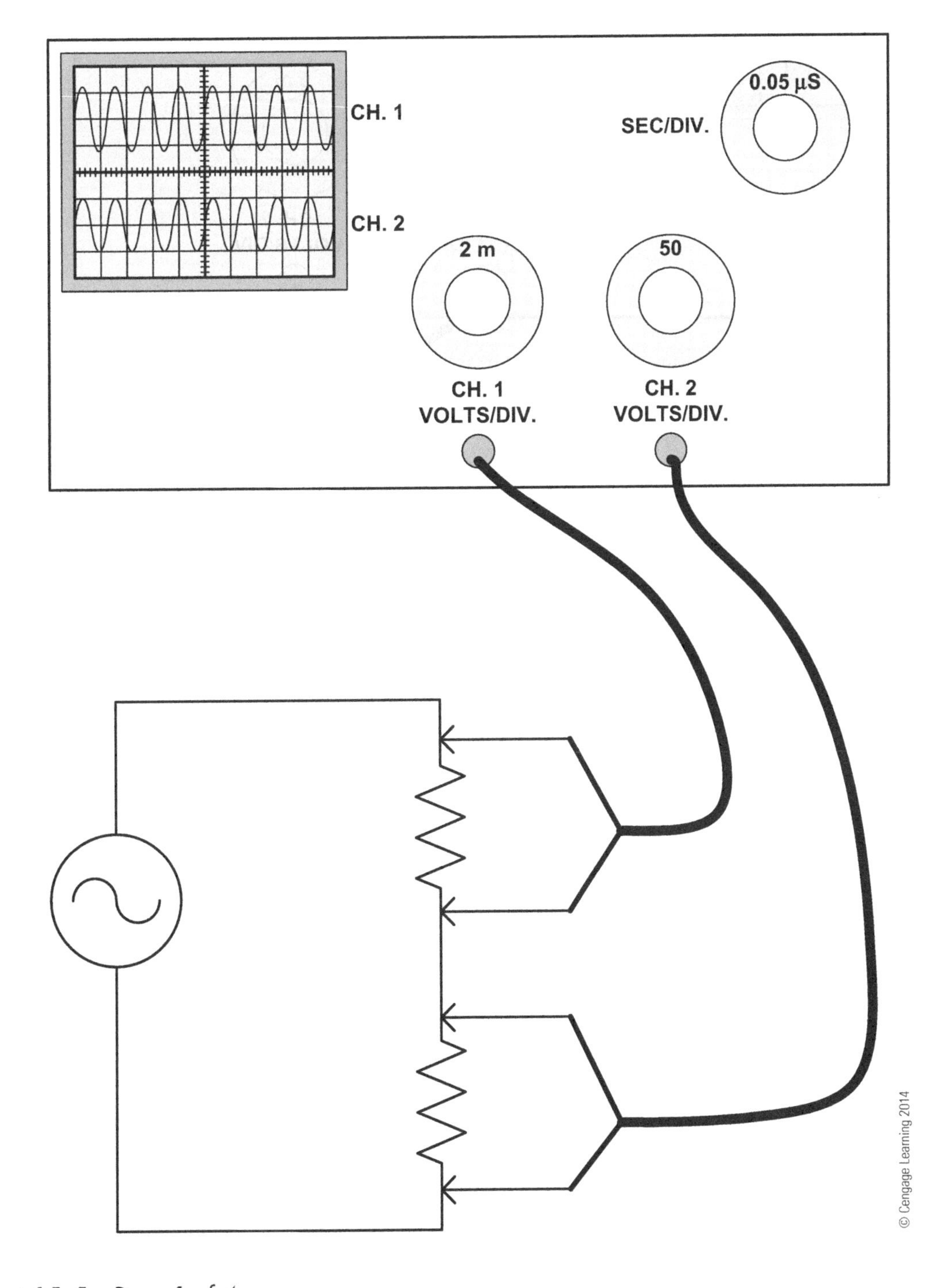

2. Given the circuit, determine the frequency of the waveform displayed on channel 1 and on channel 2 of the oscilloscope.

   a. Channel 1 frequency ____________________

   b. Channel 2 frequency ____________________

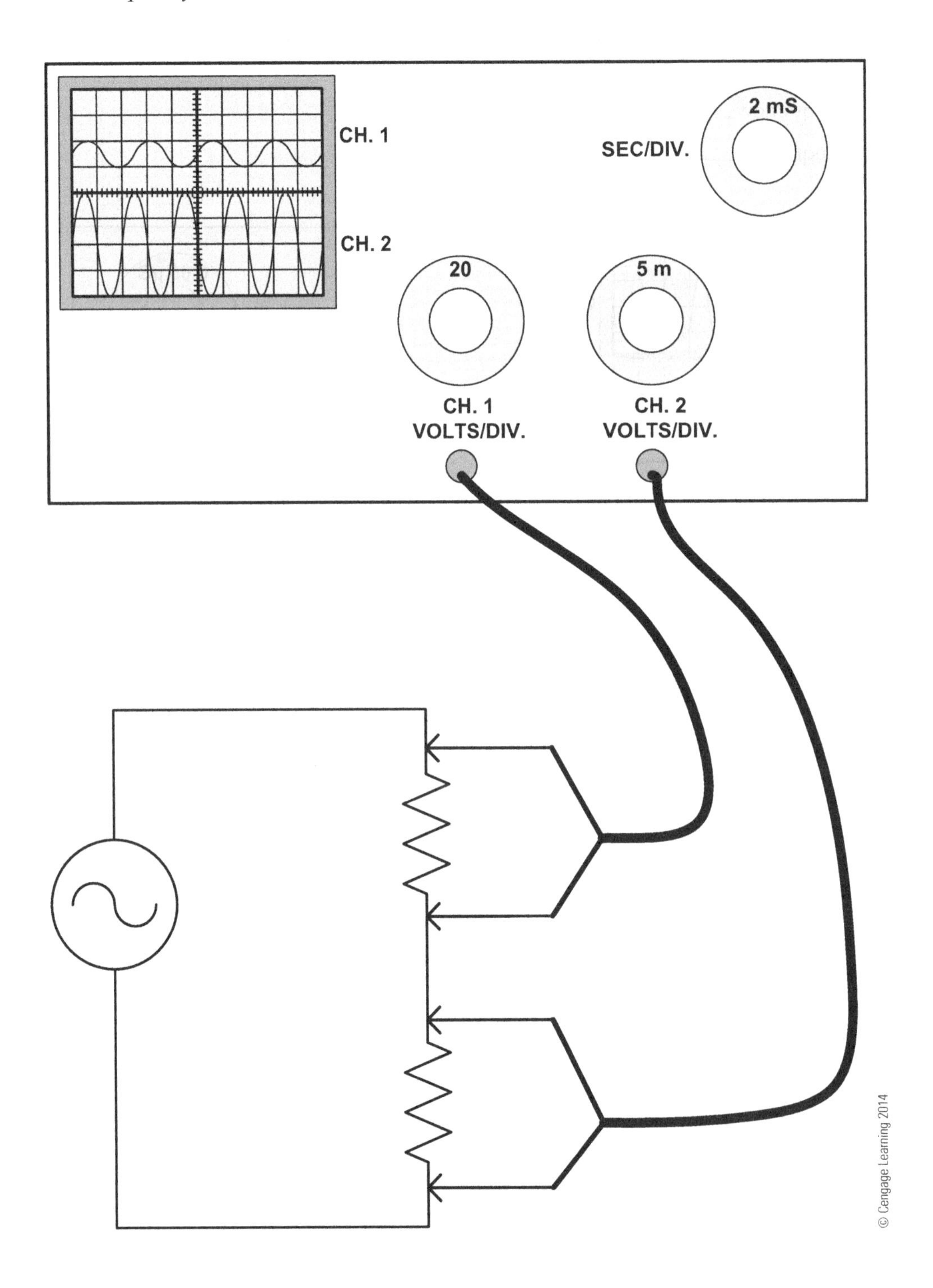

3. Given the circuit, determine the frequency of the waveform displayed on channel 1 and on channel 2 of the oscilloscope.

   a. Channel 1 frequency ____________________

   b. Channel 2 frequency ____________________

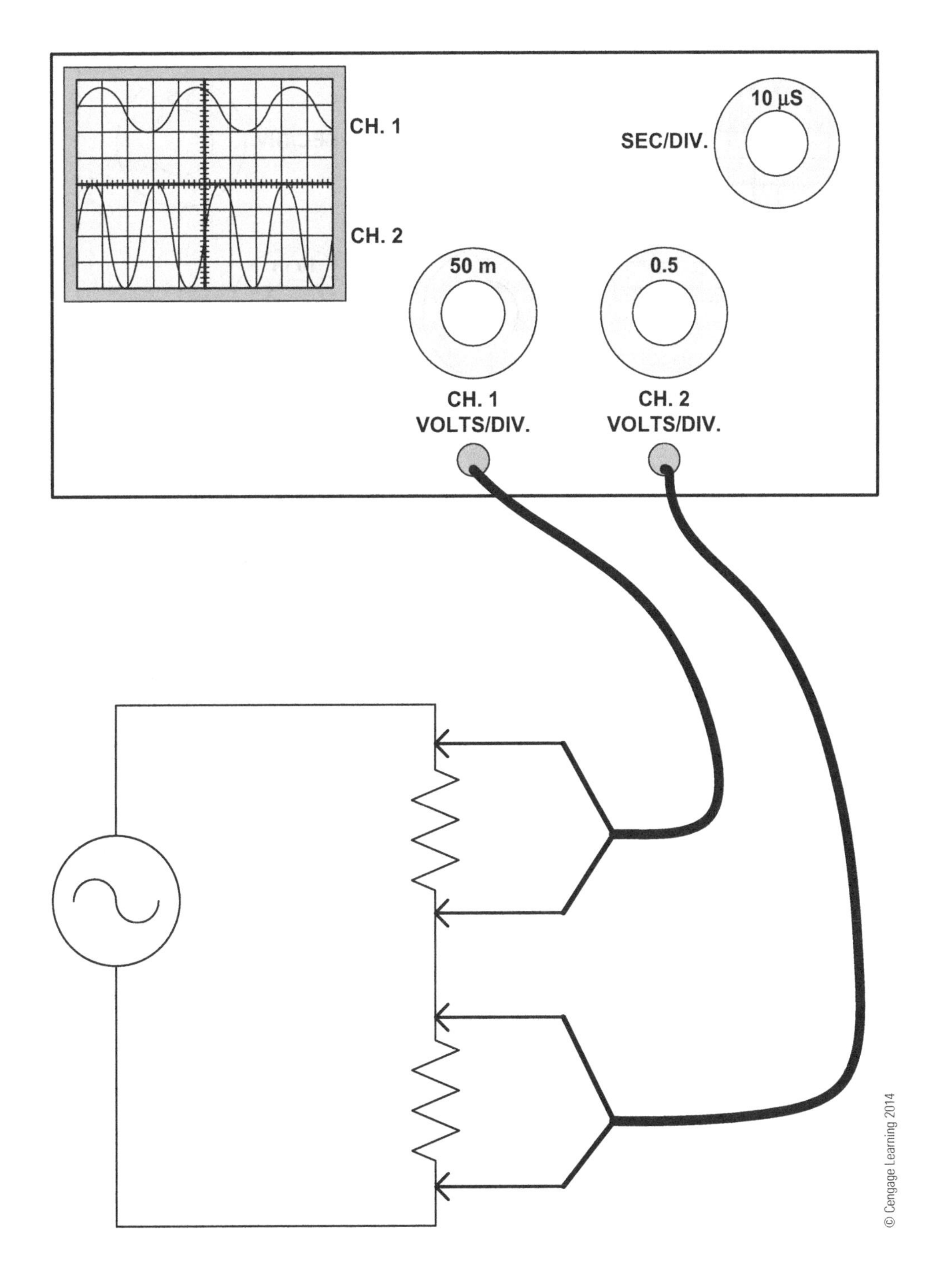

4. Given the circuit, determine the frequency of the waveform displayed on channel 1 and on channel 2 of the oscilloscope.
   a. Channel 1 frequency ____________________
   b. Channel 2 frequency ____________________

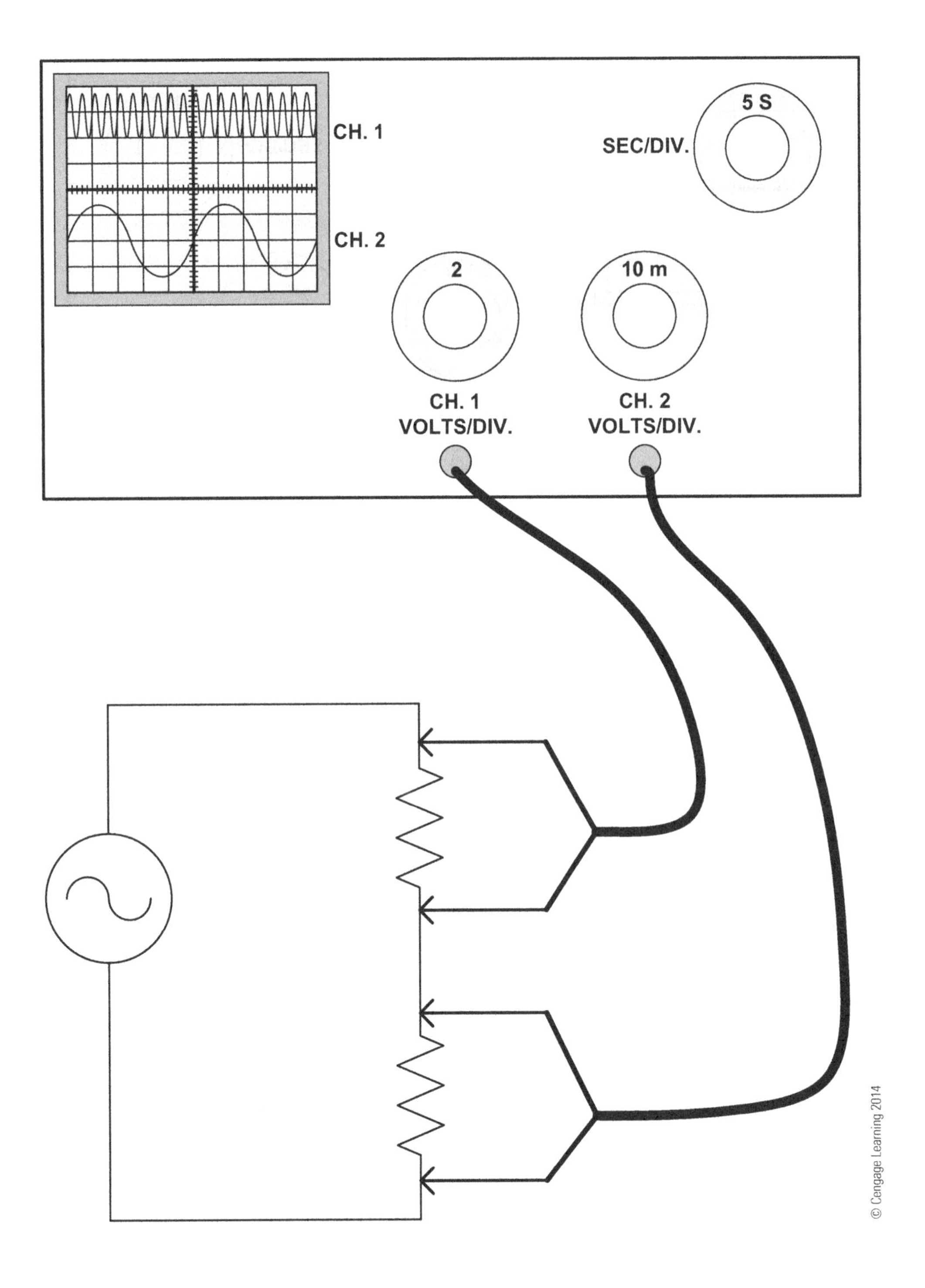

Name: ______________________ Date: ______________

SERIES CIRCUITS—PART 1

1. Given the schematic diagram, connect the components so that they electrically match the schematic.

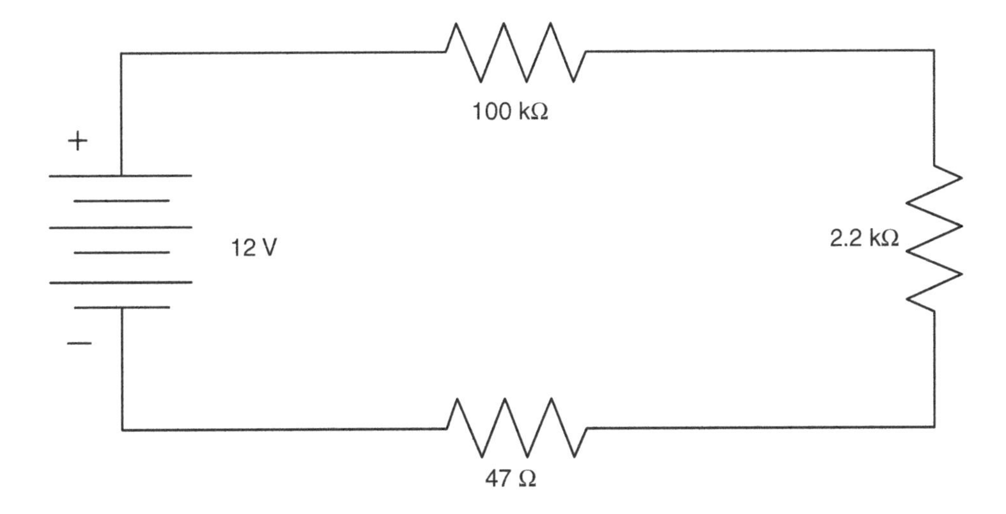

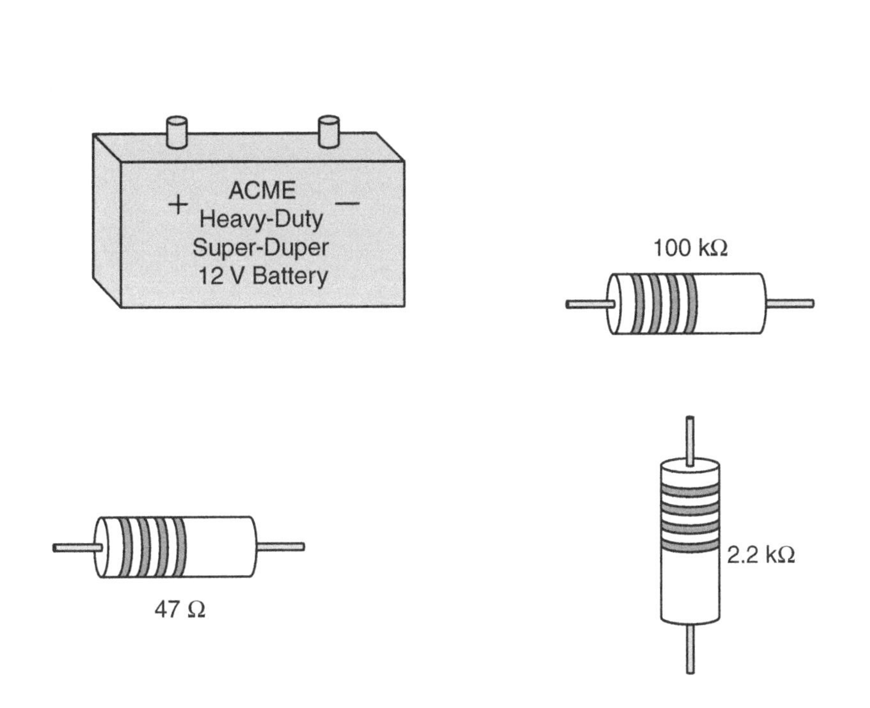

2. Given the schematic diagram, connect the components so that they electrically match the schematic.

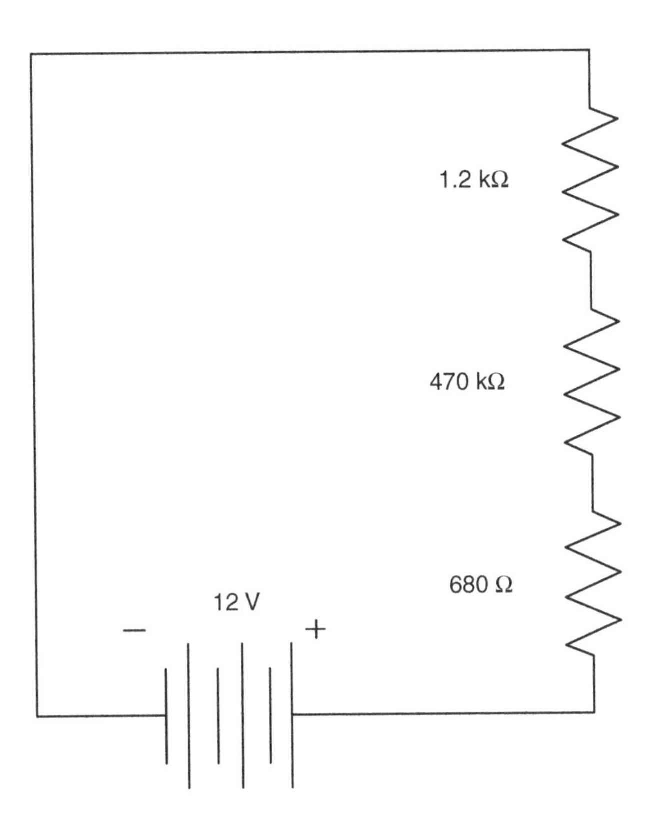

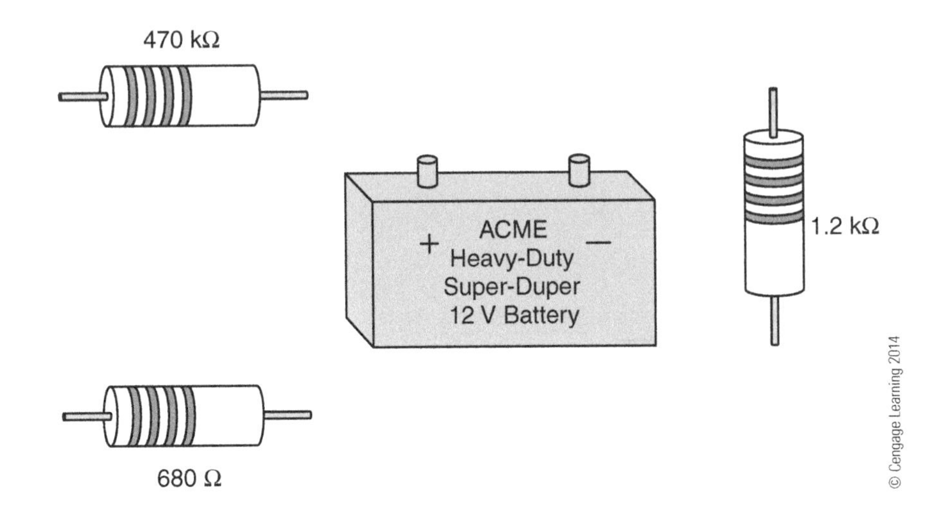

3. Given the schematic diagram, connect the components so that they electrically match the schematic.

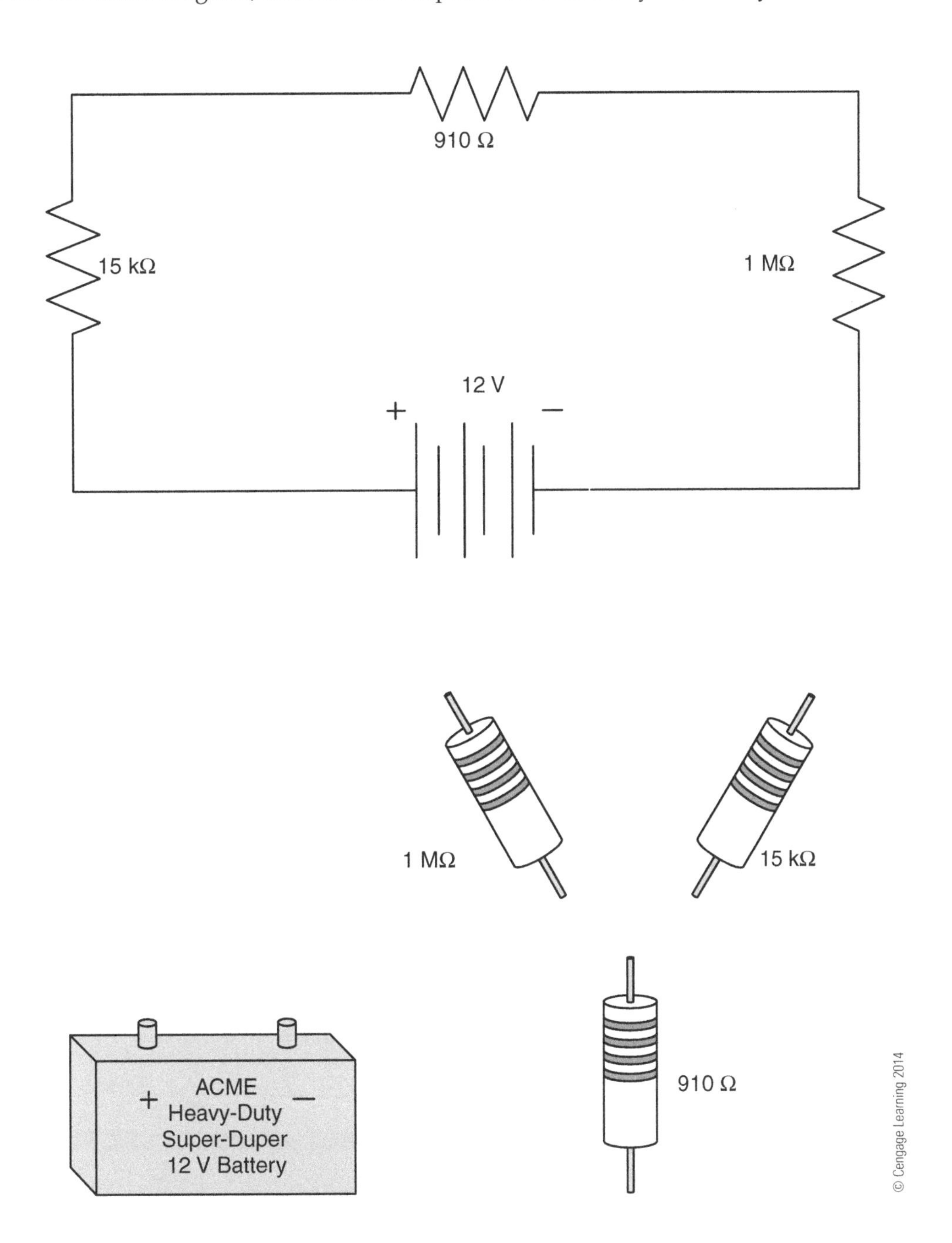

Name: ______________________________ Date: ______________________

SERIES CIRCUITS—PART 2

1. Given the circuit, find the missing values.

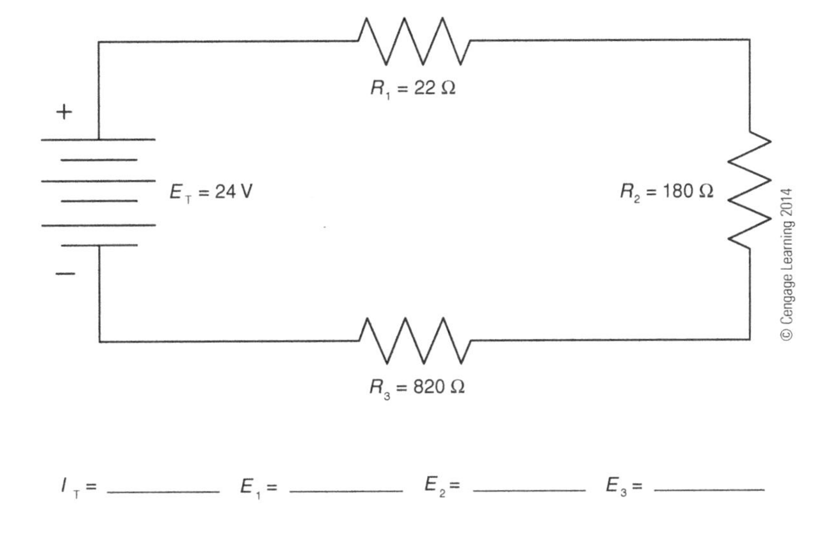

I_T = ________ E_1 = ________ E_2 = ________ E_3 = ________

R_T = ________ I_1 = ________ I_2 = ________ I_3 = ________

P_T = ________ P_1 = ________ P_2 = ________ P_3 = ________

2. Given the circuit, find the missing values.

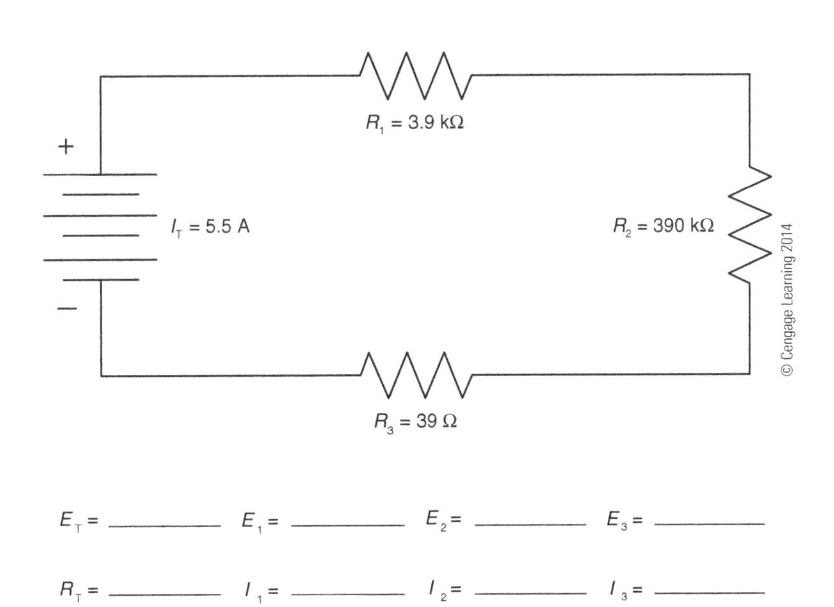

E_T = ________ E_1 = ________ E_2 = ________ E_3 = ________

R_T = ________ I_1 = ________ I_2 = ________ I_3 = ________

P_T = ________ P_1 = ________ P_2 = ________ P_3 = ________

3. Given the circuit, find the missing values.

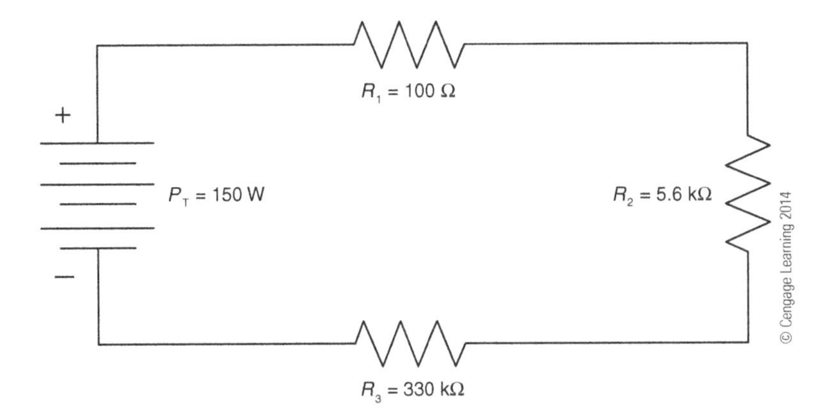

E_T = ________ E_1 = ________ E_2 = ________ E_3 = ________

I_T = ________ I_1 = ________ I_2 = ________ I_3 = ________

R_T = ________ P_1 = ________ P_2 = ________ P_3 = ________

Name: ______________________ Date: ______________

SERIES CIRCUITS—PART 3

1. Given the circuit, find the missing values.

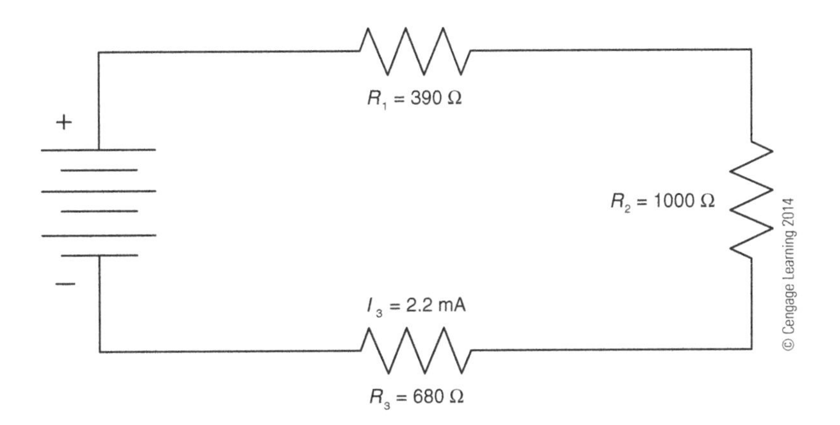

E_T = ________ E_1 = ________ E_2 = ________ E_3 = ________

I_T = ________ I_1 = ________ I_2 = ________ I_3 = ________

R_T = ________ P_1 = ________ P_2 = ________ P_3 = ________

P_T = ________

2. Given the circuit, find the missing values.

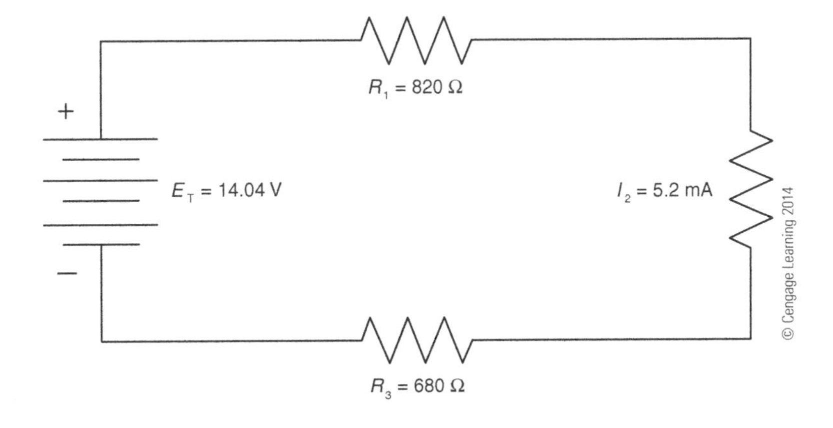

I_T = ________ E_1 = ________ E_2 = ________ E_3 = ________

R_T = ________ I_1 = ________ R_2 = ________ I_3 = ________

P_T = ________ P_1 = ________ P_2 = ________ P_3 = ________

3. Given the circuit, find the missing values.

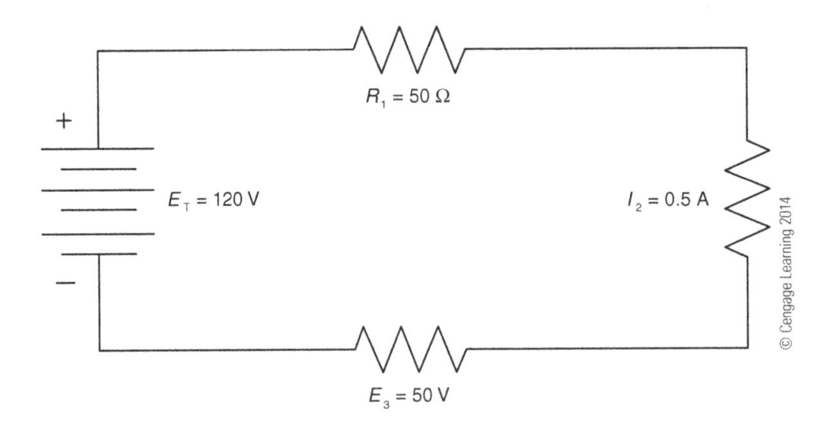

I_T = ________ E_1 = ________ E_2 = ________ I_3 = ________

R_T = ________ I_1 = ________ R_2 = ________ R_3 = ________

P_T = ________ P_1 = ________ P_2 = ________ P_3 = ________

Name: ______________________ Date: ______________

PARALLEL CIRCUITS—PART 1

1. Given the schematic diagram, connect the components so that they electrically match the schematic.

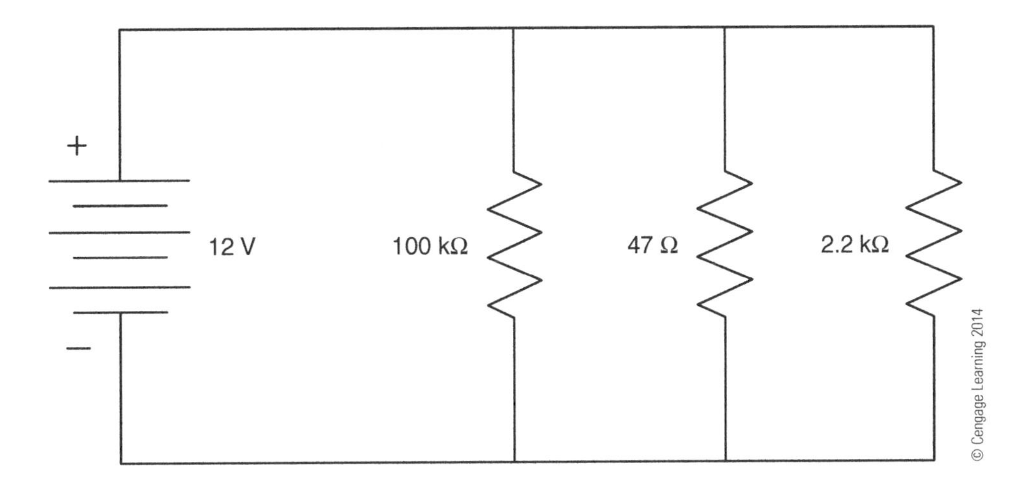

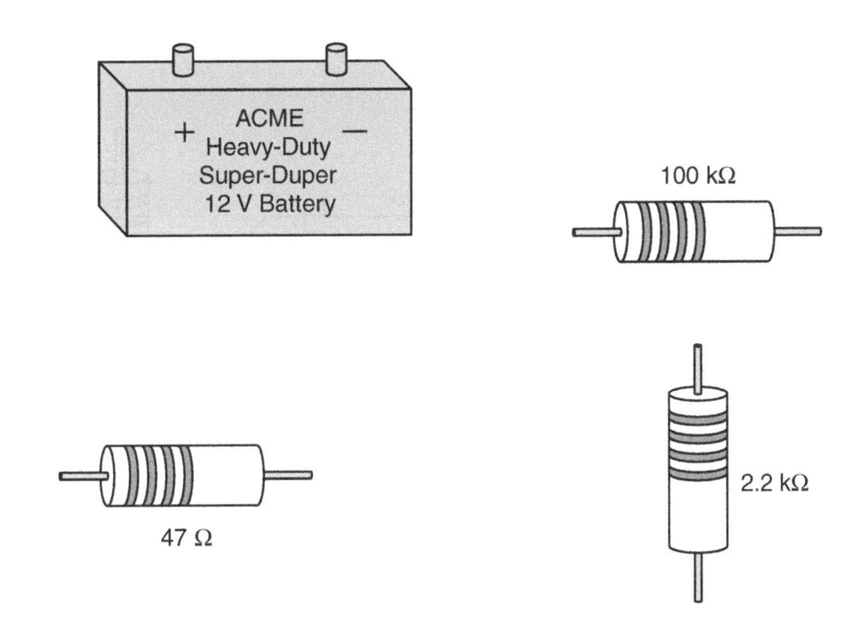

2. Given the schematic diagram, connect the components so that they electrically match the schematic.

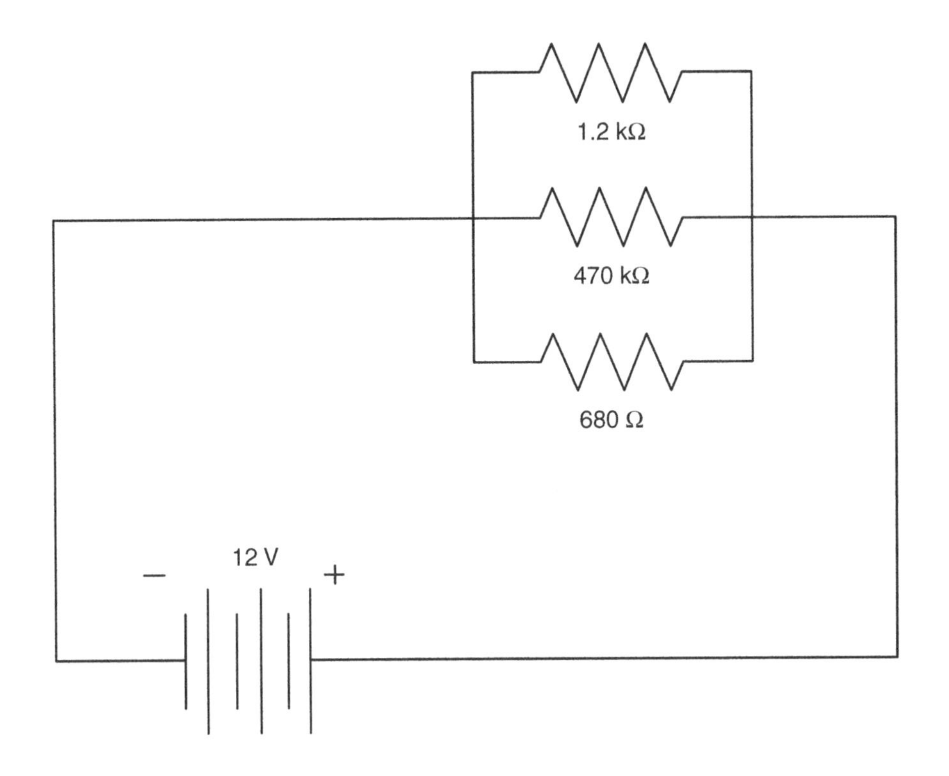

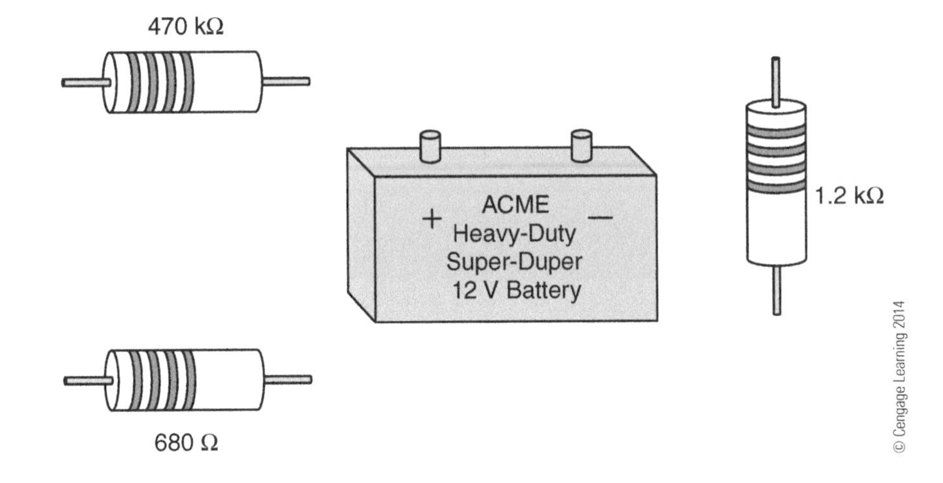

Name: ______________________________ Date: ____________________

3. Given the schematic diagram, connect the components so that they electrically match the schematic.

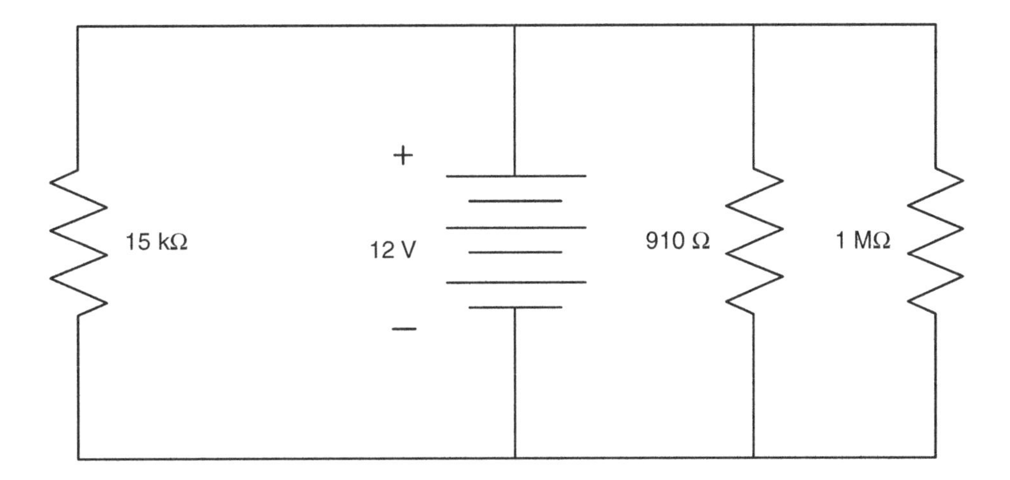

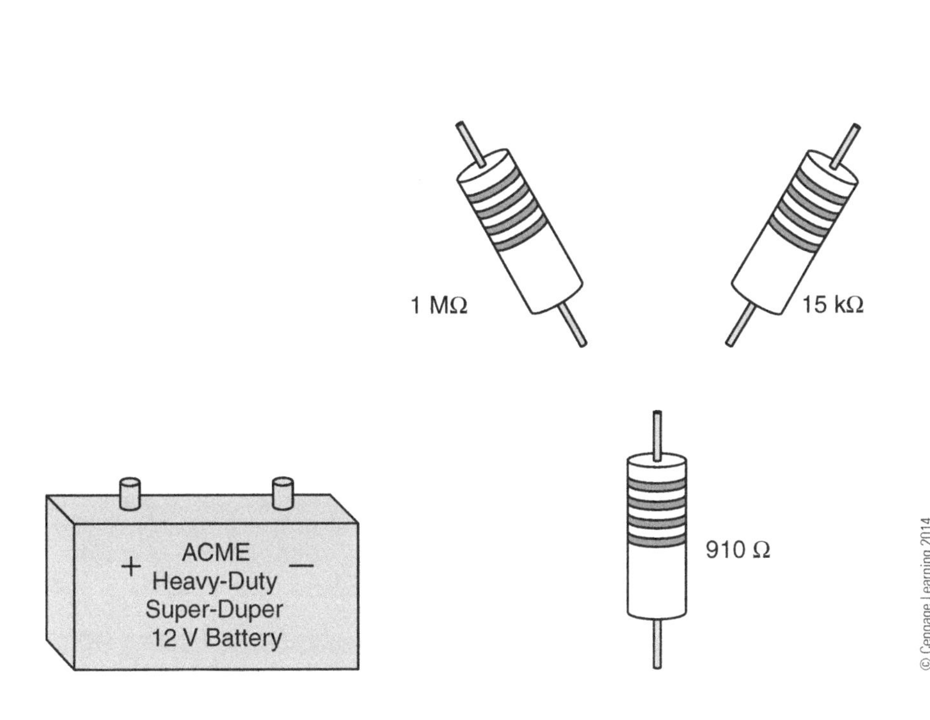

Name: ______________________________ Date: ____________________

PARALLEL CIRCUITS—PART 2

1. Given the circuit, find the missing values.

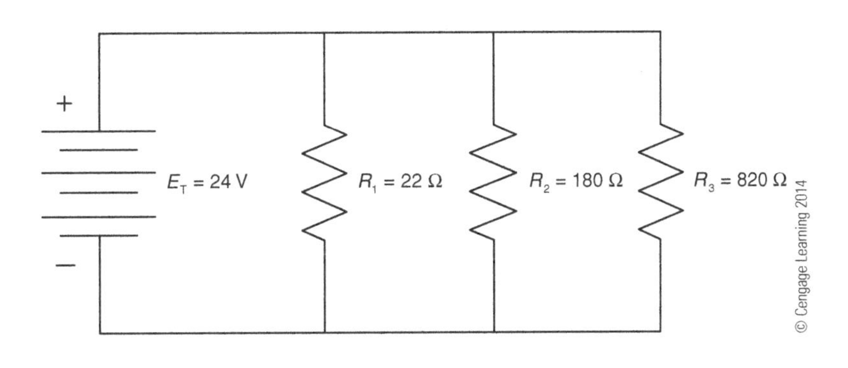

I_T = ________ E_1 = ________ E_2 = ________ E_3 = ________

R_T = ________ I_1 = ________ I_2 = ________ I_3 = ________

P_T = ________ P_1 = ________ P_2 = ________ P_3 = ________

2. Given the circuit, find the missing values.

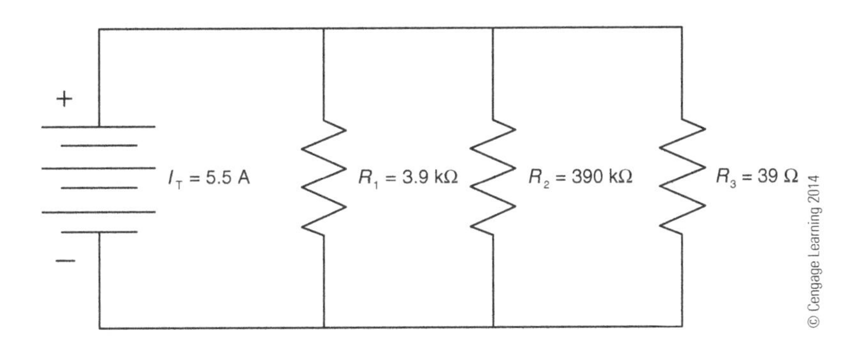

E_T = ________ E_1 = ________ E_2 = ________ E_3 = ________

R_T = ________ I_1 = ________ I_2 = ________ I_3 = ________

P_T = ________ P_1 = ________ P_2 = ________ P_3 = ________

3. Given the circuit, find the missing values.

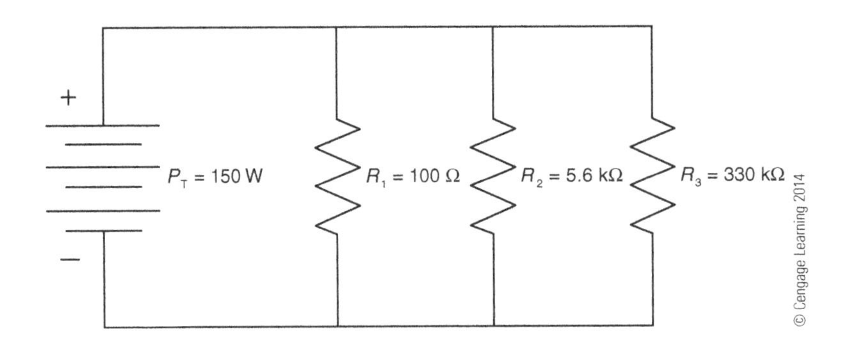

E_T = ________ E_1 = ________ E_2 = ________ E_3 = ________

I_T = ________ I_1 = ________ I_2 = ________ I_3 = ________

R_T = ________ P_1 = ________ P_2 = ________ P_3 = ________

Name: ______________________ Date: ______________

PARALLEL CIRCUITS—PART 3

1. Given the circuit, find the missing values.

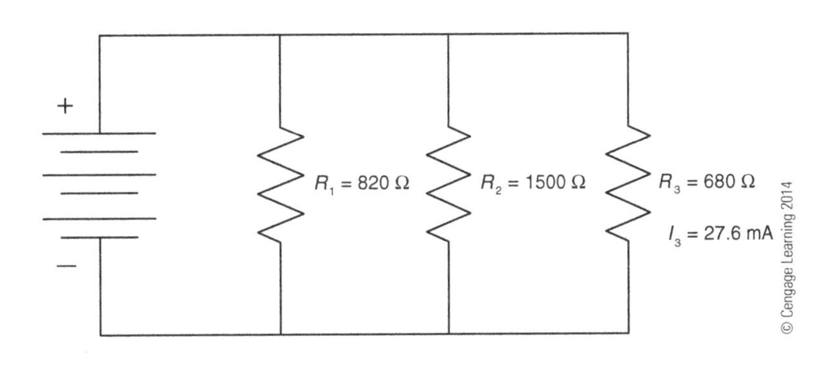

E_T = ________ E_1 = ________ E_2 = ________ E_3 = ________

I_T = ________ I_1 = ________ I_2 = ________ P_3 = ________

R_T = ________ P_1 = ________ P_2 = ________

P_T = ________

2. Given the circuit, find the missing values.

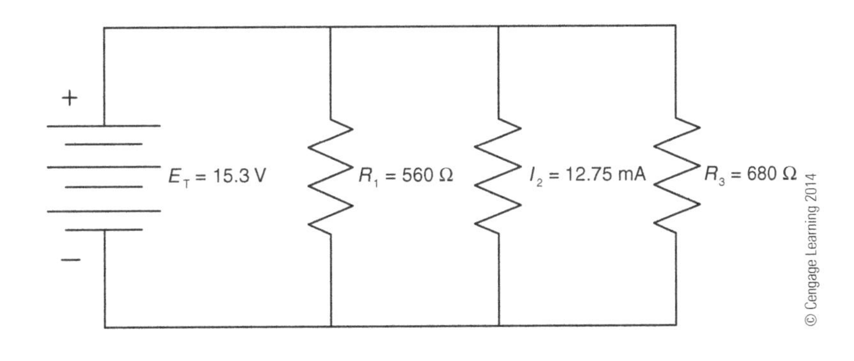

I_T = ________ E_1 = ________ E_2 = ________ E_3 = ________

R_T = ________ I_1 = ________ R_2 = ________ I_3 = ________

P_T = ________ P_1 = ________ P_2 = ________ P_3 = ________

3. Given the circuit, find the missing values.

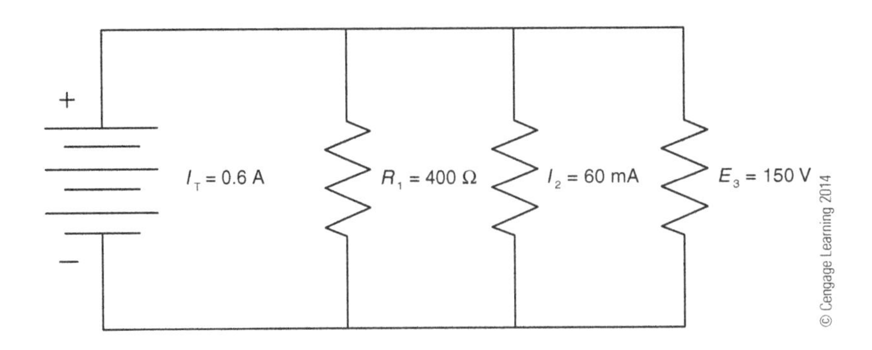

E_T = ________ E_1 = ________ E_2 = ________ I_3 = ________

R_T = ________ I_1 = ________ R_2 = ________ R_3 = ________

P_T = ________ P_1 = ________ P_2 = ________ P_3 = ________

Name: ______________________ Date: ______________________

COMBINATION CIRCUITS—PART 1

1. Given the schematic diagram, connect the components so that they electrically match the schematic.

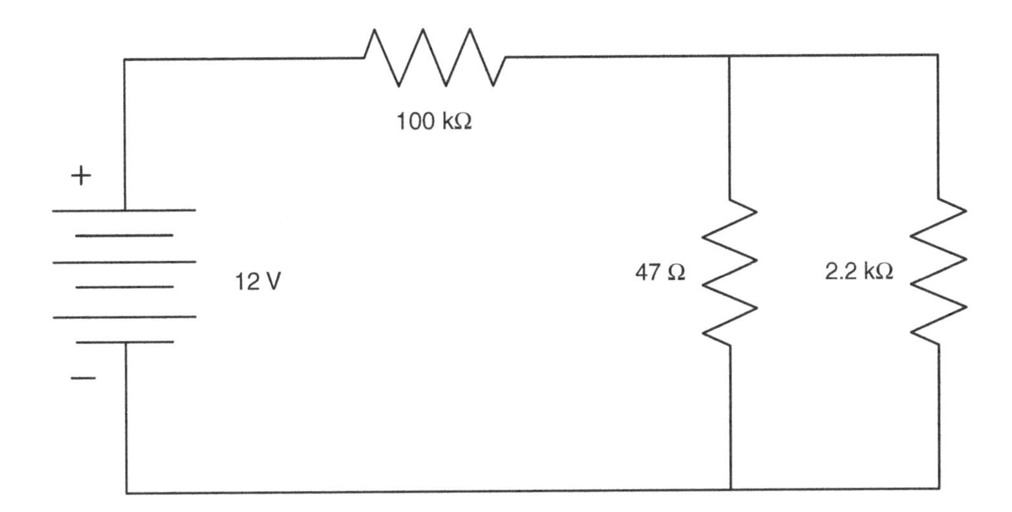

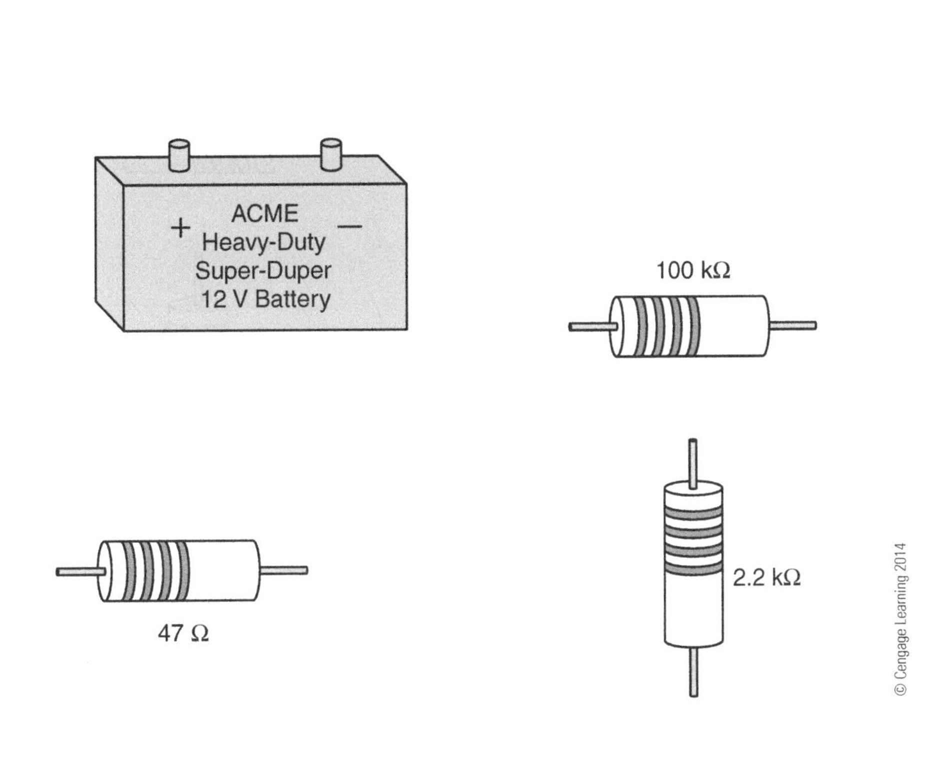

2. Given the schematic diagram, connect the components so that they electrically match the schematic.

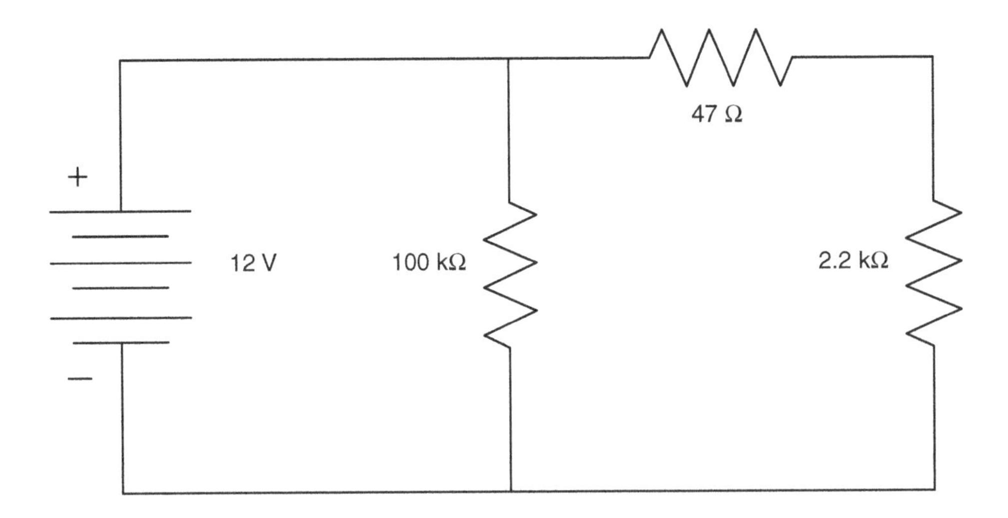

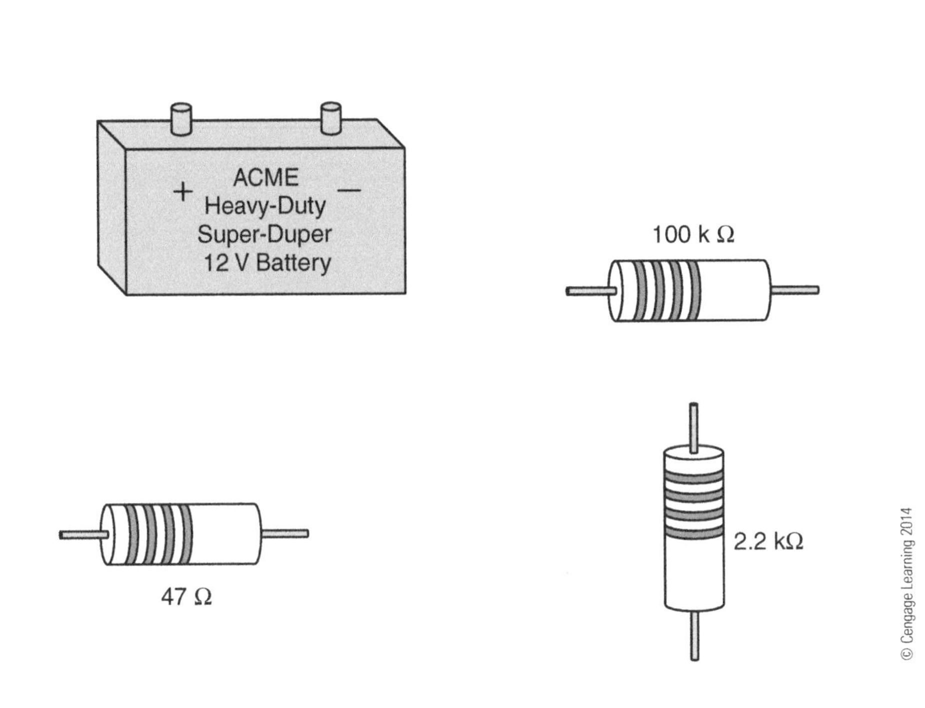

Name: ______________________ Date: ______________

3. Given the schematic diagram, connect the components so that they electrically match the schematic.

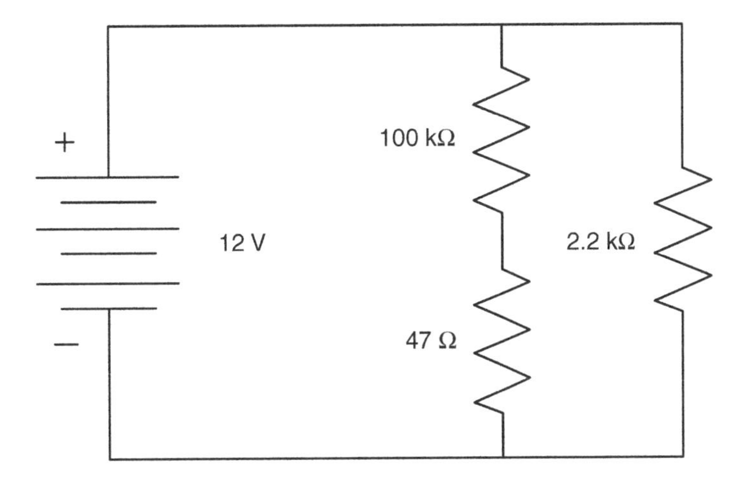

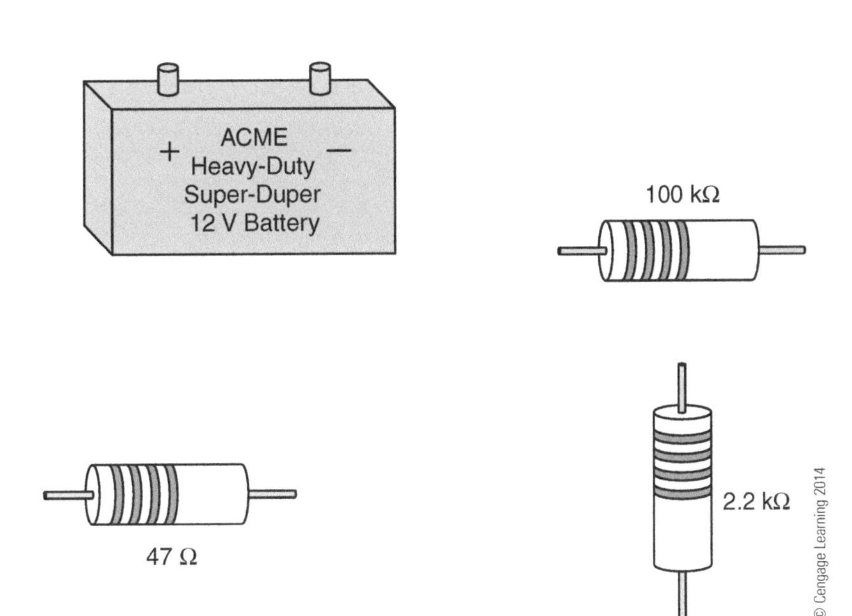

Name: ______________________________ Date: ______________

COMBINATION CIRCUITS—PART 2

1. Given the circuit, find the missing values.

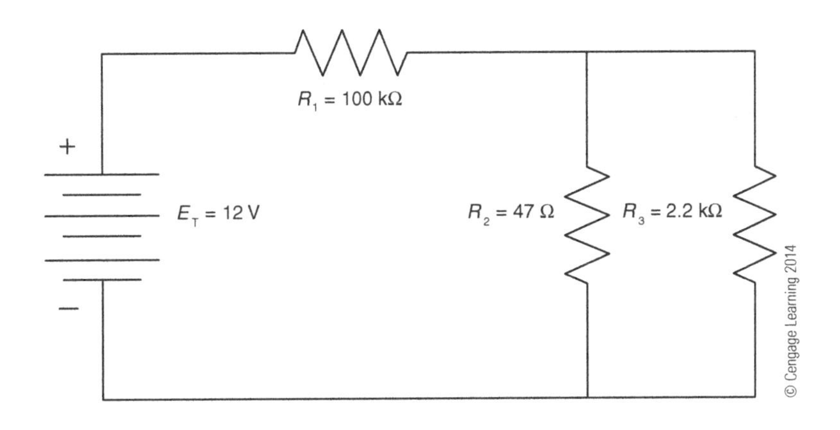

I_T = ________ E_1 = ________ E_2 = ________ E_3 = ________

R_T = ________ I_1 = ________ I_2 = ________ I_3 = ________

P_T = ________ P_1 = ________ P_2 = ________ P_3 = ________

2. Given the circuit, find the missing values.

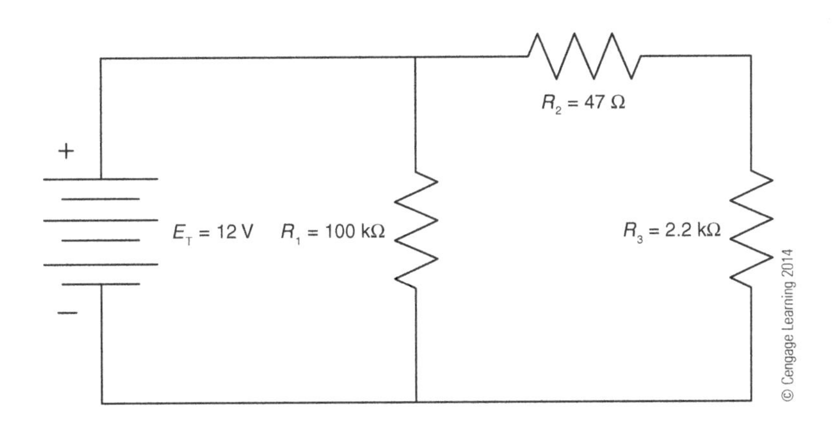

I_T = ________ E_1 = ________ E_2 = ________ E_3 = ________

R_T = ________ I_1 = ________ I_2 = ________ I_3 = ________

P_T = ________ P_1 = ________ P_2 = ________ P_3 = ________

3. Given the circuit, find the missing values.

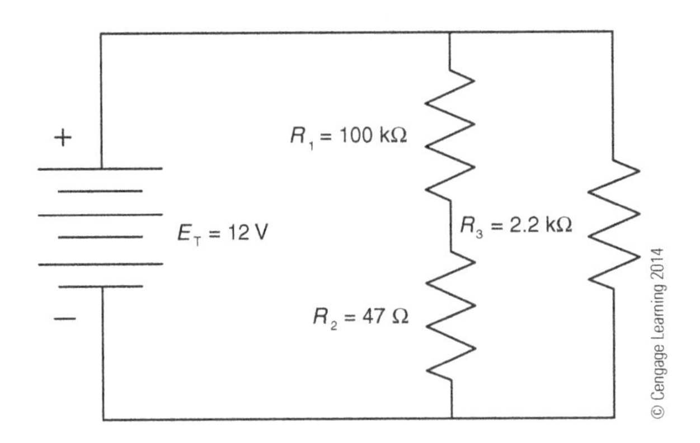

I_T = ________ E_1 = ________ E_2 = ________ E_3 = ________

R_T = ________ I_1 = ________ I_2 = ________ I_3 = ________

P_T = ________ P_1 = ________ P_2 = ________ P_3 = ________

Name: ______________________ Date: ______________

COMBINATION CIRCUITS—PART 3

1. Given the circuit, find the missing values.

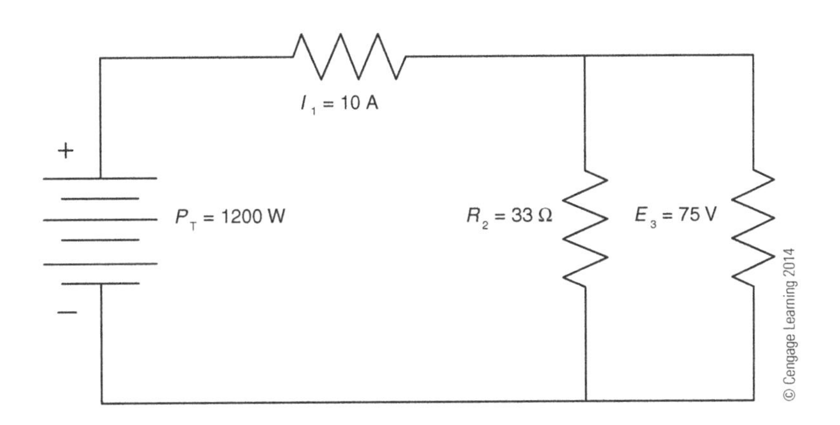

E_T = ______	E_1 = ______	E_2 = ______	I_3 = ______
I_T = ______	R_1 = ______	I_2 = ______	R_3 = ______
R_T = ______	P_1 = ______	P_2 = ______	P_3 = ______

2. Given the circuit, find the missing values.

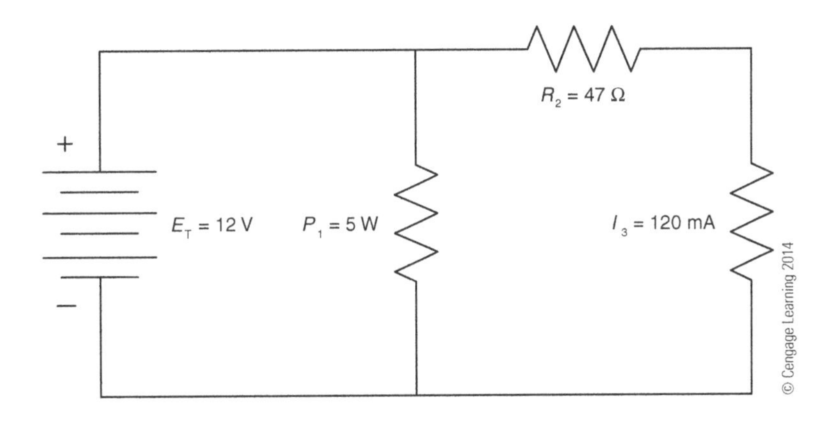

I_T = ______	E_1 = ______	E_2 = ______	E_3 = ______
R_T = ______	I_1 = ______	I_2 = ______	R_3 = ______
P_T = ______	R_1 = ______	P_2 = ______	P_3 = ______

3. Given the circuit, find the missing values.

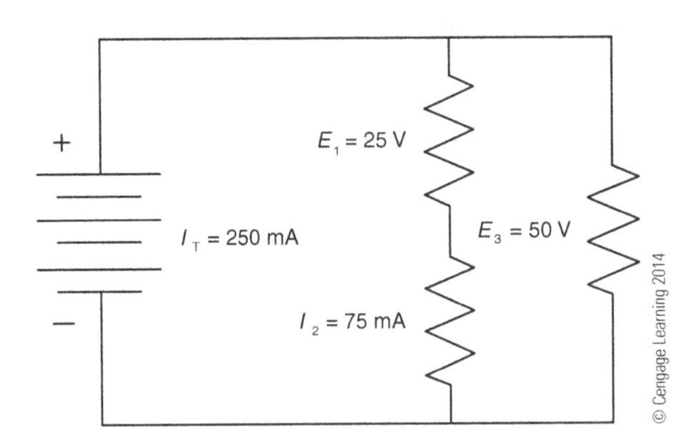

E_T = ________ I_1 = ________ E_2 = ________ I_3 = ________

R_T = ________ R_1 = ________ R_2 = ________ R_3 = ________

P_T = ________ P_1 = ________ P_2 = ________ P_3 = ________

Name: ______________________ **Date:** ______________

R-L SERIES CIRCUITS

1. Given the circuit, find the missing values.

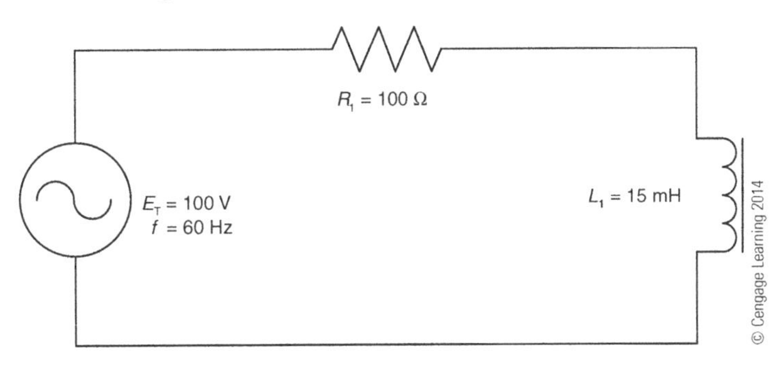

I_T = ________	E_1 = ________	E_{L1} = ________
Z_T = ________	I_1 = ________	I_{L1} = ________
VA = ________	P_1 = ________	X_{L1} = ________
PF = ________		$VARs_{L1}$ = ________
$\angle\theta$ = ________		

Choose one: leading, lagging, or in phase

2. Given the circuit, find the missing values.

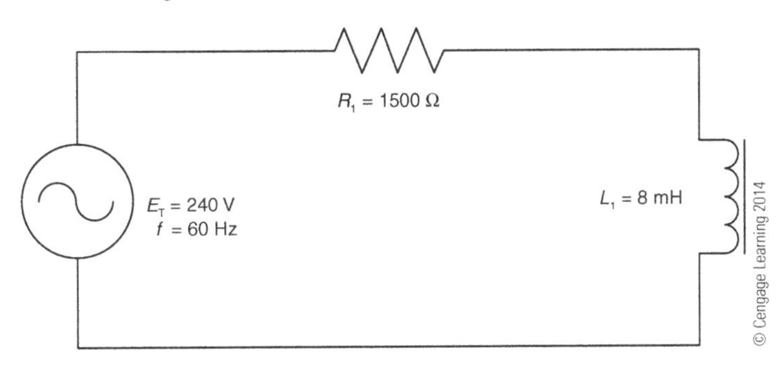

I_T = ________	E_1 = ________	E_{L1} = ________
Z_T = ________	I_1 = ________	I_{L1} = ________
VA = ________	P_1 = ________	X_{L1} = ________
PF = ________		$VARs_{L1}$ = ________
$\angle\theta$ = ________		

Choose one: leading, lagging, or in phase

3. Given the circuit, find the missing values.

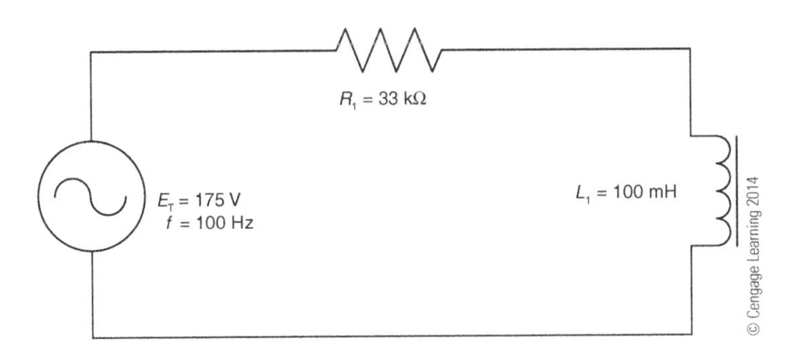

I_T = ________ E_1 = ________ E_{L1} = ________

Z_T = ________ I_1 = ________ I_{L1} = ________

VA = ________ P_1 = ________ X_{L1} = ________

PF = ________ $VARs_{L1}$ = ________

$\angle\theta$ = ________

Choose one: leading, lagging, or in phase

Name: ______________________ **Date:** ______________

R-L PARALLEL CIRCUITS

1. Given the circuit, find the missing values.

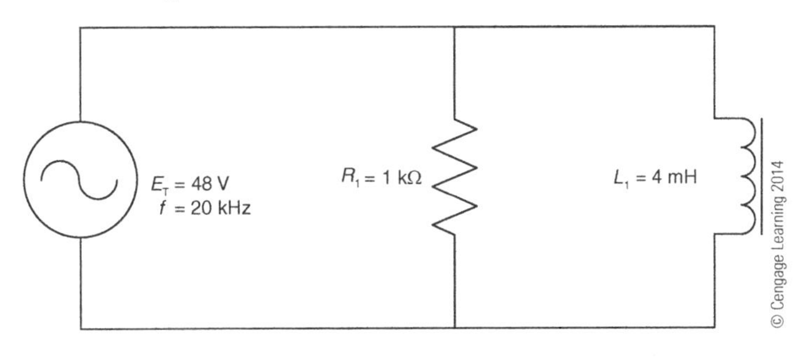

I_T = ________ E_1 = ________ E_{L1} = ________

Z_T = ________ I_1 = ________ I_{L1} = ________

VA = ________ P_1 = ________ X_{L1} = ________

PF = ________ $VARs_{L1}$ = ________

$\angle\theta$ = ________

Choose one: leading, lagging, or in phase

2. Given the circuit, find the missing values.

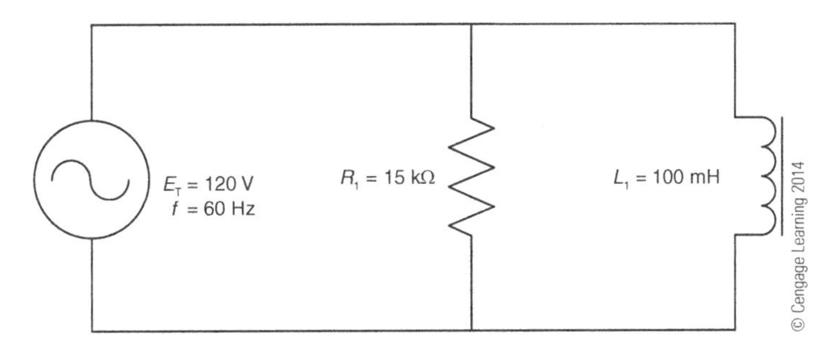

I_T = ________ E_1 = ________ E_{L1} = ________

Z_T = ________ I_1 = ________ I_{L1} = ________

VA = ________ P_1 = ________ X_{L1} = ________

PF = ________ $VARs_{L1}$ = ________

$\angle\theta$ = ________

Choose one: leading, lagging, or in phase

3. Given the circuit, find the missing values.

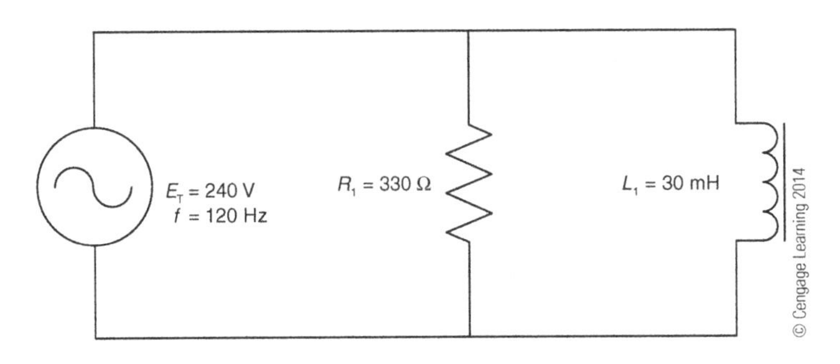

I_T = ________ E_1 = ________ E_{L1} = ________

Z_T = ________ I_1 = ________ I_{L1} = ________

VA = ________ P_1 = ________ X_{L1} = ________

PF = ________ $VARs_{L1}$ = ________

$\angle\theta$ = ________

Choose one: leading, lagging, or in phase

Name: ______________________ **Date:** ______________

R-C SERIES CIRCUITS

1. Given the circuit, find the missing values.

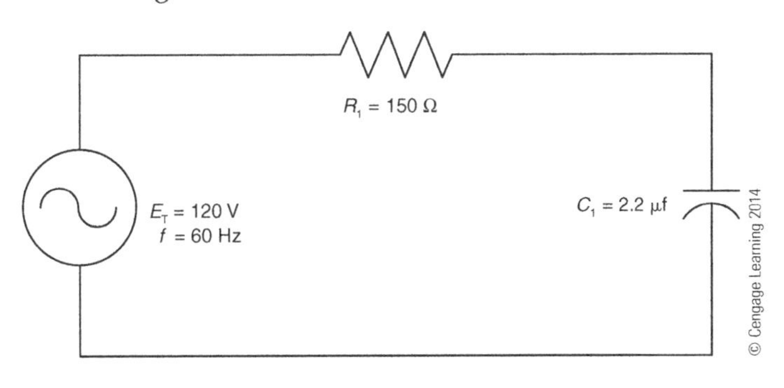

I_T = ________ E_1 = ________ E_{C1} = ________

Z_T = ________ I_1 = ________ I_{C1} = ________

VA = ________ P_1 = ________ X_{C1} = ________

PF = ________ $VARs_{C1}$ = ________

$\angle\theta$ = ________

Choose one: leading, lagging, or in phase

2. Given the circuit, find the missing values.

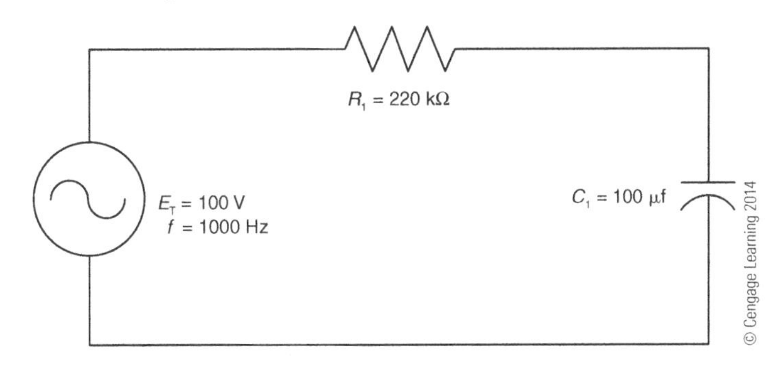

I_T = ________ E_1 = ________ E_{C1} = ________

Z_T = ________ I_1 = ________ I_{C1} = ________

VA = ________ P_1 = ________ X_{C1} = ________

PF = ________ $VARs_{C1}$ = ________

$\angle\theta$ = ________

Choose one: leading, lagging, or in phase

3. Given the circuit, find the missing values.

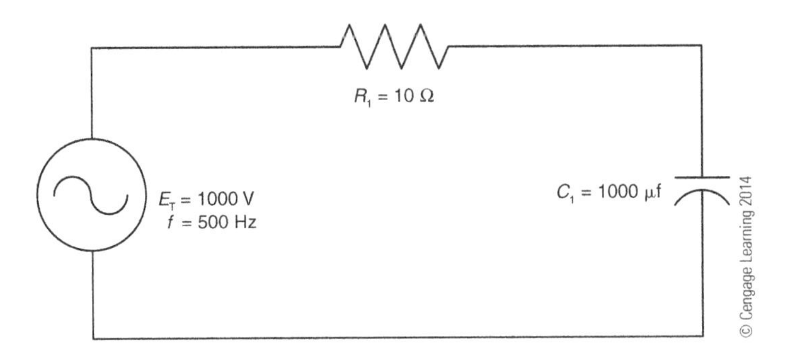

I_T = ____________ E_1 = ____________ E_{C1} = ____________

Z_T = ____________ I_1 = ____________ I_{C1} = ____________

VA = ____________ P_1 = ____________ X_{C1} = ____________

PF = ____________ $VARs_{C1}$ = ____________

$\angle\theta$ = ____________

Choose one: leading, lagging, or in phase

Name: ____________________ Date: ____________________

R-C PARALLEL CIRCUITS

1. Given the circuit, find the missing values.

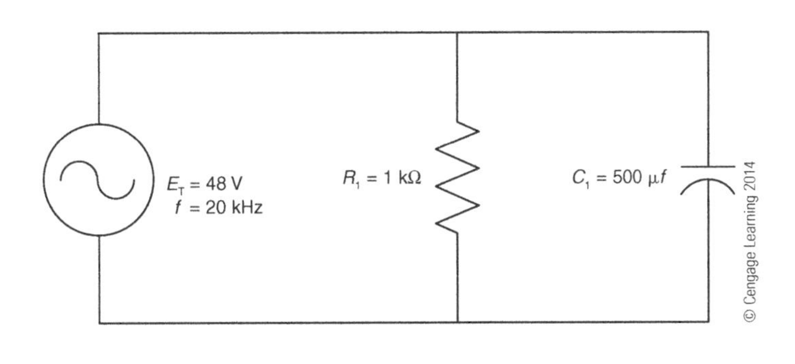

I_T = ________ E_1 = ________ E_{C1} = ________

Z_T = ________ I_1 = ________ I_{C1} = ________

VA = ________ P_1 = ________ X_{C1} = ________

PF = ________ $VARs_{C1}$ = ________

$\angle\theta$ = ________

Choose one: leading, lagging, or in phase

2. Given the circuit, find the missing values.

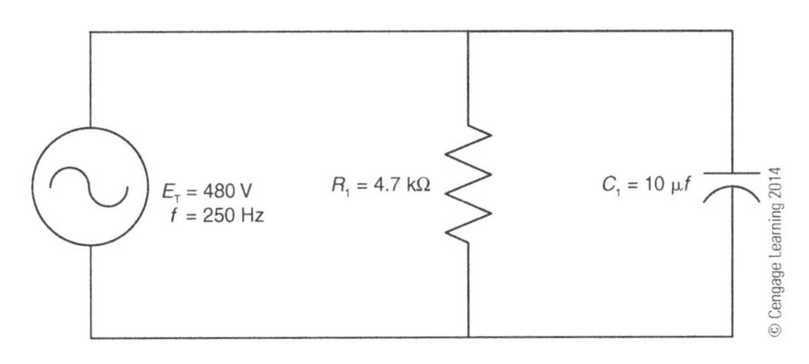

I_T = ________ E_1 = ________ E_{C1} = ________

Z_T = ________ I_1 = ________ I_{C1} = ________

VA = ________ P_1 = ________ X_{C1} = ________

PF = ________ $VARs_{C1}$ = ________

$\angle\theta$ = ________

Choose one: leading, lagging, or in phase

3. Given the circuit, find the missing values.

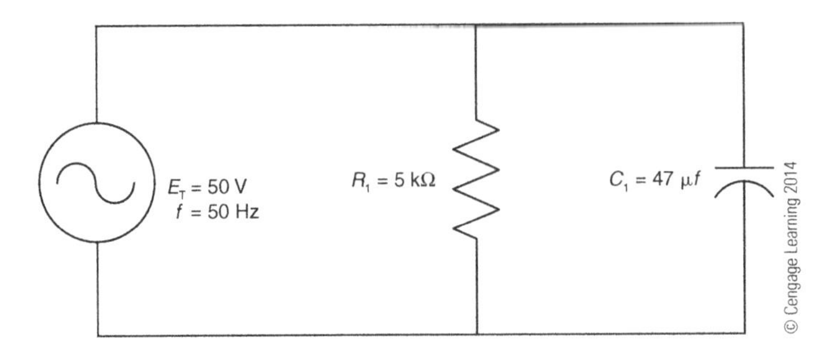

I_T = ________	E_1 = ________	E_{C1} = ________
Z_T = ________	I_1 = ________	I_{C1} = ________
VA = ________	P_1 = ________	X_{C1} = ________
PF = ________		$VARs_{C1}$ = ________
$\angle\theta$ = ________		

Choose one: leading, lagging, or in phase

Name: ______________________ Date: ______________

R-L-C SERIES CIRCUITS

1. Given the circuit, find the missing values.

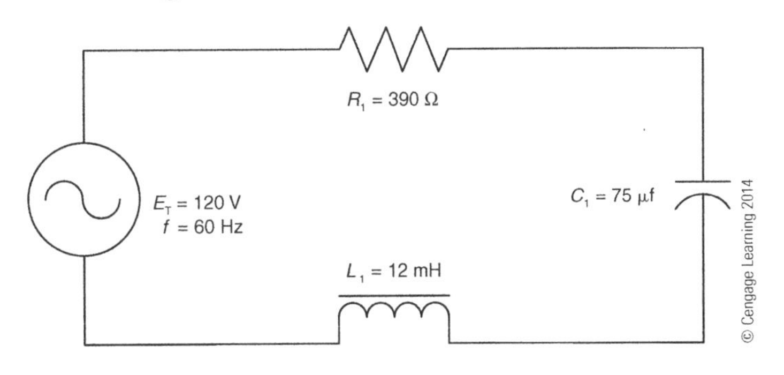

I_T = ______	E_1 = ______	E_{C1} = ______	E_{L1} = ______
Z_T = ______	I_1 = ______	I_{C1} = ______	I_{L1} = ______
VA = ______	P_1 = ______	X_{C1} = ______	X_{L1} = ______
PF = ______		$VARs_{C1}$ = ______	$VARs_{L1}$ = ______
$\angle\theta$ = ______			

Choose one: leading, lagging, or in phase

2. Given the circuit, find the missing values.

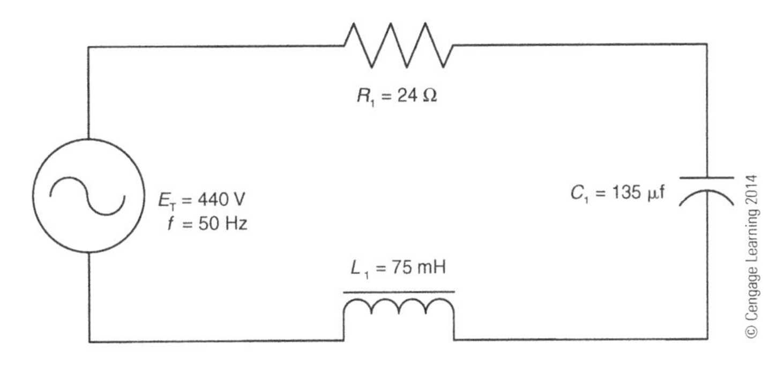

I_T = ______	E_1 = ______	E_{C1} = ______	E_{L1} = ______
Z_T = ______	I_1 = ______	I_{C1} = ______	I_{L1} = ______
VA = ______	P_1 = ______	X_{C1} = ______	X_{L1} = ______
PF = ______		$VARs_{C1}$ = ______	$VARs_{L1}$ = ______
$\angle\theta$ = ______			

Choose one: leading, lagging, or in phase

3. Given the circuit, find the missing values.

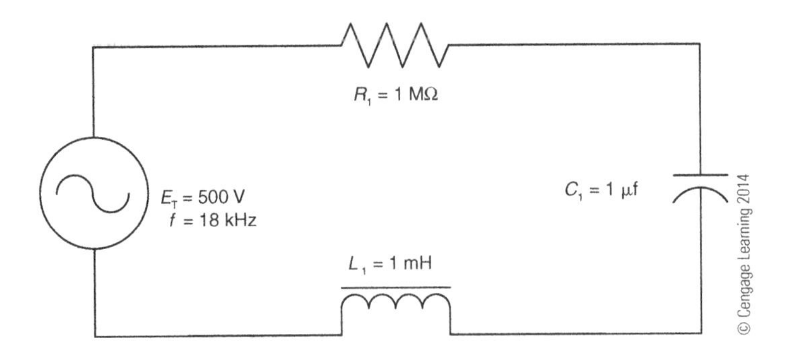

I_T = ________ E_1 = ________ E_{C1} = ________ E_{L1} = ________

Z_T = ________ I_1 = ________ I_{C1} = ________ I_{L1} = ________

VA = ________ P_1 = ________ X_{C1} = ________ X_{L1} = ________

PF = ________ $VARs_{C1}$ = ________ $VARs_{L1}$ = ________

$\angle\theta$ = ________

Choose one: leading, lagging, or in phase

Name: ______________________ **Date:** ______________

R-L-C PARALLEL CIRCUITS

1. Given the circuit, find the missing values.

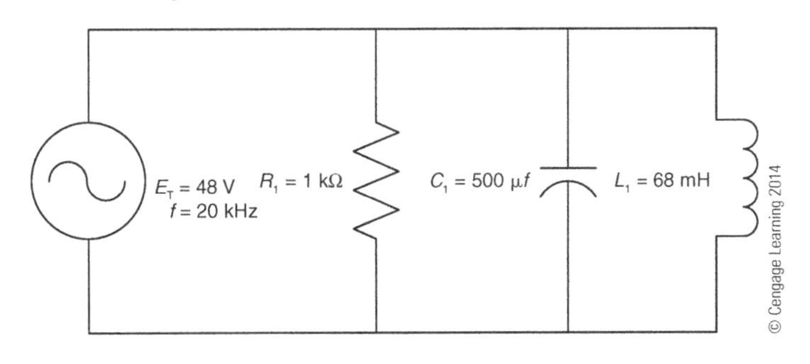

I_T = ________ E_1 = ________ E_{C1} = ________ E_{L1} = ________

Z_T = ________ I_1 = ________ I_{C1} = ________ I_{L1} = ________

VA = ________ P_1 = ________ X_{C1} = ________ X_{L1} = ________

PF = ________ $VARs_{C1}$ = ________ $VARs_{L1}$ = ________

$\angle\theta$ = ________

Choose one: leading, lagging, or in phase

2. Given the circuit, find the missing values.

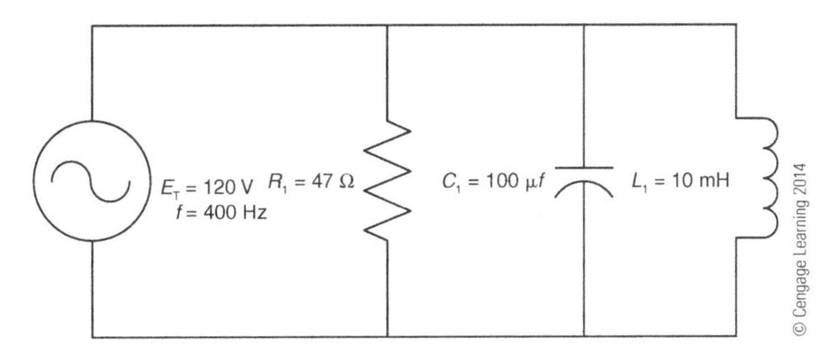

I_T = ________ E_1 = ________ E_{C1} = ________ E_{L1} = ________

Z_T = ________ I_1 = ________ I_{C1} = ________ I_{L1} = ________

VA = ________ P_1 = ________ X_{C1} = ________ X_{L1} = ________

PF = ________ $VARs_{C1}$ = ________ $VARs_{L1}$ = ________

$\angle\theta$ = ________

Choose one: leading, lagging, or in phase

3. Given the circuit, find the missing values.

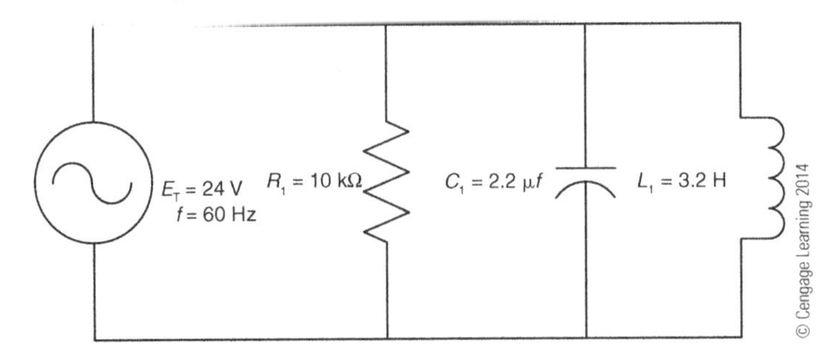

I_T = ________ E_1 = ________ E_{C1} = ________ E_{L1} = ________

Z_T = ________ I_1 = ________ I_{C1} = ________ I_{L1} = ________

VA = ________ P_1 = ________ X_{C1} = ________ X_{L1} = ________

PF = ________ $VARs_{C1}$ = ________ $VARs_{L1}$ = ________

$\angle\theta$ = ________

Choose one: leading, lagging, or in phase

Name: ______________________ Date: ______________

POWER FACTOR CORRECTION

1. Given the motor circuit, find the capacitor value that could be added into the circuit to correct the motor power factor to 91%.

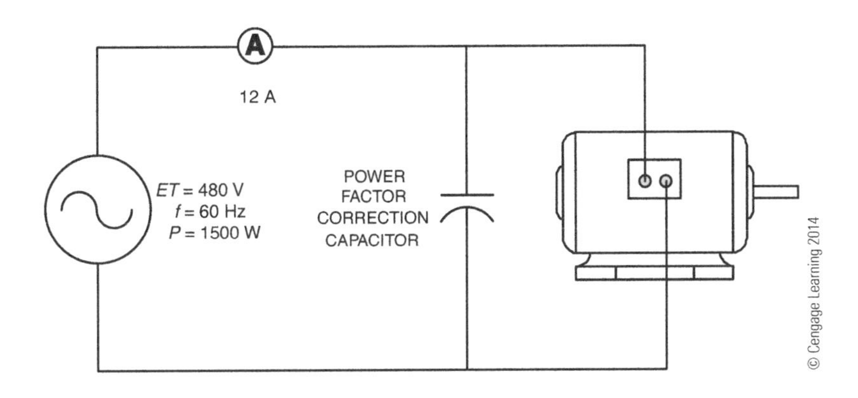

2. Given the motor circuit, find the capacitor value that could be added into the circuit to correct the motor power factor to 82%.

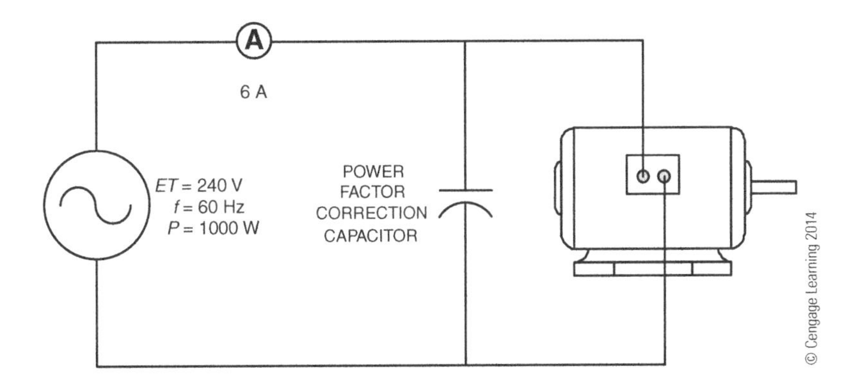

3. Given the motor circuit, find the capacitor value that could be added into the circuit to correct the motor power factor to 87%.

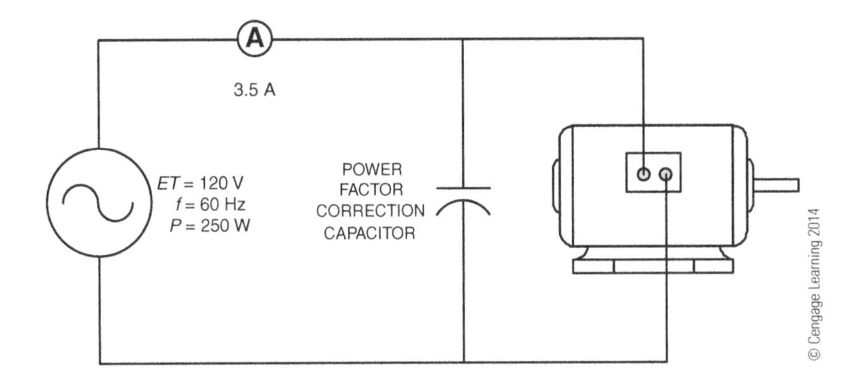

Name: ______________________ Date: ______________

THREE-PHASE POWER FACTOR CORRECTION

1. Given the motor circuit, find the capacitor value that could be added into the circuit to correct the motor power factor to 80%.

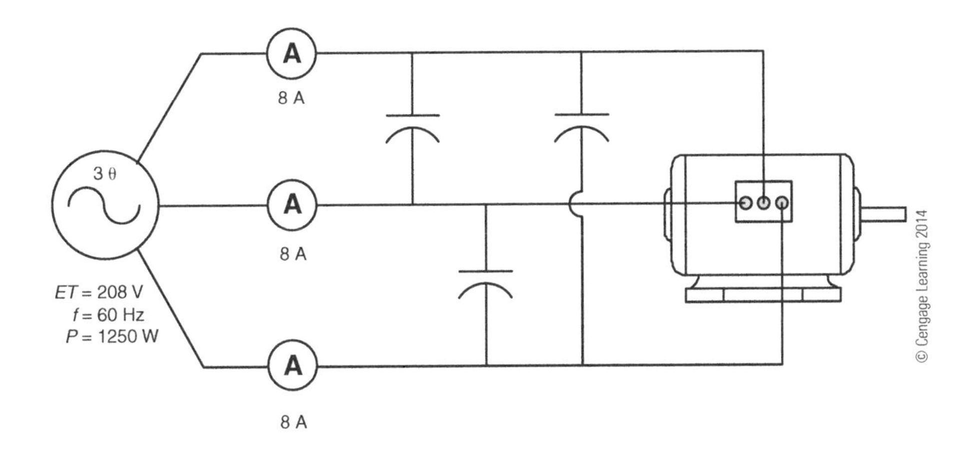

2. Given the motor circuit, find the capacitor value that could be added into the circuit to correct the motor power factor to 85%.

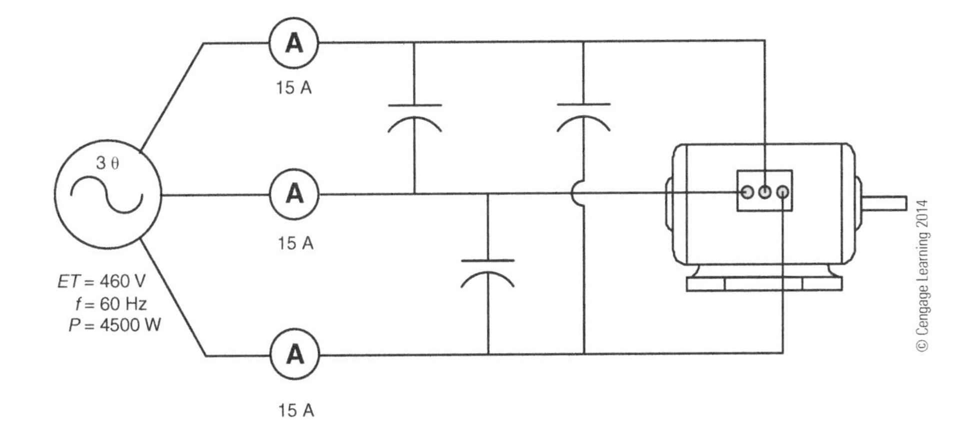

3. Given the motor circuit, find the capacitor value that could be added into the circuit to correct the motor power factor to 96%.

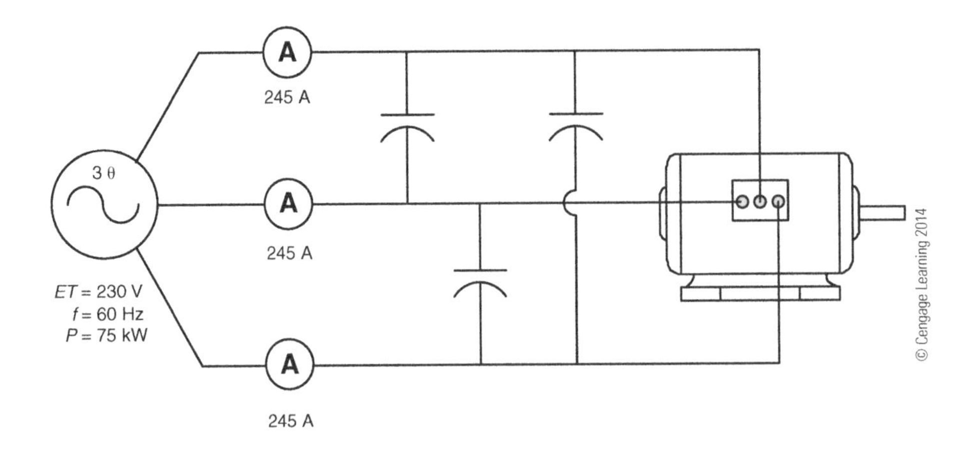

Name: ______________________________ Date: ______________________

CONDUCTOR SIZING

1. Using the tables found in the Appendices of the textbook, determine the minimum size of Type UF cable (rated at 60°C, or 140°F) copper conductors used to supply fifty 277-volt HID (high-intensity discharge) luminaires as shown, located in a supermarket parking lot.

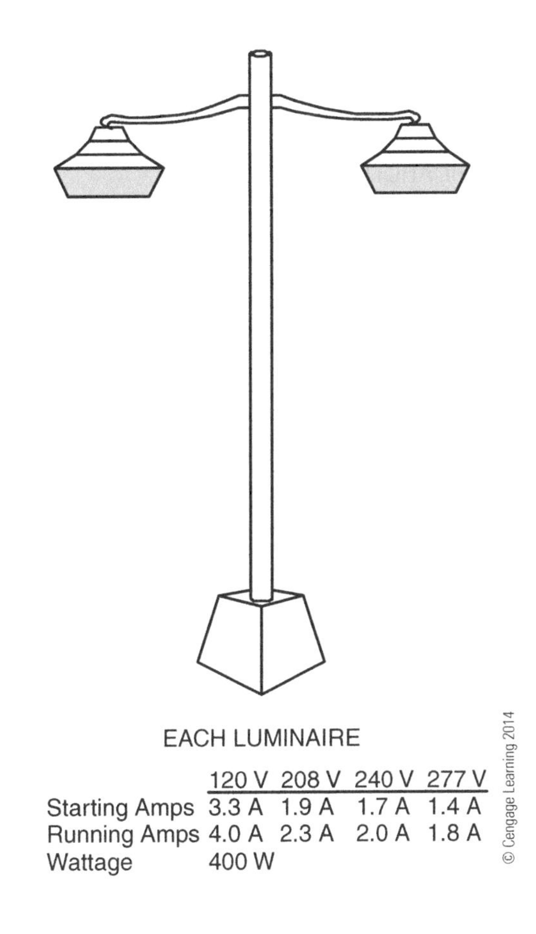
EACH LUMINAIRE

	120 V	208 V	240 V	277 V
Starting Amps	3.3 A	1.9 A	1.7 A	1.4 A
Running Amps	4.0 A	2.3 A	2.0 A	1.8 A
Wattage	400 W			

2. Using the tables found in the Appendices of the textbook, determine the minimum size of Type THWN copper conductors, contained within a nonmetallic-sheathed cable, used to supply 24 floodlights of the type shown, to provide perimeter security lighting at a hardware store.

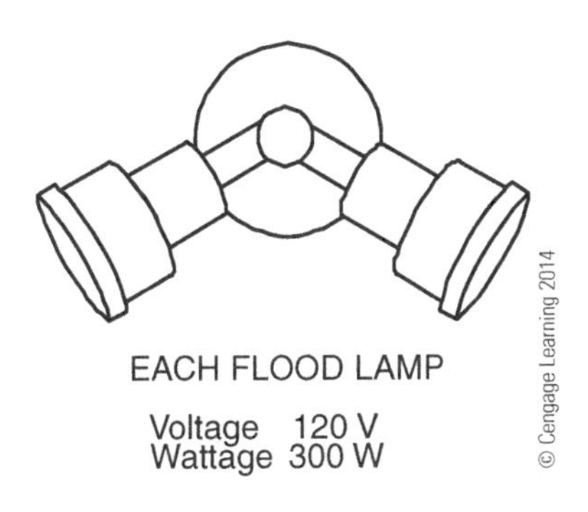

3. Using the tables found in the Appendices of the textbook, determine the minimum size of Type THW-2 (rated at 90°C, or 194°F) copper conductors used in a multiple outlet extension cord to supply the equipment shown.

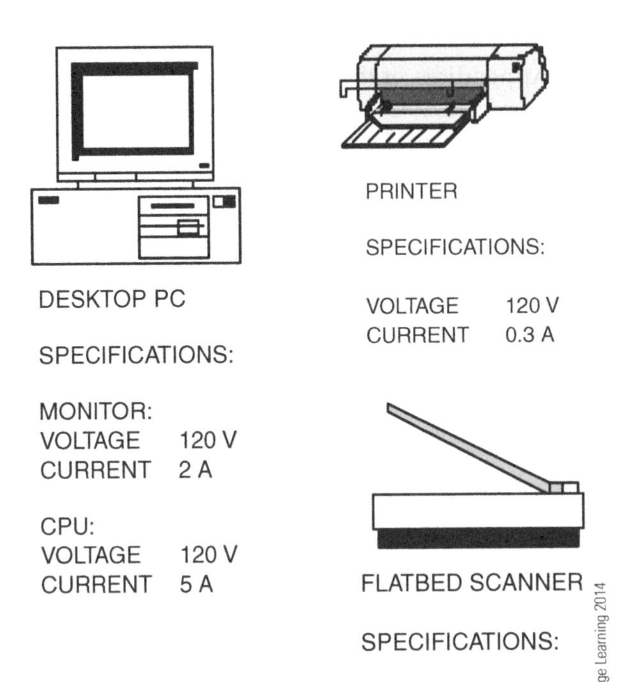

Name: ______________________ Date: ______________

CONDUCTOR COLOR CODE

1. Using the standard color code, label the correct color for each conductor.

Conductor A color ____________

Conductor B color ____________

Conductor C color ____________

Conductor D color ____________

Conductor E color ____________

Conductor F color ____________

Conductor G color ____________

Conductor H color ____________

Conductor I color ____________

Conductor J color ____________

Conductor K color ____________

Conductor L color ____________

Conductor M color ____________

Conductor N color ____________

Conductor O color ____________

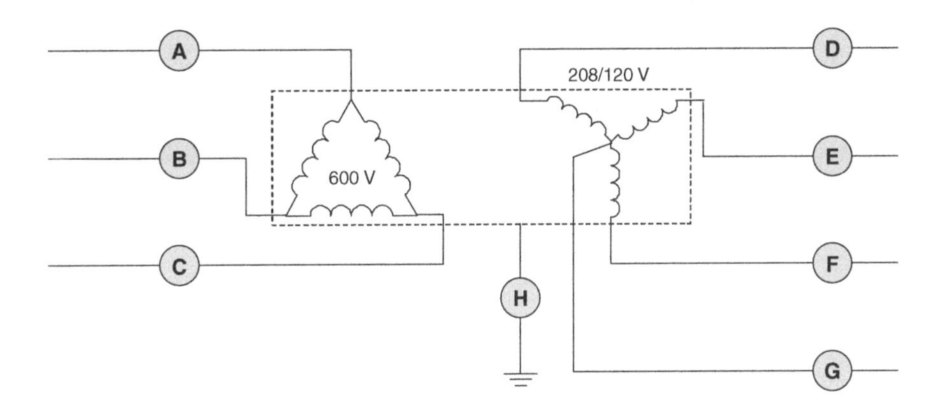

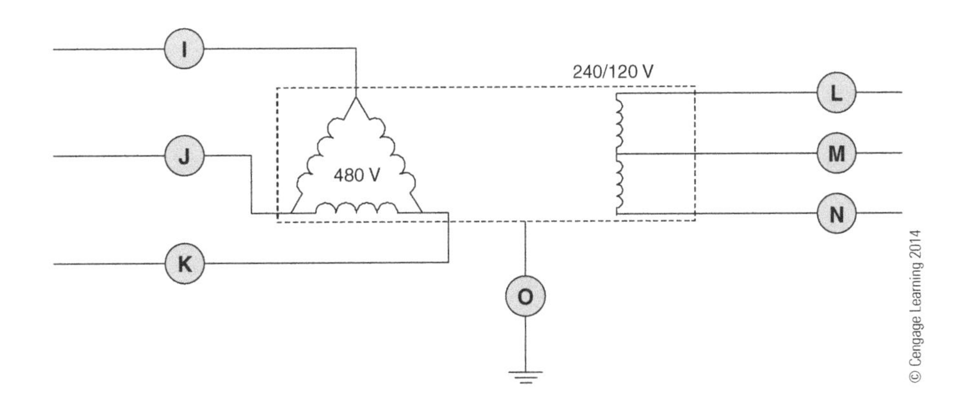

Name: ______________________ Date: ______________

RACEWAY SIZING

Use the ladder diagram to determine the wiring of components within the raceway for questions 1 through 4.

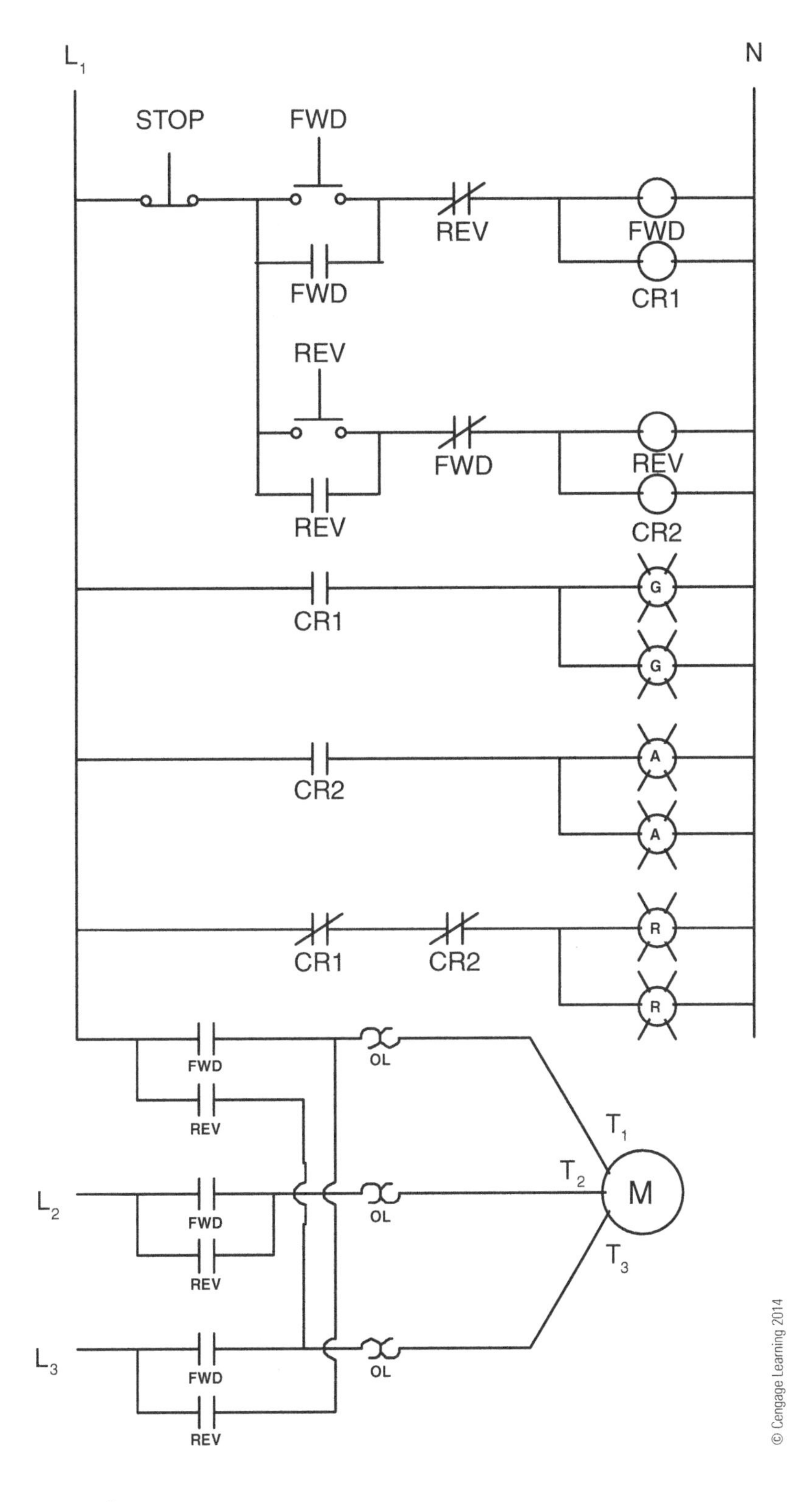

1. Power is fed through the switch enclosure. Show the wiring of the components. Then determine the minimum trade size of electrical metallic tubing (EMT) raceway required at each location.

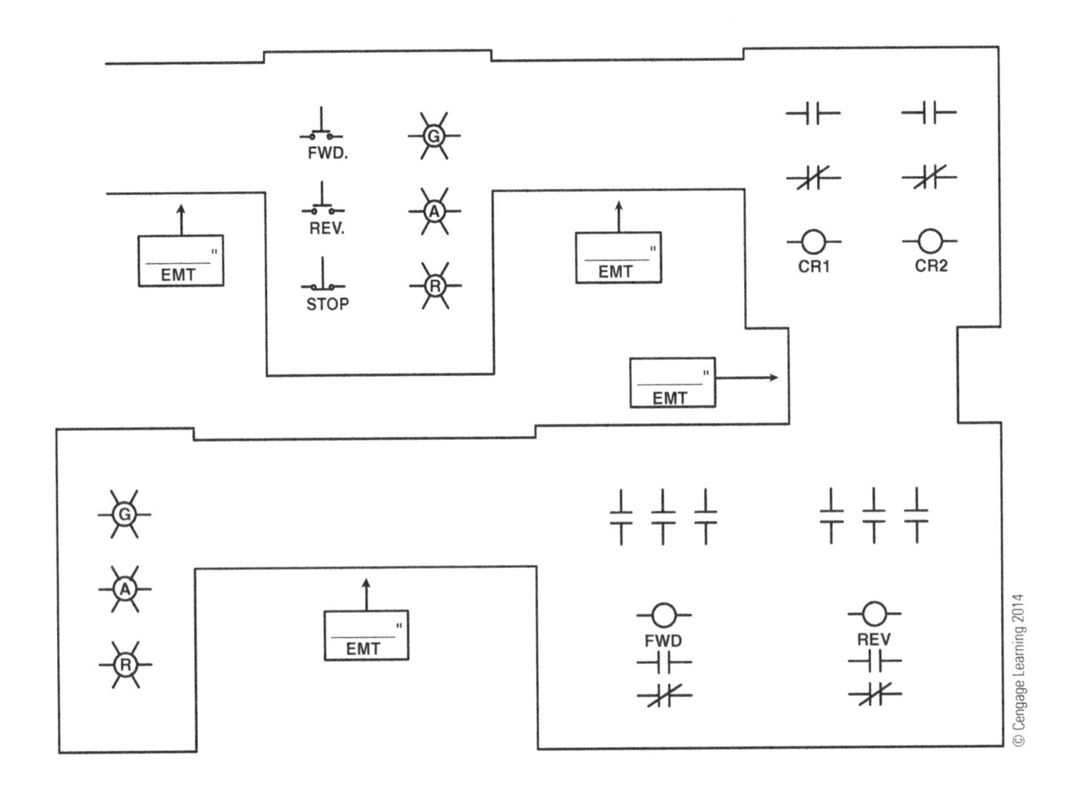

2. Power is fed through the control relay enclosure. Show the wiring of the components. Then determine the minimum trade size of liquidtight flexible metal conduit (LFMC) raceway required at each location.

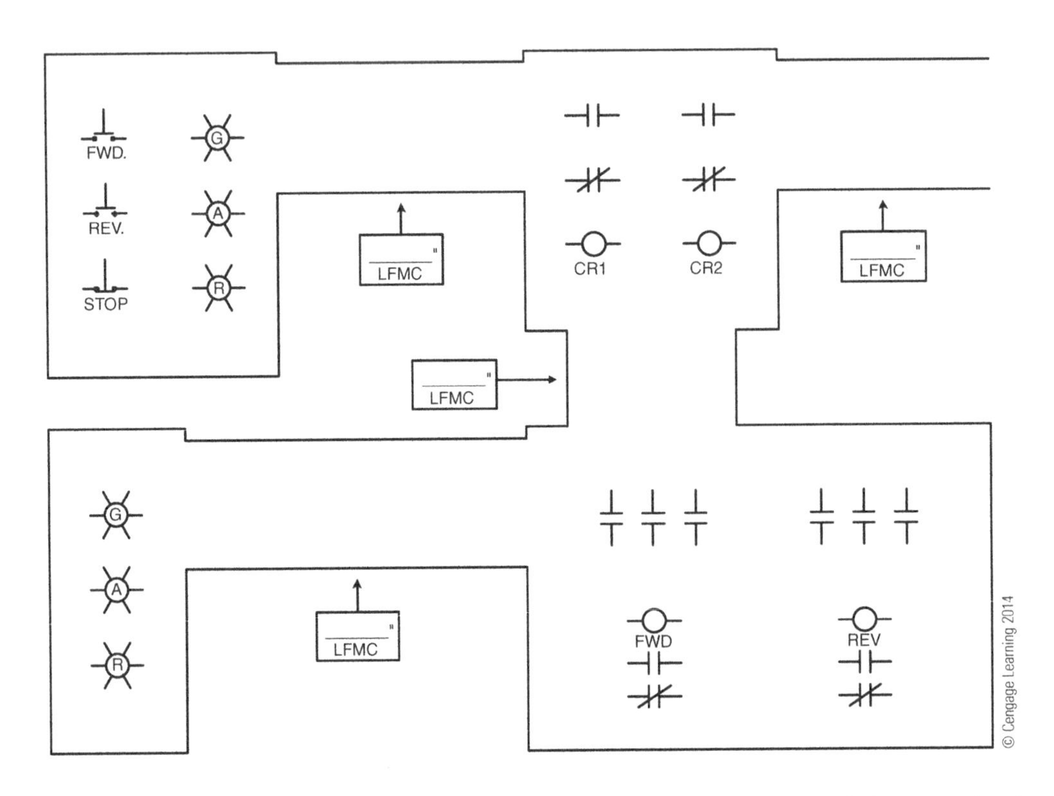

3. Power is fed through the motor starter enclosure. Show the wiring of the components. Then determine the minimum trade size of rigid metal conduit (RMC) raceway required at each location.

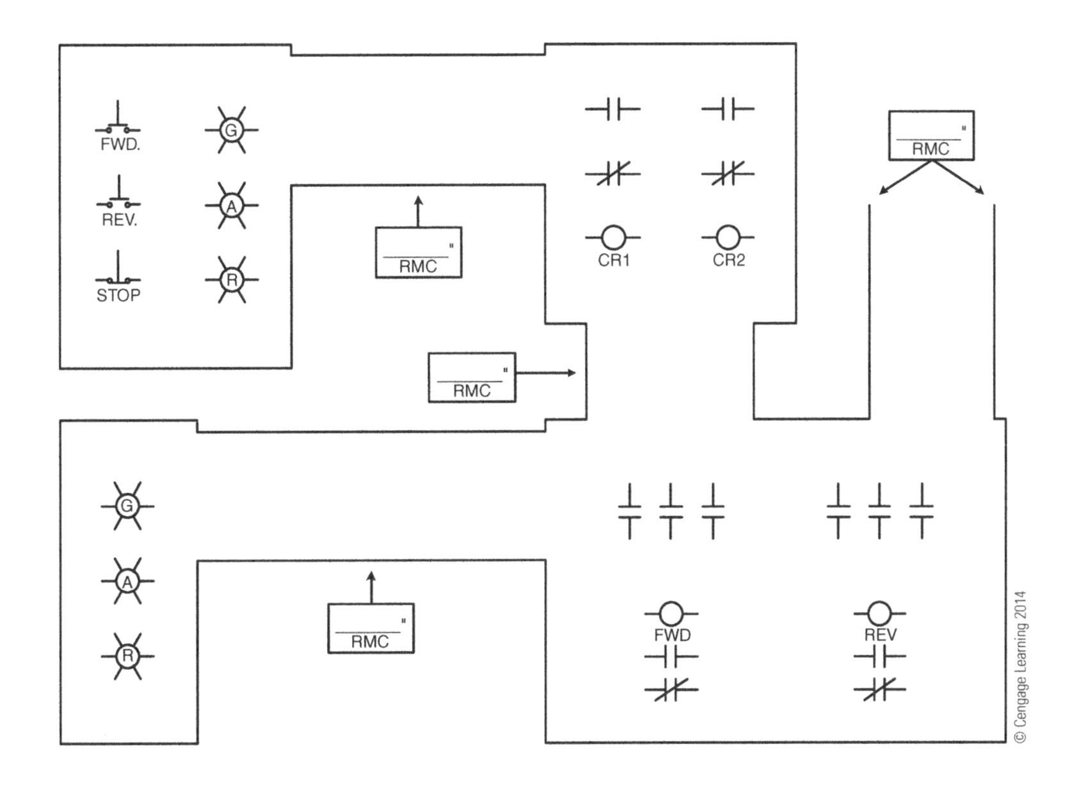

4. Power is fed through the indicator lamp enclosure. Show the wiring of the components. Then determine the minimum trade size of rigid nonmetallic conduit (RNC) (Schedule 80) raceway required at each location.

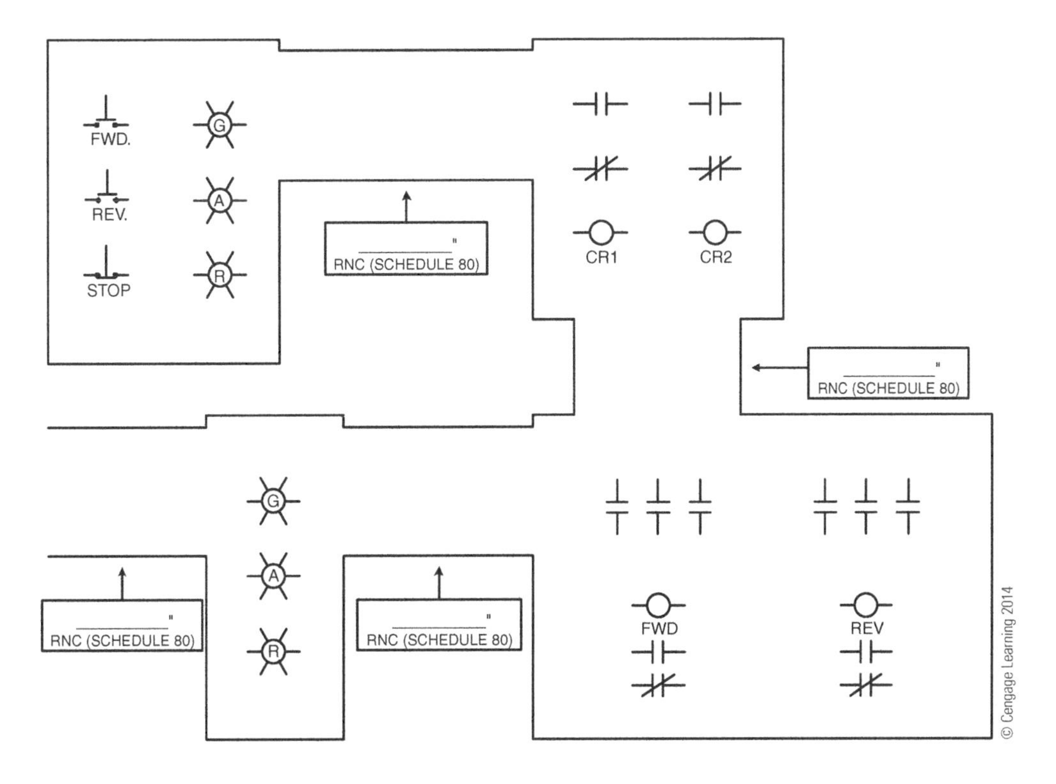

Name: ______________________________ Date: ____________________

TRANSFORMERS

1. Match the transformer to the correct drawing.

_______ Step-down autotransformer

_______ Isolation transformer

_______ Step-up transformer

_______ Center-tapped transformer

_______ Step-up autotransformer

_______ Transformer with multiple taps

_______ Step-down transformer

_______ Autotransformer with multiple taps

_______ Transformer with multiple windings

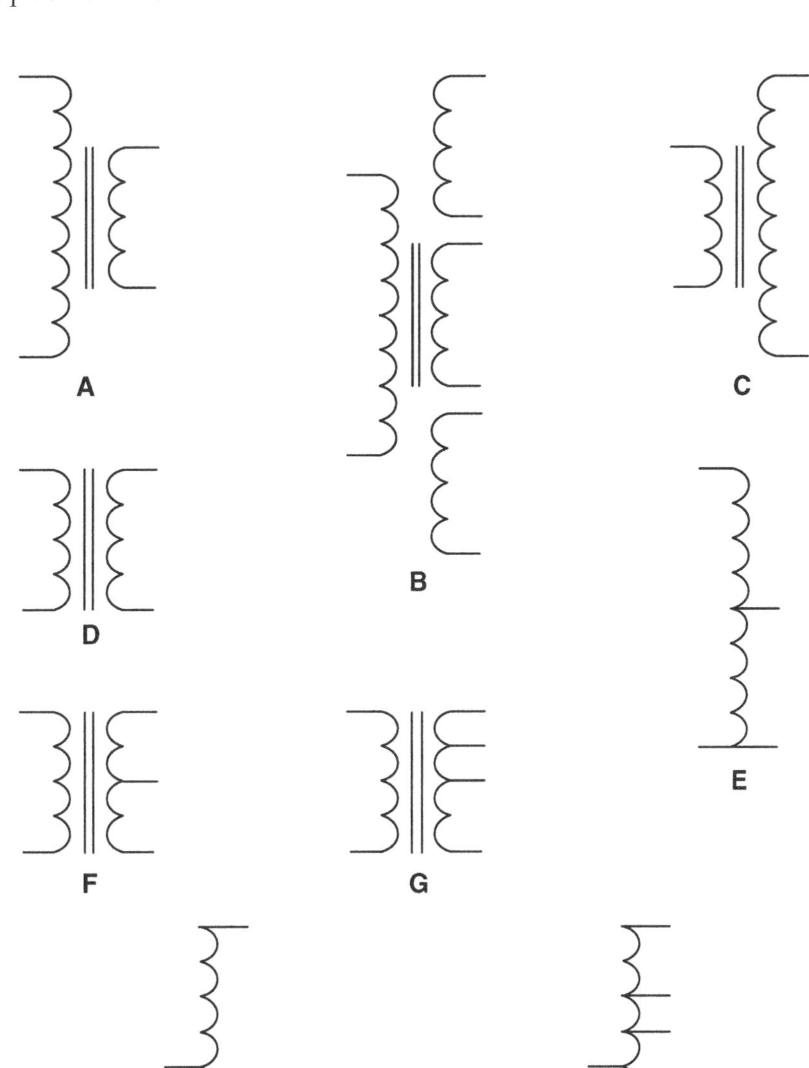

Name: ______________________ Date: ______________

TRANSFORMER CALCULATIONS

1. Given the transformer circuit, determine the unknown values.

   a. _______ Primary current

   b. _______ Primary power

   c. _______ Secondary voltage

   d. _______ Secondary power

   e. _______ Voltage ratio

   f. _______ Current ratio

   g. _______ Power ratio

   h. _______ Turns ratio

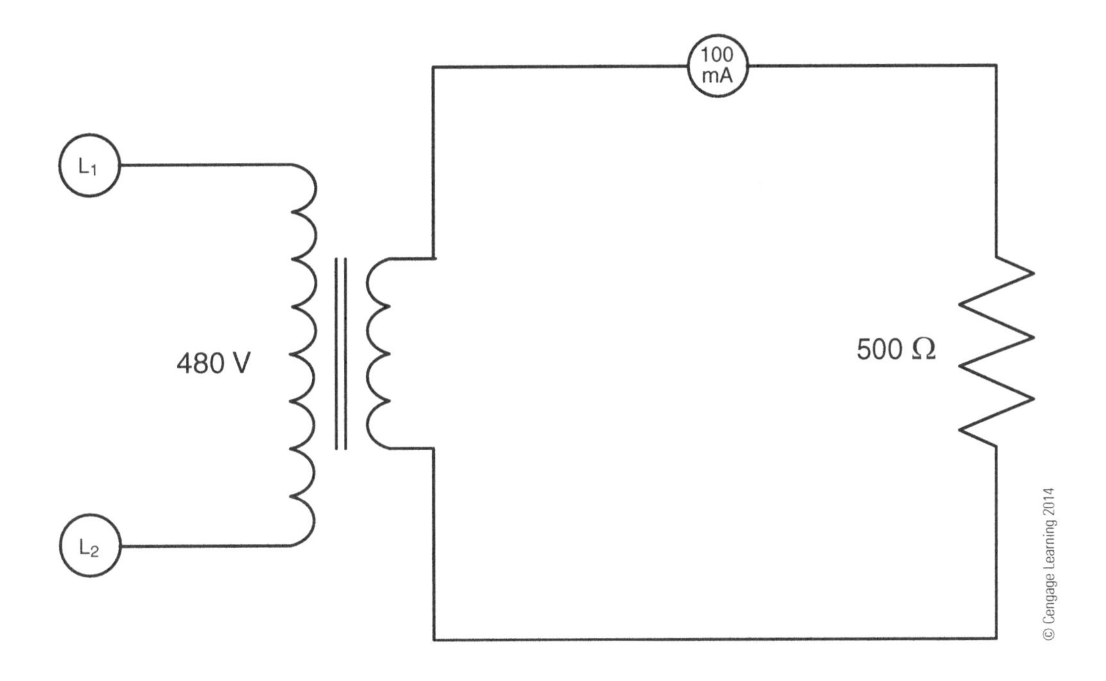

2. Given the transformer circuit, determine the unknown values.

   a. _______ Primary current

   b. _______ Primary power

   c. _______ Secondary voltage

   d. _______ Secondary current

   e. _______ Secondary power

   f. _______ Current ratio

   g. _______ Power ratio

   h. _______ Turns ratio

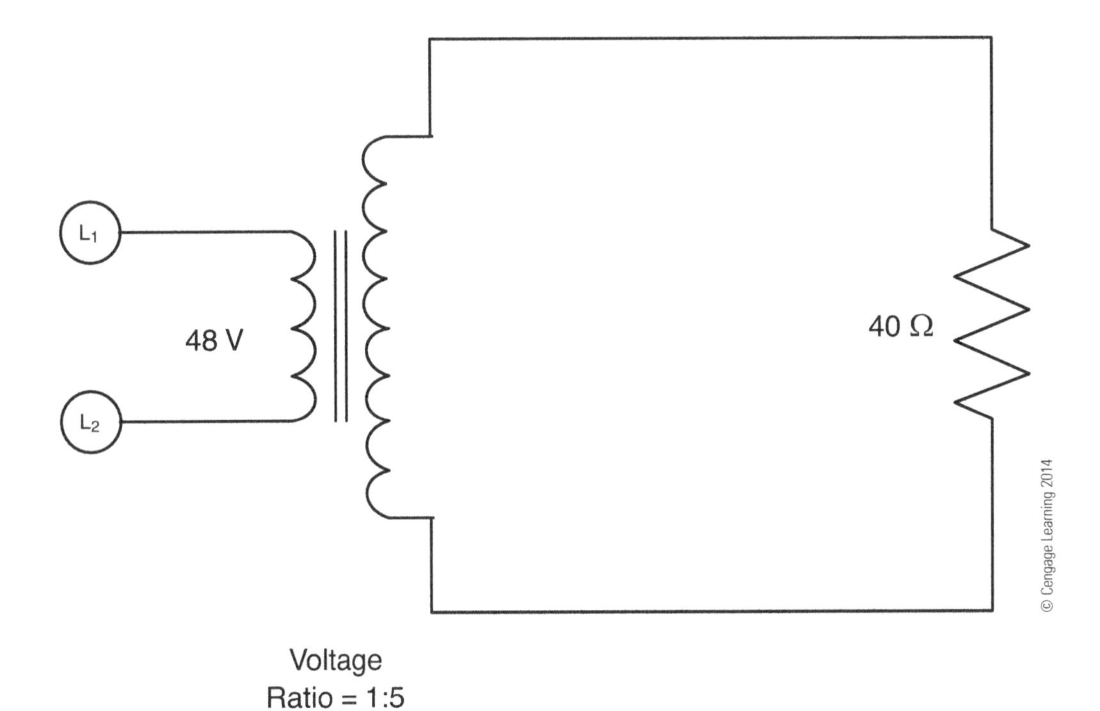

Name: ______________________ Date: ______________

TRANSFORMER CONNECTIONS

1. With a source voltage of 240 V, connect the transformer windings needed to produce 360 V.

2. With a source voltage of 240 V, connect the transformer windings needed to produce 176 V.

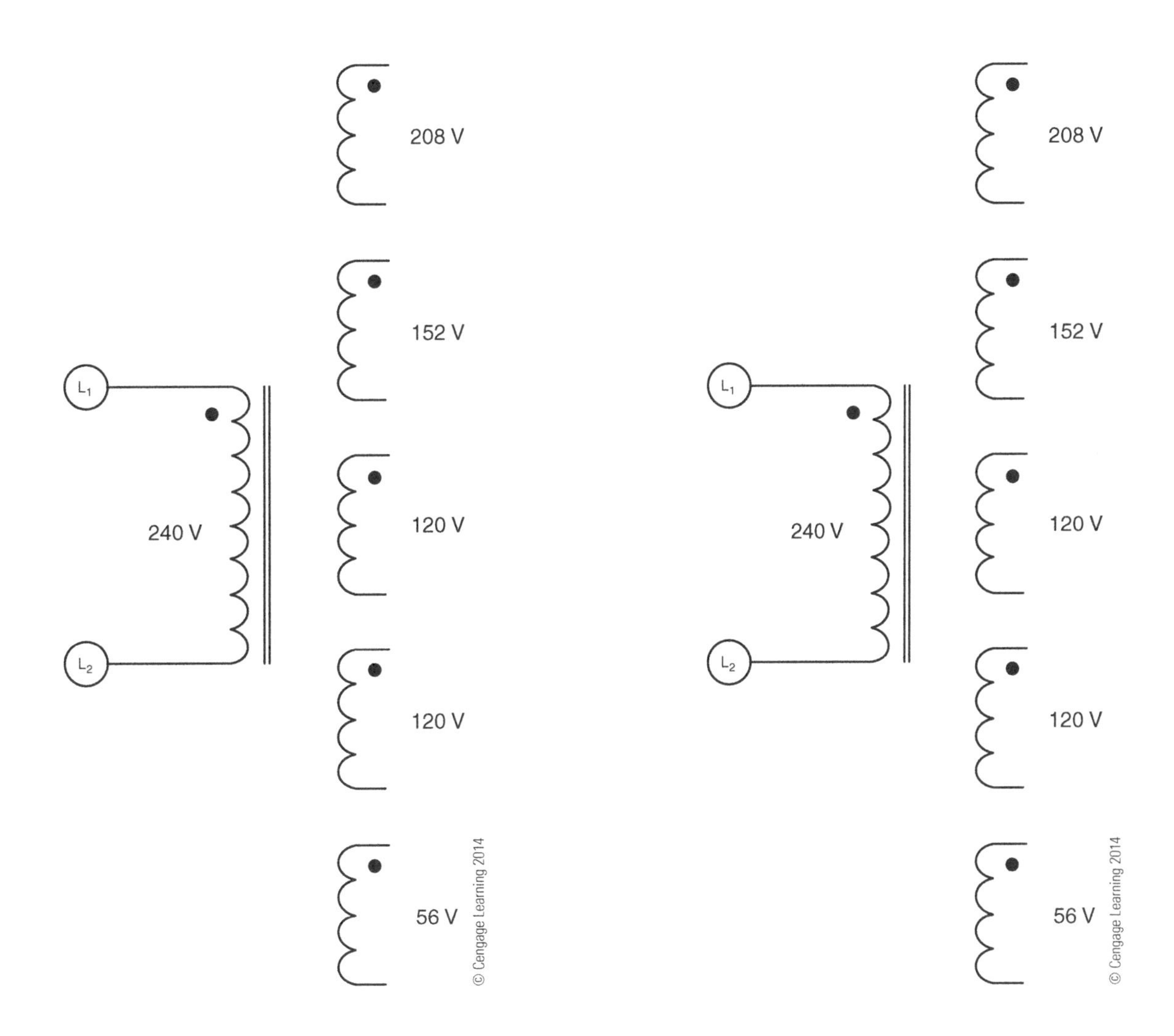

3. With a source voltage of 240 V, connect the transformer windings needed to produce 480 V.

4. With a source voltage of 240 V, connect the transformer windings needed to produce 184 V.

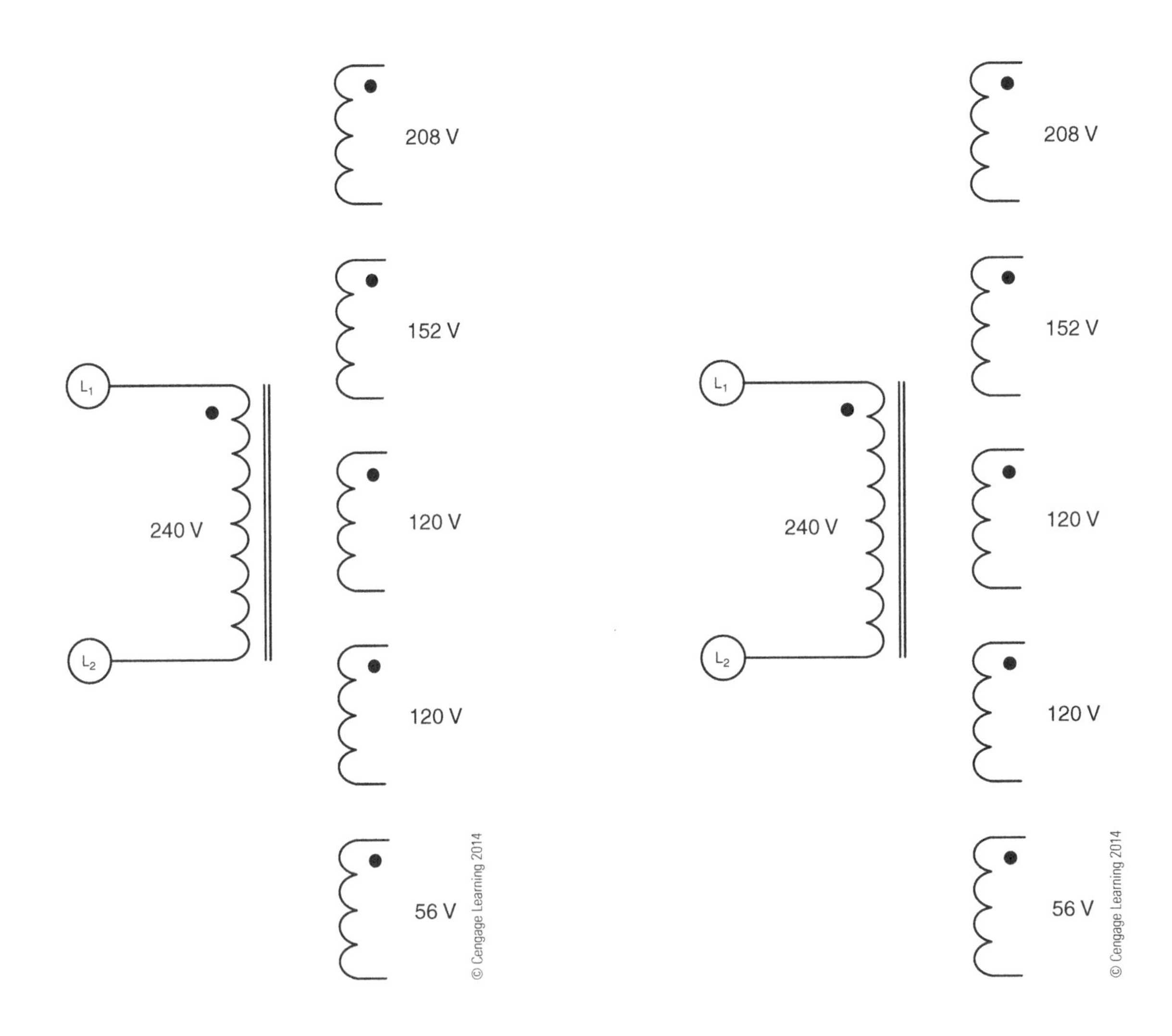

Name: ______________________ Date: ______________

THREE-PHASE TRANSFORMERS

1. Connect the three single-phase transformers to form a three-phase delta–delta connection.

L_1 ______________________

L_2 ______________________

L_3 ______________________

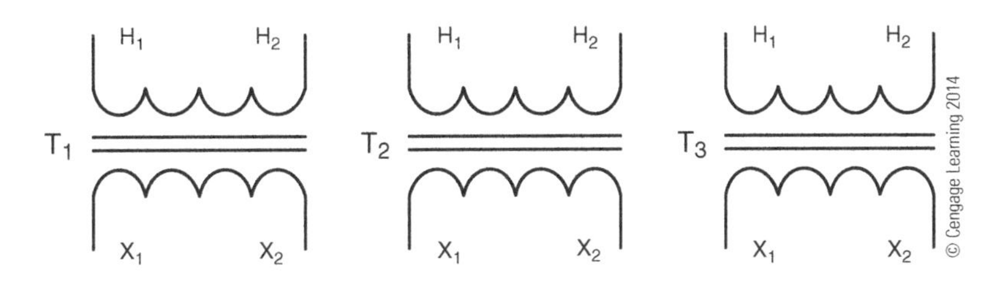

L_1 A ______________________

L_2 A ______________________

L_3 A ______________________

2. Connect the three single-phase transformers to form a three-phase delta–wye connection.

L_1 ______________________

L_2 ______________________

L_3 ______________________

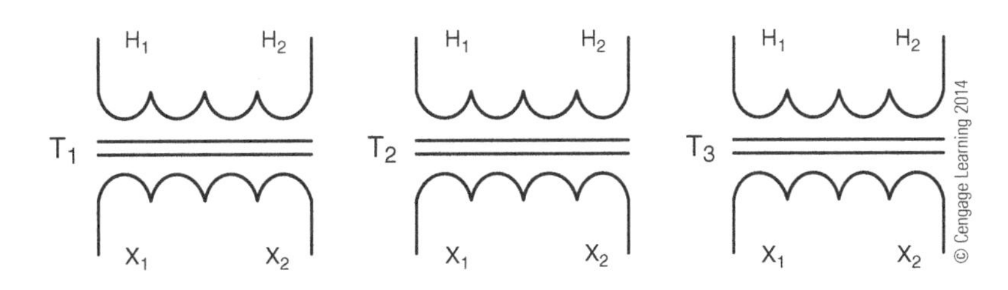

L_1 A ______________________

L_2 A ______________________

L_3 A ______________________

3. Connect the three single-phase transformers to form a three-phase wye–wye connection.

L_1

L_2

L_3

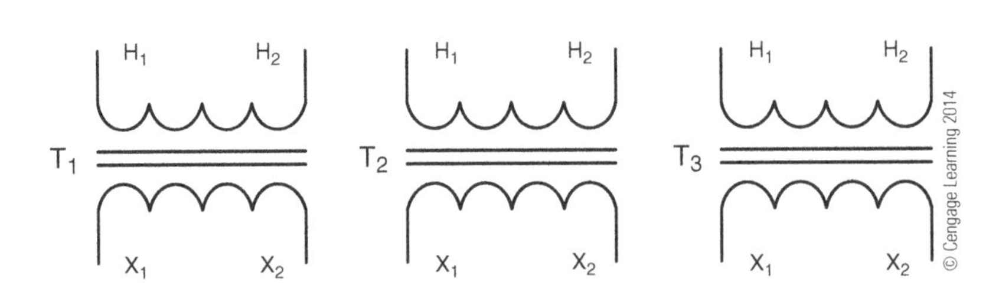

L_1 A

L_2 A

L_3 A

4. Connect the three single-phase transformers to form a three-phase wye–delta connection.

L_1

L_2

L_3

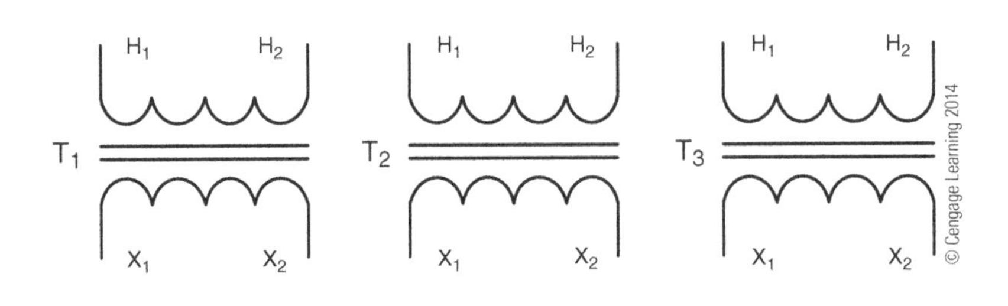

L_1 A

L_2 A

L_3 A

Name: ______________________________ Date: ______________________

CONSUMER DISTRIBUTION SYSTEM

1. Connect each load to the service panel.

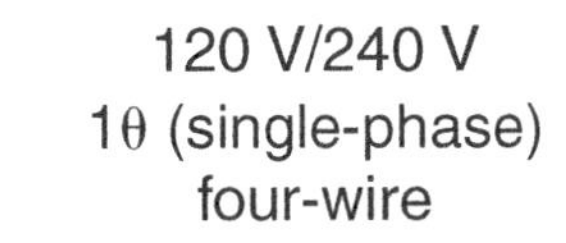

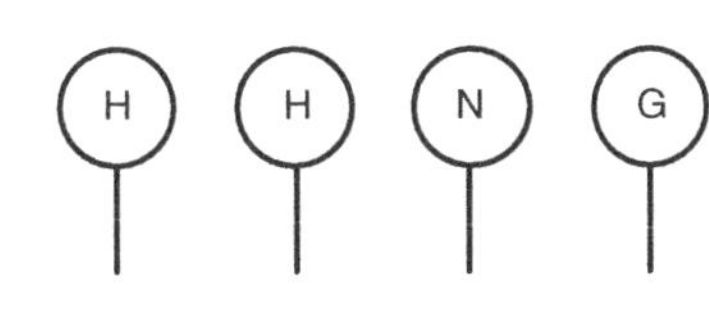

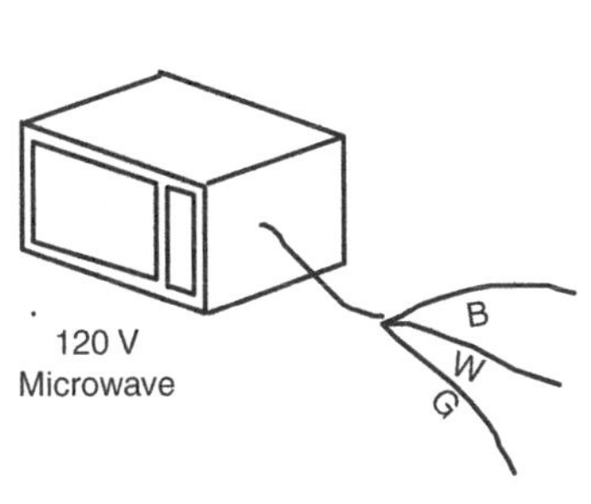

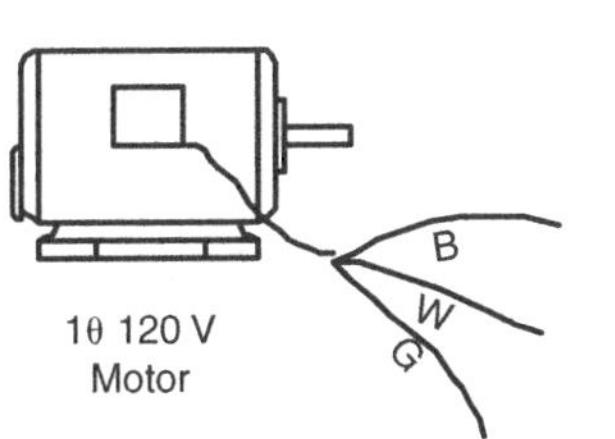

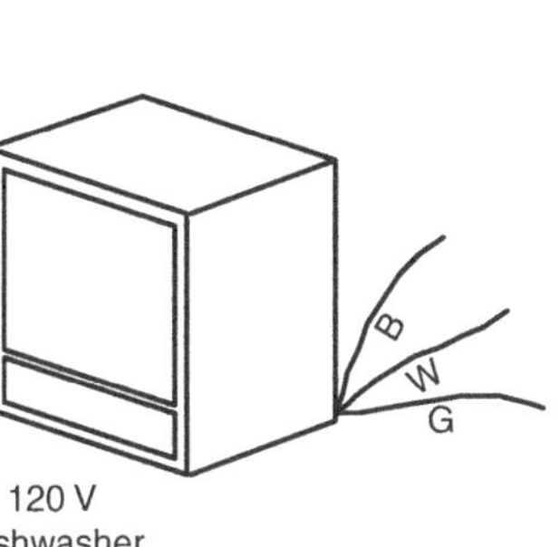

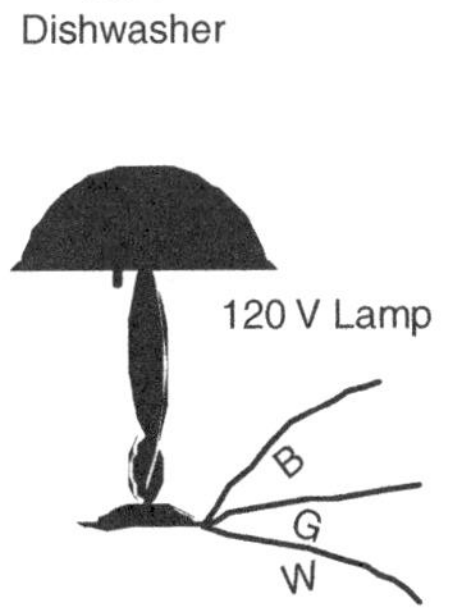

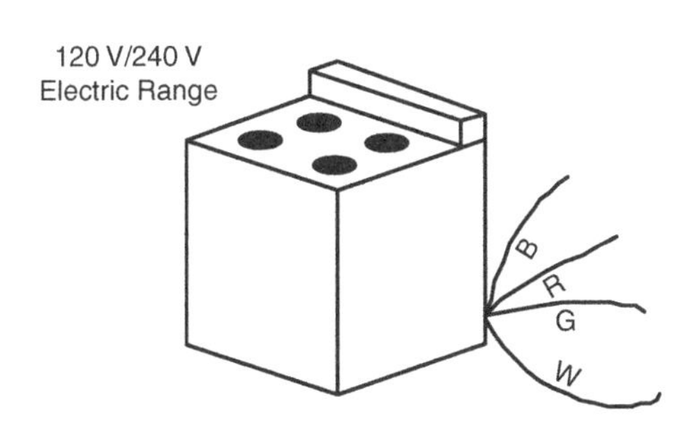

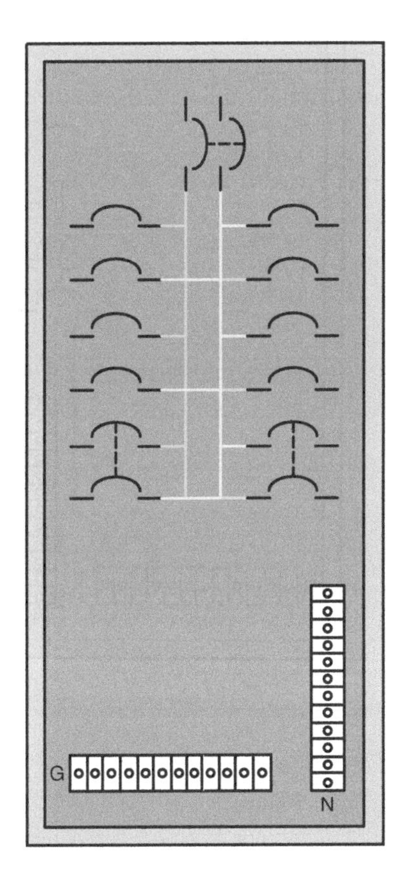

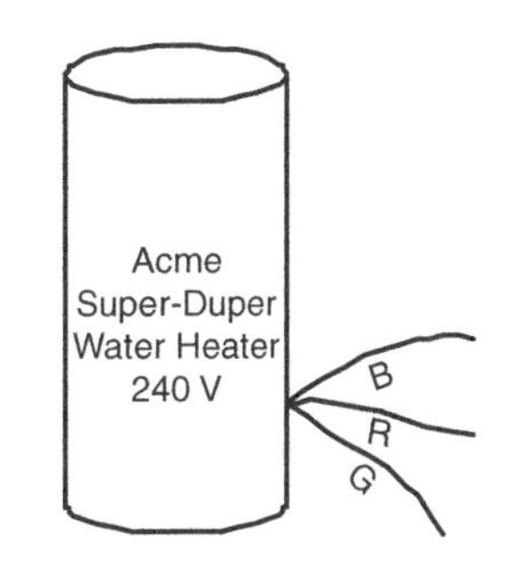

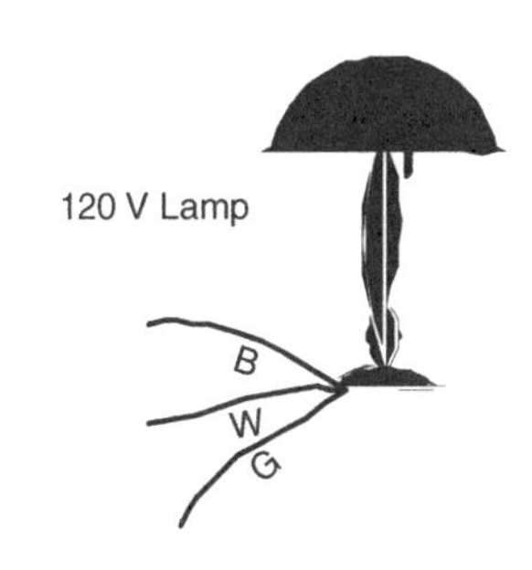

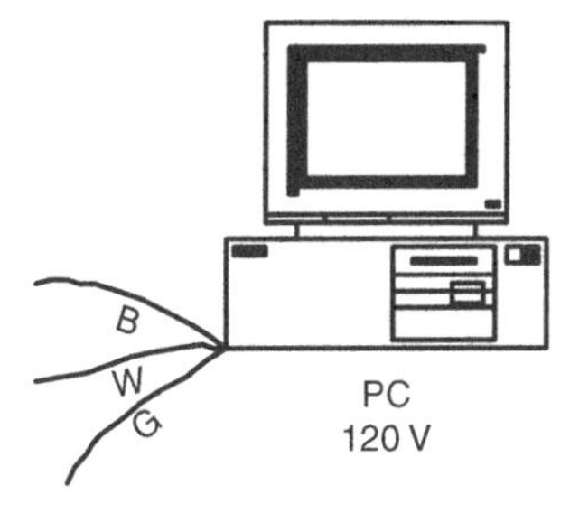

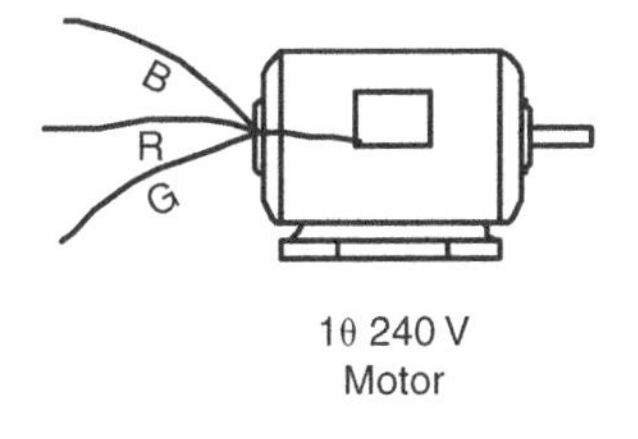

1θ 240 V
Motor

2. Connect each load to the service panel.

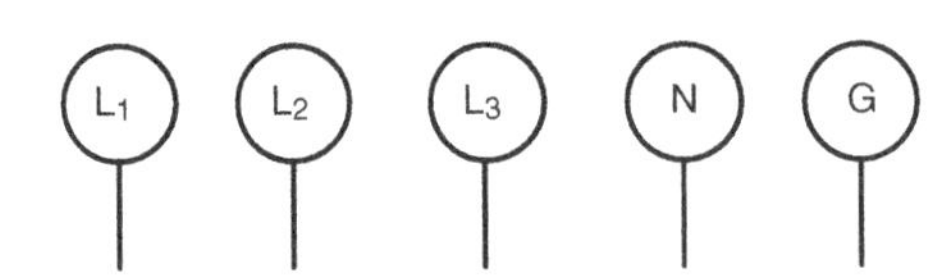

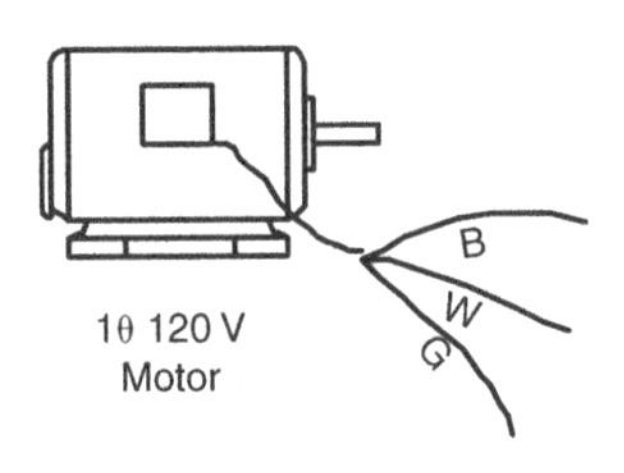

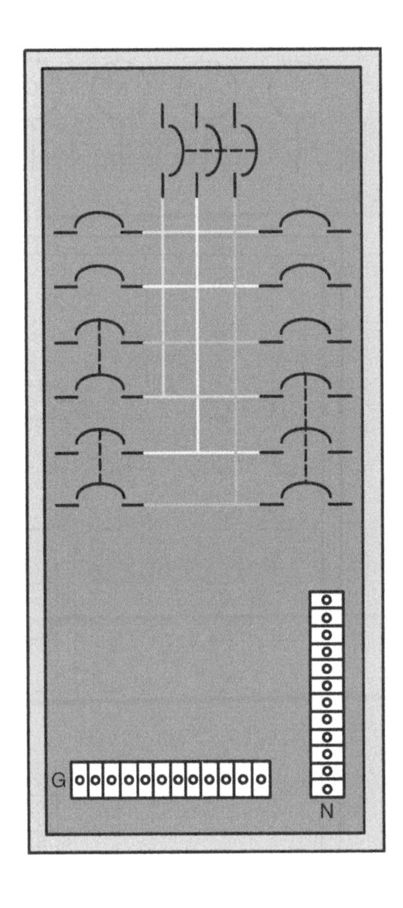

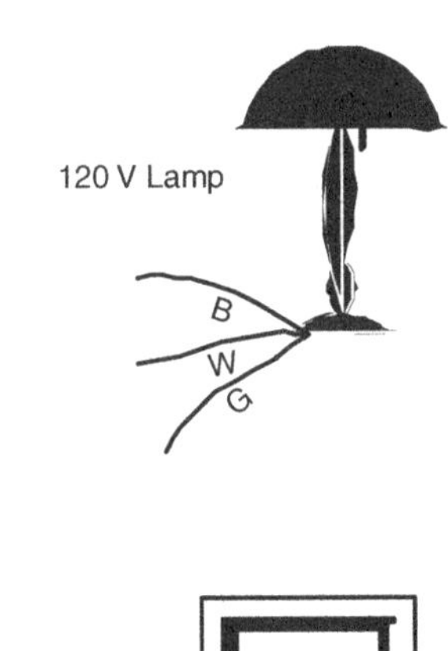

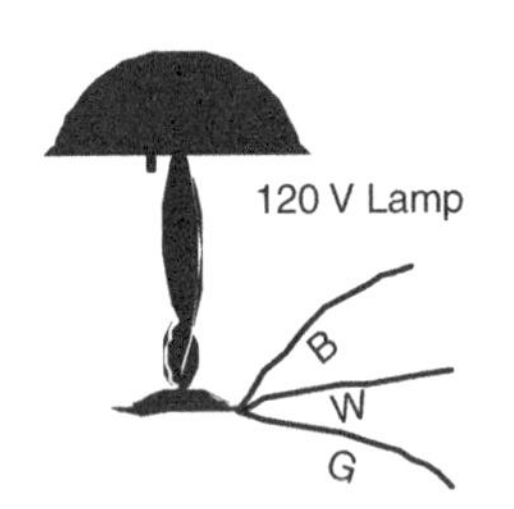

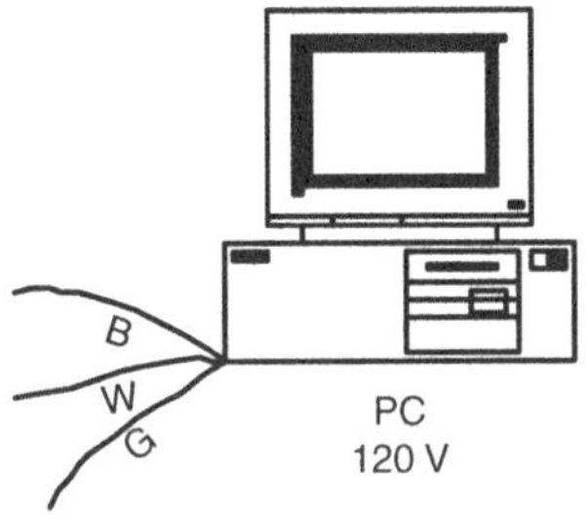

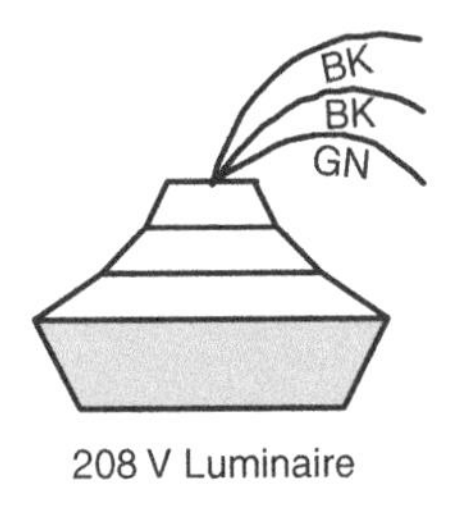

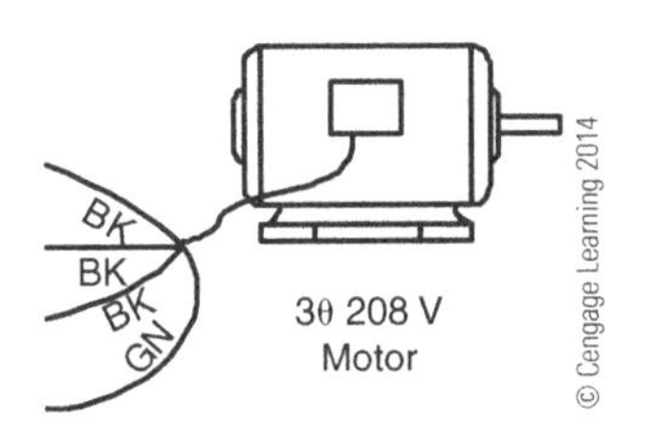

Name: ______________________ Date: ______________

DC GENERATORS

1. Given the schematic diagram, complete the connections necessary for a series-wound DC generator.

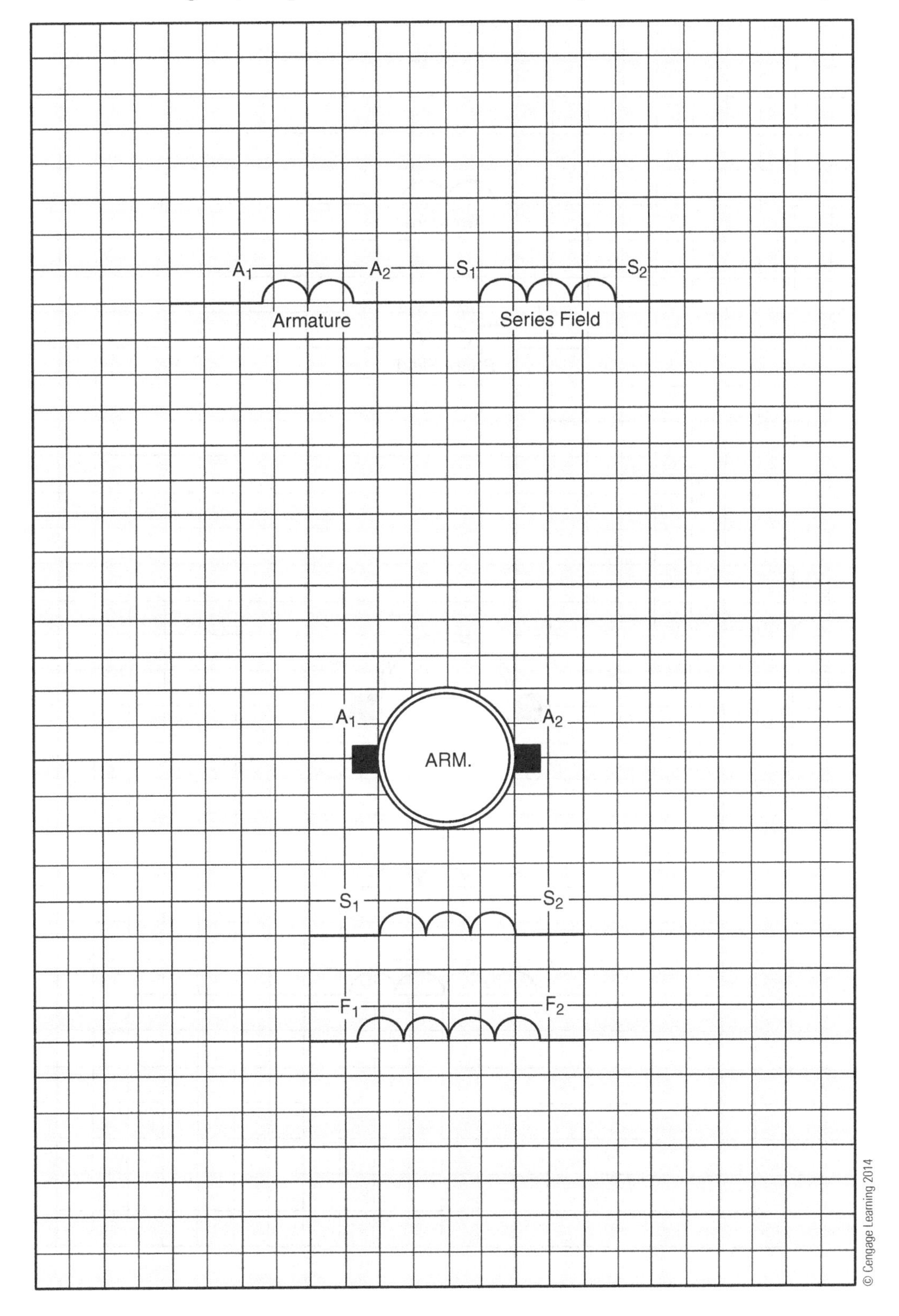

2. Given the schematic diagram, complete the connections necessary for a self-excited, shunt-wound DC generator.

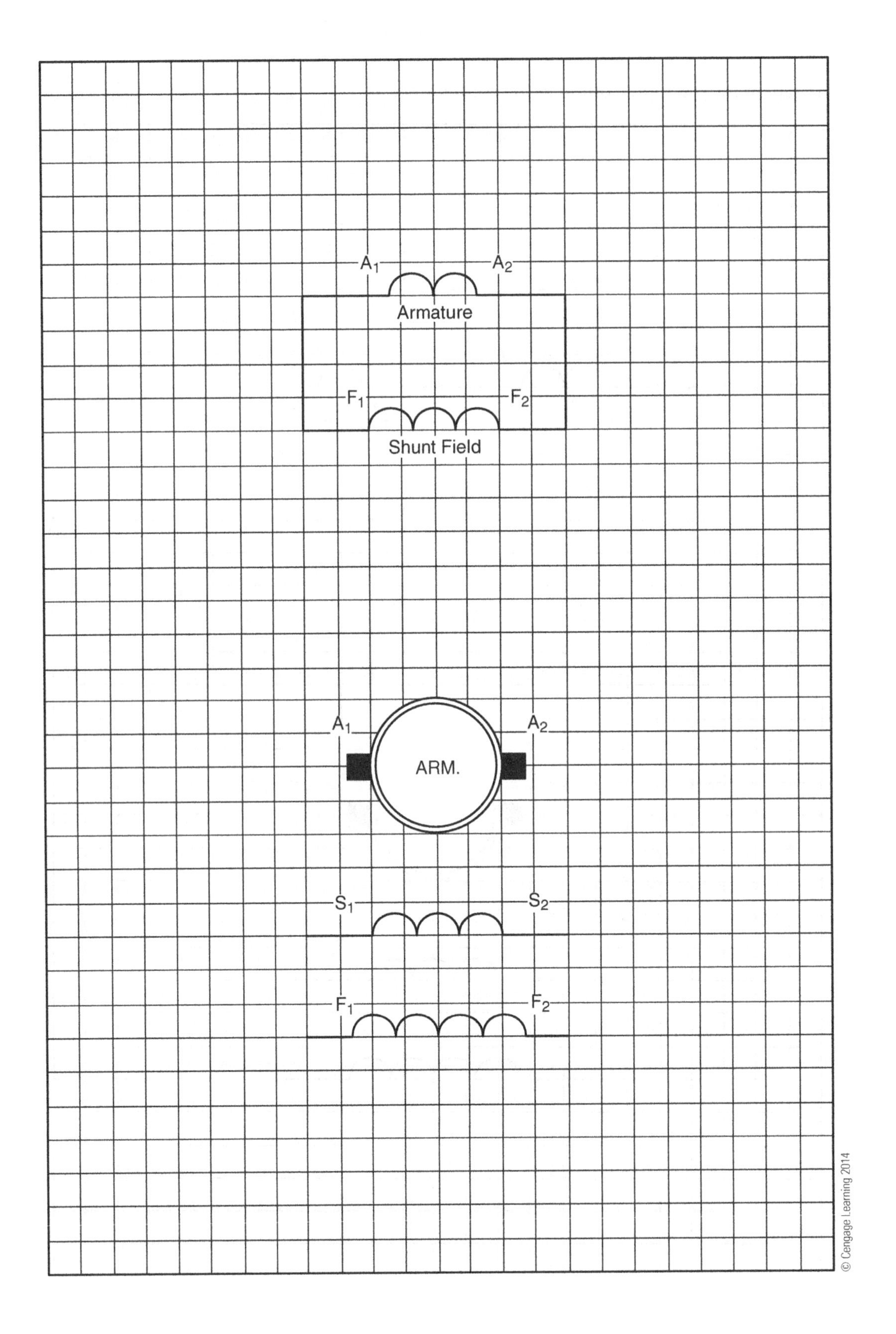

3. Given the schematic diagram, complete the connections necessary for a separately excited, shunt-wound DC generator.

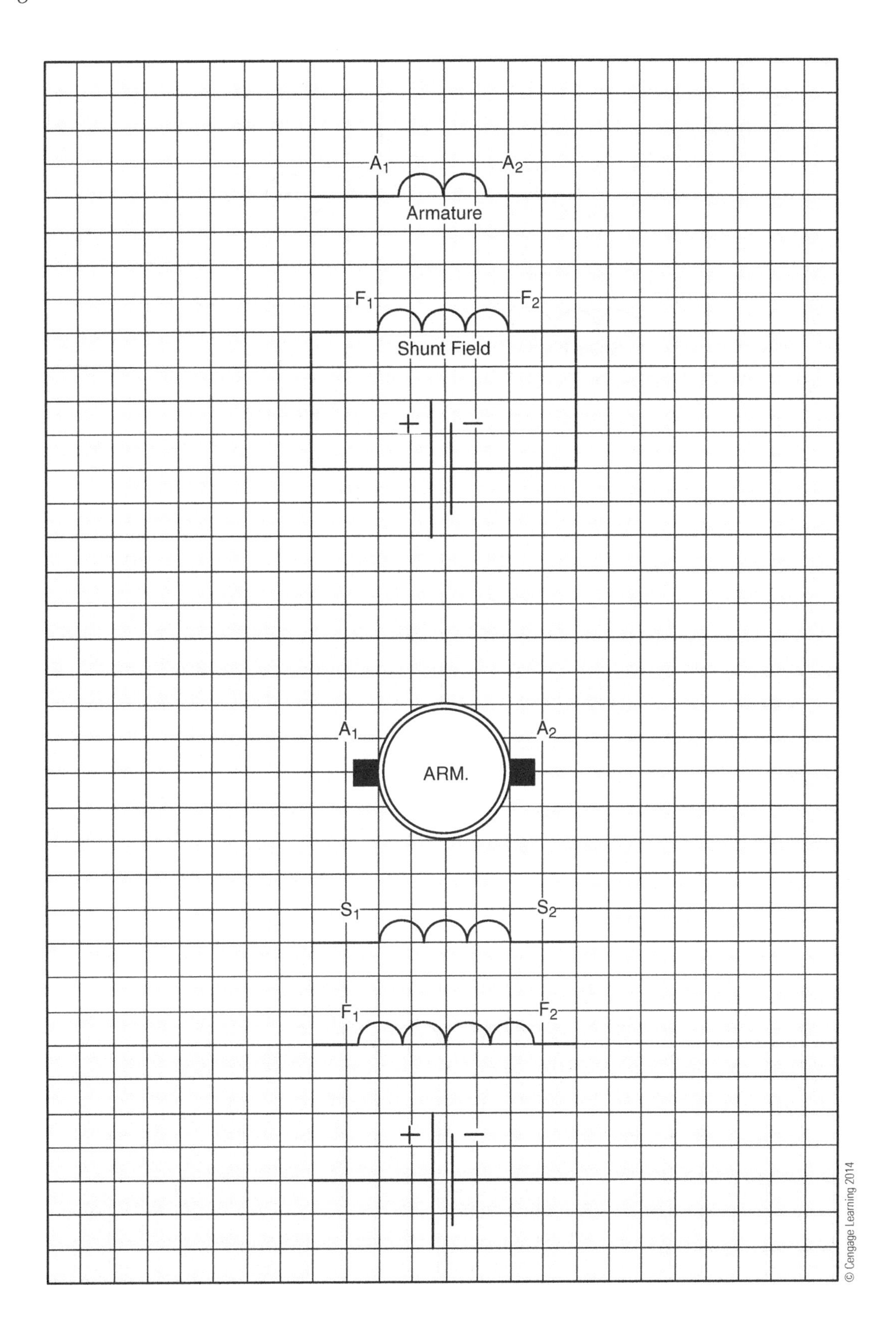

4. Given the schematic diagram, complete the connections necessary for a cumulative compound-wound DC generator.

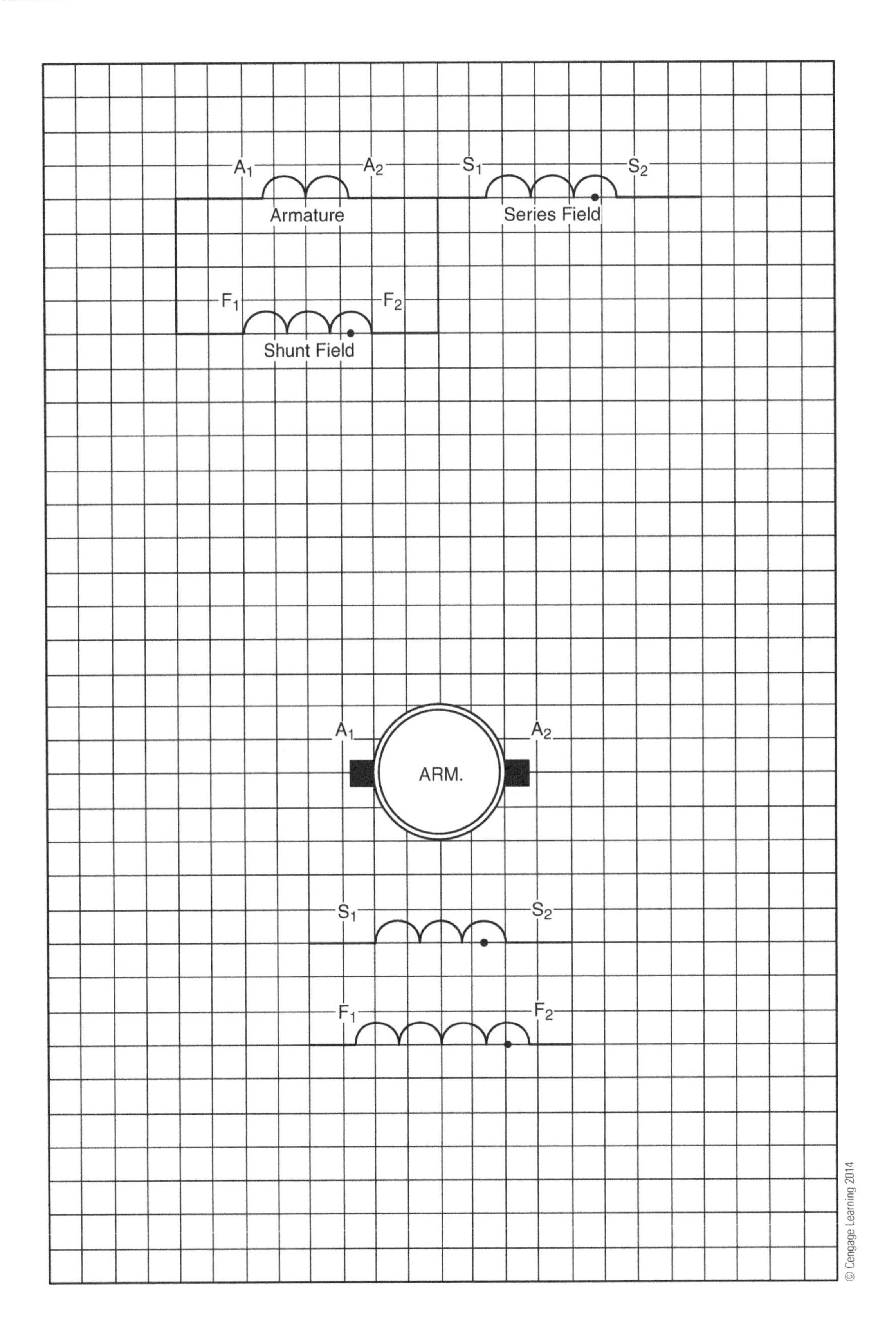

5. Given the schematic diagram, complete the connections necessary for a differential-compound-wound DC generator.

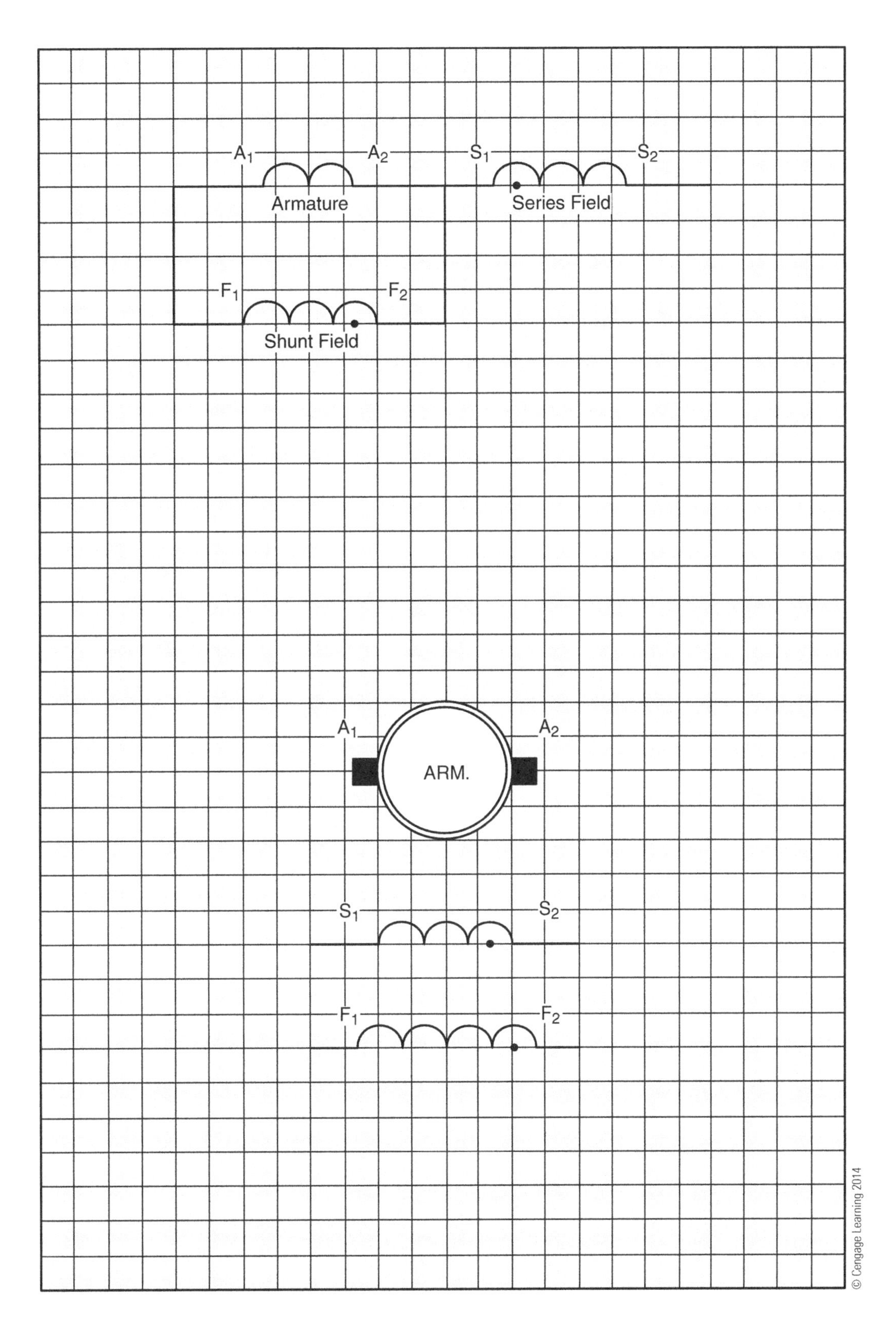

Name: ______________________ Date: ______________

DC MOTORS

1. Given the schematic diagram, complete the connections necessary for a series-wound DC motor.

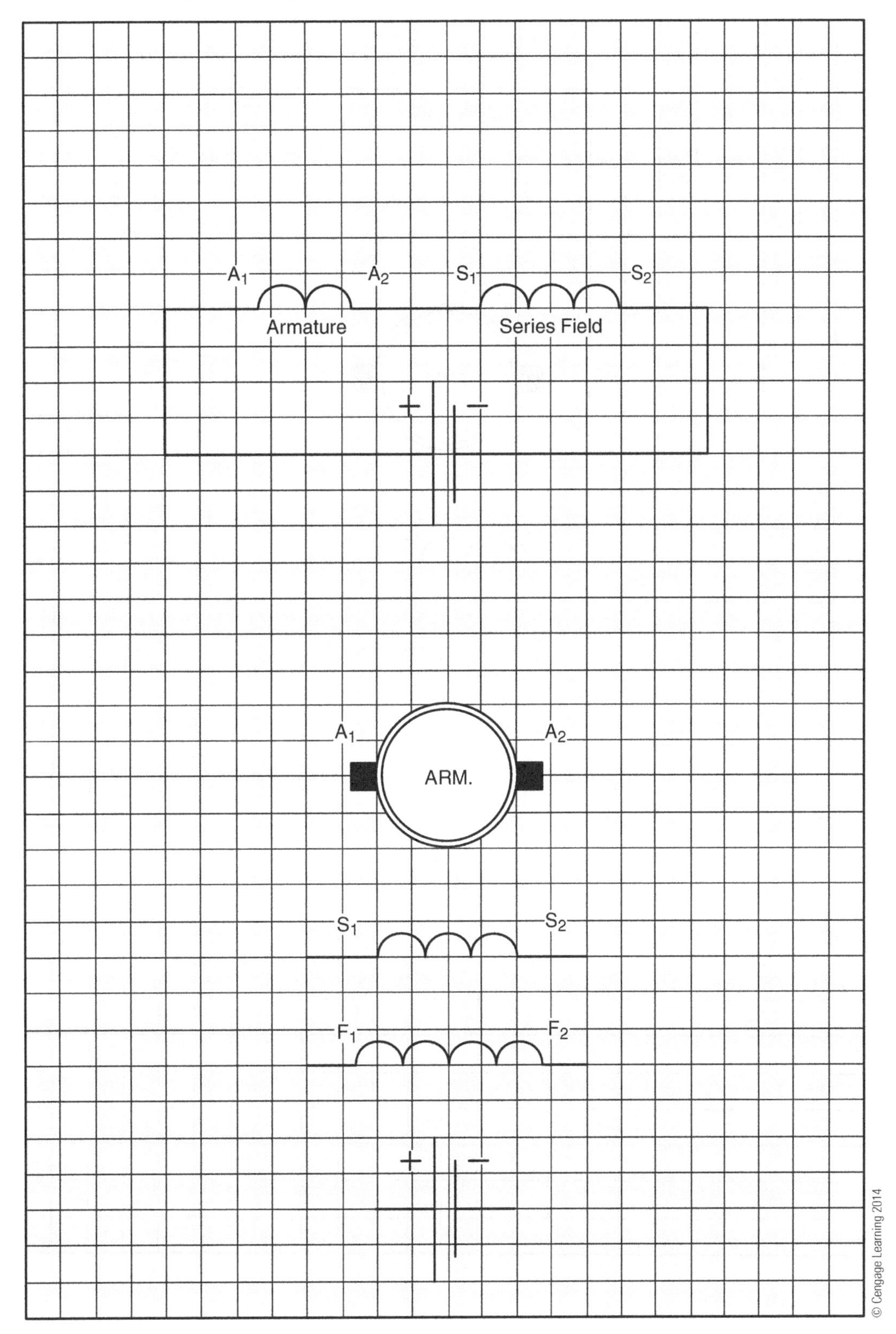

2. Show the connections necessary to reverse the direction of rotation of the series-wound DC motor that you showed the wiring for in question 1.

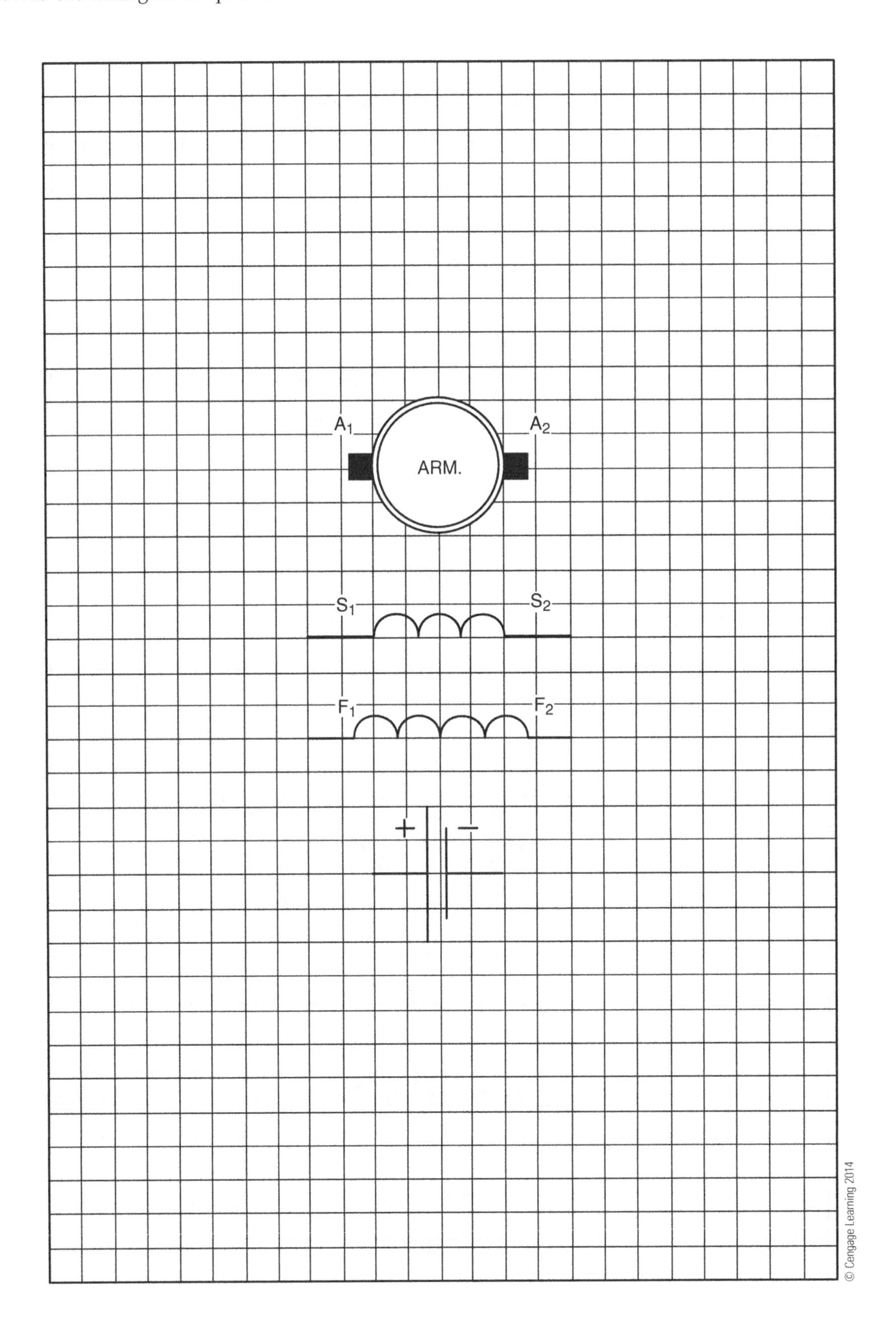

3. Given the schematic diagram, complete the connections necessary for a shunt-wound DC motor.

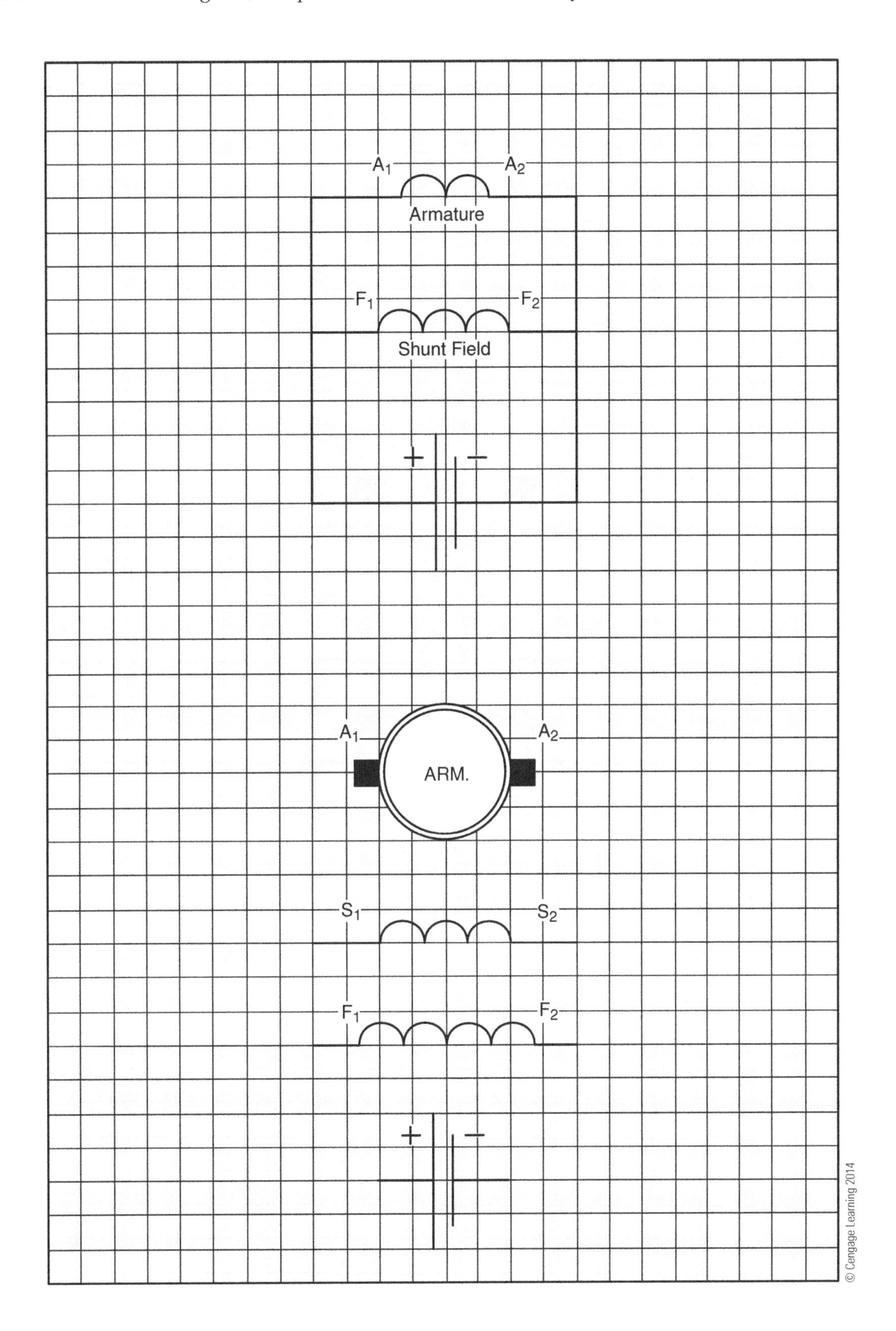

4. Show the connections necessary to reverse the direction of rotation of the shunt-wound DC motor that you showed the wiring for in question 3.

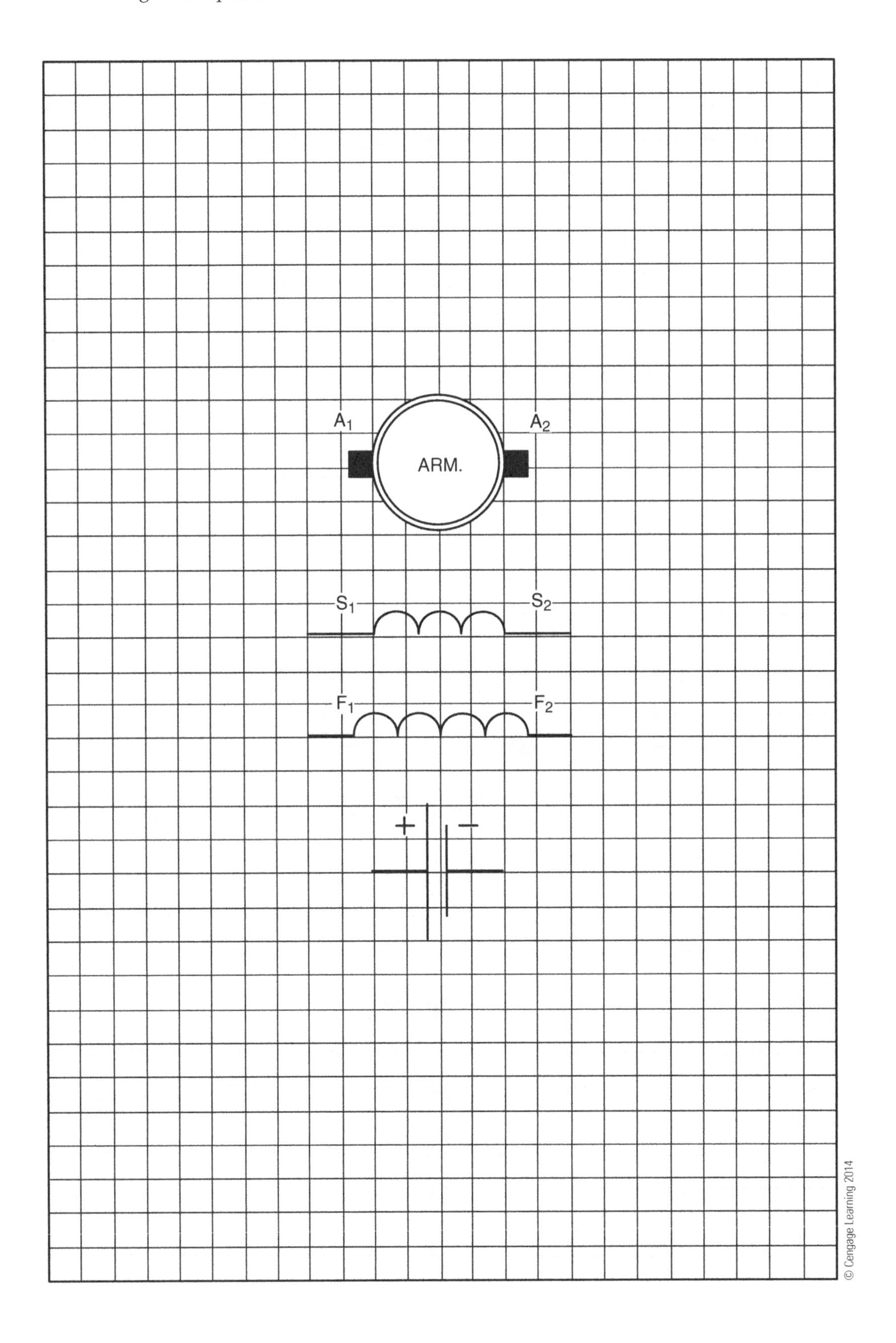

5. Given the schematic diagram, complete the connections necessary for a separately excited, shunt-wound DC motor.

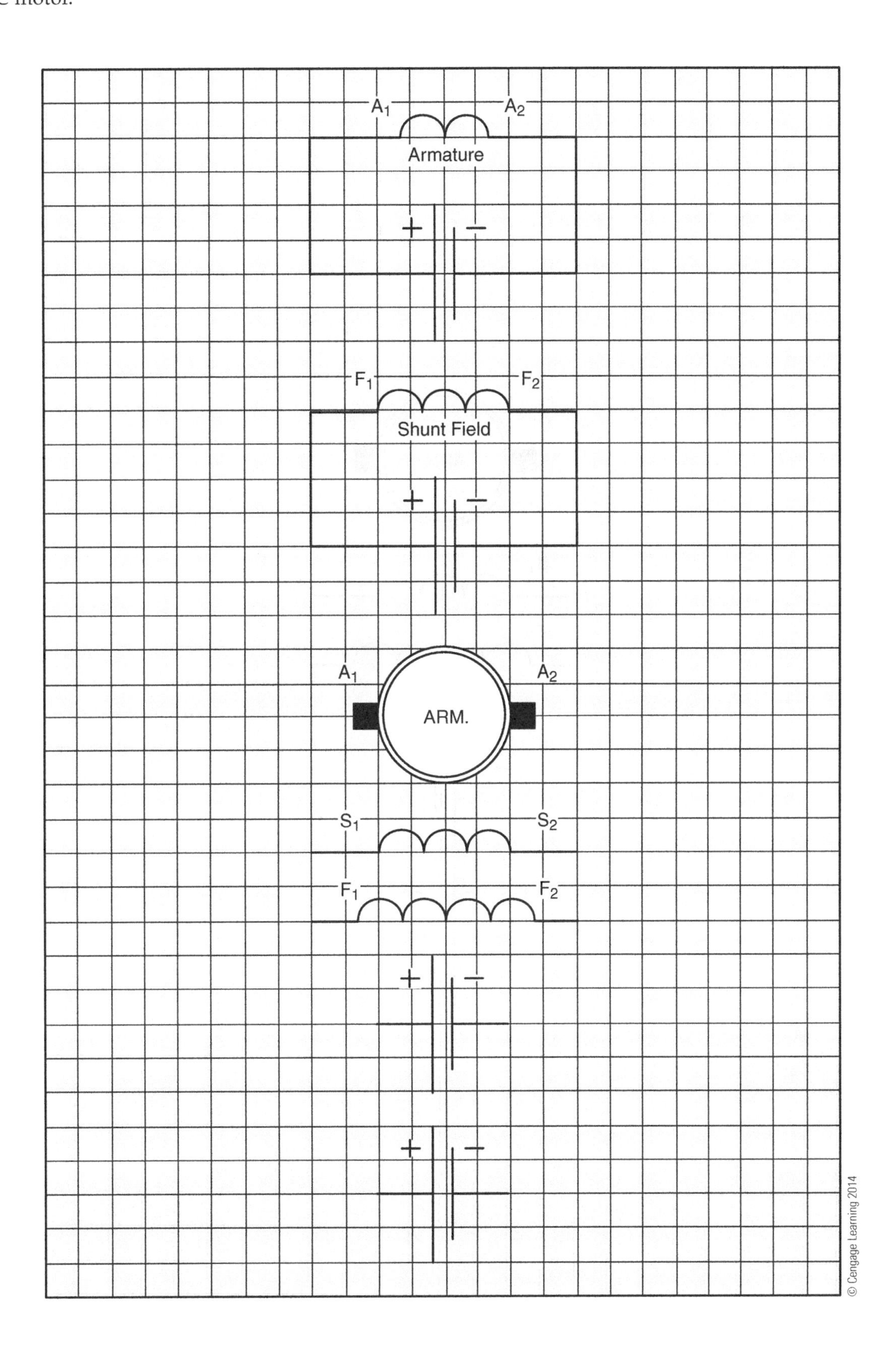

6. Show the connections necessary to reverse the direction of rotation of the separately excited, shunt-wound DC motor that you showed the wiring for in question 5.

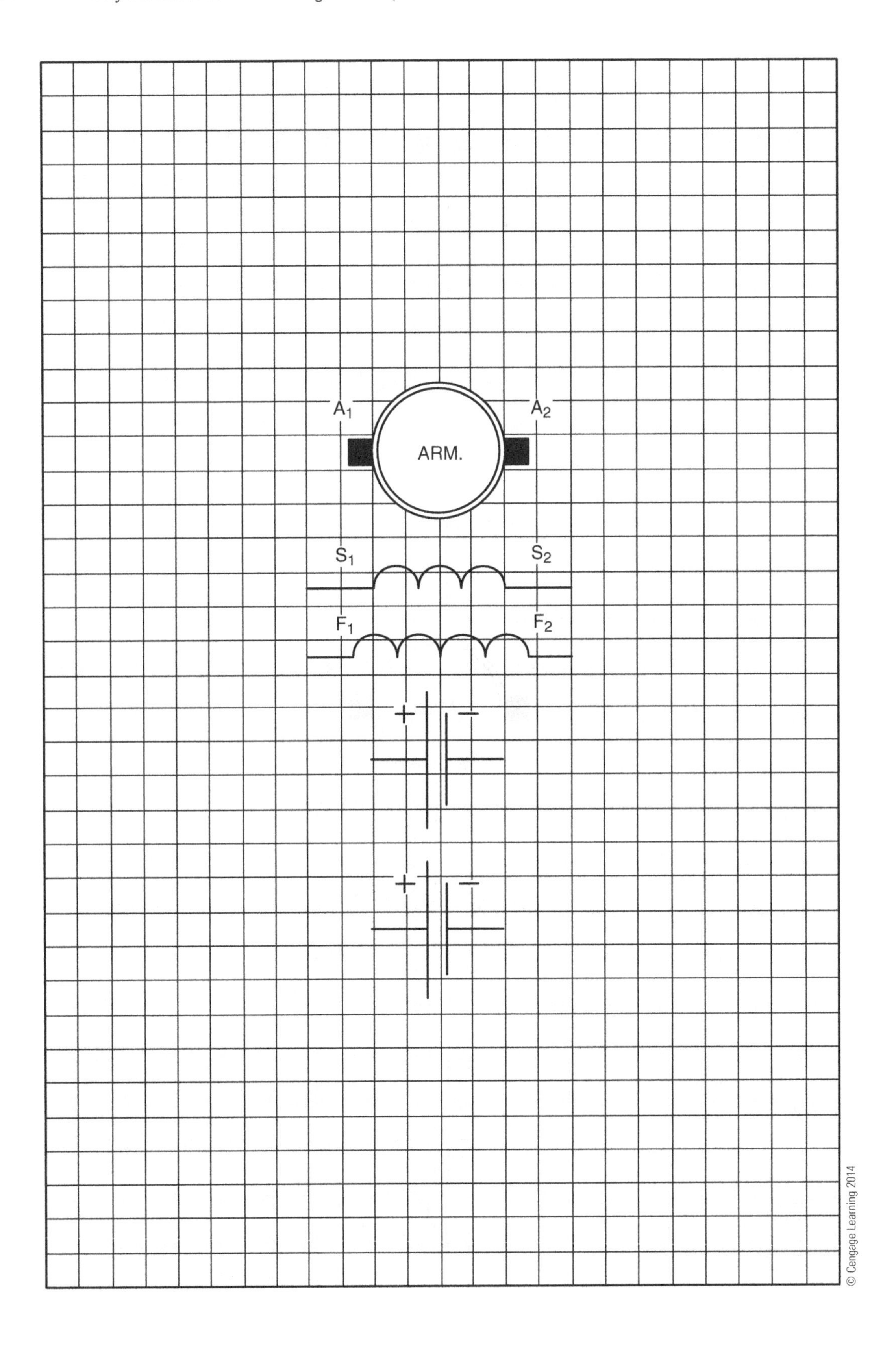

7. Given the schematic diagram, complete the connections necessary for a compound-wound DC motor.

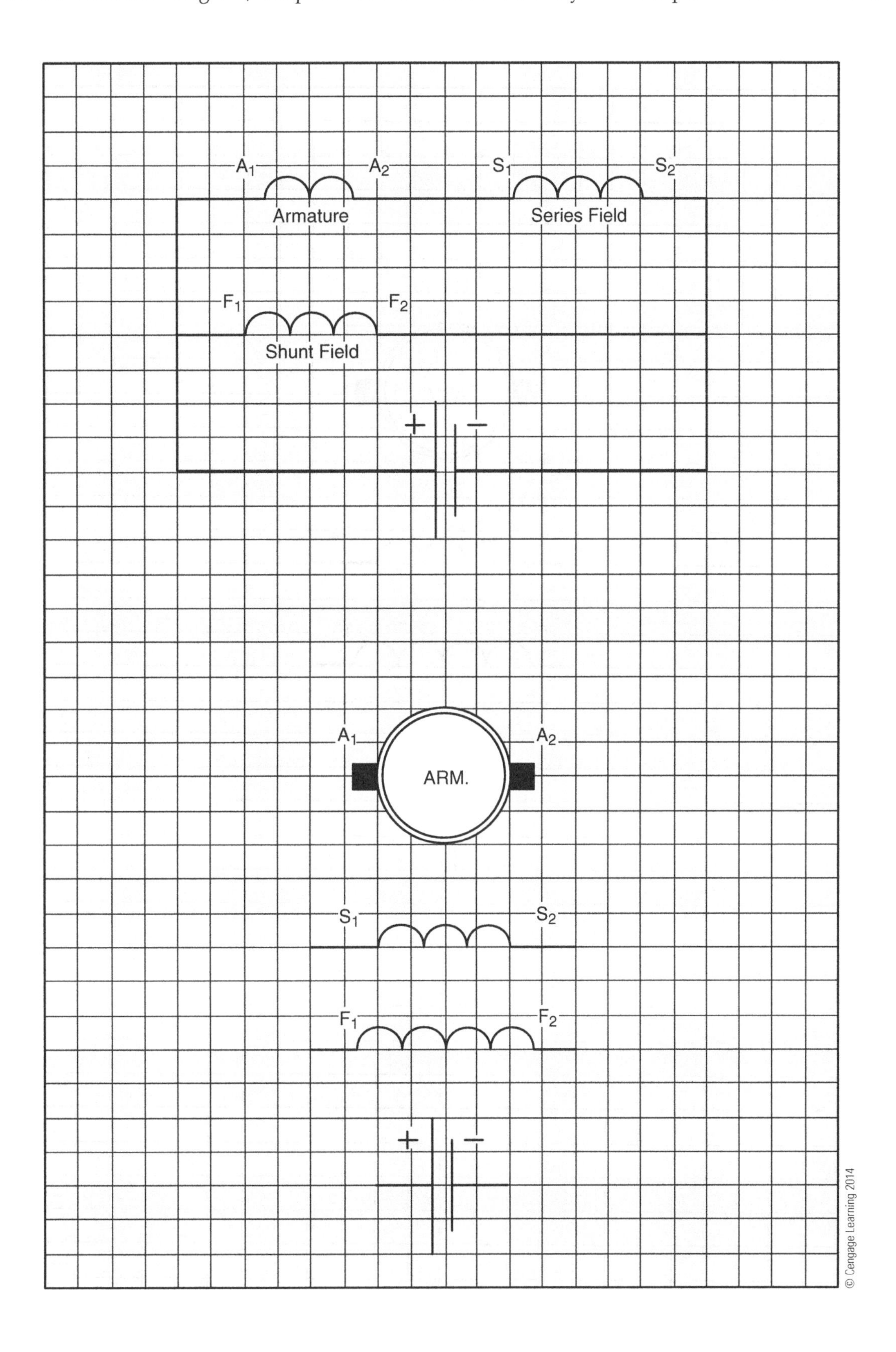

8. Show the connections necessary to reverse the direction of rotation of the compound-wound DC motor that you showed the wiring for in question 7.

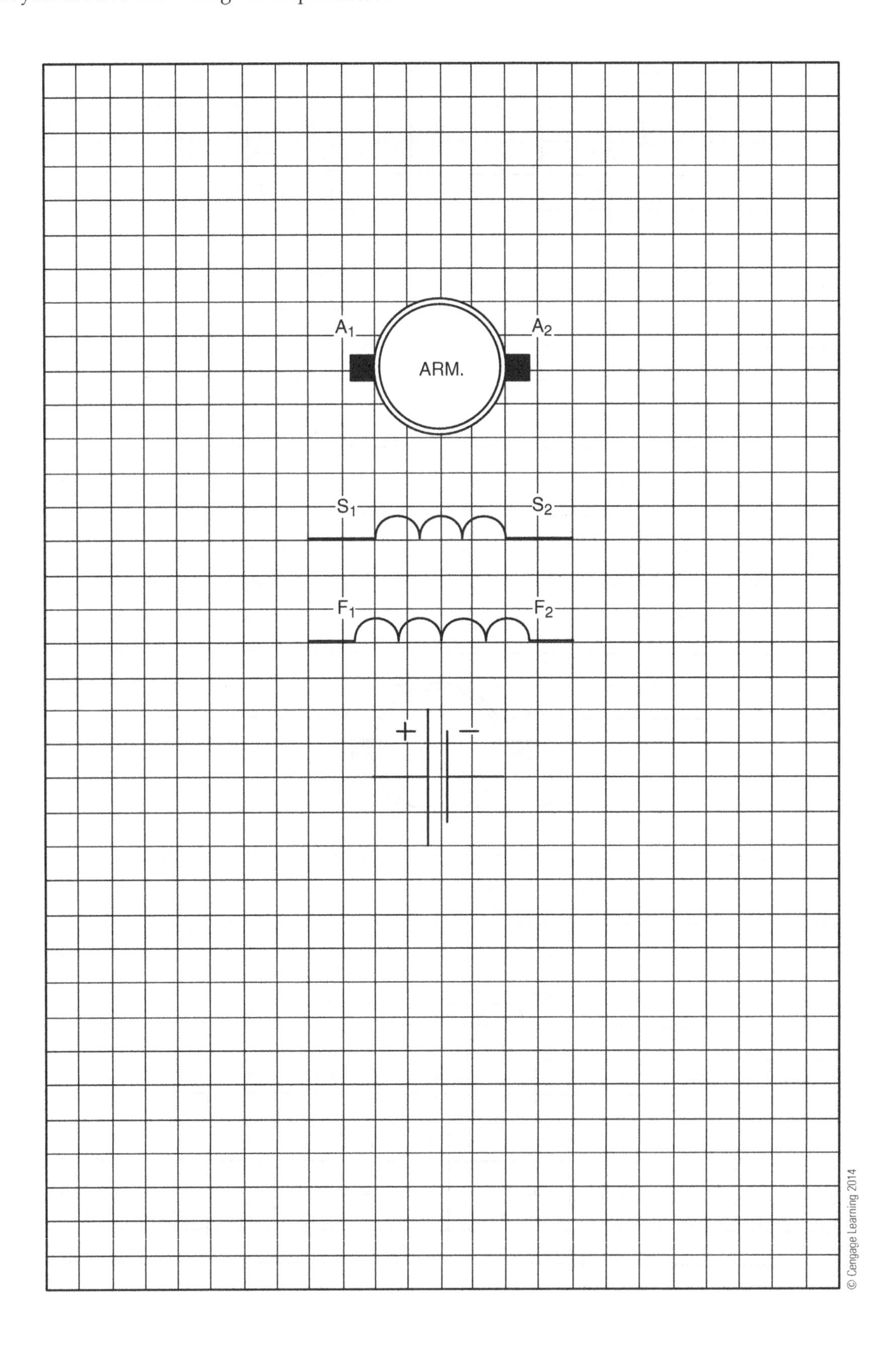

Name: ______________________________ Date: ____________________

THREE-PHASE MOTORS

1. Given the schematic diagram, complete the connections necessary for a squirrel-cage induction motor.

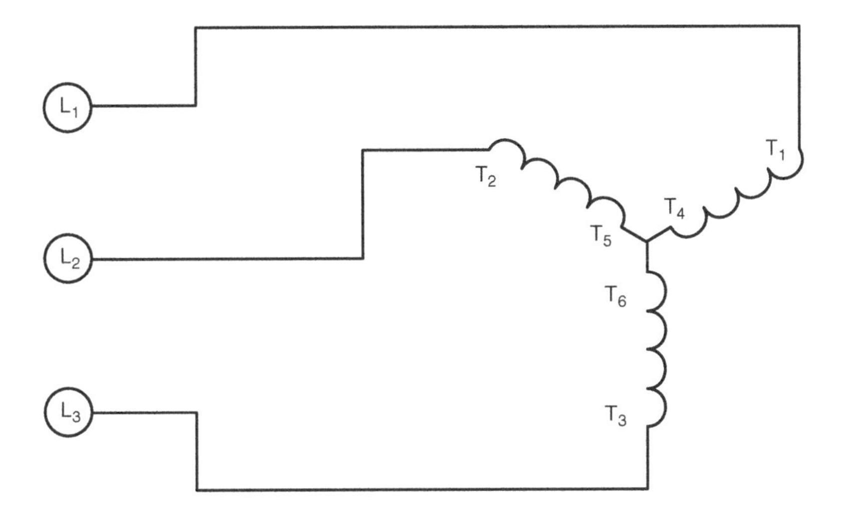

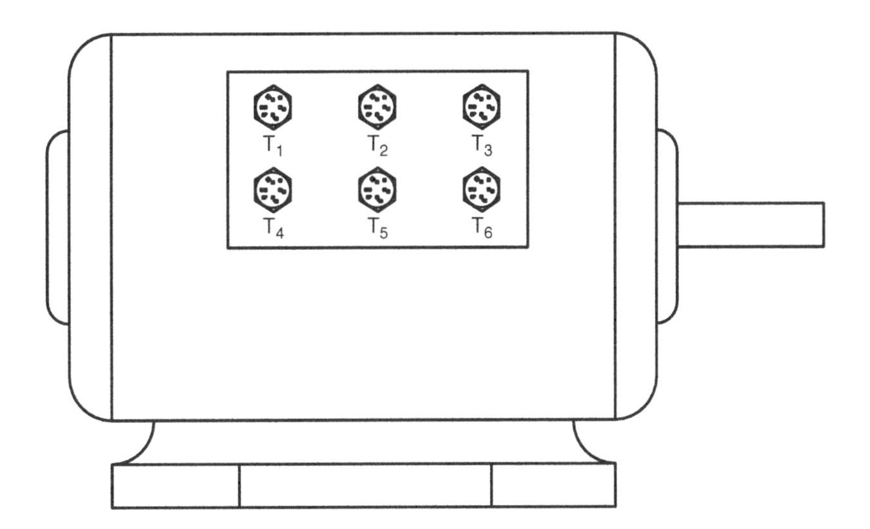

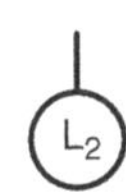

2. Show the connections necessary to reverse the direction of rotation of the squirrel-cage induction motor that you showed the wiring for in question 1.

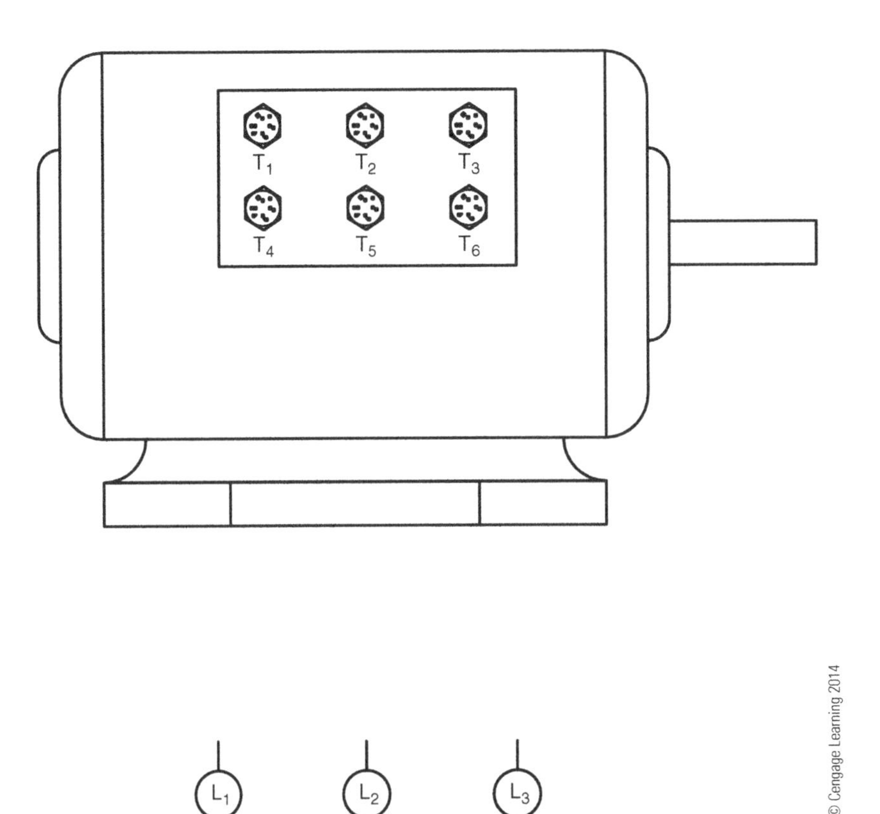

3. Given the schematic diagram, complete the connections necessary for a wound-rotor induction motor.

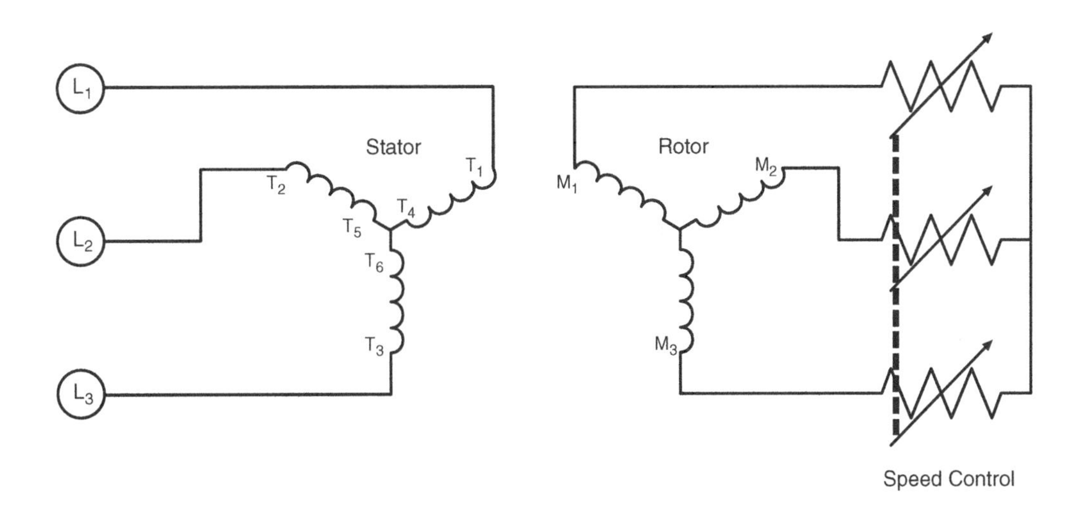

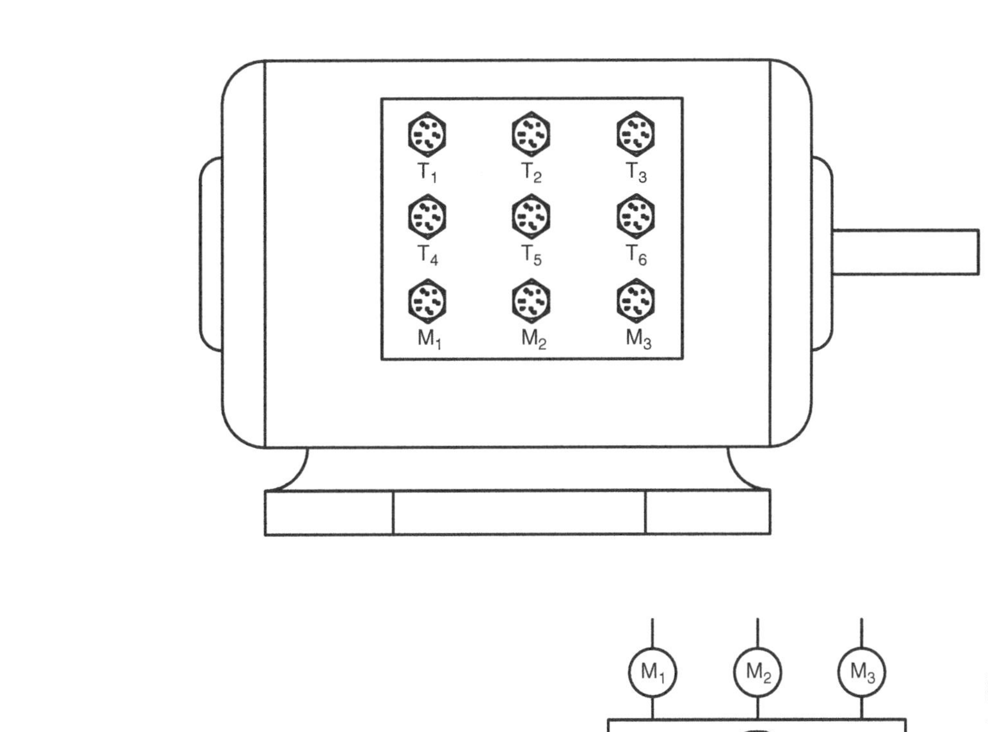

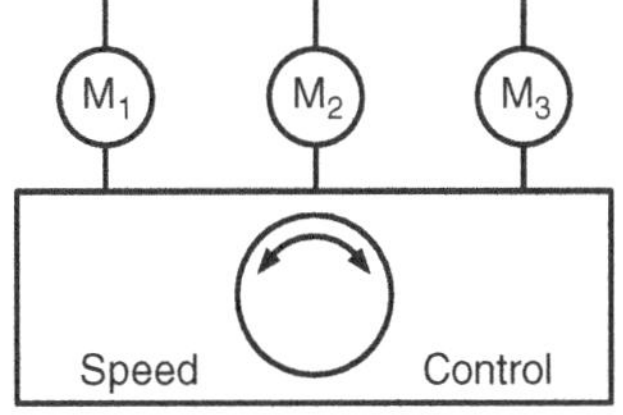

4. Show the connections necessary to reverse the direction of rotation of the wound-rotor induction motor that you showed the wiring for in question 3.

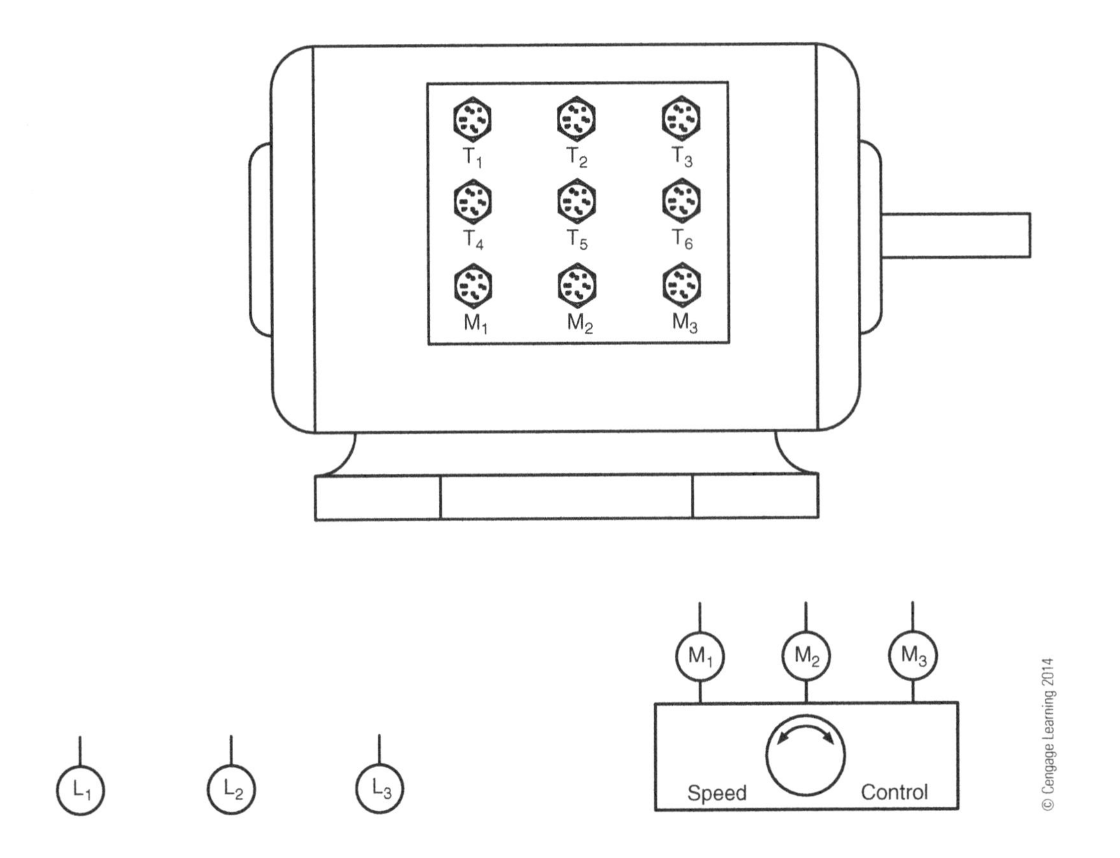

5. Given the schematic diagram, complete the connections necessary for a synchronous motor.

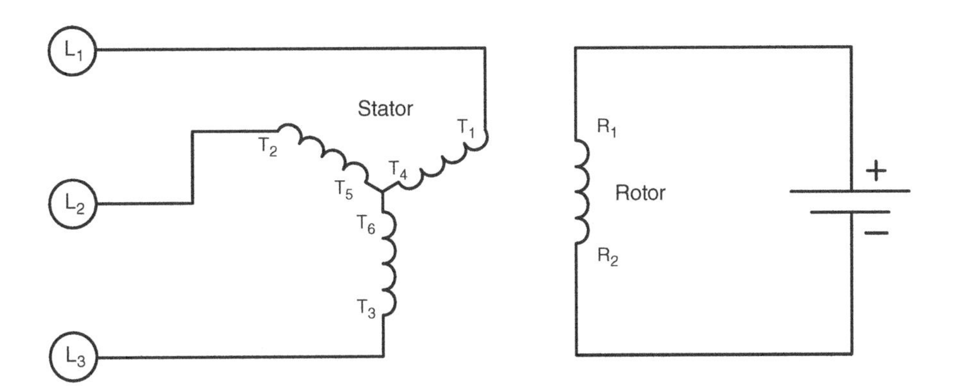

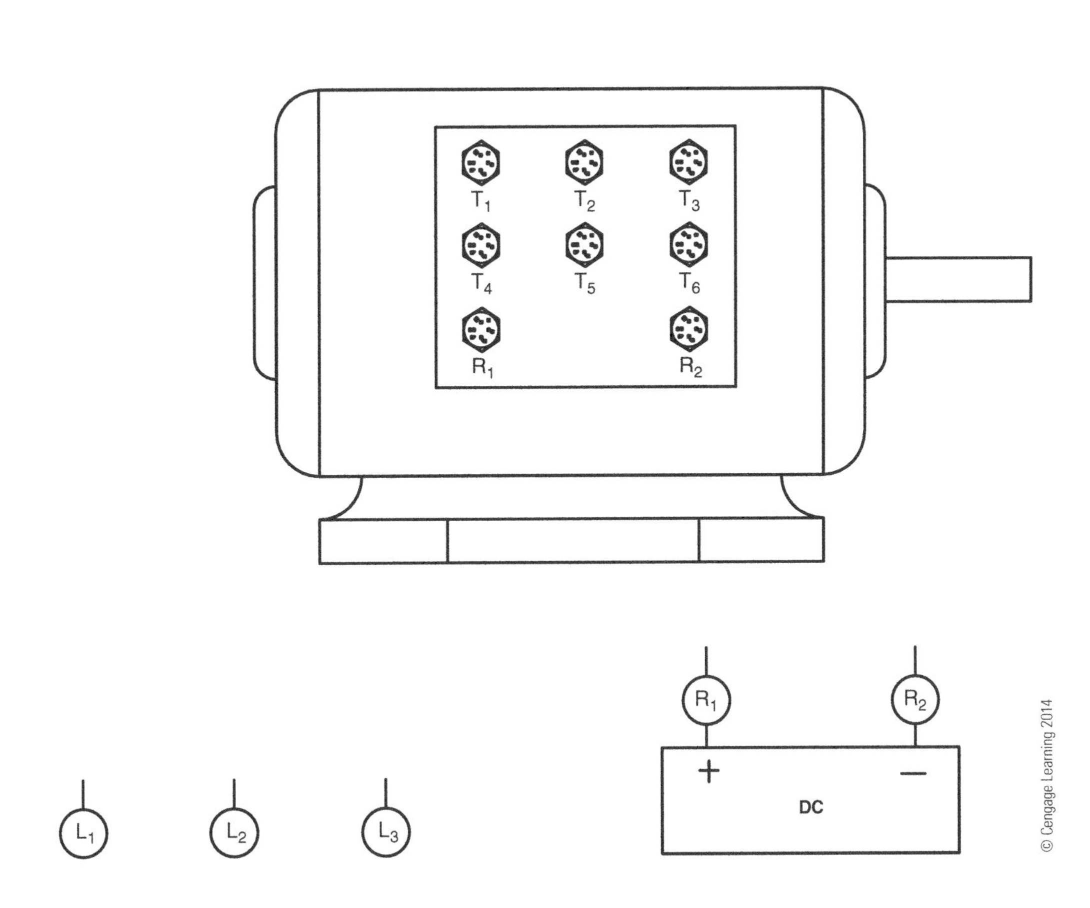

6. Show the connections necessary to reverse the direction of rotation of the synchronous motor that you showed the wiring for in question 5.

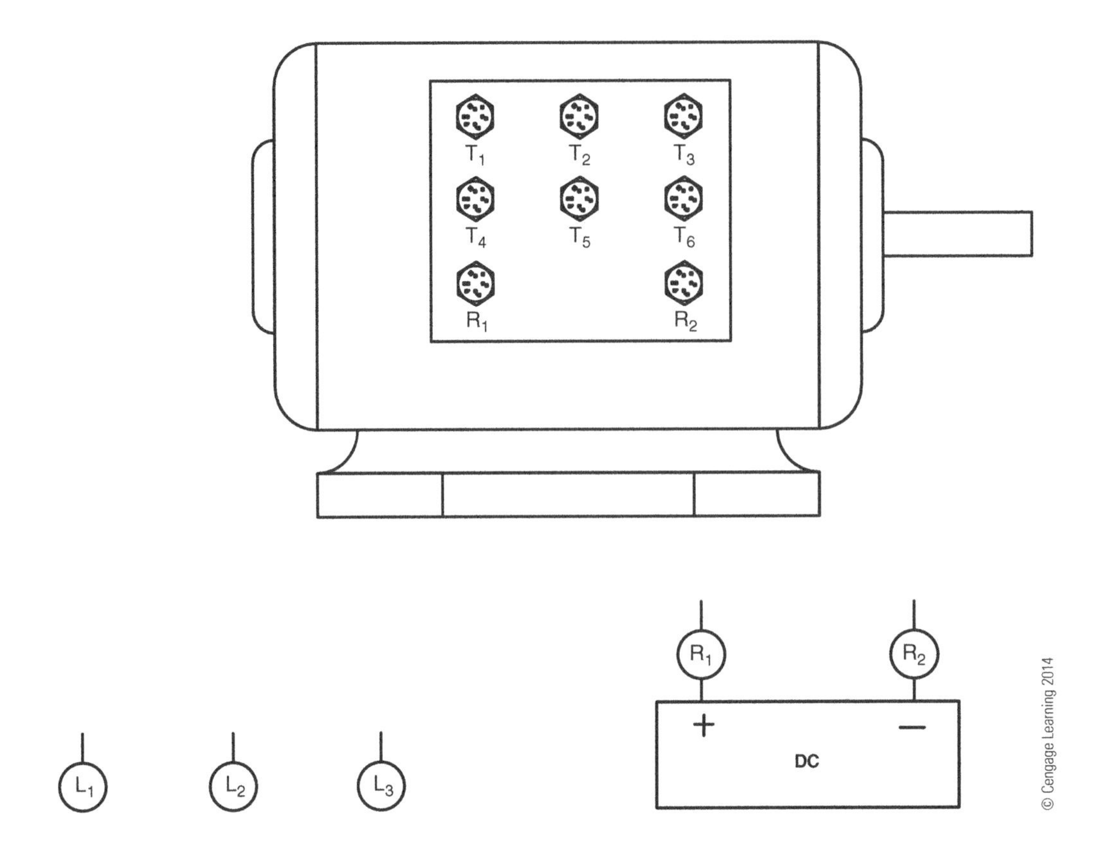

Name: ______________________ Date: ______________

SINGLE-PHASE MOTORS

1. Given the schematic diagram, complete the connections necessary for a shaded-pole motor.

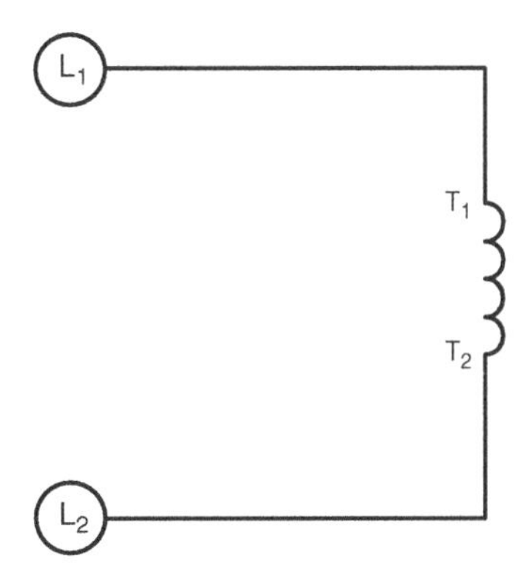

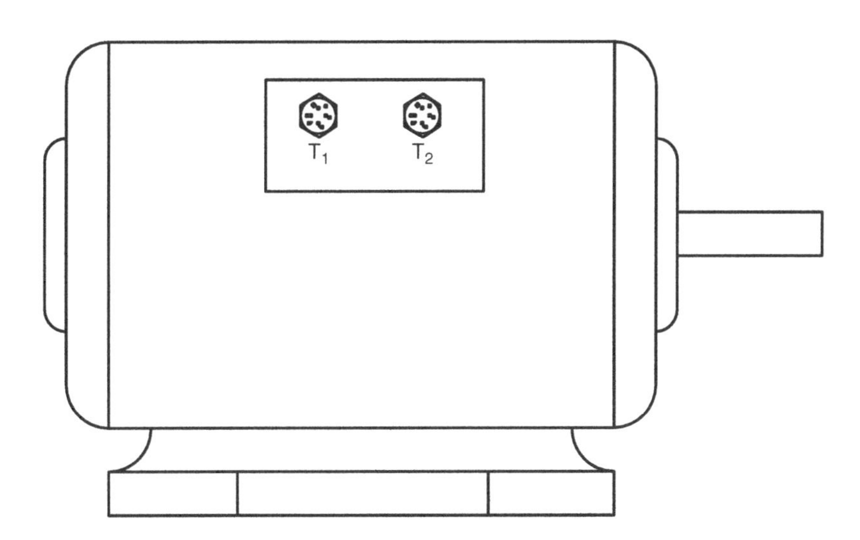

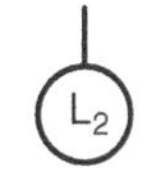

2. Given the schematic diagram, complete the connections necessary for a split-phase motor.

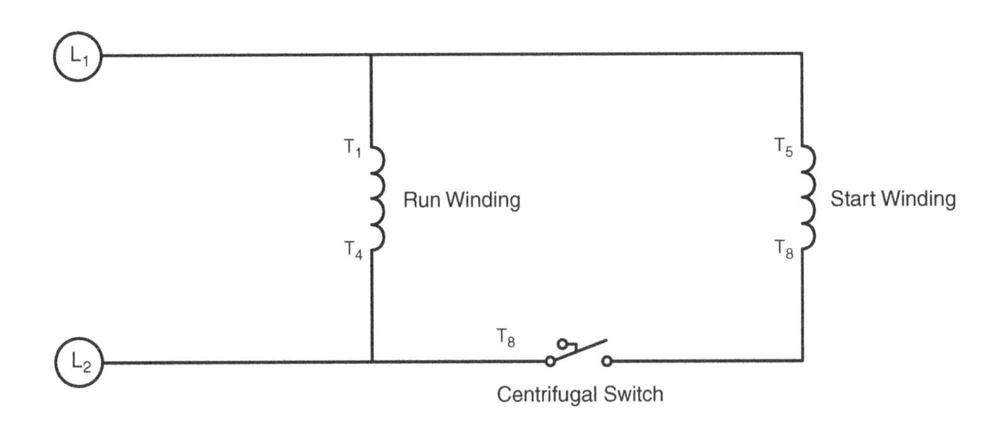

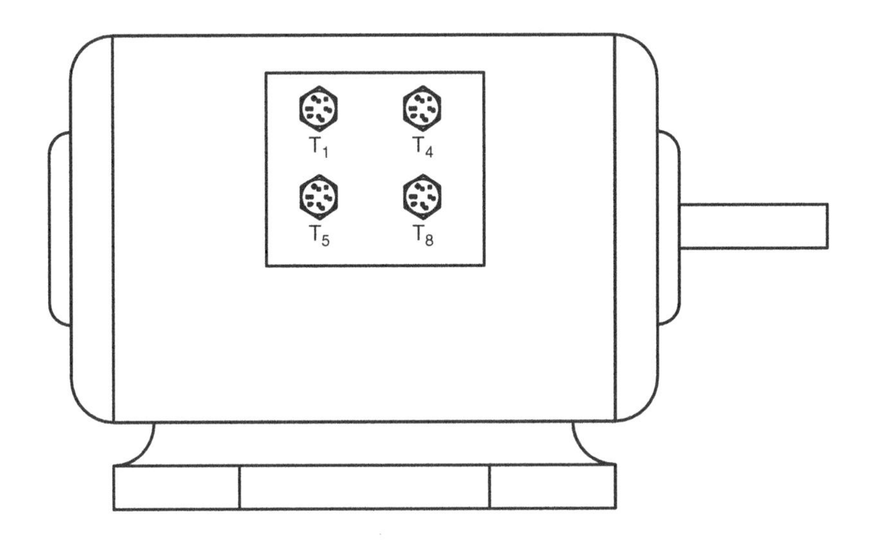

3. Show the connections necessary to reverse the direction of rotation of the split-phase motor that you showed the wiring for in question 2.

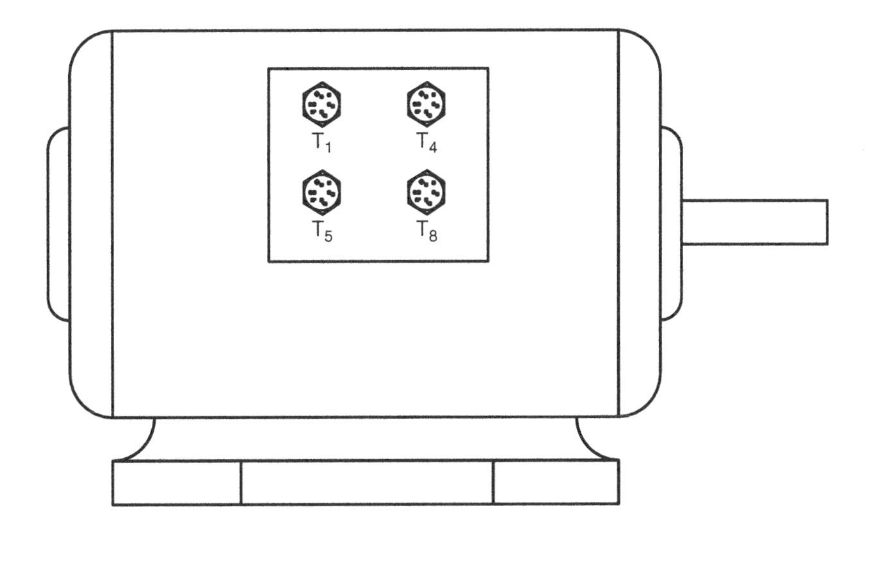

4. Given the schematic diagram, complete the connections necessary for a capacitor-start motor.

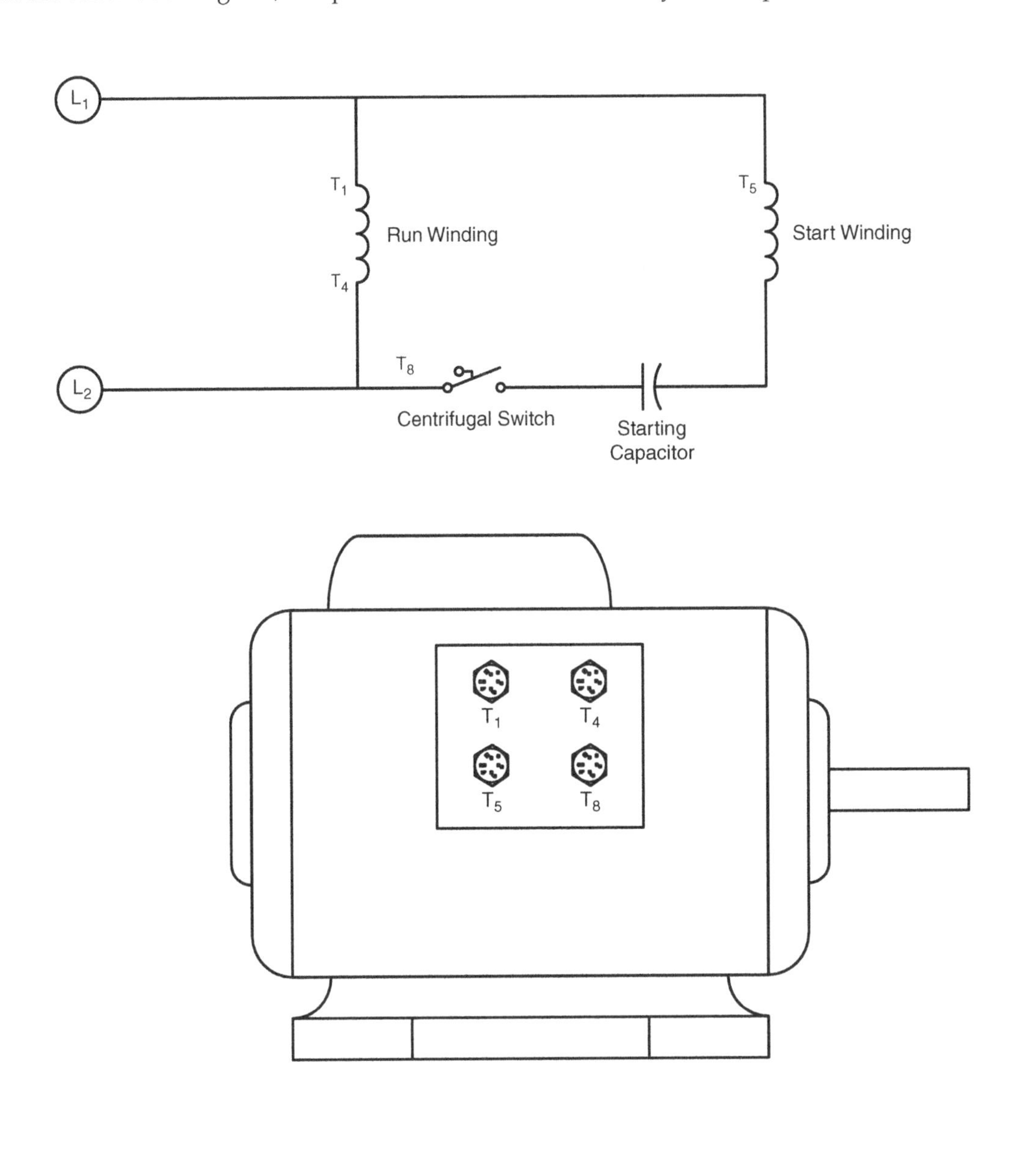

5. Show the connections necessary to reverse the direction of rotation of the capacitor-start motor that you showed the wiring for in question 4.

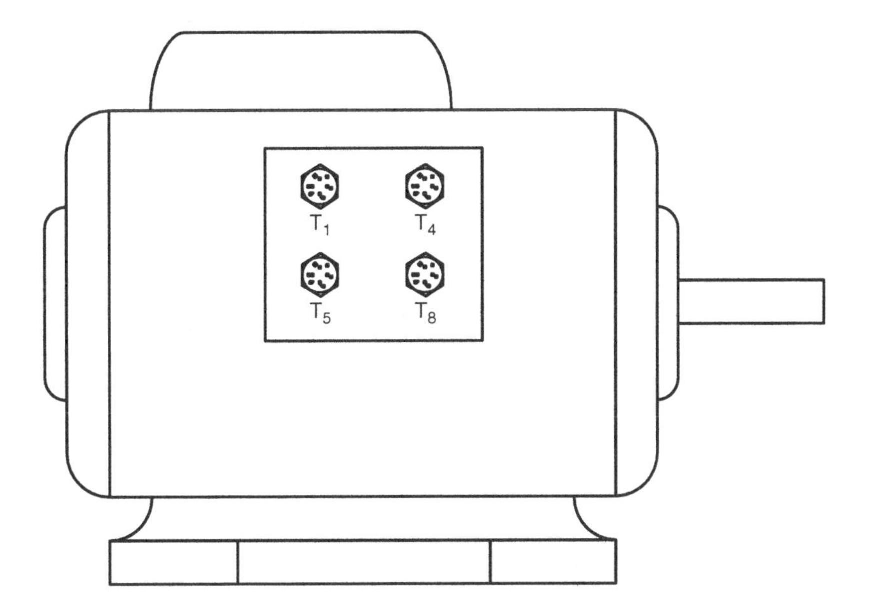

6. Given the schematic diagram, complete the connections necessary for a capacitor-run motor.

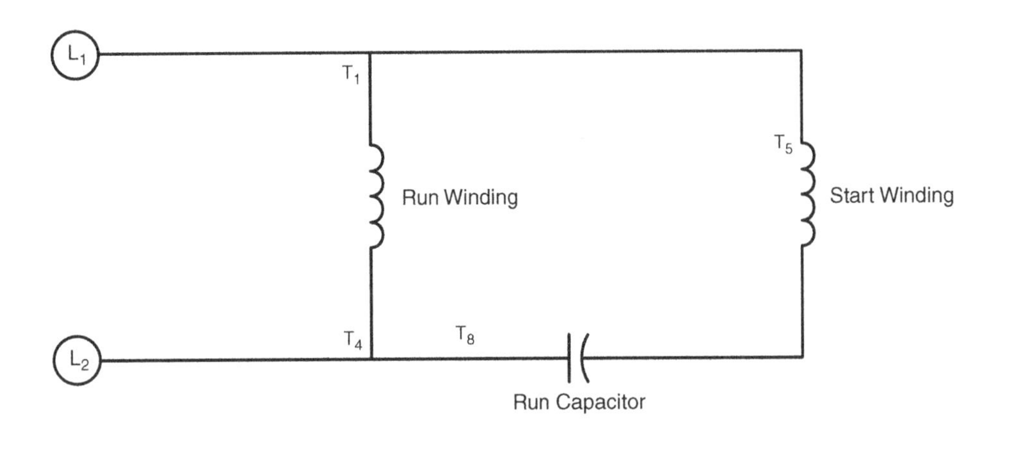

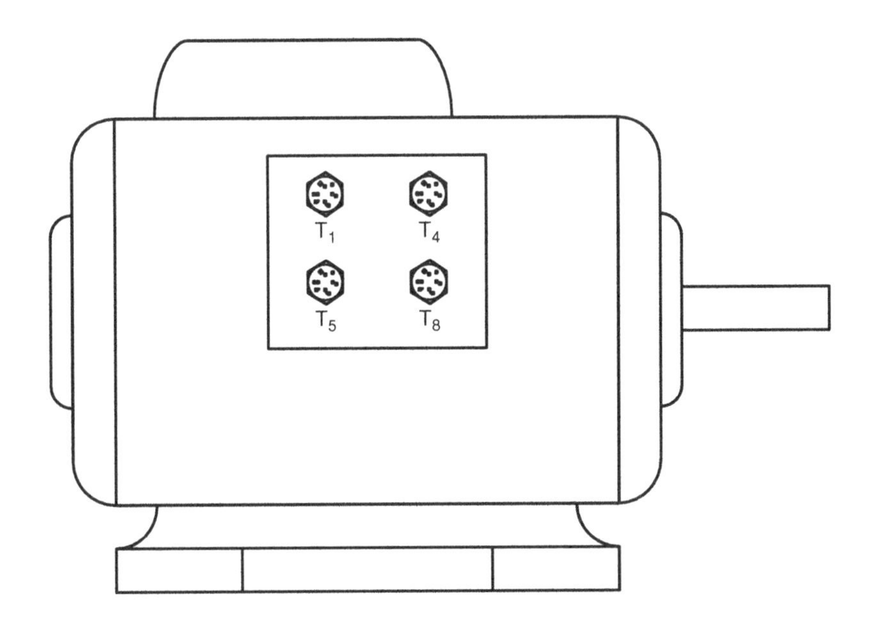

7. Show the connections necessary to reverse the direction of rotation of the capacitor-run motor that you showed the wiring for in question 6.

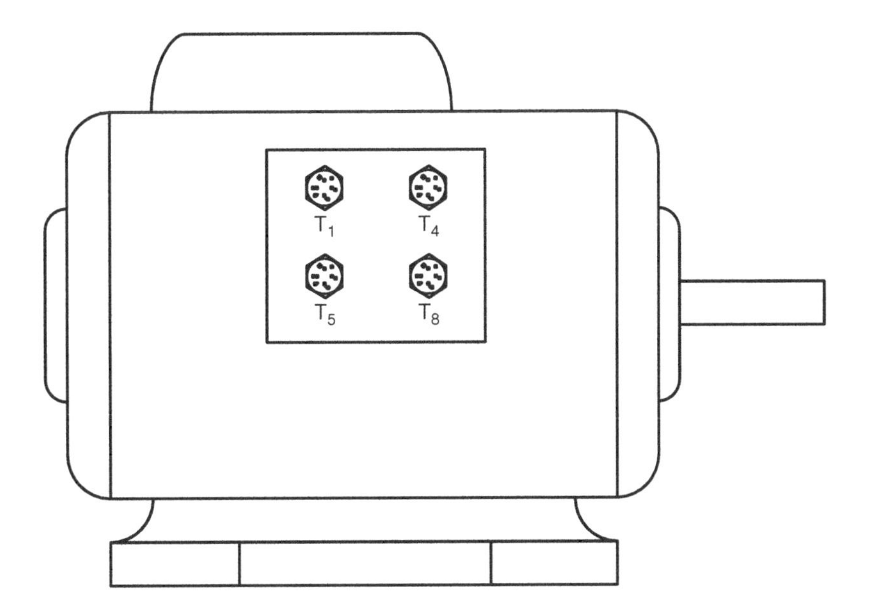

8. Given the schematic diagram, complete the connections necessary for a capacitor-start/capacitor-run motor.

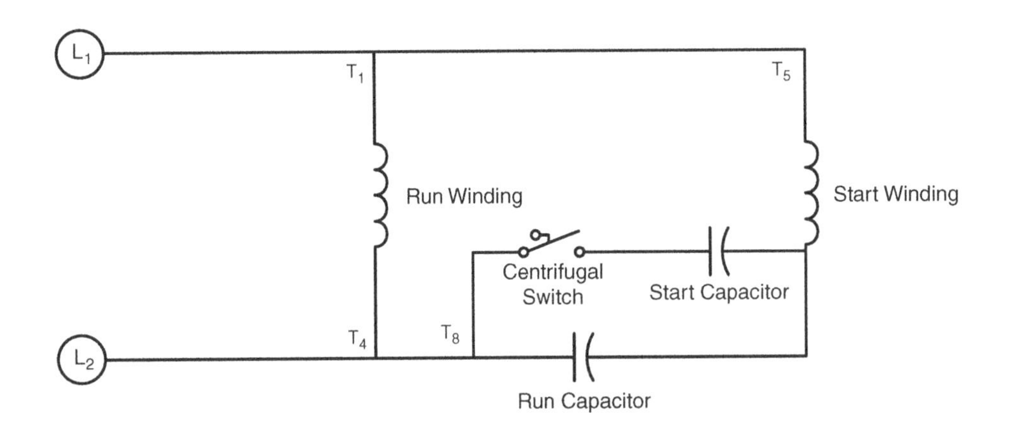

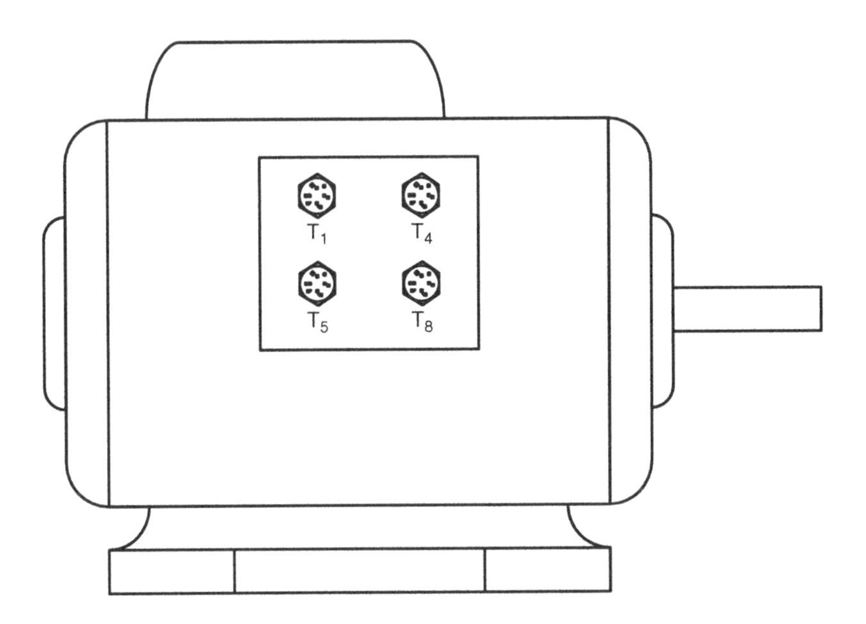

9. Show the connections necessary to reverse the direction of rotation of the capacitor-start/capacitor-run motor that you showed the wiring for in question 8.

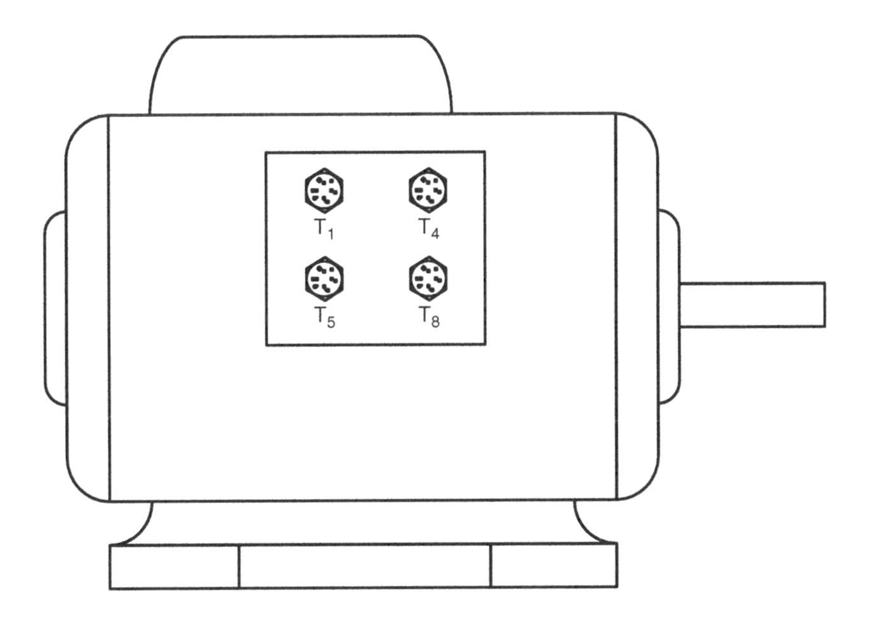

Name: ______________________________ Date: ____________________

CONTROL DEVICES: MANUAL

1. Match the object to its schematic symbol.

E-stop pushbutton _____

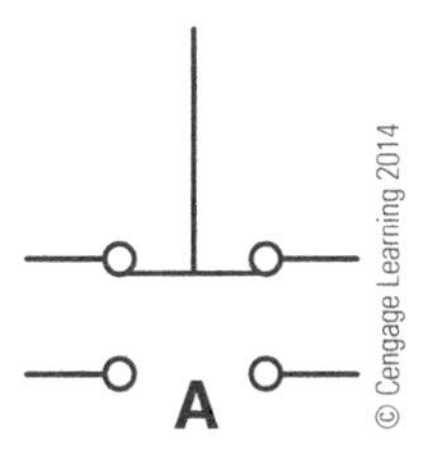

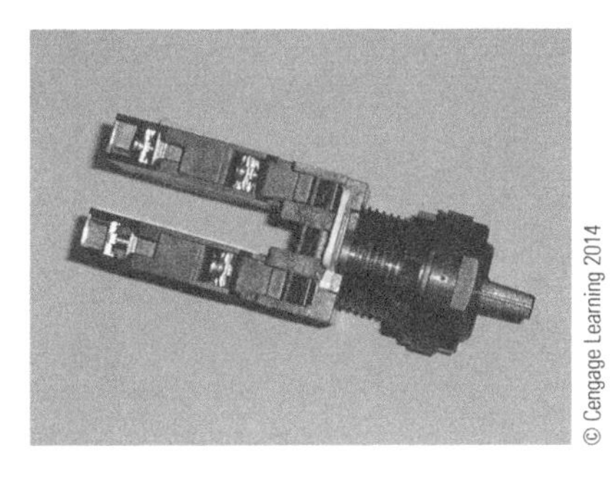

Selector switch _____

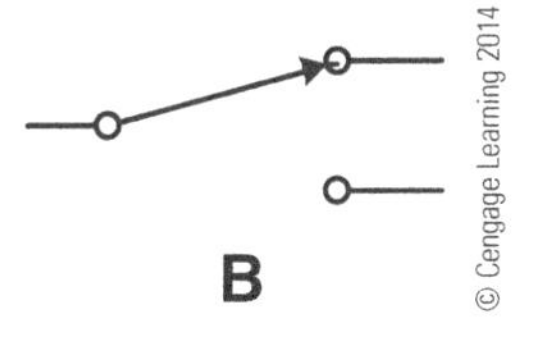

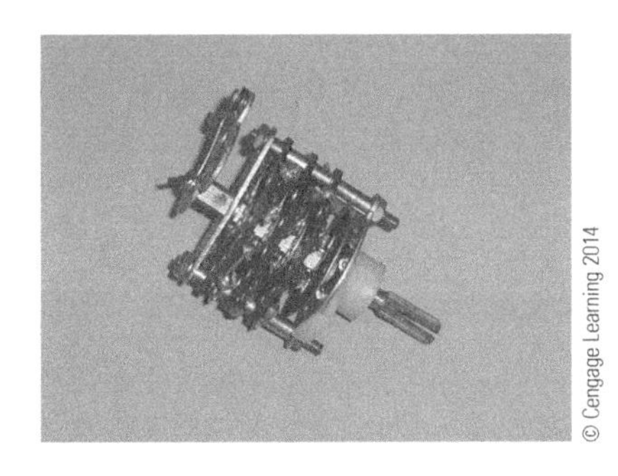

Multideck selector switch _____

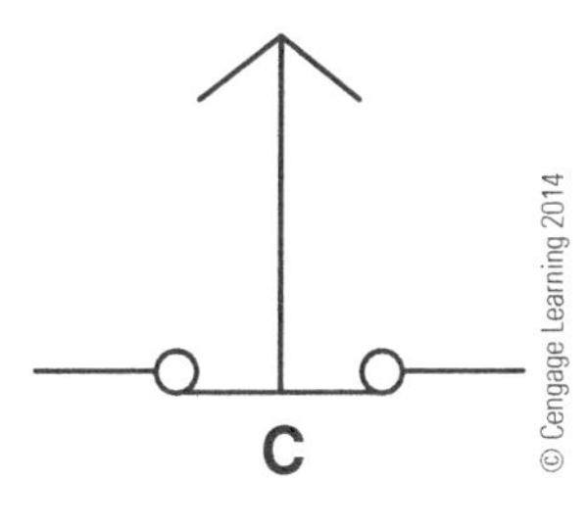

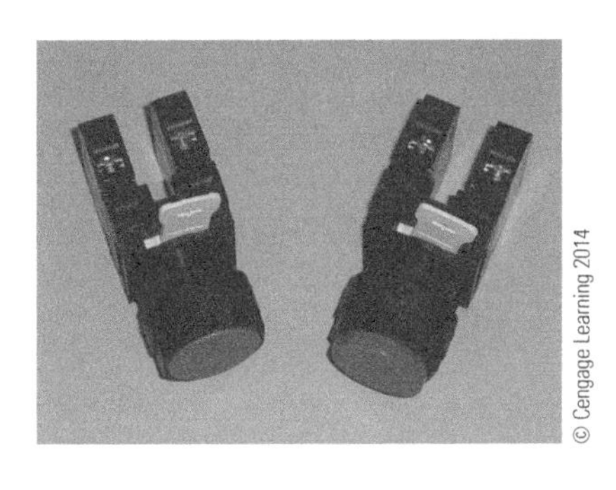

Momentary pushbuttons _____

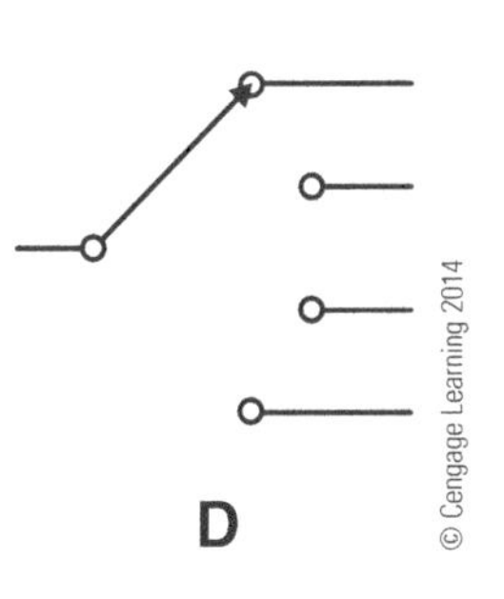

Name: ______________________________ Date: ______________

CONTROL DEVICES: AUTOMATIC

1. Match the object to its schematic symbol.

Limit switches ______

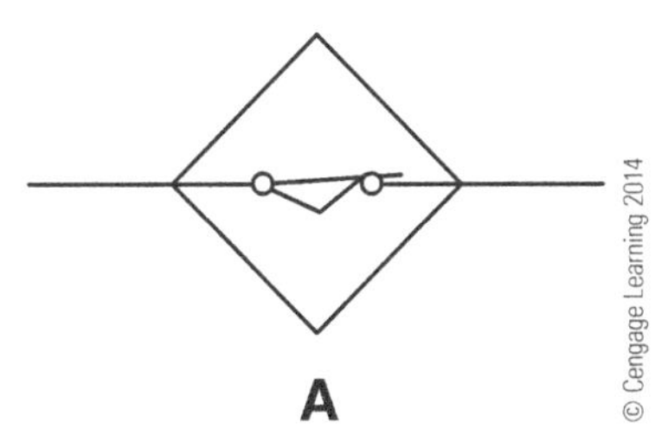

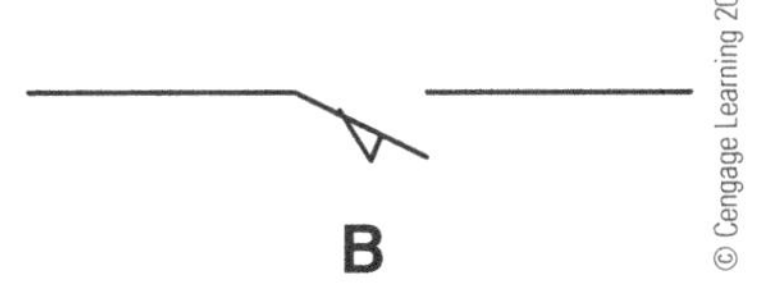

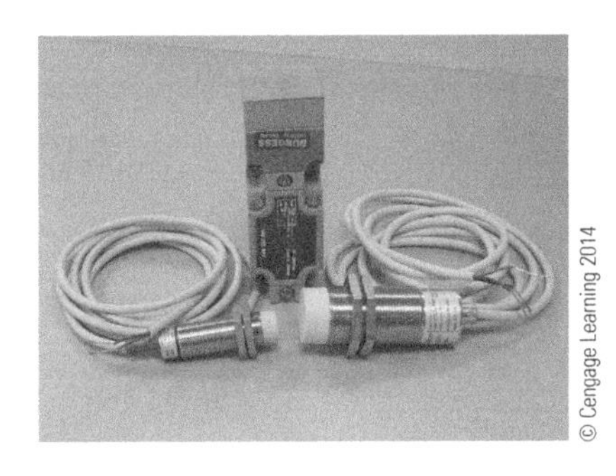

Proximity switches ______

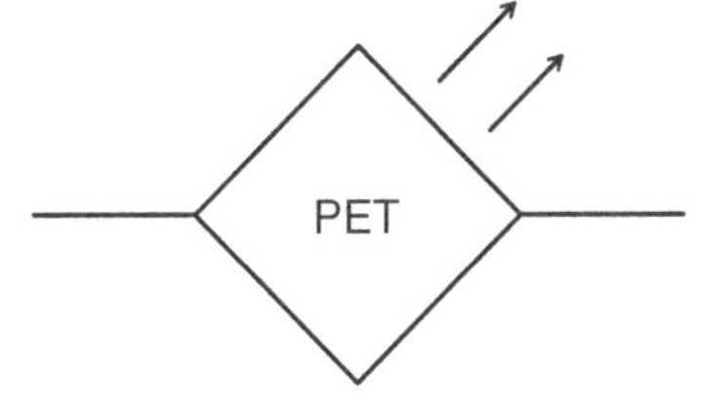

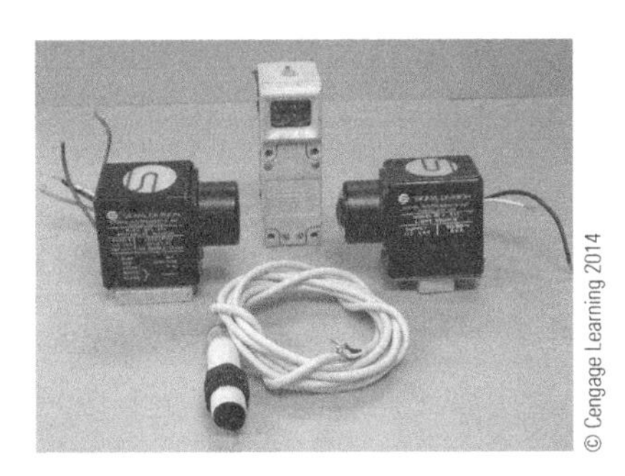

Photoelectric switches ______

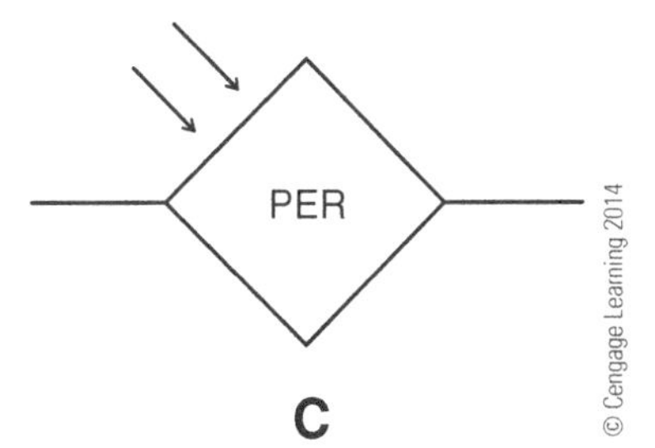

Name: ______________________________ Date: ______________

CONTROLLED DEVICES

1. Match the object to its schematic symbol.

Relays ______

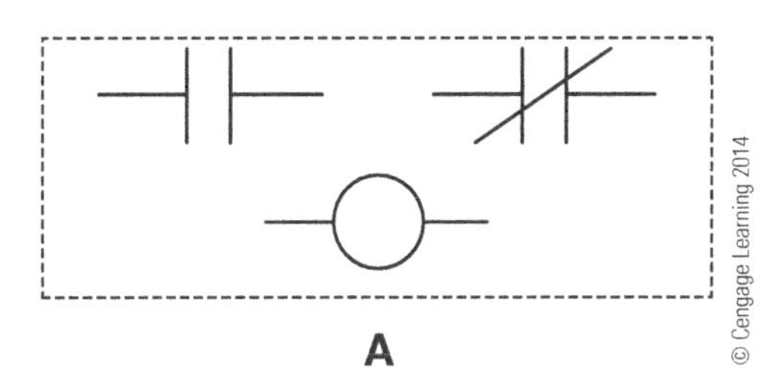

A

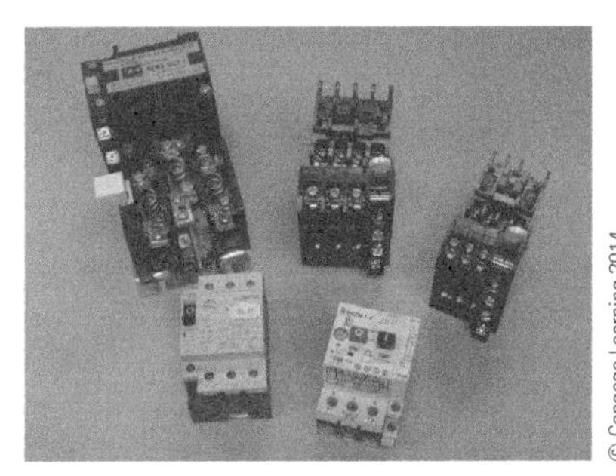

Motor starters ______

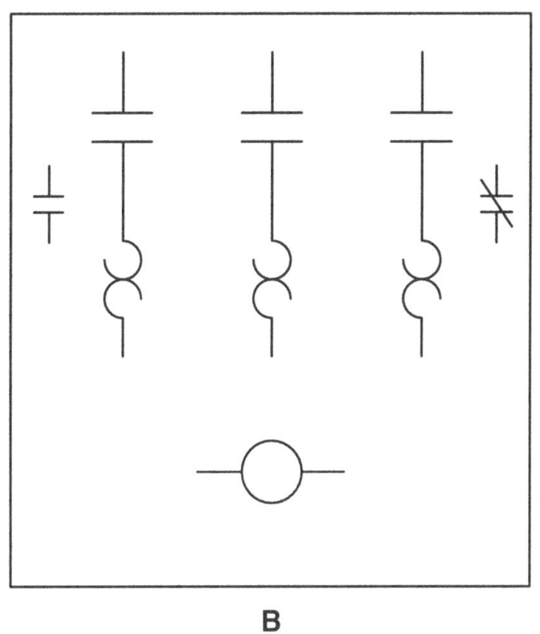

B

Name: ______________________________ Date: ______________

TWO-WIRE CONTROLS

1. Draw a ladder diagram for a circuit that uses a thermostat to control the operation of a fan.

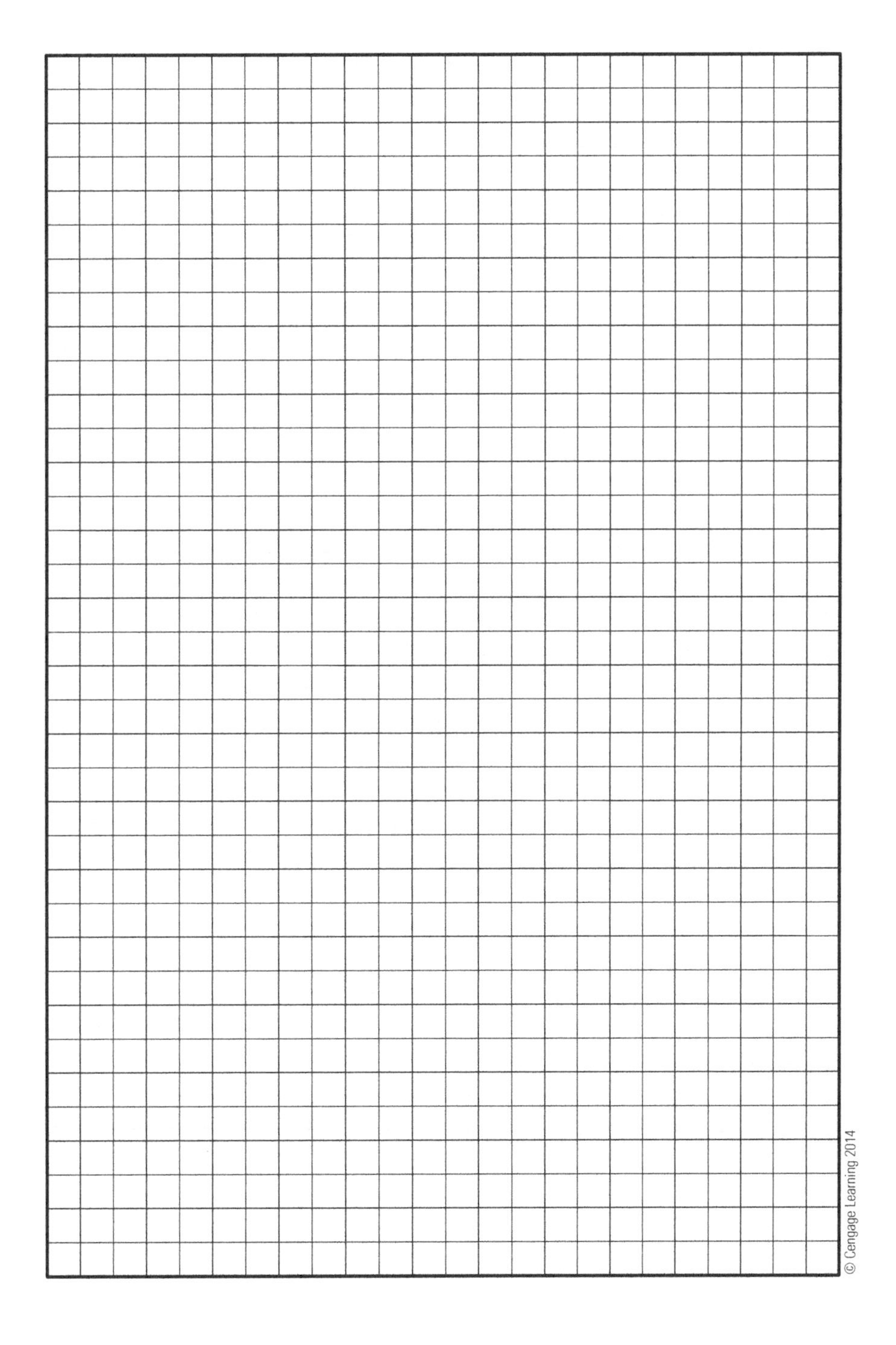

2. Modify the circuit in question 1 to include a red pilot light, which indicates that the fan is not running, and a green pilot light, which indicates that the fan is running.

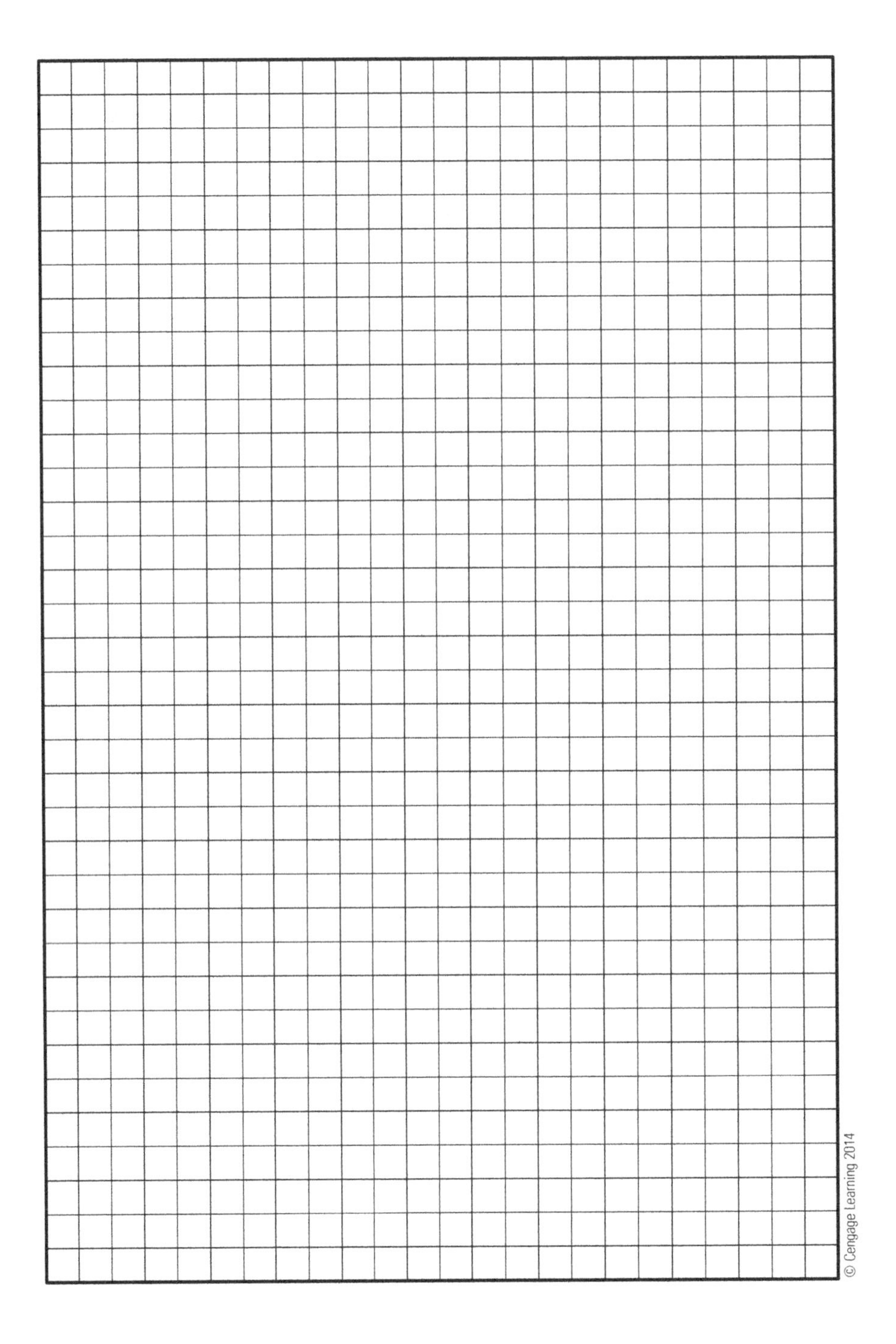

3. Draw a ladder diagram for a circuit that uses a toggle switch to control the operation of a blower.

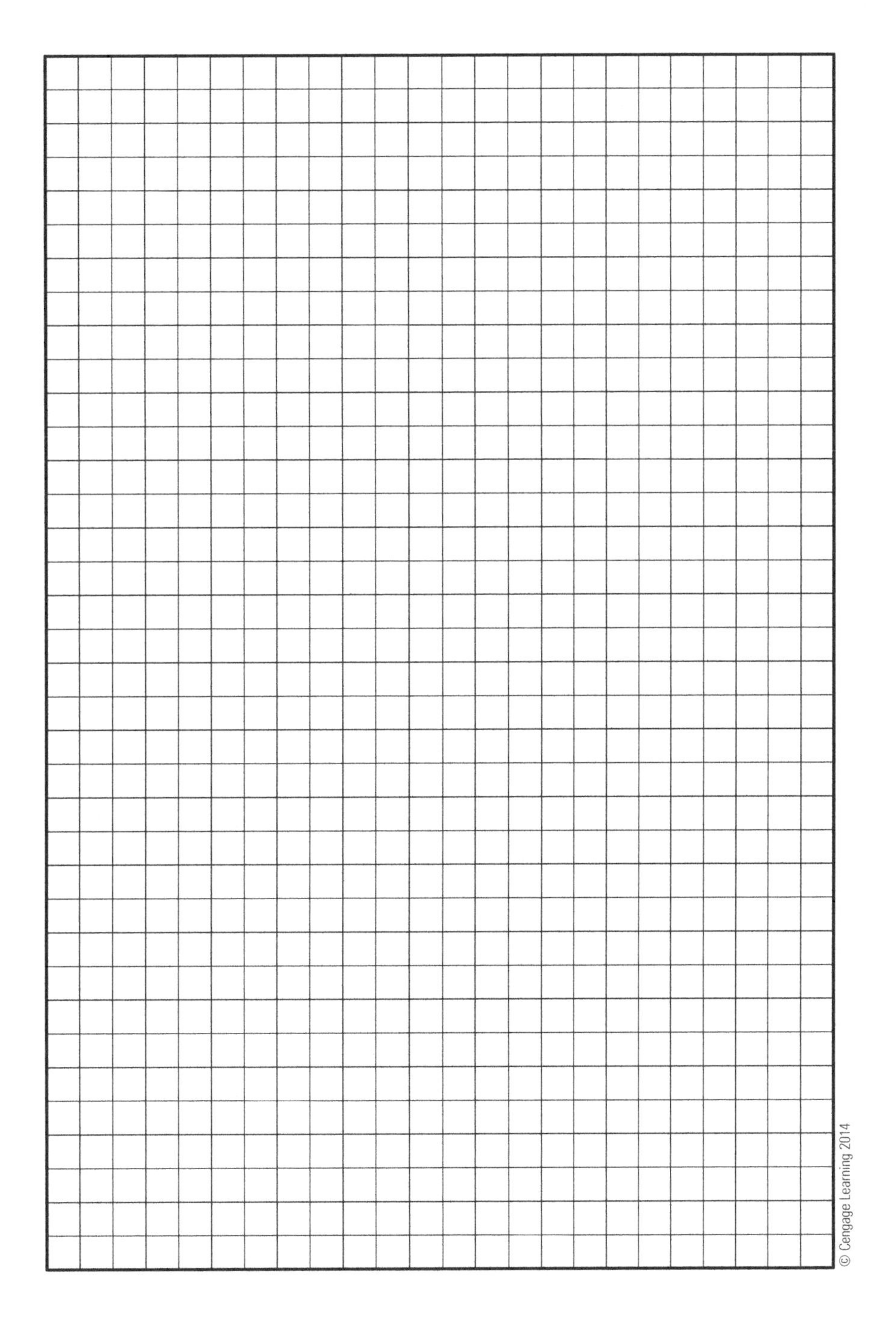

4. Modify the circuit in question 3 to include a flow switch to detect airflow through a filter. Include a red pilot light, which indicates a dirty filter (reduced airflow), and a green pilot light, which indicates normal airflow.

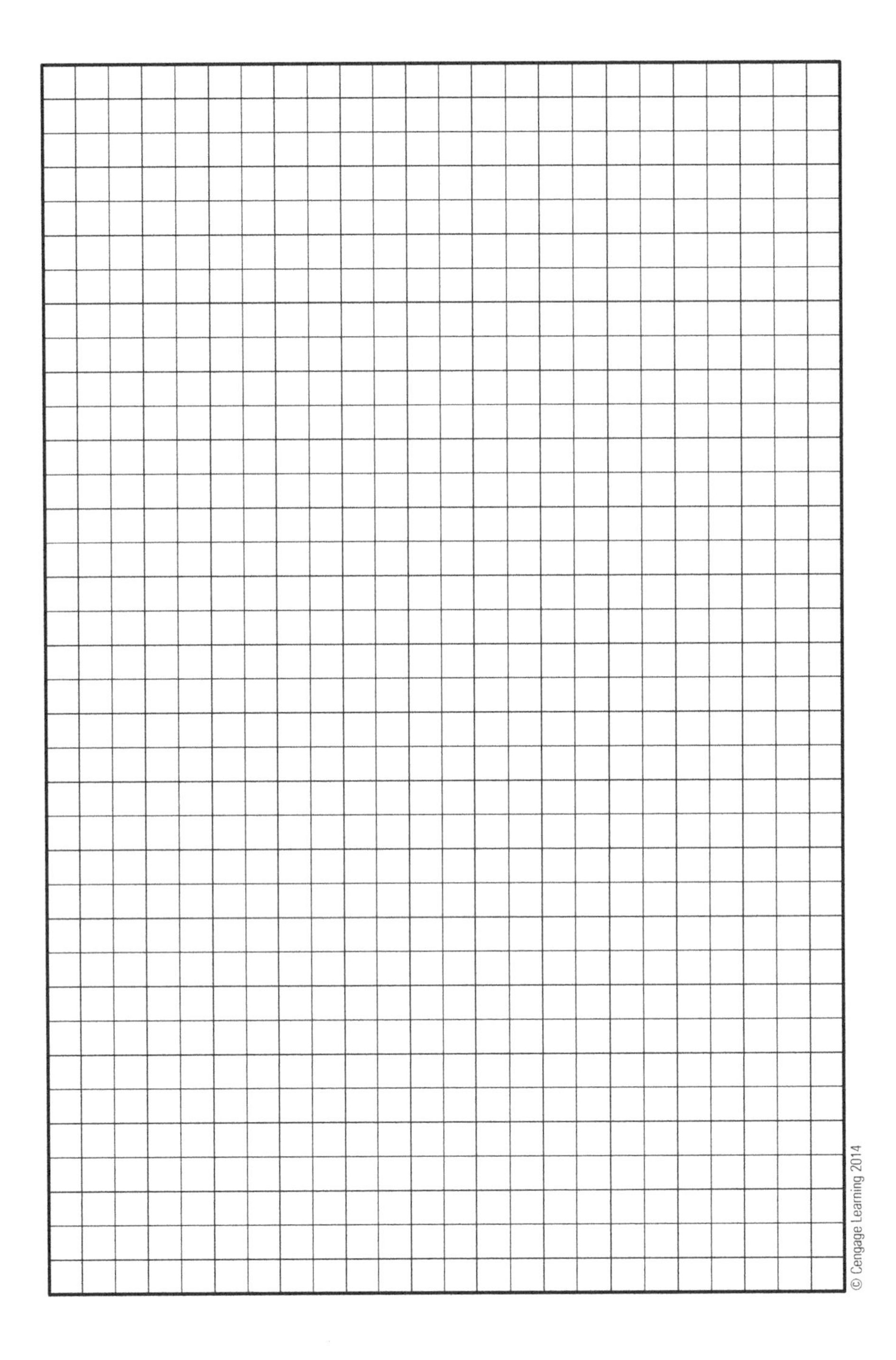

Name: ______________________ Date: ______________

THREE-WIRE CONTROLS

1. Draw a ladder diagram for a circuit that performs as follows:
   a. Pressing and releasing a momentary pushbutton causes a cylinder to advance, even after the pushbutton is released.
   b. When the cylinder strikes a limit switch, the cylinder's motion is stopped.

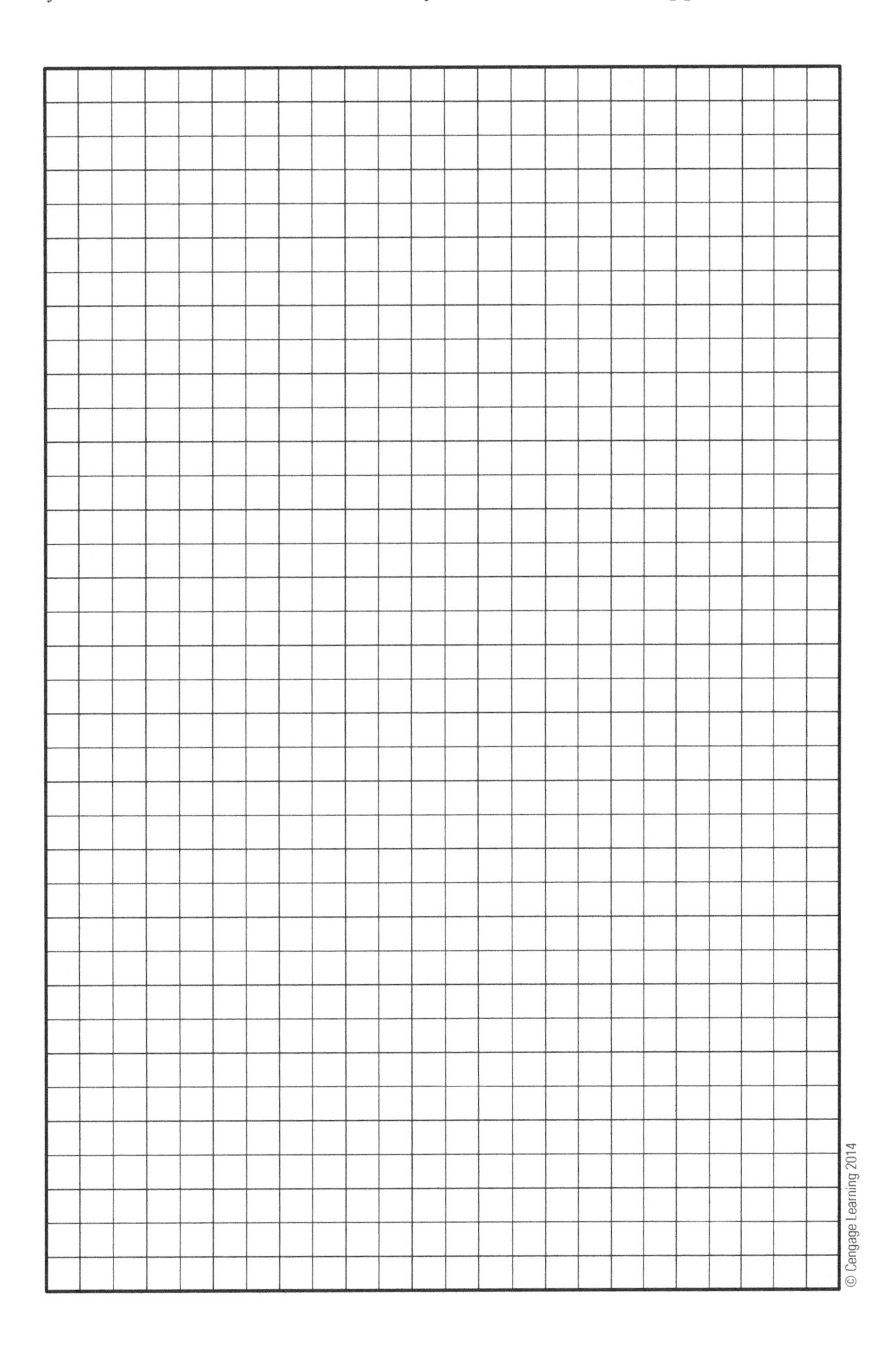

2. Modify the circuit in question 1 to include an alarm that sounds for 5 seconds before the cylinder advances.

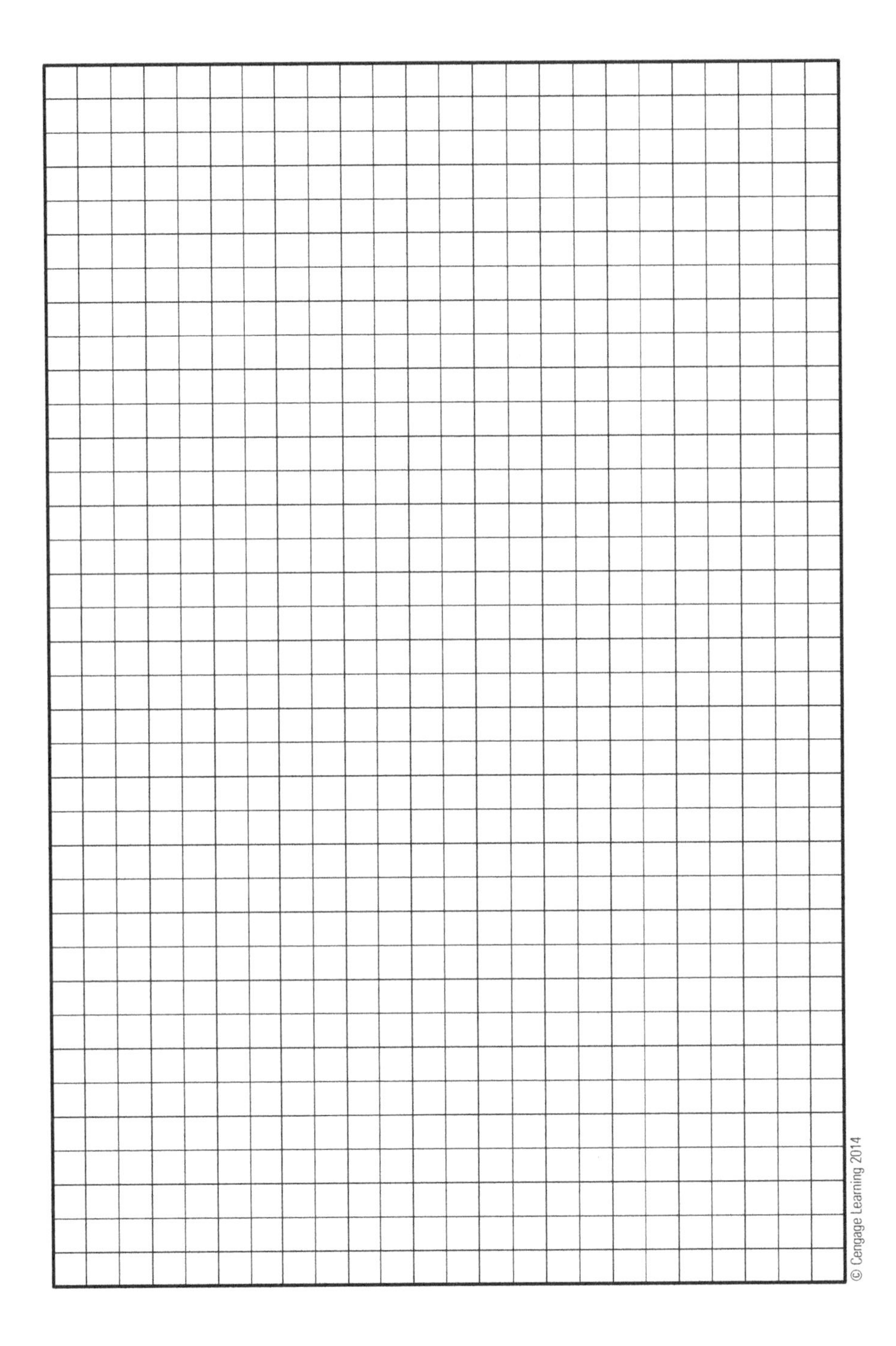

Name: ______________________ Date: ______________

MULTIPLE START/STOP STATIONS

1. Draw a ladder diagram for a circuit that performs as follows:
   a. A conveyor contains three start/stop stations. A start/stop station is mounted at each end of the conveyor and in the middle.
   b. The conveyor must be able to be started or stopped from any one of the three locations.

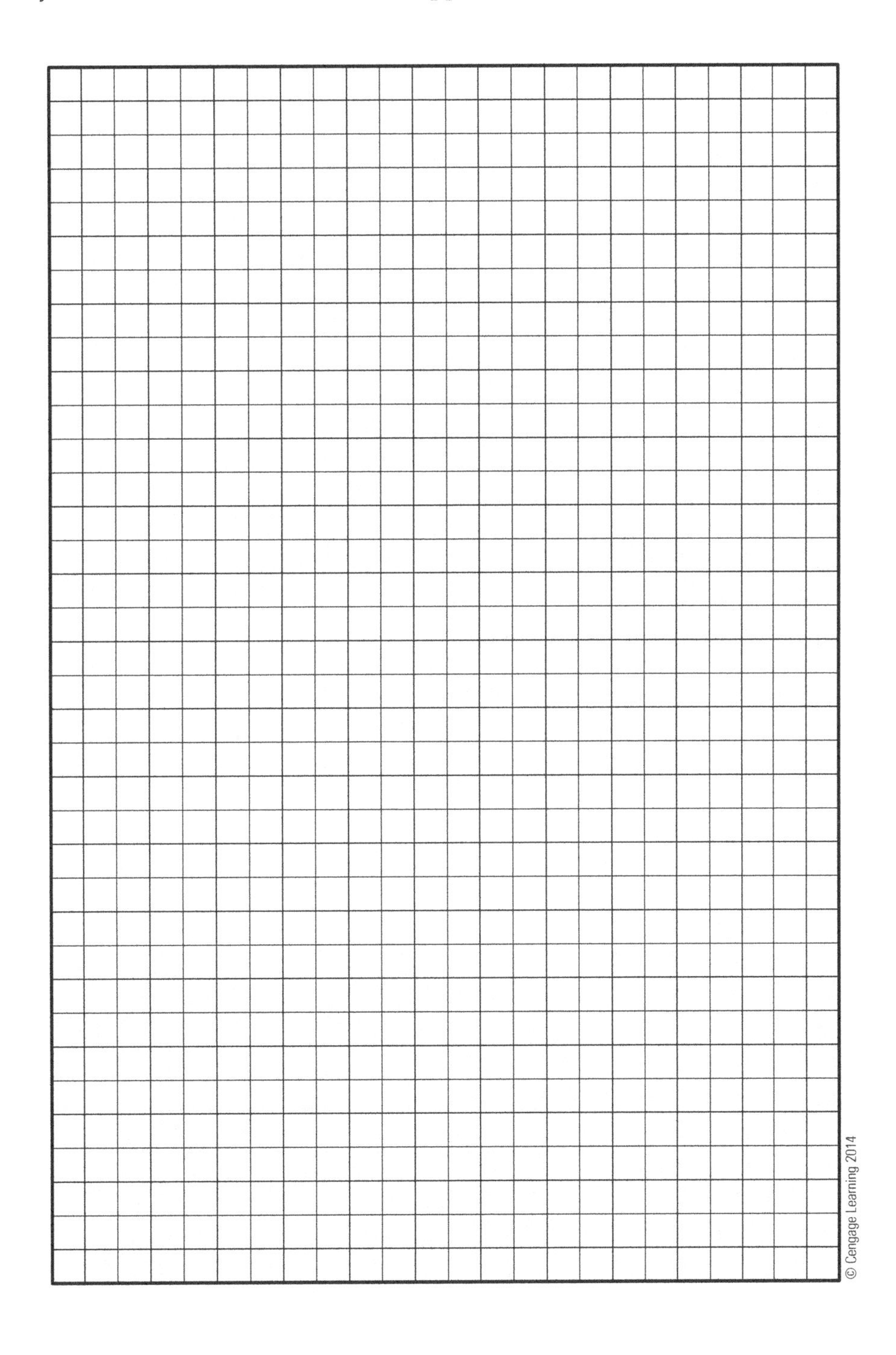

2. A drill press is started by an operator. To ensure the safety of the operator, the operator must simultaneously press one start pushbutton with the left hand and one pushbutton with the right hand. The pushbuttons are located on opposite sides of the drill table to ensure that the operator's hands are not in the drilling space when the machine is started. The drill press can be stopped by pressing a stop pushbutton located near each start pushbutton. Draw a ladder diagram for a circuit that performs this operation.

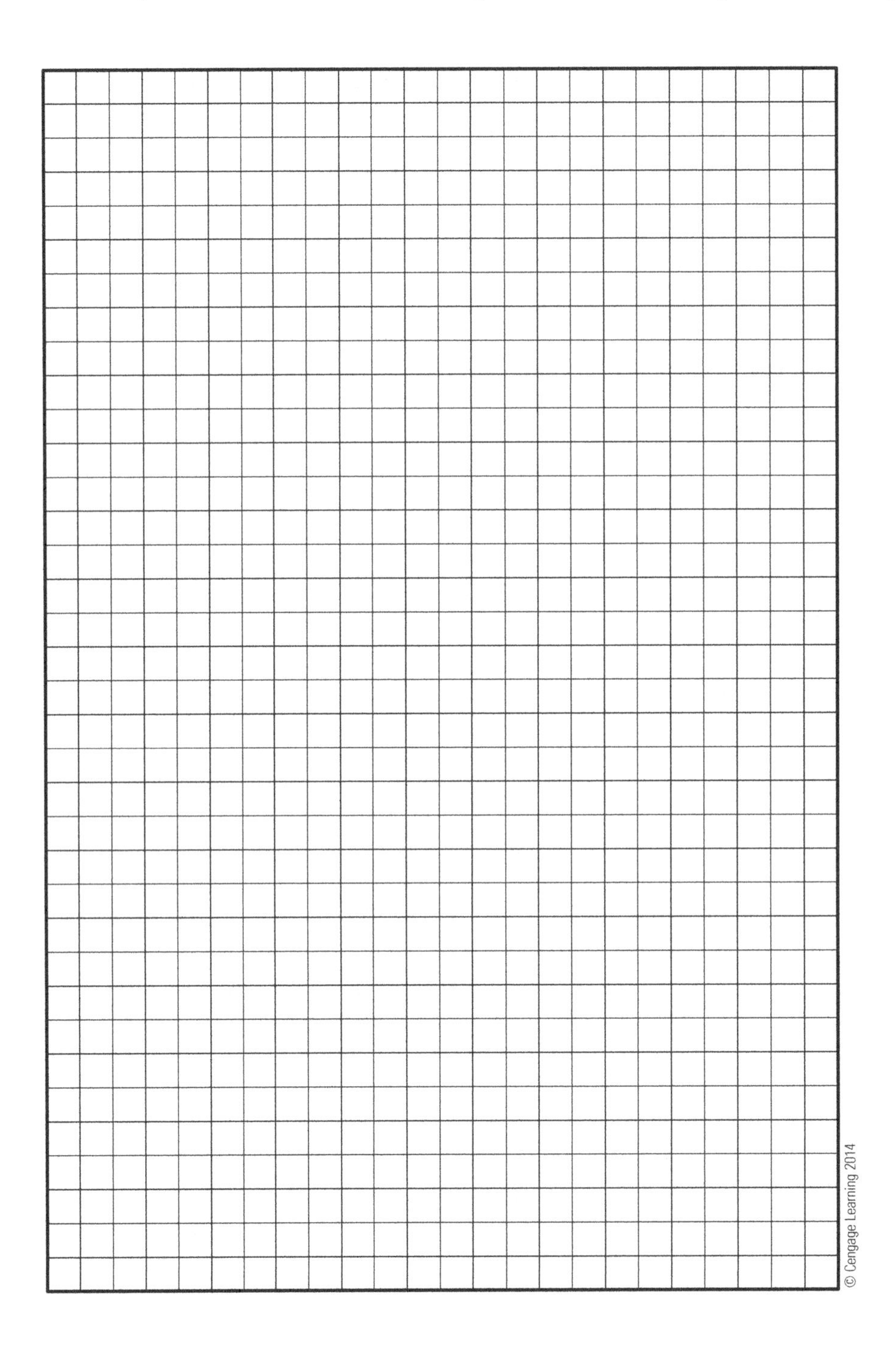

Name: ______________________________ Date: ______________

FORWARD/REVERSE CONTROLS

1. Draw a ladder diagram for a circuit that performs as follows:
   a. A momentary start pushbutton is used to cause a cylinder to extend.
   b. When the cylinder extends, it contacts a limit switch, causing the cylinder to retract.
   c. When the cylinder retracts, it contacts a limit switch, causing the cylinder to extend.
   d. The cylinder cycles back and forth until the stop pushbutton is pressed.

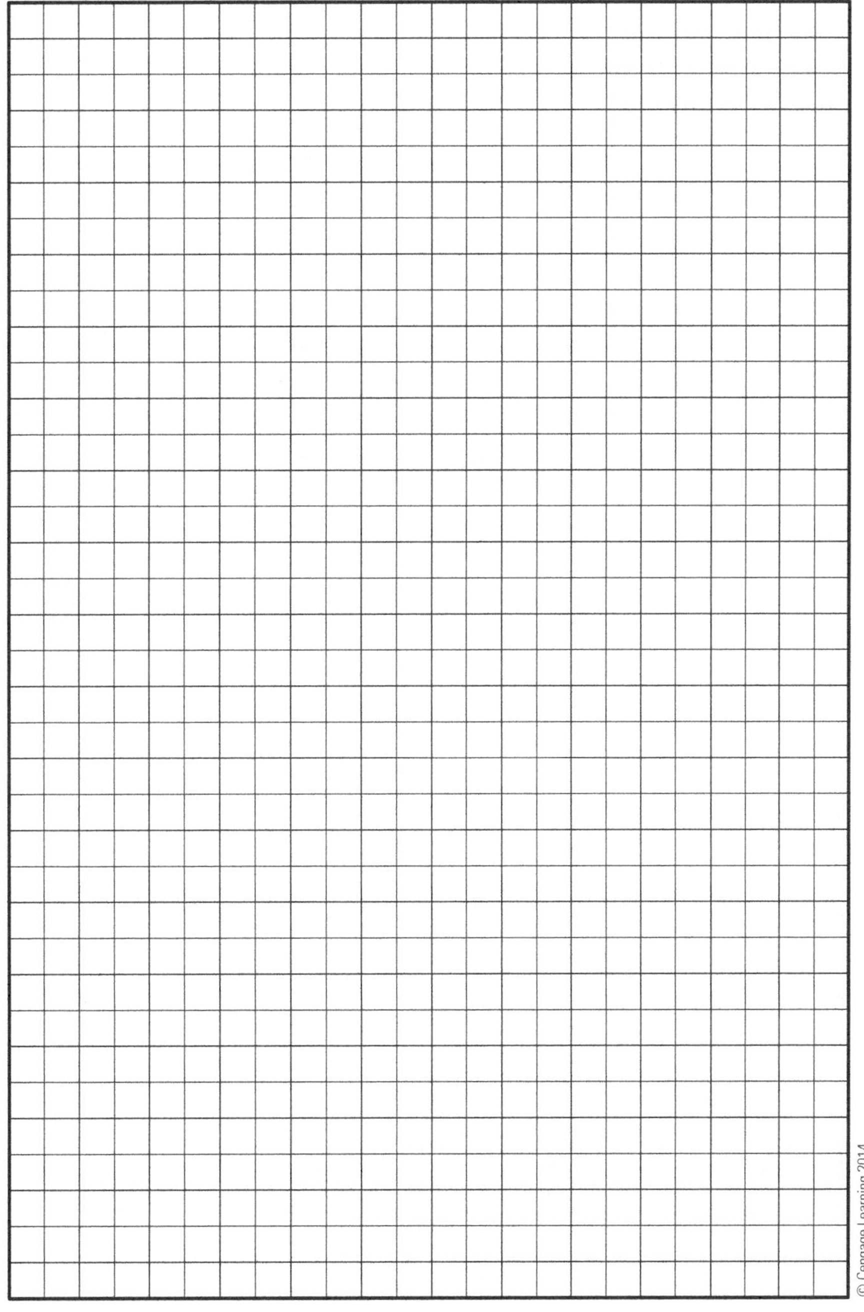

2. Draw a ladder diagram for a circuit that performs as follows:
   a. Pressing a momentary pushbutton causes a motor to run in the forward direction.
   b. After 10 seconds, the motor stops.
   c. After 5 seconds, the motor runs in the reverse direction.
   d. After 10 seconds, the motor stops.
   e. After 5 seconds, the motor again runs in the forward direction.
   f. The process repeats until the stop pushbutton is pressed.

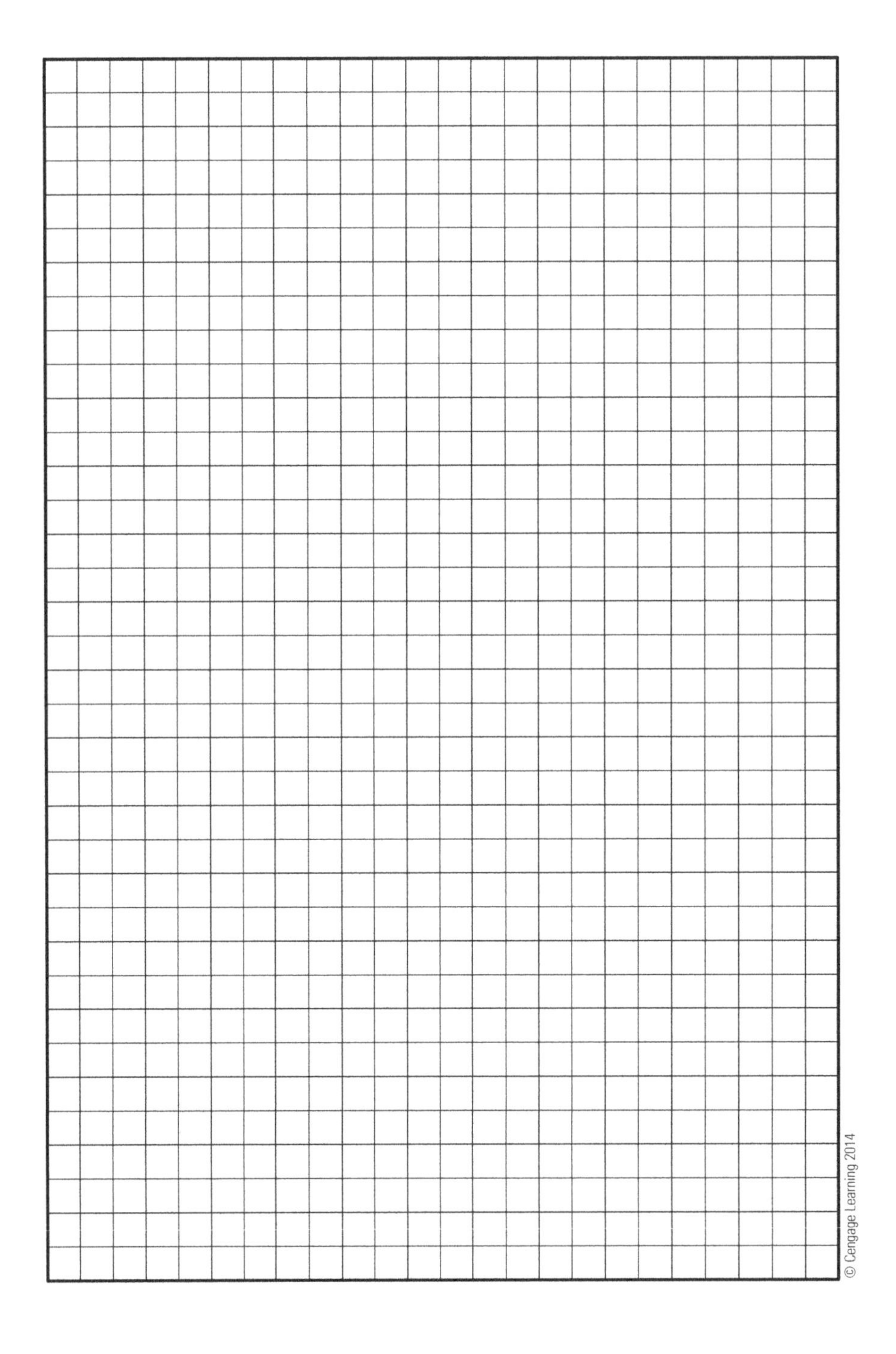

Name: ______________________ Date: ______________

SPEED CONTROL

1. Draw a ladder diagram for a circuit that performs as follows:
   a. Three momentary pushbuttons are used to select slow, medium, or fast speed.
   b. The motor must be started with the slow speed pushbutton.
      i. The motor cannot be started with the medium speed pushbutton.
      ii. The motor cannot be started with the fast speed pushbutton.
   c. Once the motor is started in slow speed, its speed may be changed directly to either medium or fast.
   d. A stop pushbutton is used to stop the motor at any time.

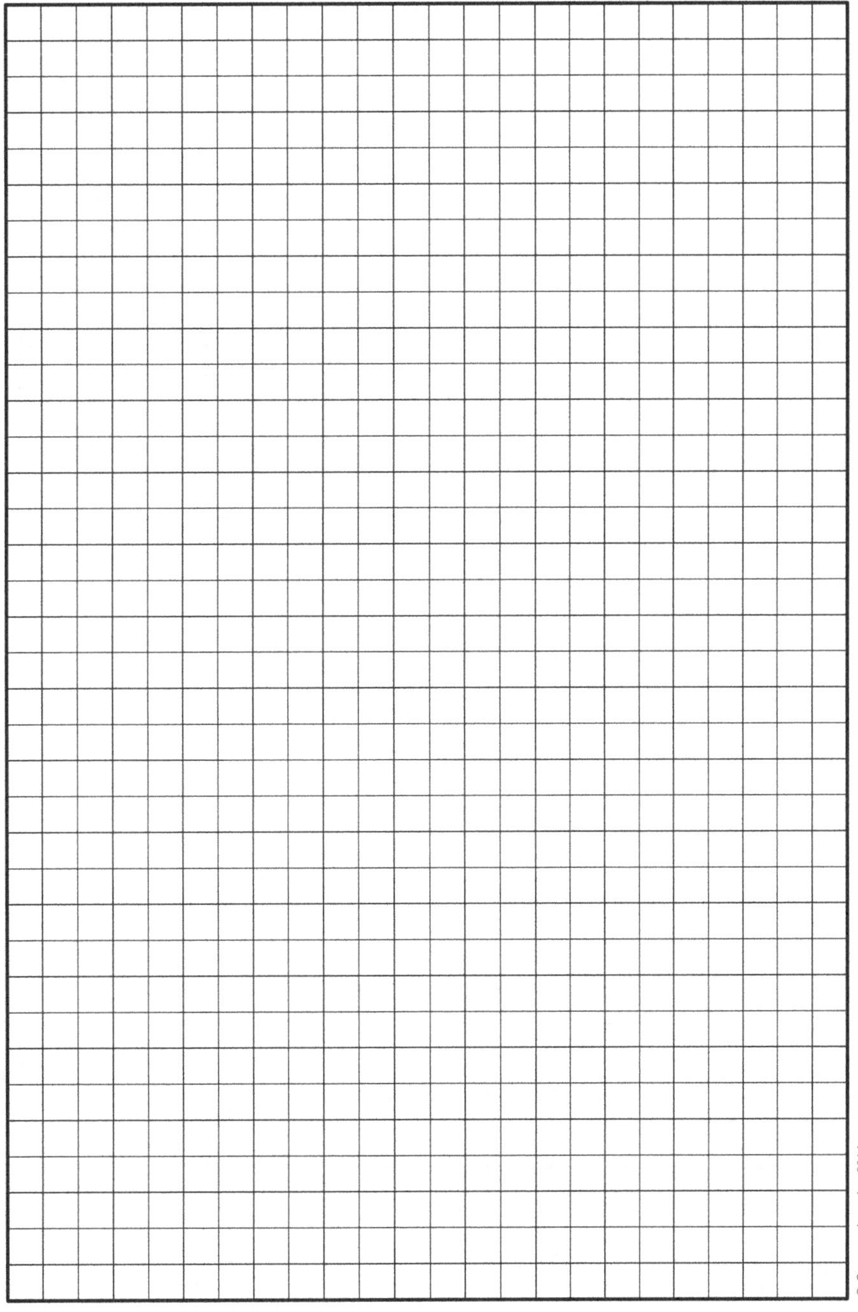

2. Draw a ladder diagram for a circuit that performs as follows:
   a. A motor is started by pressing a momentary pushbutton switch.
   b. The motor starts in low speed.
   c. After 5 seconds, the motor accelerates to medium speed.
   d. After 5 seconds, the motor accelerates to fast speed.
   e. A stop pushbutton is used to stop the motor at any time.

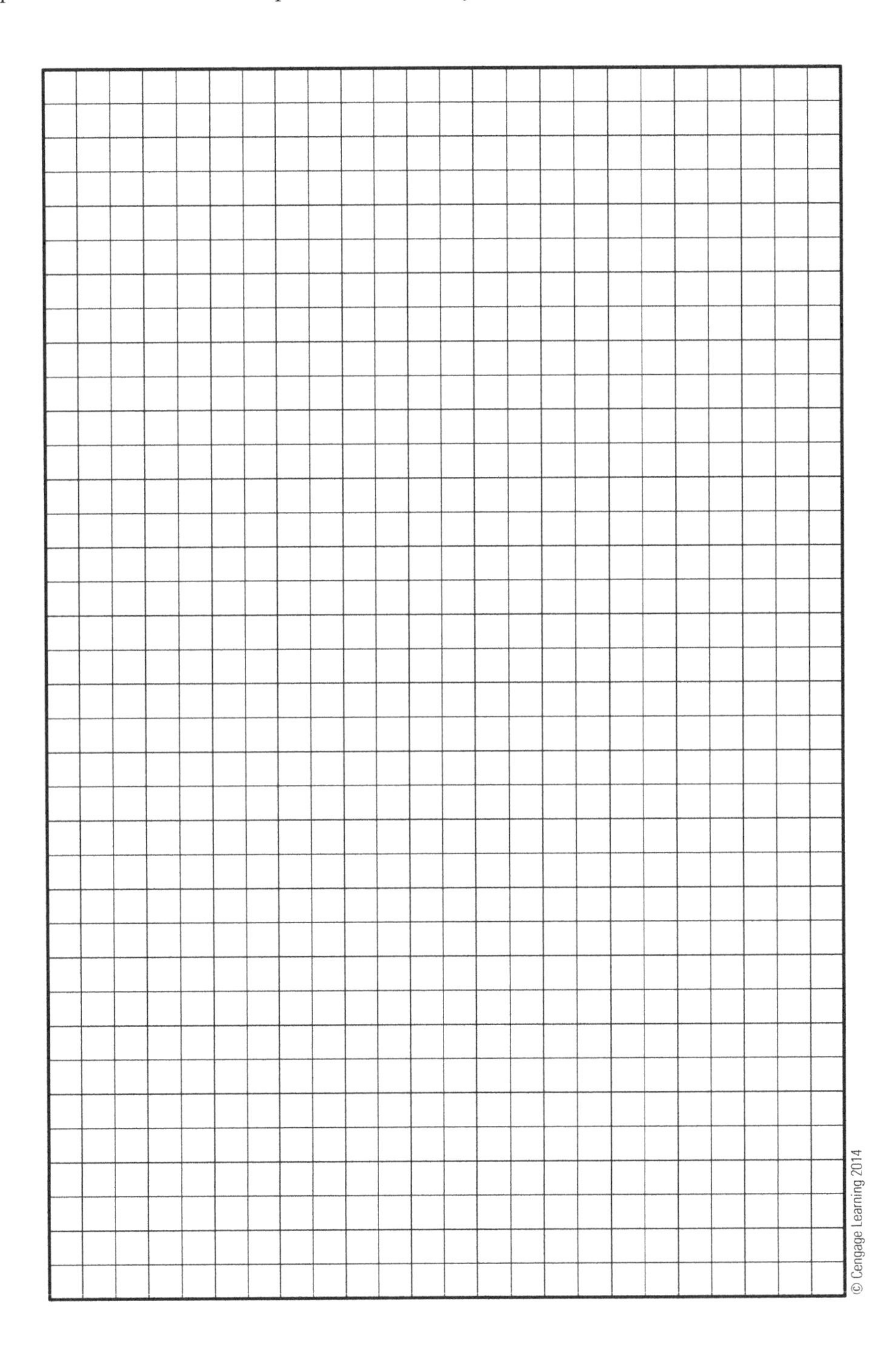

Name: ______________________ Date: ______________

JOG CONTROL

1. Draw a ladder diagram for a circuit that performs as follows:
   a. Use three pushbuttons to perform the functions of stop, run, and jog.

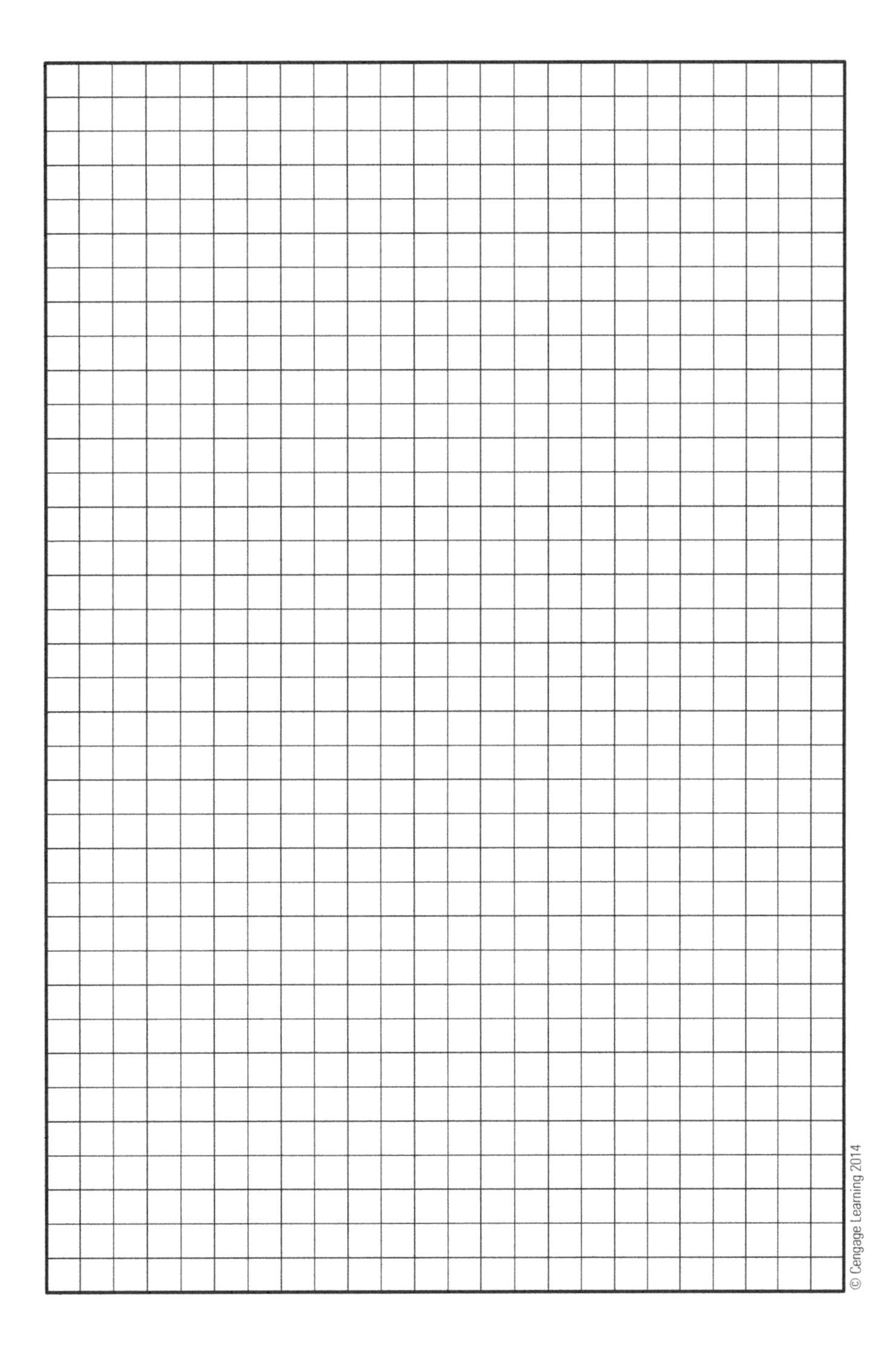

2. Draw a ladder diagram for a circuit that performs the following:
   a. Use two pushbuttons (red and green) and a selector switch.
      i. With the selector switch in the "run" position, pressing the green pushbutton causes the motor to run continuously. Pressing the red pushbutton stops the motor.
      ii. With the selector switch in the "jog" position, pressing the green pushbutton jogs the motor.

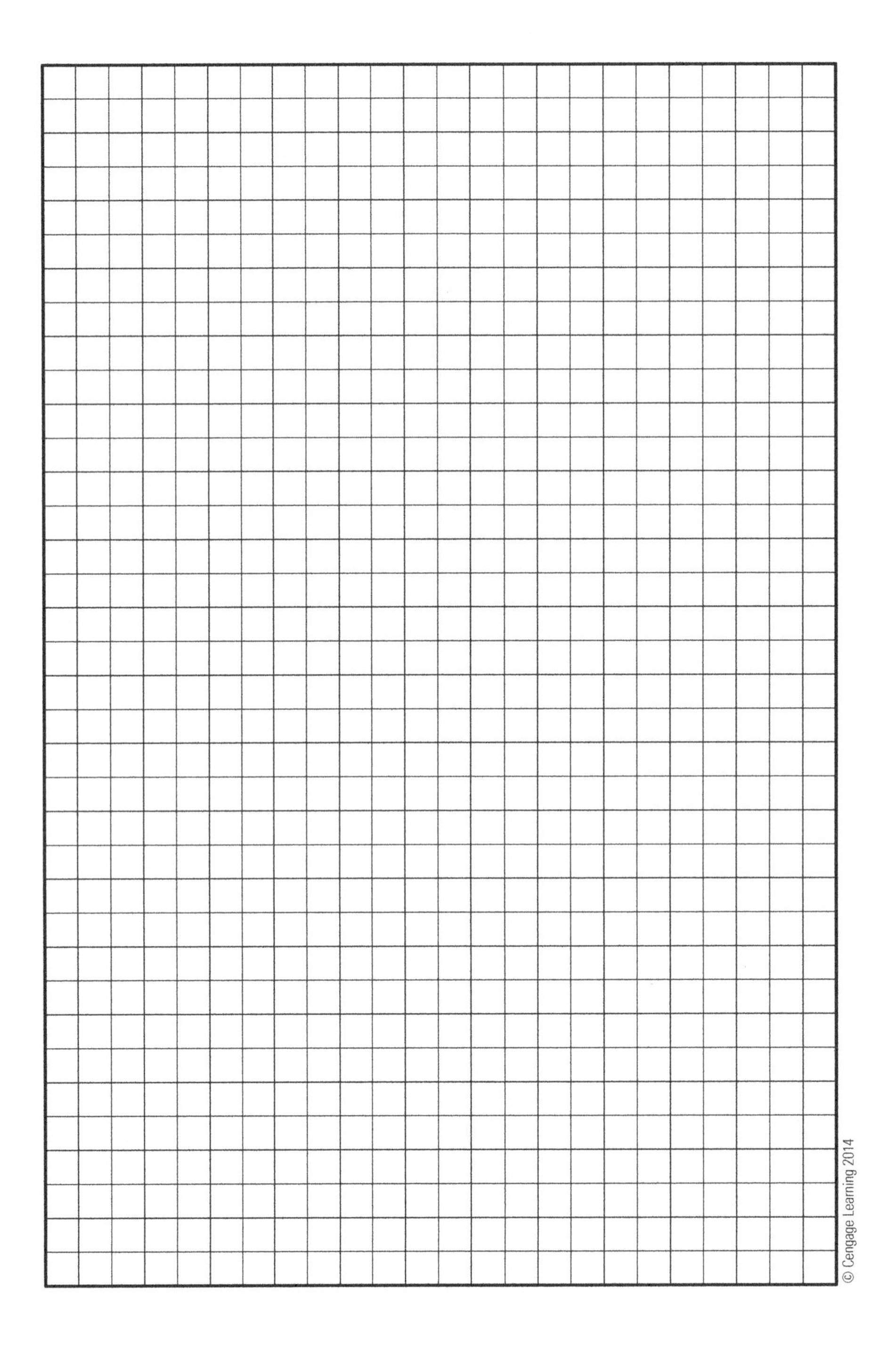

Name: ______________________ Date: ______________

HAND-OFF-AUTOMATIC CONTROL

1. Draw a ladder diagram for a circuit that performs as follows:
   a. A selector switch is used to select between manual, off, and automatic modes in a residential heating system.
   b. With the selector switch in the off position:
      i. The circuit is de-energized.
   c. With the selector switch in the manual position:
      i. The heater runs continuously.
   d. With the selector switch in the automatic position:
      i. The heater runs whenever a thermostatic switch closes.
      ii. The heater is off whenever a thermostatic switch is open.

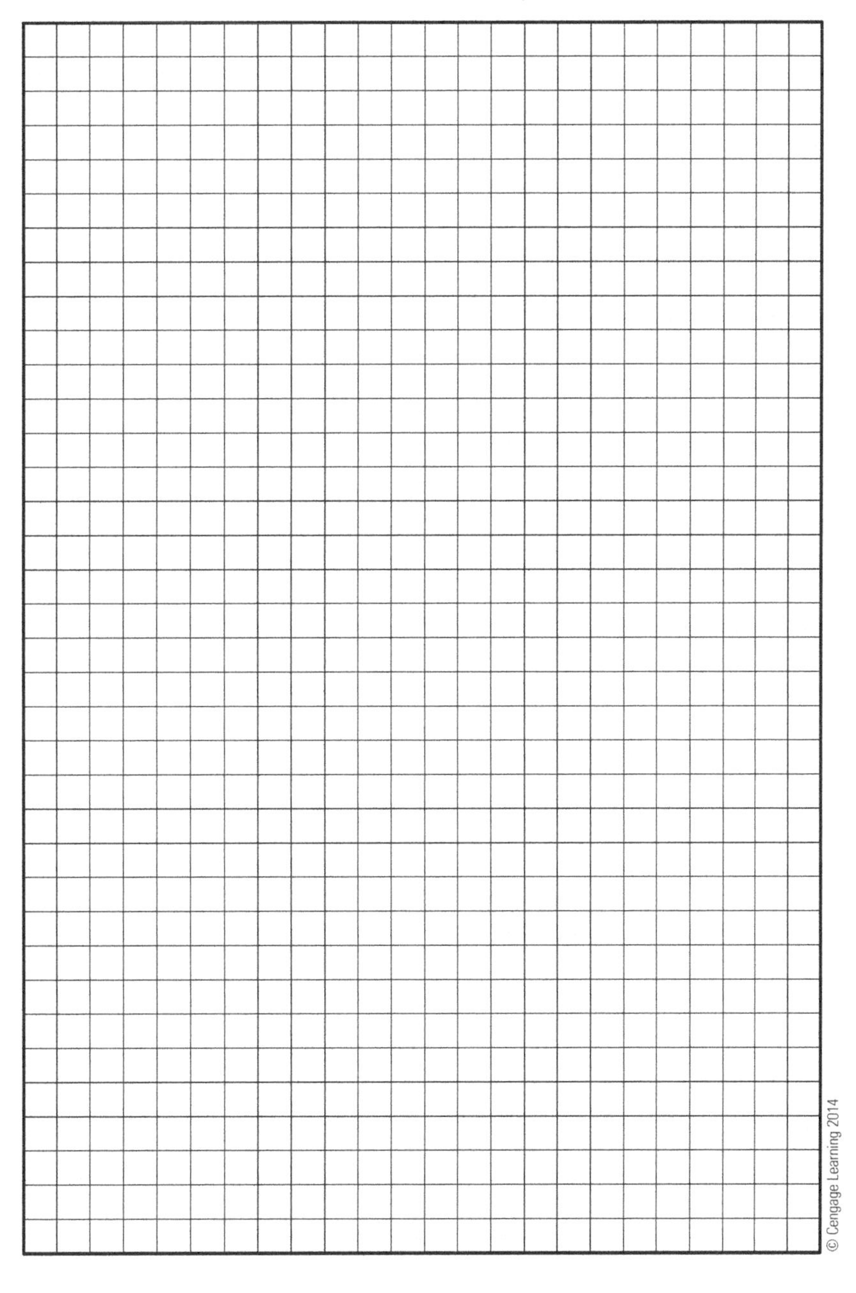

2. Modify the circuit in question 1 so that a circulating fan turns on 5 seconds after the heater turns on.
   a. When the heater turns off, the circulating fan continues to run for 5 seconds.

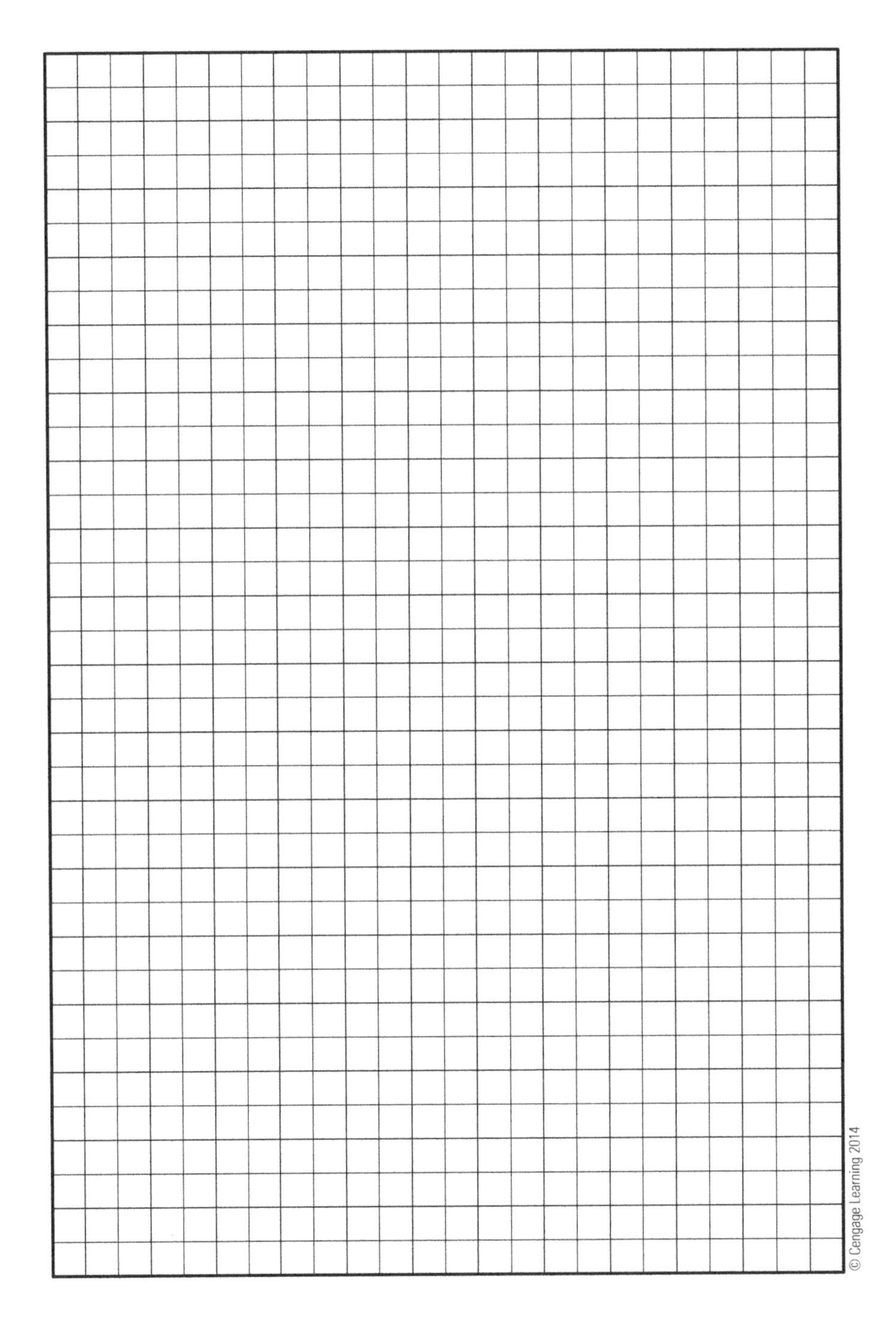

Name: ______________________ Date: ______________

MULTIPLE MOTOR STARTER CONTROL

1. Draw a ladder diagram for a circuit that performs as follows:
   a. Pressing a start pushbutton causes three conveyors to start simultaneously.
   b. Pressing a stop pushbutton stops all three conveyors simultaneously.
   c. Should an overload occur on conveyor 1, conveyor 2 and conveyor 3 will not run.
   d. Should an overload occur on conveyor 2, conveyor 1 will run, but conveyor 3 will not run.
   e. Should an overload occur on conveyor 3, conveyor 1 and conveyor 2 will run.

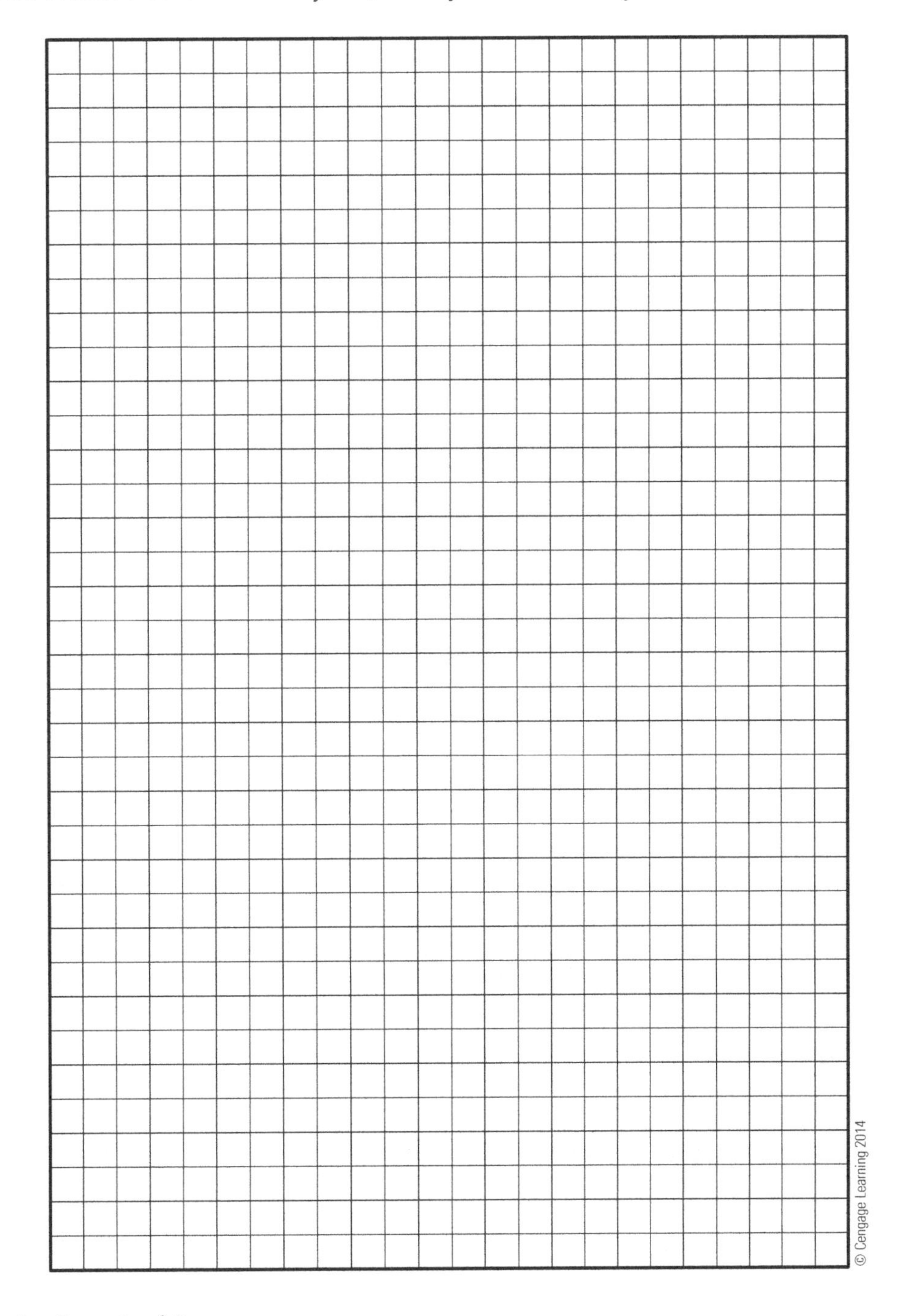

2. Draw a ladder diagram for a control circuit that uses a master stop pushbutton and three separate motors, each with its own start/stop station.

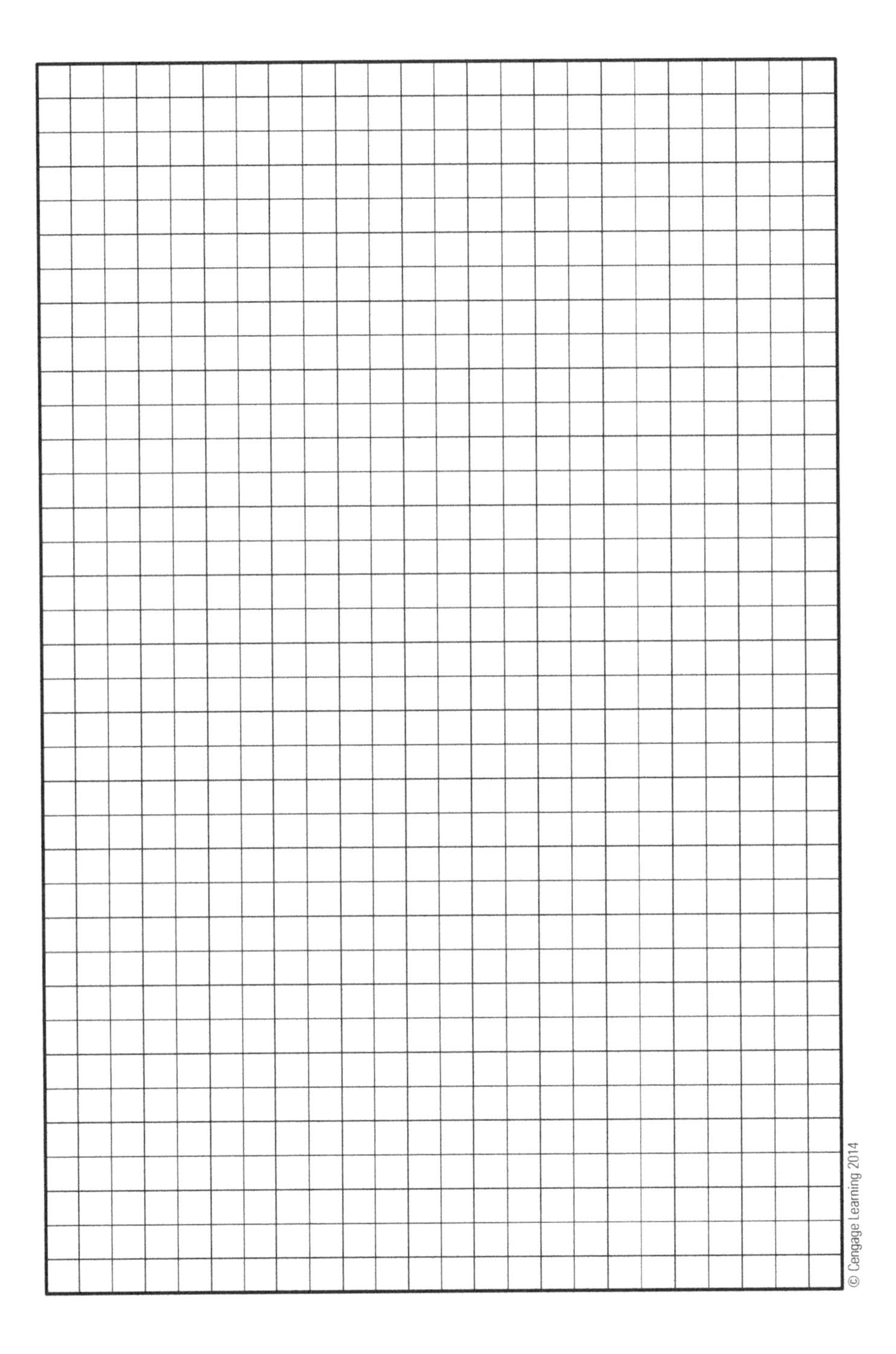

Name: ____________________ Date: ____________________

SEQUENTIAL STARTING CONTROL

1. Draw a ladder diagram for a circuit that performs the following:
   a. A start pushbutton is pressed to start a series of three conveyors:
      i. Conveyor 3 must start first.
      ii. 5 seconds after conveyor 3 starts, conveyor 2 starts.
      iii. 5 seconds after conveyor 2 starts, conveyor 1 starts.
   b. If conveyor 3 is not running, conveyor 2 and conveyor 1 will not run.
   c. If conveyor 2 is not running, conveyor 3 may run, but conveyor 1 will not run.
   d. If conveyor 1 is not running, conveyor 2 and conveyor 3 may run.
   e. An overload on any of the conveyors will shut down all conveyors.
   f. Pressing the stop pushbutton stops all conveyors.

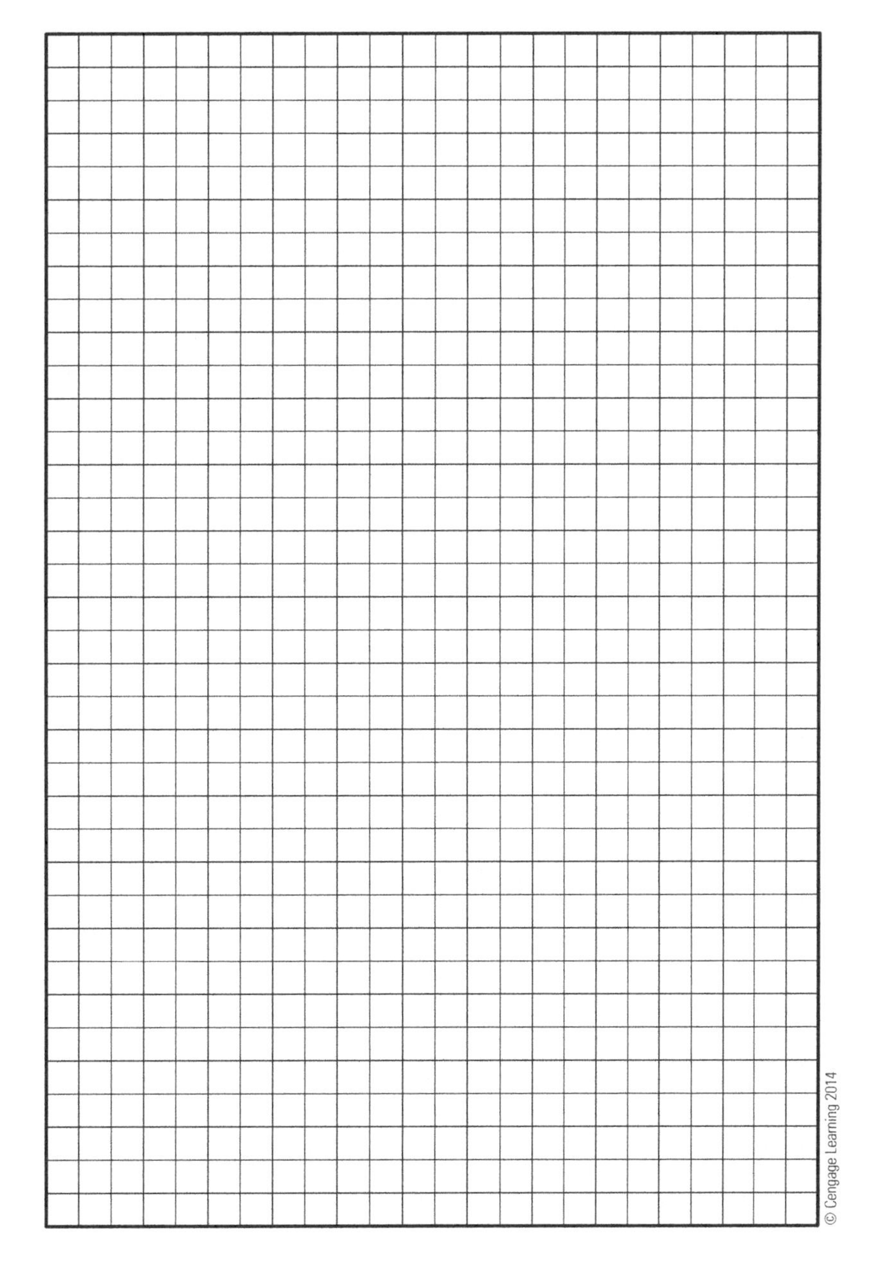

2. A grinding machine consists of a coolant pump and a grinding motor. Draw a ladder diagram for a circuit that will perform the following:

   a. When the start pushbutton is pressed, an alarm sounds for 5 seconds.

   b. After the alarm has sounded, a coolant pump begins to pump coolant to the grinding wheel. A pressure switch detects whether there is sufficient coolant pressure for the operation to continue.

      i. If there is insufficient pressure, the grinding operation cannot go forth.

      ii. If there is sufficient pressure, the grinding wheel begins to turn.

   c. If at any time during the operation a loss of coolant pressure is detected, the grinding operation is stopped.

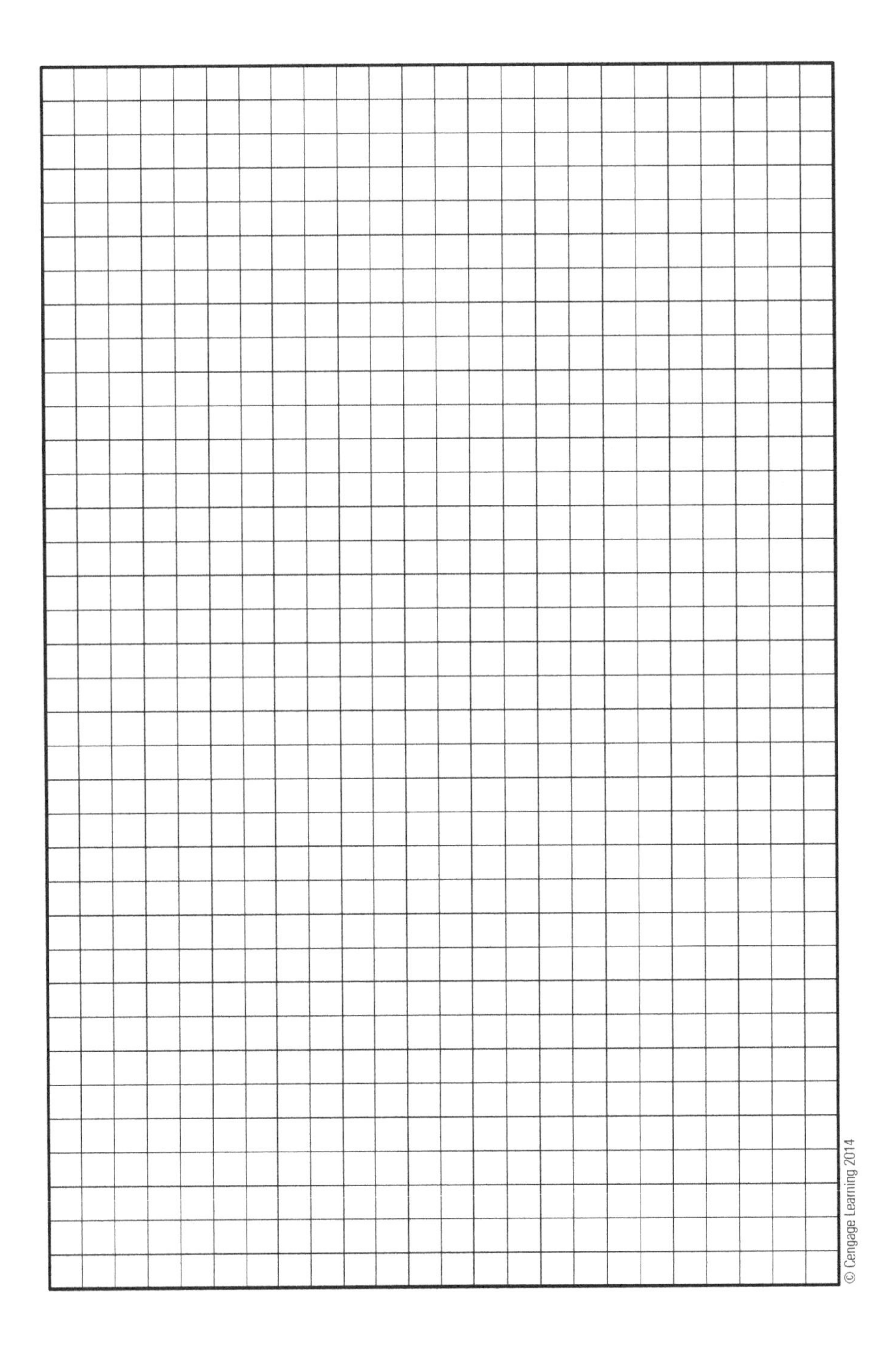

Name: ______________________ Date: ______________

VARIOUS STARTING METHODS

1. Draw a ladder diagram for a circuit that performs a primary resistor starting function.

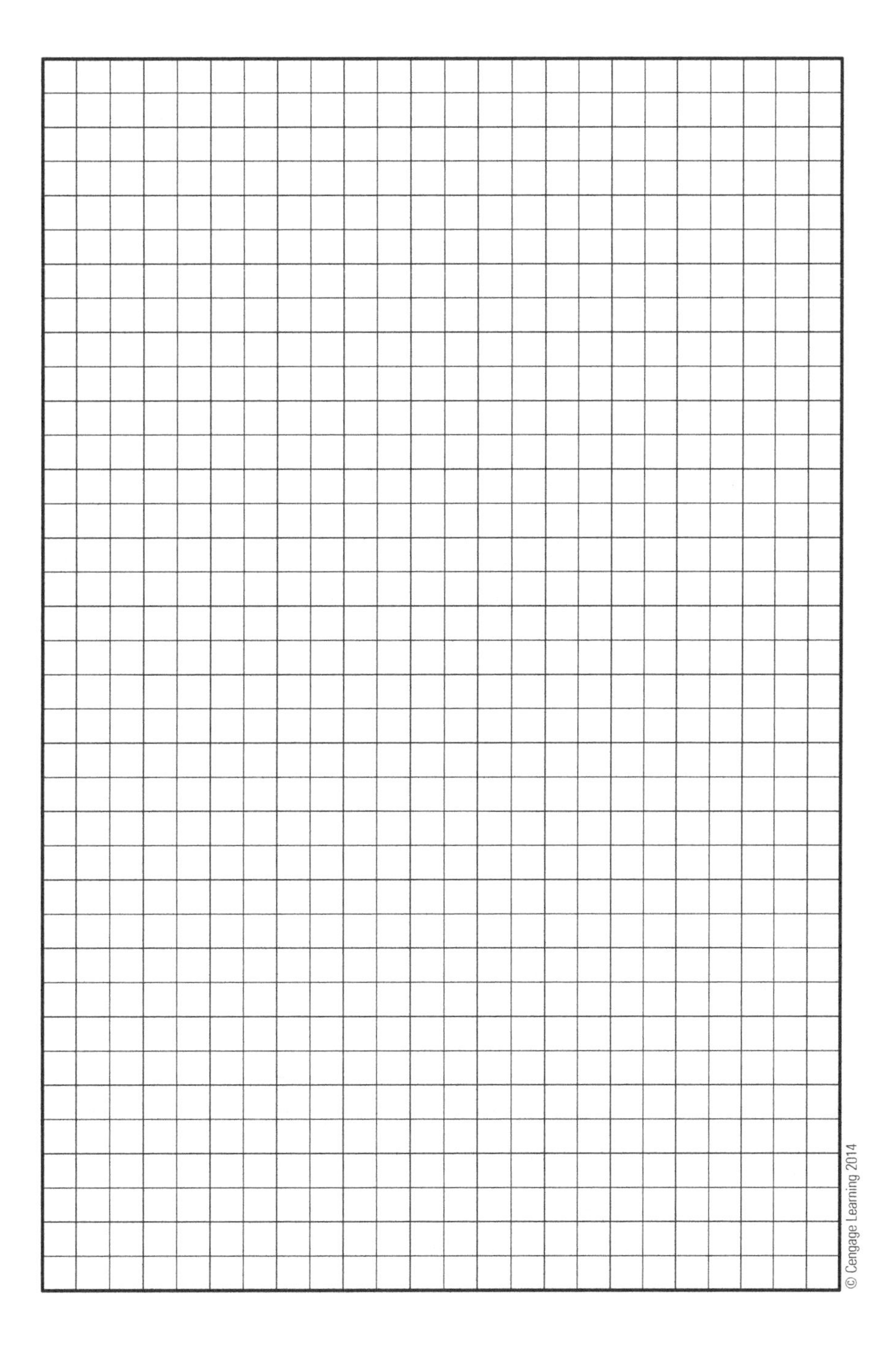

2. Draw a ladder diagram for a circuit that performs a two-step part-winding starting function.

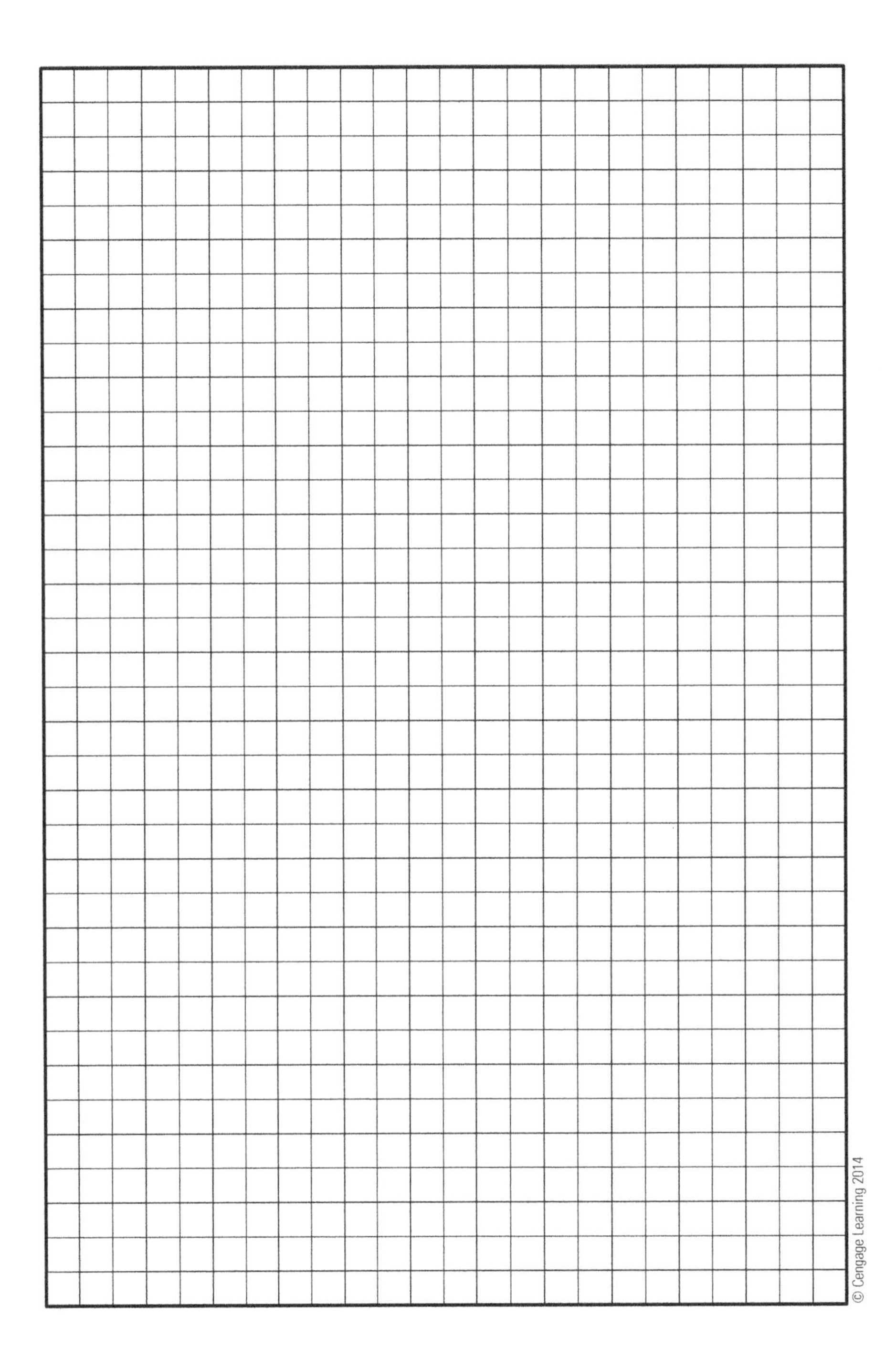

3. Draw a ladder diagram for a circuit that performs a wye–delta starting function.

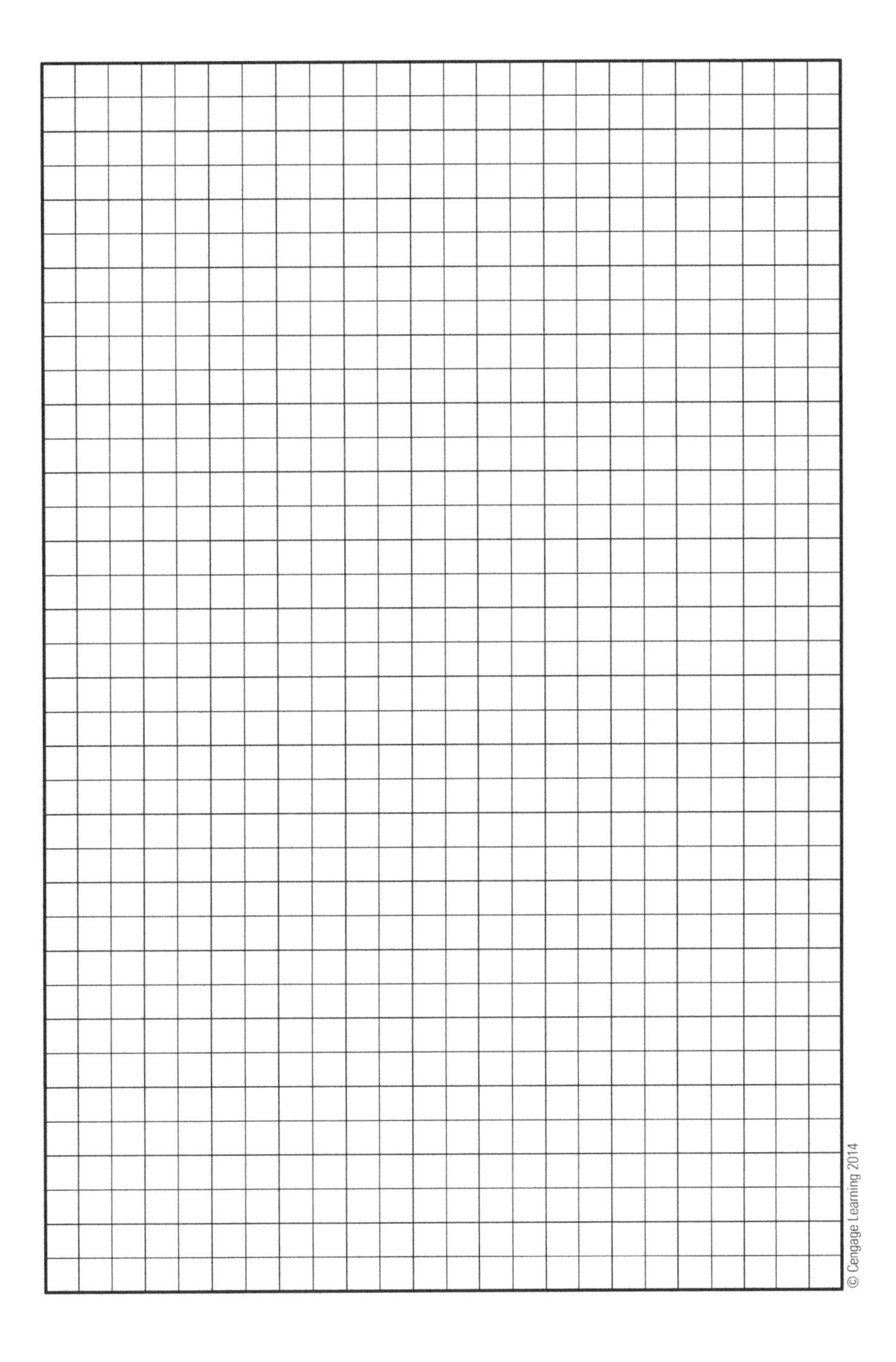

Name: ______________________ Date: ______________

BRAKING

1. Draw a ladder diagram for a circuit that provides braking of a three-phase motor by plugging.

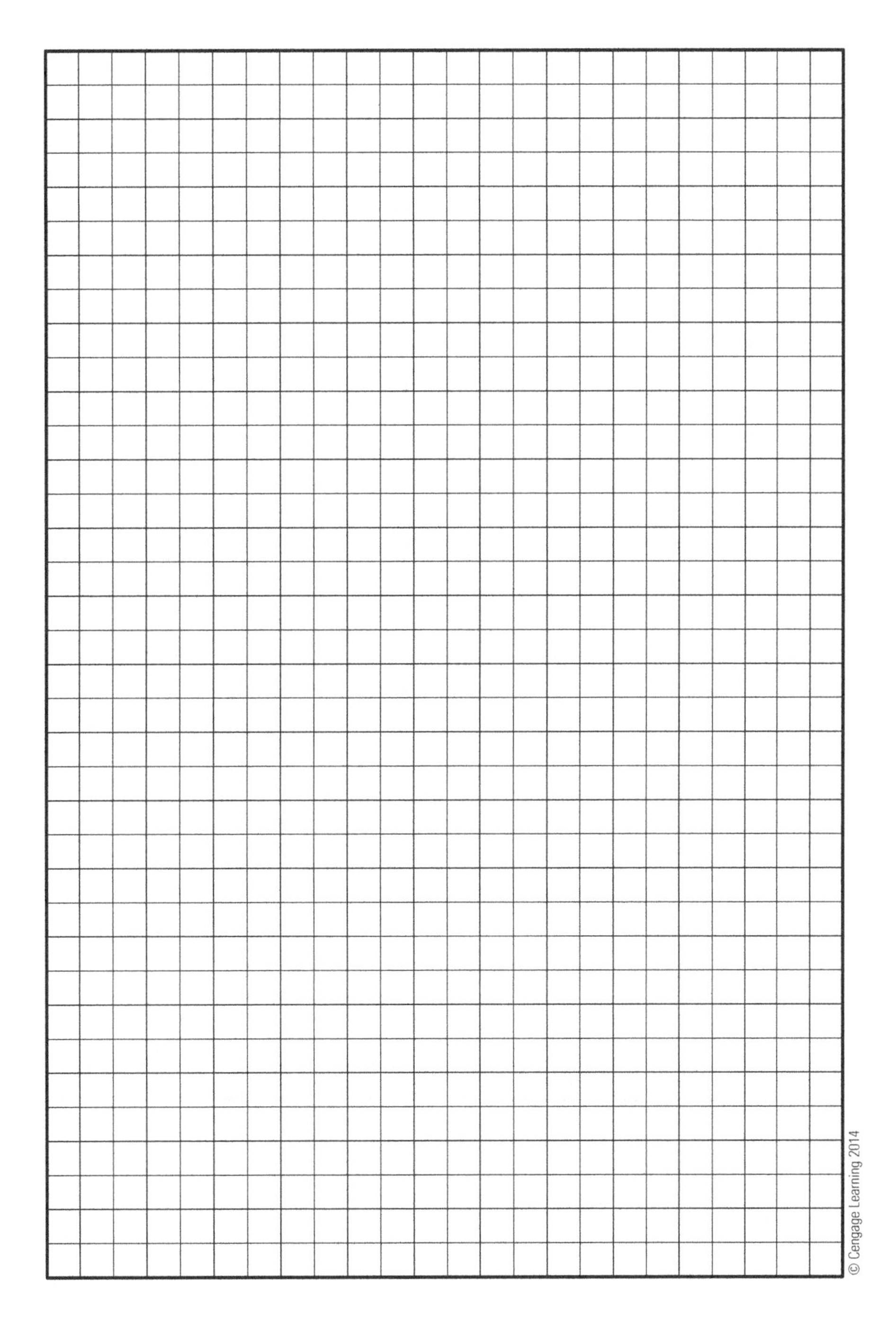

2. Draw a ladder diagram for a circuit that provides dynamic braking of a DC motor.

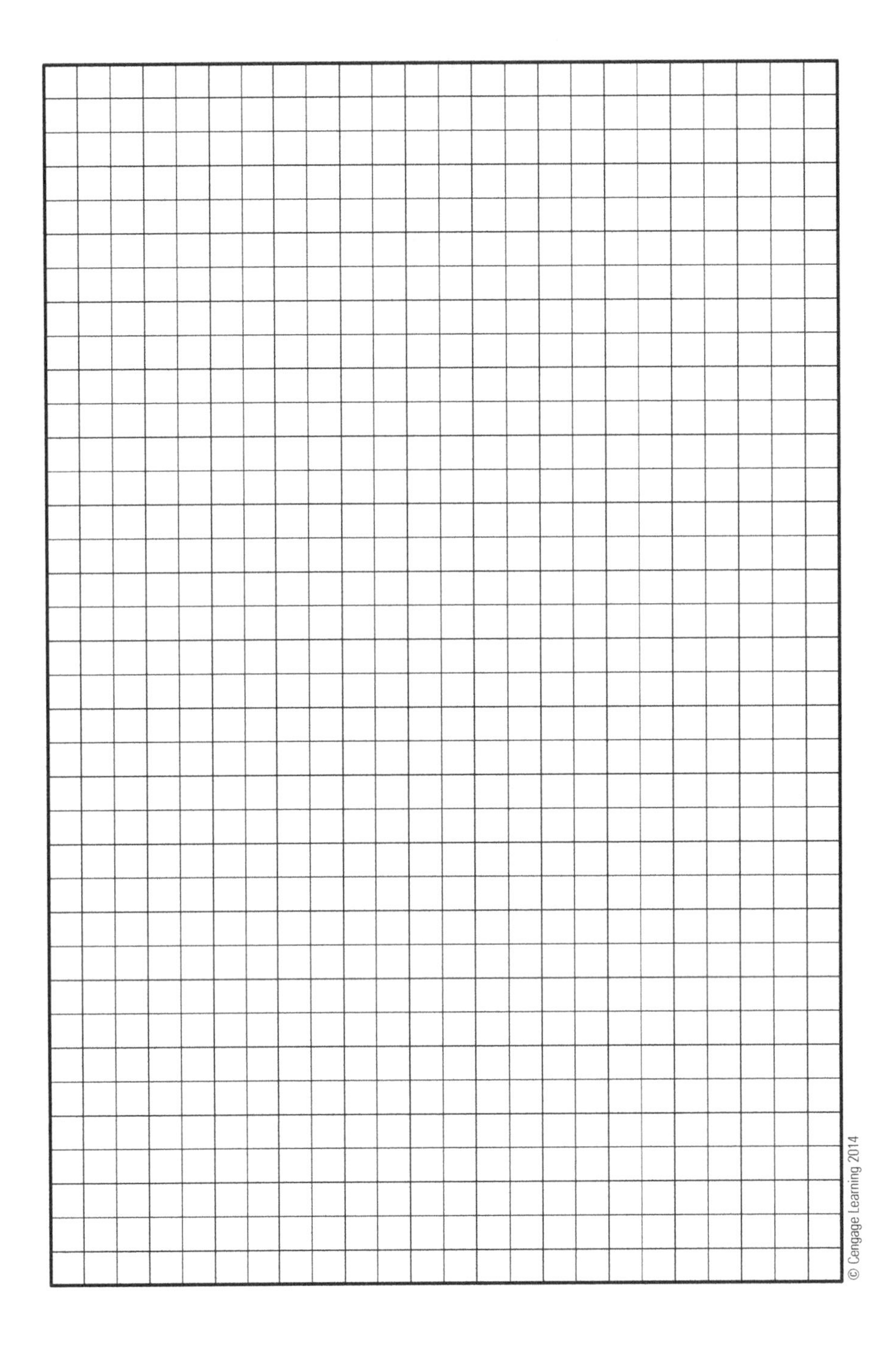

Name: ______________________________ Date: ______________________

BASIC INDUSTRIAL ELECTRONICS

1. Match the device to its schematic symbol.

_______ Diode

_______ Bridge rectifier

_______ Zener diode

_______ LED

_______ Photo diode

_______ NPN transistor

_______ PNP transistor

_______ Darlington transistor

_______ IGBT

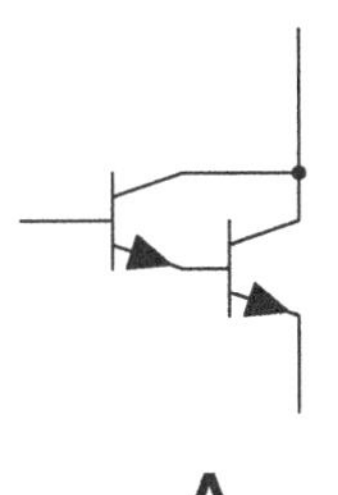

A

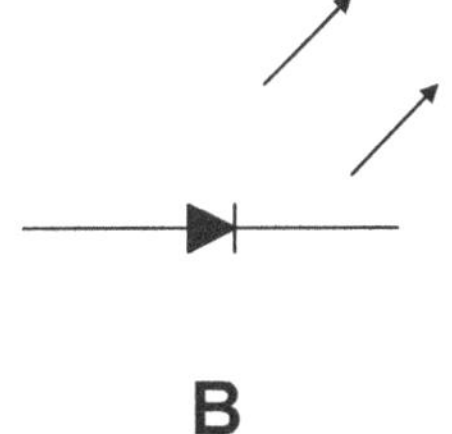

B

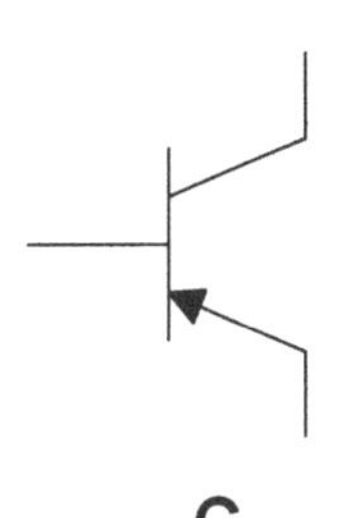

C

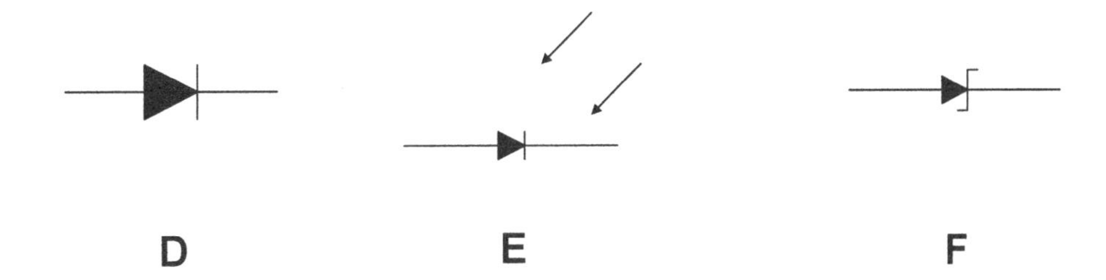

D E F

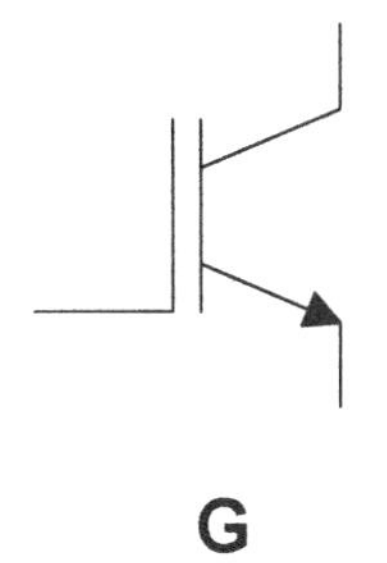

G

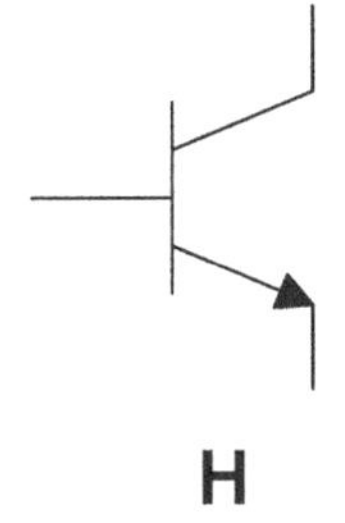

H

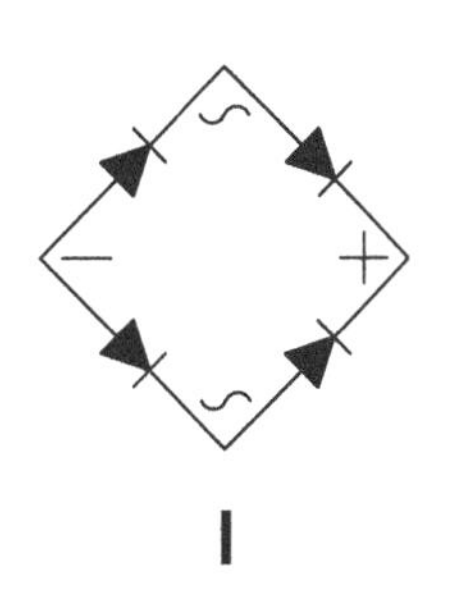

I

2. Match the device to its schematic symbol.

______	N-channel JFET	______	P-channel E-MOSFET
______	P-channel JFET	______	UJT
______	N-channel DE-MOSFET	______	SCR
______	P-channel DE-MOSFET	______	Diac
______	N-channel E-MOSFET	______	Triac

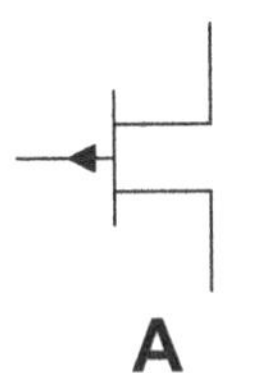
A

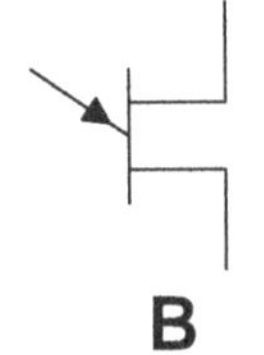
B

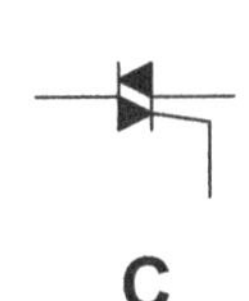
C

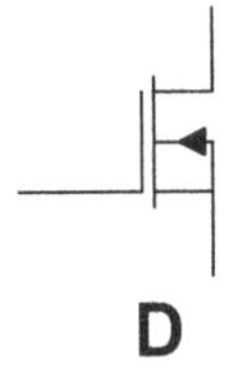
D

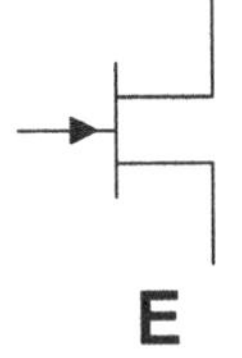
E

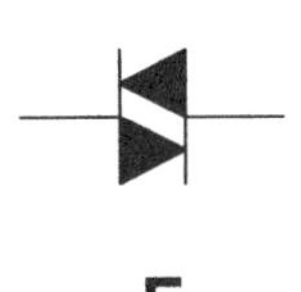
F

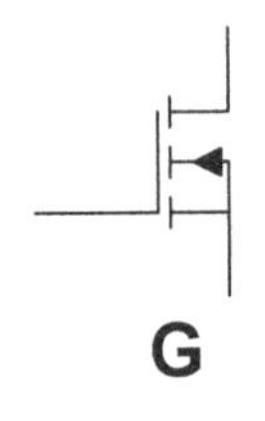
G

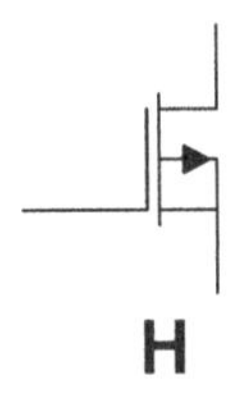
H

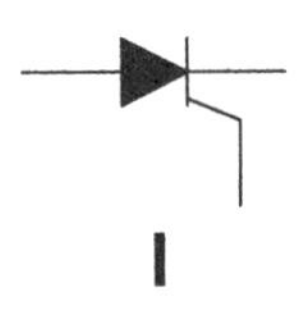
I

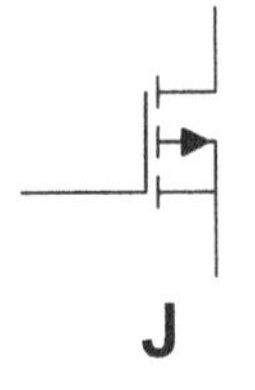
J

Name: ______________________________ Date: ______________________

TESTING ELECTRONIC DEVICES: DIODES

1. Show the connections that would allow you to test a diode in the forward-biased condition. Be sure to indicate the proper position of the selector switch on the digital multimeter (DMM).

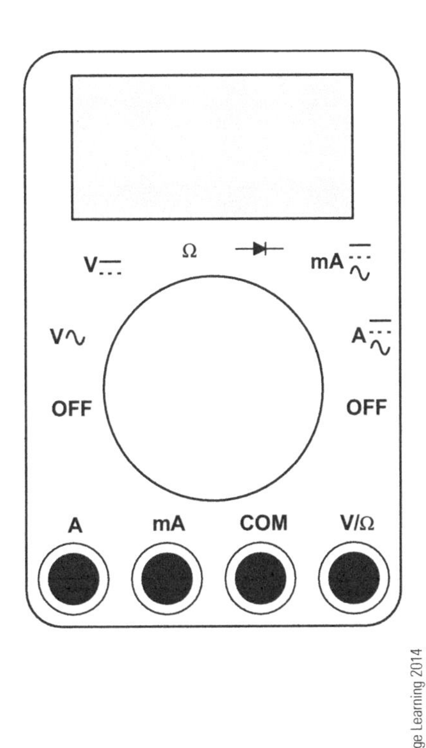

2. Show the connections that would allow you to test the diode in question 1 in the reverse-biased condition. Be sure to indicate the proper position of the selector switch on the DMM.

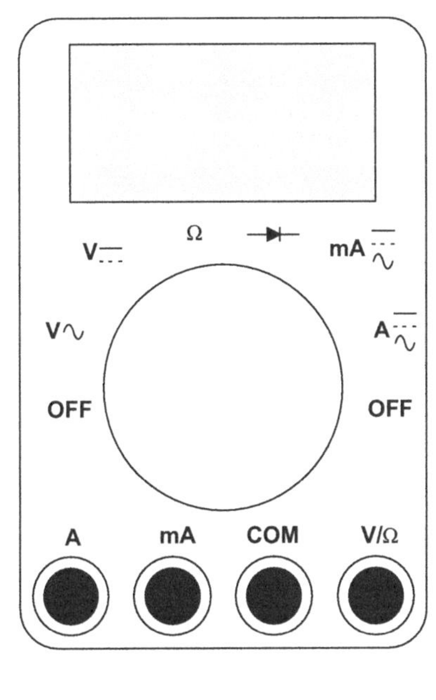

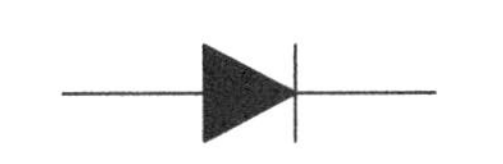

Name: ______________________________ Date: ______________

TESTING ELECTRONIC DEVICES: TRANSISTORS

1. Show the connections that would allow you to test the emitter-base junction of an NPN transistor in the forward-biased condition. Be sure to indicate the proper position of the selector switch on the DMM.

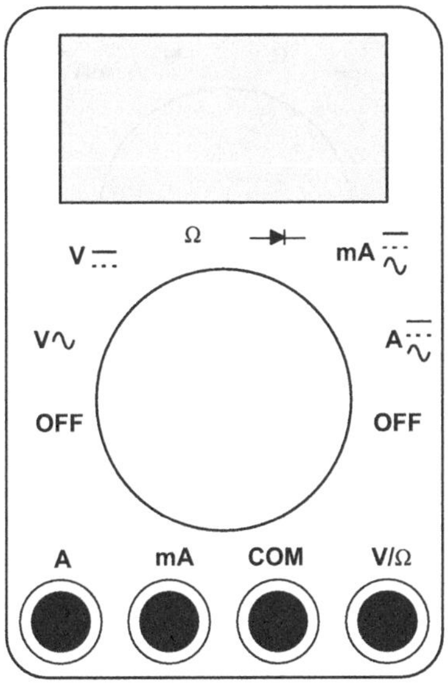

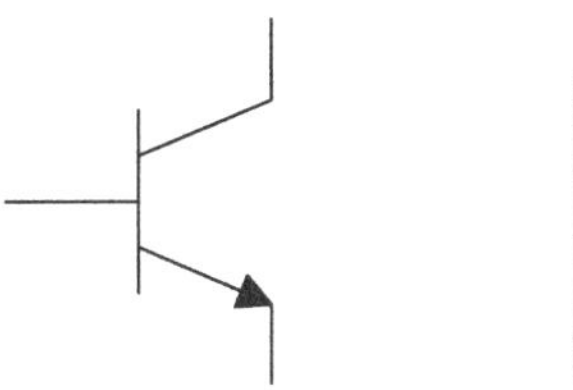

2. Show the connections that would allow you to test the emitter-base junction of the NPN transistor in question 1 in the reverse-biased condition. Be sure to indicate the proper position of the selector switch on the DMM.

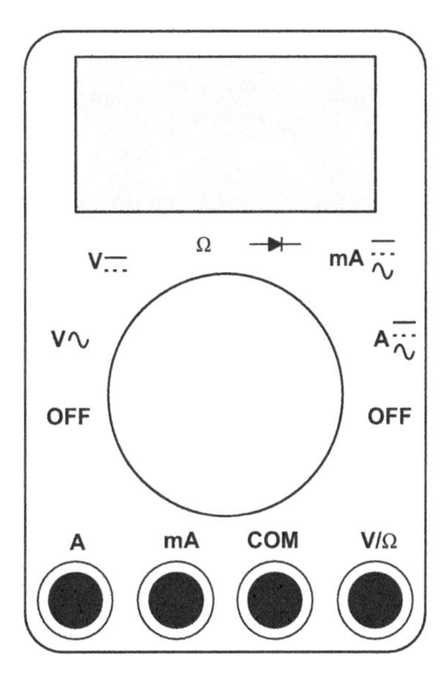

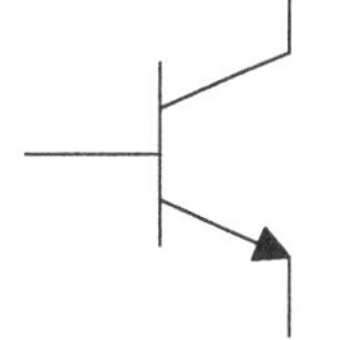

© Cengage Learning 2014

3. Show the connections that would allow you to test the collector-base junction of the NPN transistor in question 1 in the forward-biased condition. Be sure to indicate the proper position of the selector switch on the DMM.

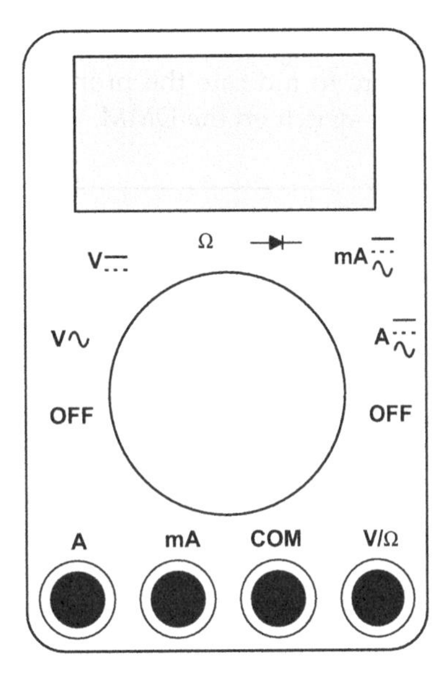

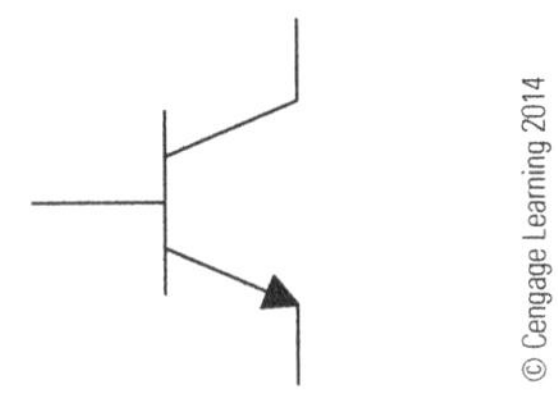

4. Show the connections that would allow you to test the collector-base junction of the NPN transistor in question 1 in the reverse-biased condition. Be sure to indicate the proper position of the selector switch on the DMM.

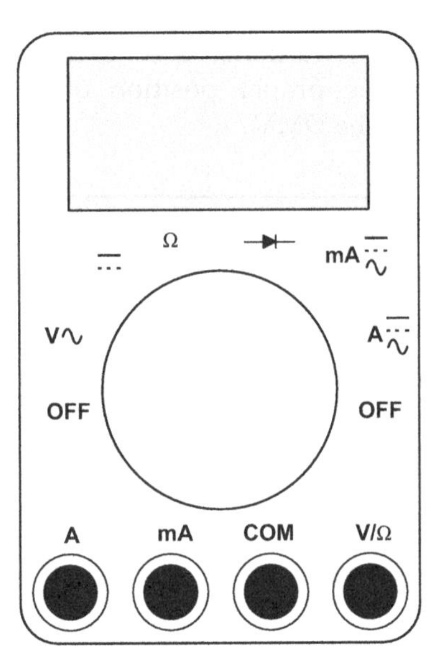

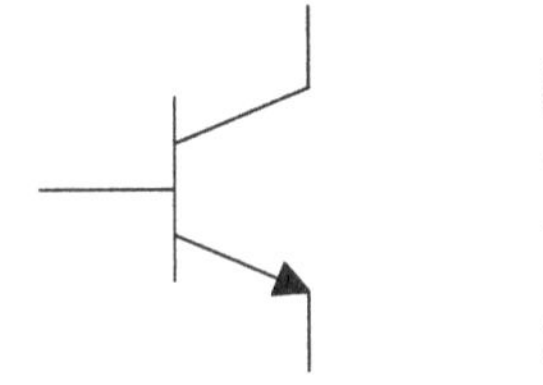

5. Show the connections that would allow you to test the collector-emitter junction of the NPN transistor in question 1 in the forward-biased condition. Be sure to indicate the proper position of the selector switch on the DMM.

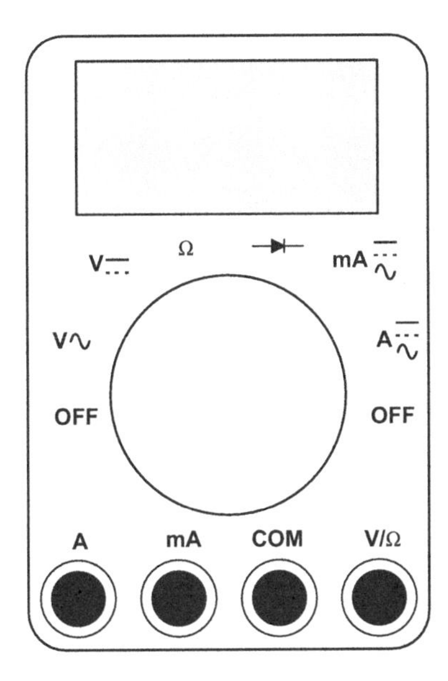

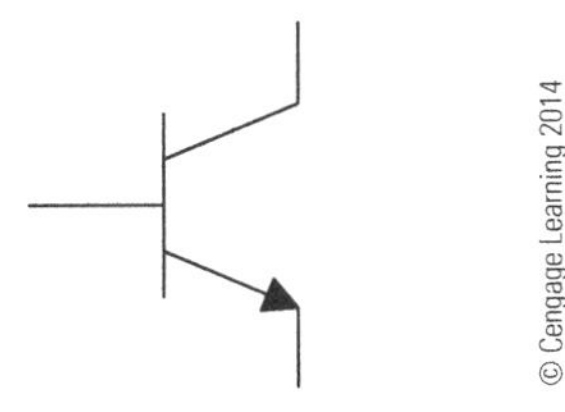

6. Show the connections that would allow you to test the collector-emitter junction of the NPN transistor in question 1 in the reverse-biased condition. Be sure to indicate the proper position of the selector switch on the DMM.

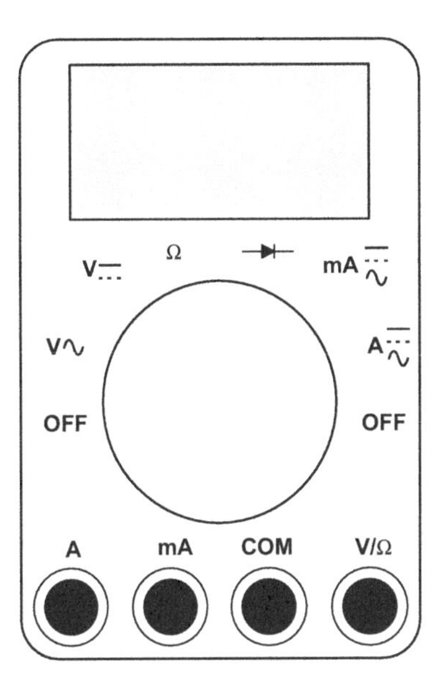

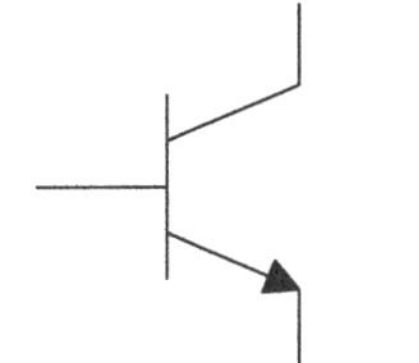

Name: ______________________________ Date: ______________

TESTING ELECTRONIC DEVICES: JFET

1. Show the connections that would allow you to test the source-gate junction of an N-channel JFET in the forward-biased condition. Be sure to indicate the proper position of the selector switch on the DMM.

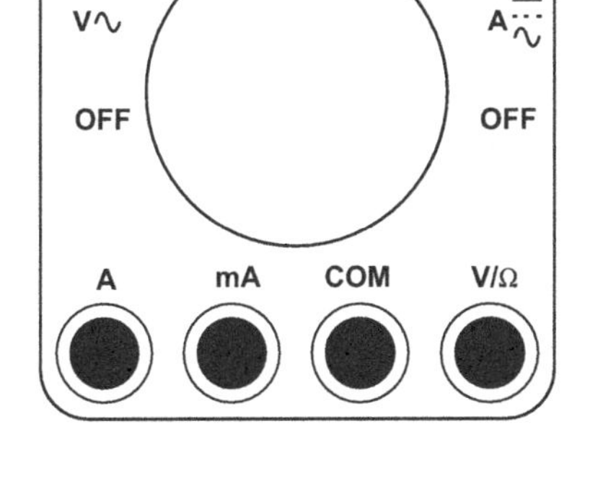

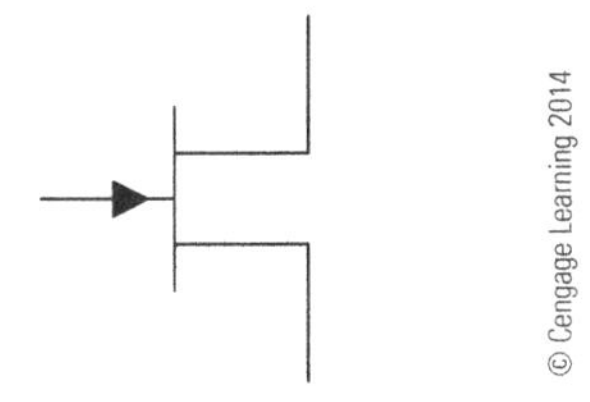

2. Show the connections that would allow you to test the source-gate junction of the N-channel JFET in question 1 in the reverse-biased condition. Be sure to indicate the proper position of the selector switch on the DMM.

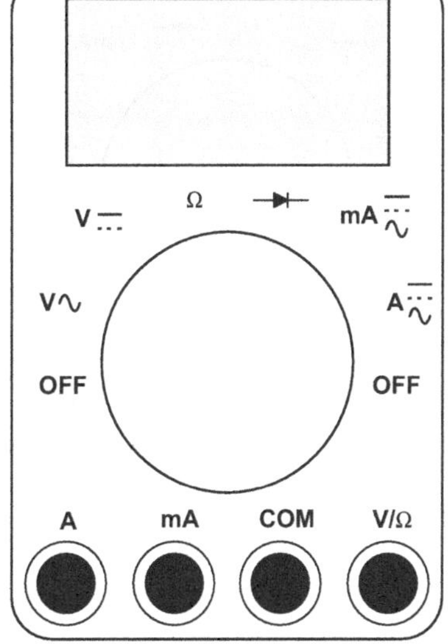

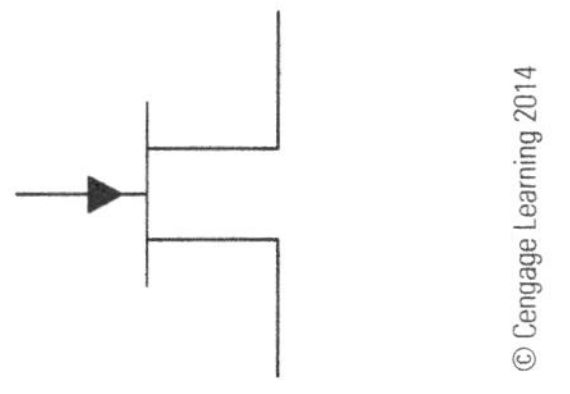

3. Show the connections that would allow you to test the drain-gate junction of an N-channel JFET in the forward-biased condition. Be sure to indicate the proper position of the selector switch on the DMM.

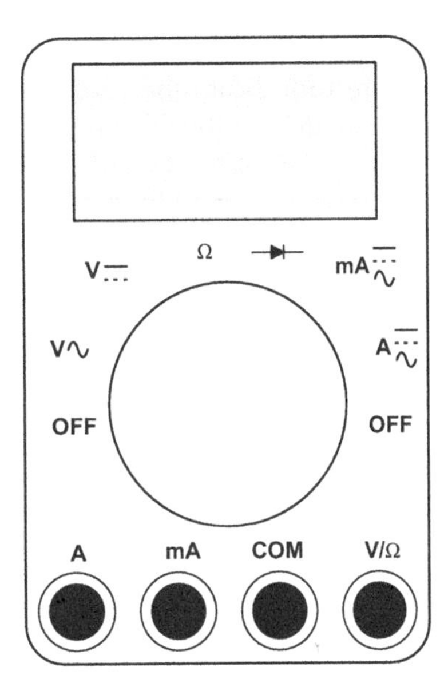

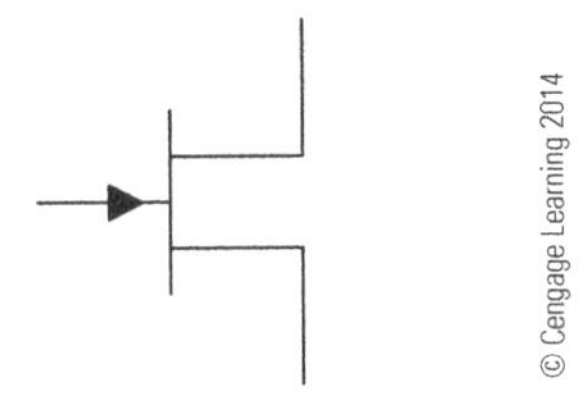

4. Show the connections that would allow you to test the drain gate junction of the N-channel JFET in question 1 in the reverse biased condition. Be sure to indicate the proper position of the selector switch on the DMM.

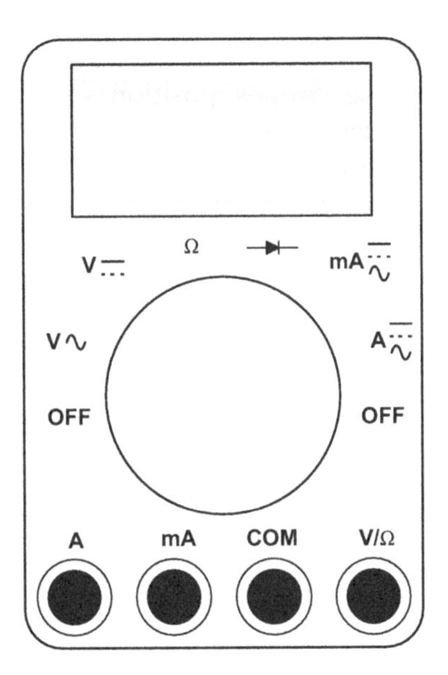

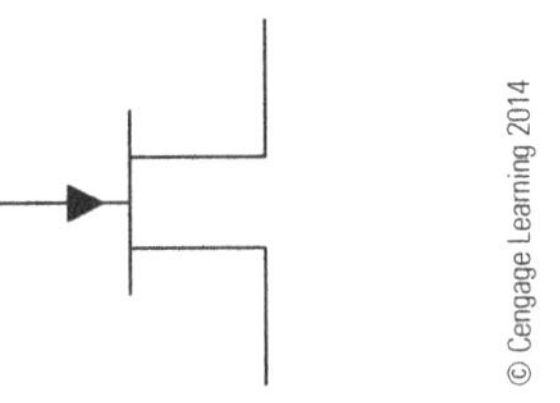

5. Show the connections that would allow you to test the drain-source junction of an N-channel JFET in the forward-biased condition. Be sure to indicate the proper position of the selector switch on the DMM.

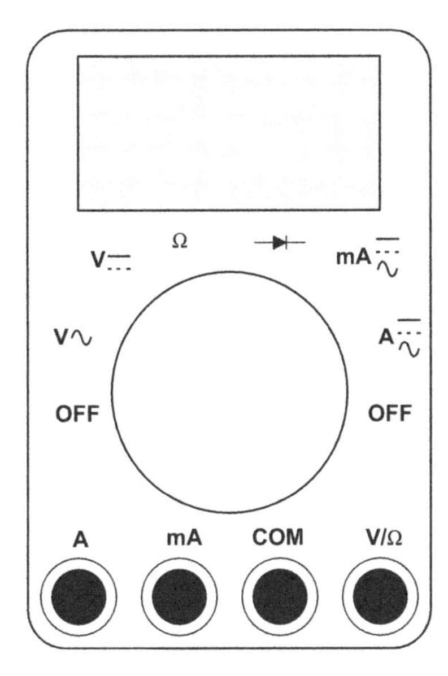

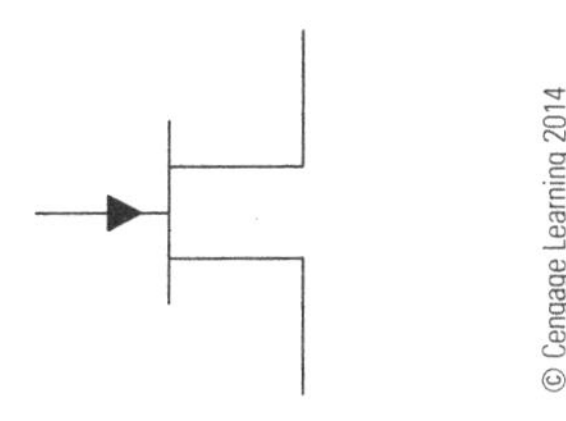

6. Show the connections that would allow you to test the drain-source junction of the N-channel JFET in question 1 in the reverse-biased condition. Be sure to indicate the proper position of the selector switch on the DMM.

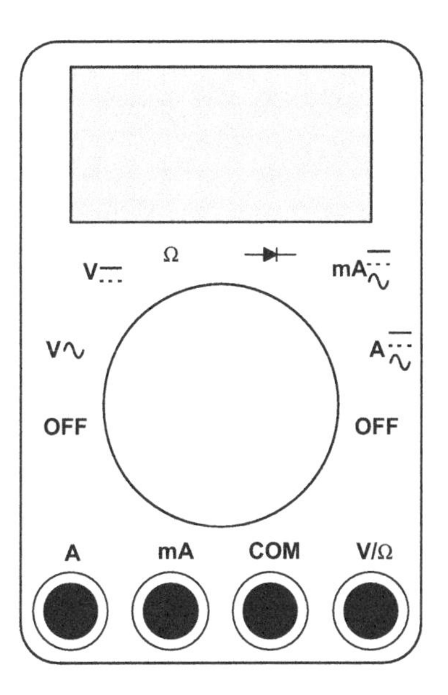

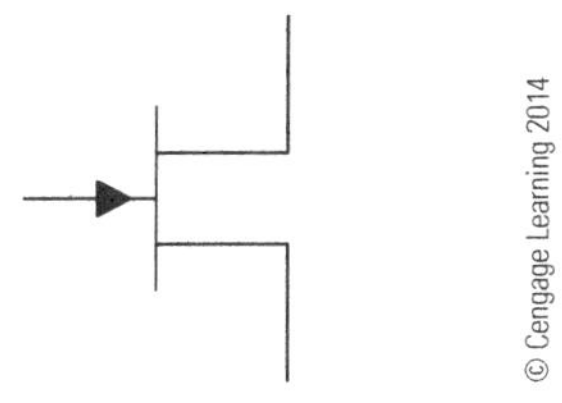

Name: ______________________ Date: ______________

TESTING ELECTRONIC DEVICES: UJT

1. Show the connections that would allow you to test the emitter–base B_1 junction of a UJT in the forward-biased condition. Be sure to indicate the proper position of the selector switch on the DMM.

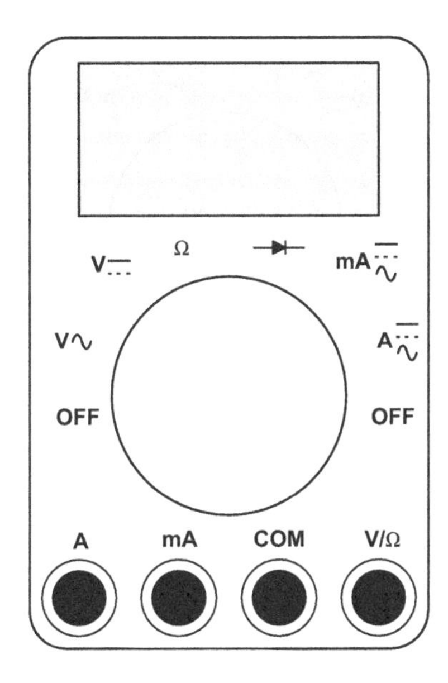

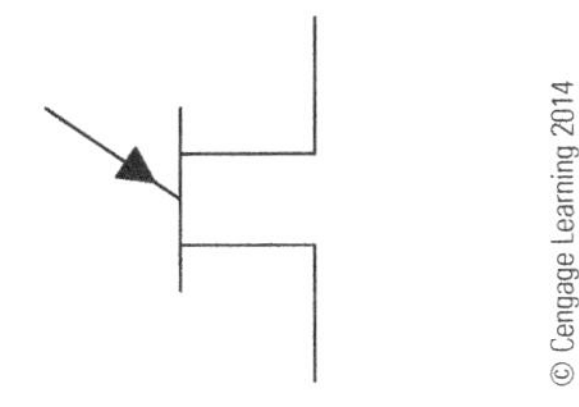

2. Show the connections that would allow you to test the emitter–base B_1 junction of the UJT in question 1 in the reverse-biased condition. Be sure to indicate the proper position of the selector switch on the DMM.

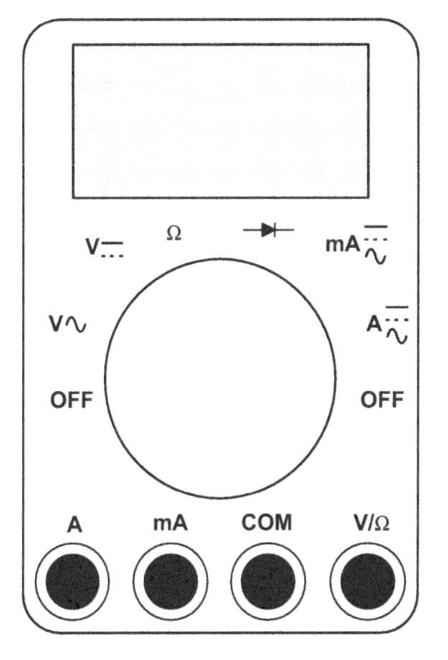

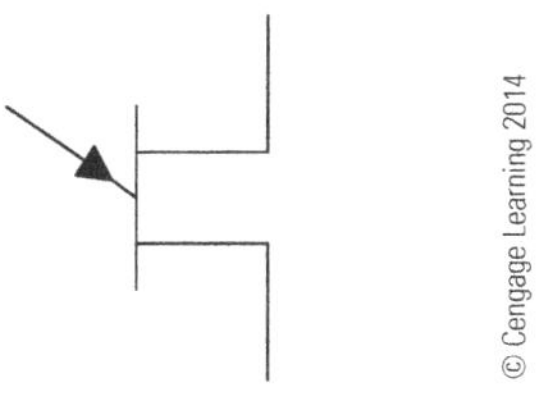

3. Show the connections that would allow you to test the emitter–base B_2 junction of a UJT in the forward-biased condition. Be sure to indicate the proper position of the selector switch on the DMM.

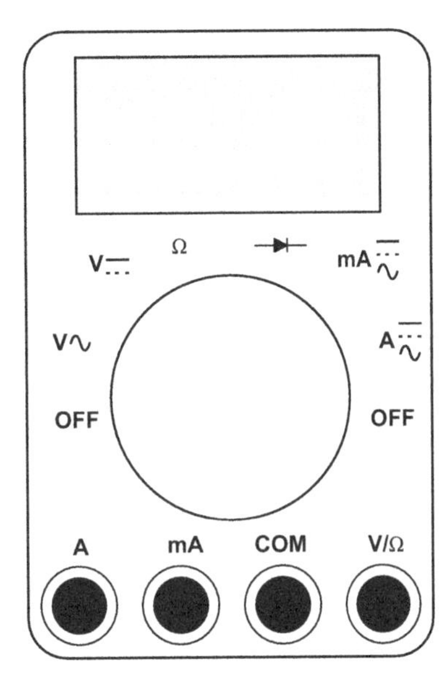

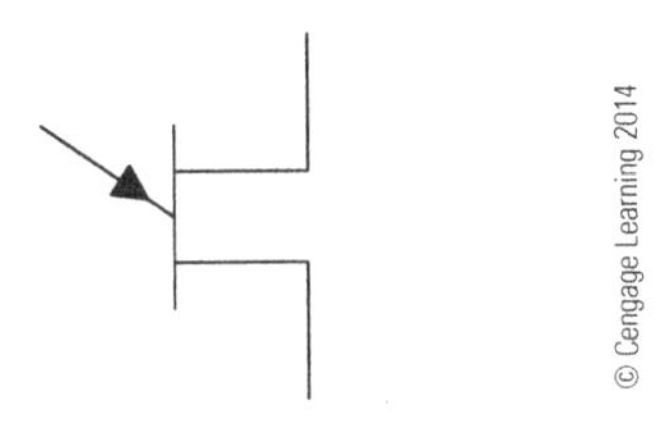

4. Show the connections that would allow you to test the emitter–base B_2 junction of the UJT in question 1 in the reverse-biased condition. Be sure to indicate the proper position of the selector switch on the DMM.

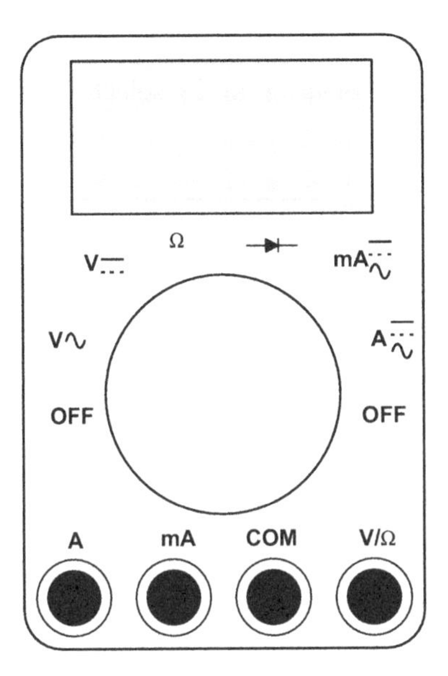

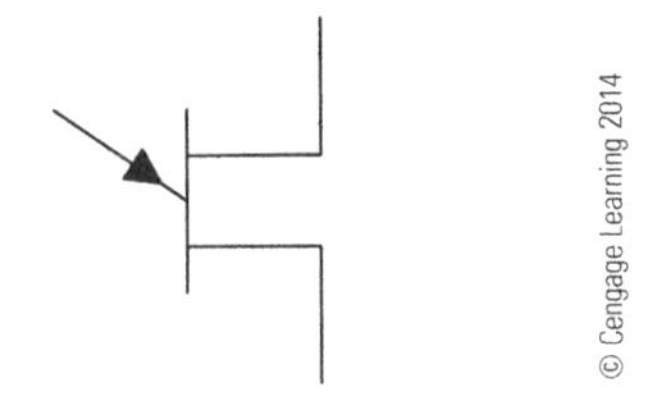

5. Show the connections that would allow you to test the base B_1–B_2 junction of a UJT in the forward-biased condition. Be sure to indicate the proper position of the selector switch on the DMM.

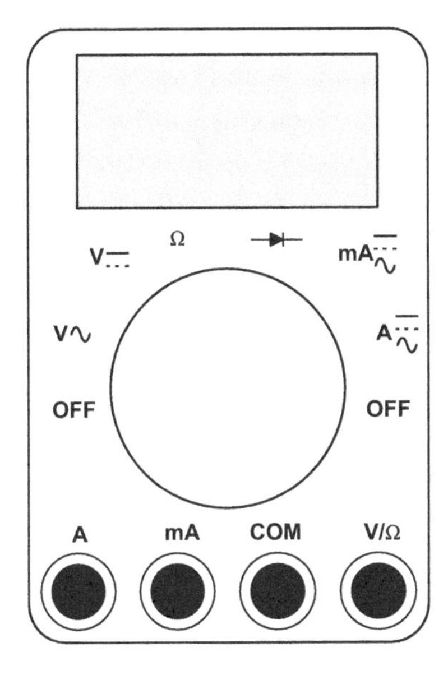

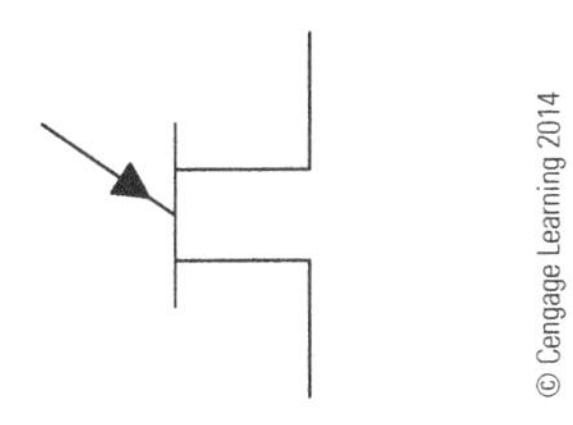

© Cengage Learning 2014

6. Show the connections that would allow you to test the base B_1–B_2 junction of the UJT in question 1 in the reverse-biased condition. Be sure to indicate the proper position of the selector switch on the DMM.

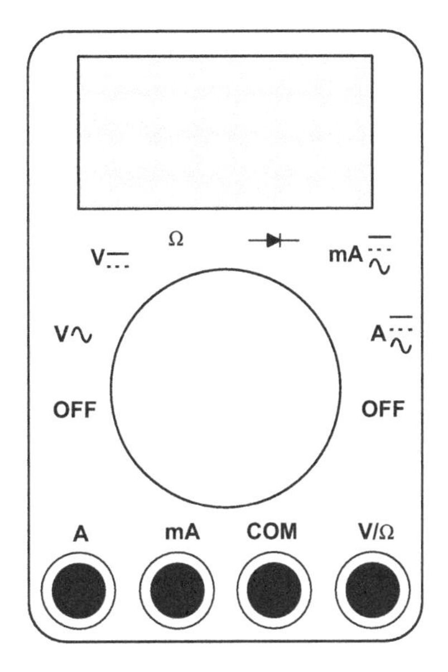

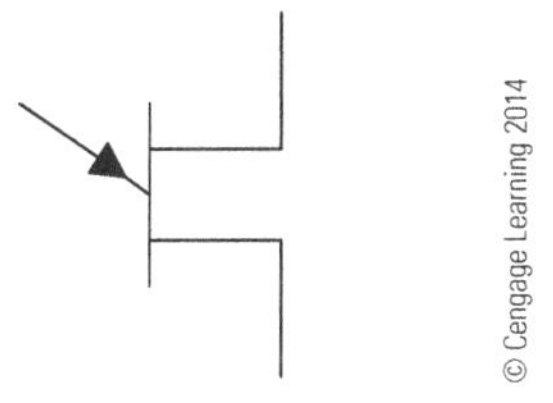

© Cengage Learning 2014

Name: ______________________ Date: ______________

TESTING ELECTRONIC DEVICES: SCR

1. Show the connections that would allow you to test the diode portion of an SCR in the forward-biased condition. Be sure to indicate the proper position of the selector switch on the DMM.

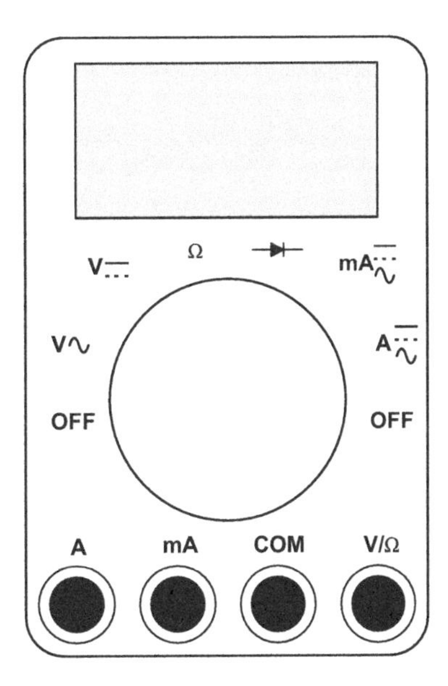

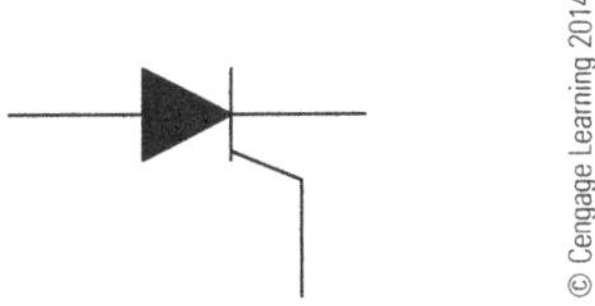

2. Show the connections that would allow you to test the diode portion of the SCR in question 1 in the reverse-biased condition. Be sure to indicate the proper position of the selector switch on the DMM.

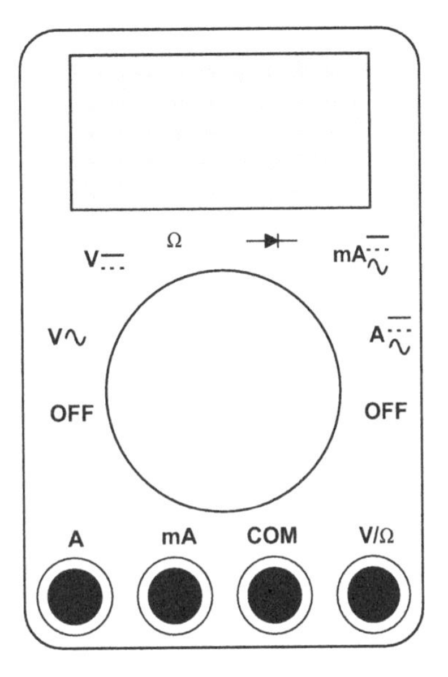

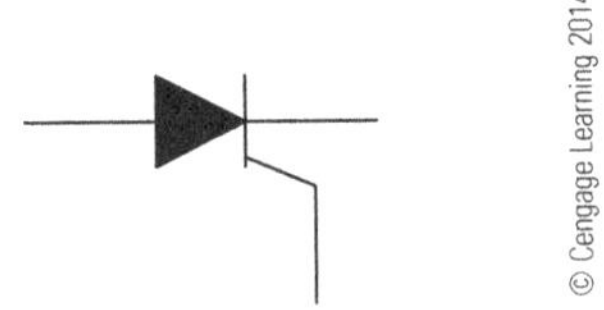

3. Show the connections that would allow you to test the gate function of the SCR in question 1 in the forward-biased condition. Be sure to indicate the proper position of the selector switch on the DMM.

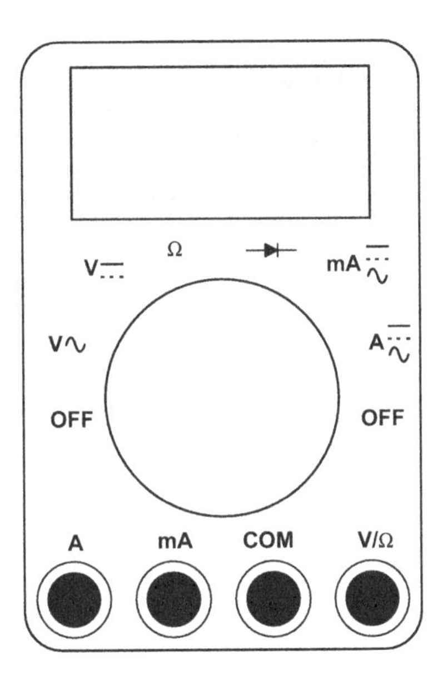

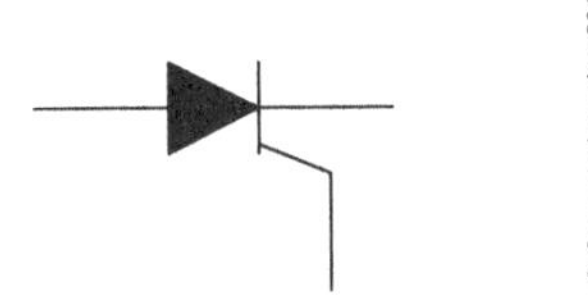

Name: ______________________________ Date: ____________________

TESTING ELECTRONIC DEVICES: DIAC

1. Show the connections that would allow you to test one portion of the diac. Be sure to indicate the proper position of the selector switch on the DMM.

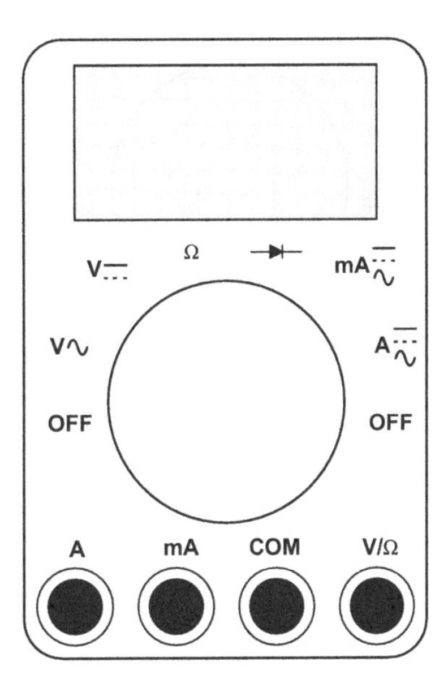

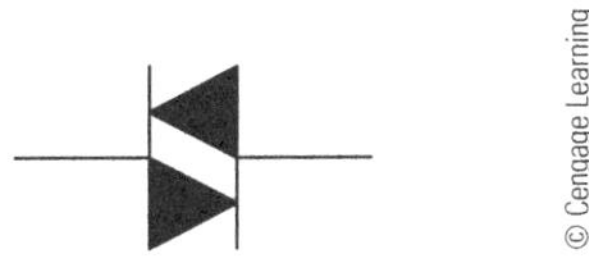

2. Show the connections that would allow you to test the remaining portion of the diac in question 1. Be sure to indicate the proper position of the selector switch on the DMM.

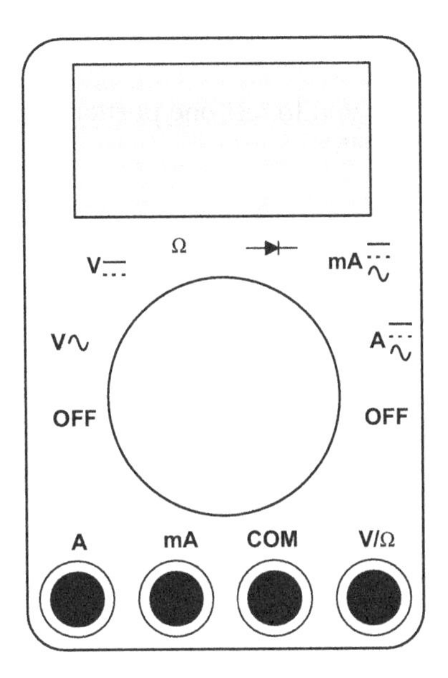

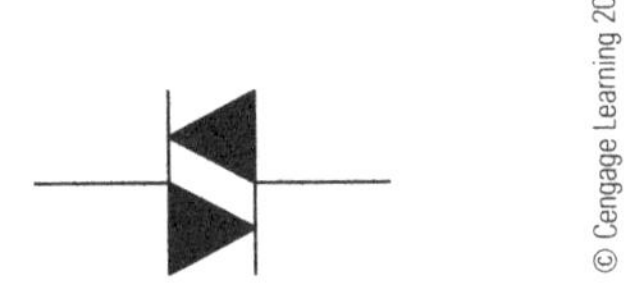

Name: ______________________ Date: ______________

TESTING ELECTRONIC DEVICES: TRIAC

1. Show the connections that would allow you to test one portion of the diode portion of a triac. Be sure to indicate the proper position of the selector switch on the DMM.

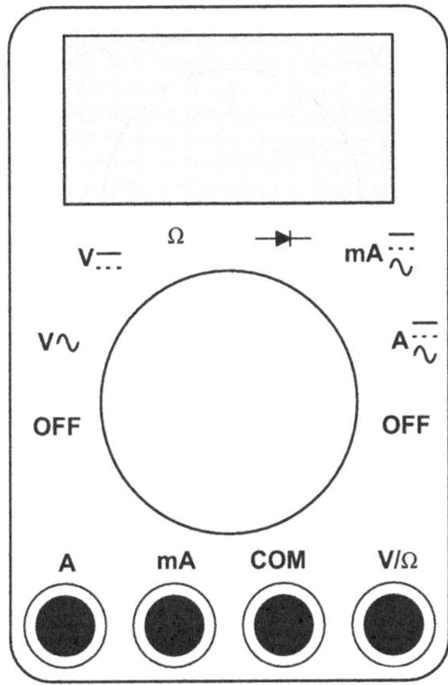

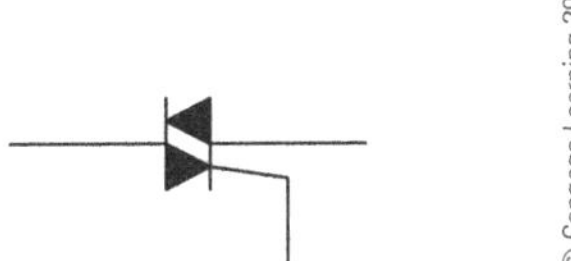

2. Show the connections that would allow you to test the remaining diode portion of the triac in question 1. Be sure to indicate the proper position of the selector switch on the DMM.

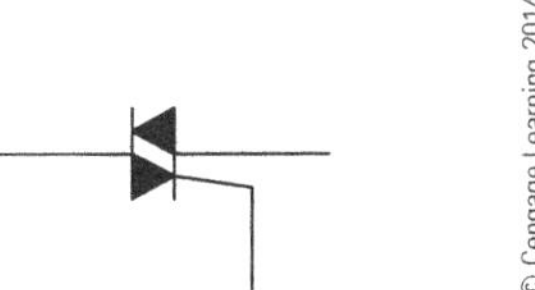

3. Show the connections that would allow you to test the gate function of one portion of the triac in question 1. Be sure to indicate the proper position of the selector switch on the DMM.

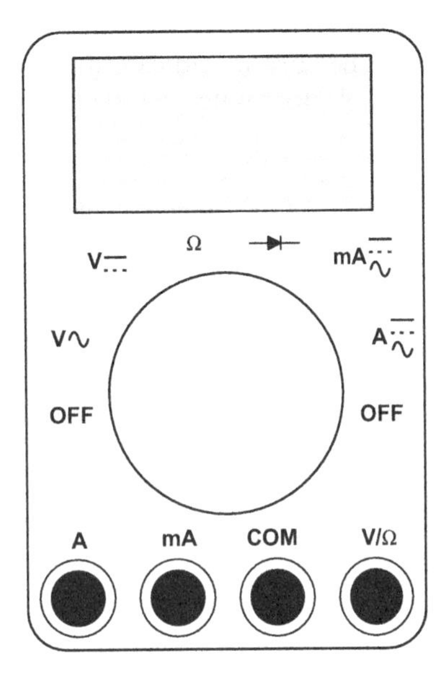

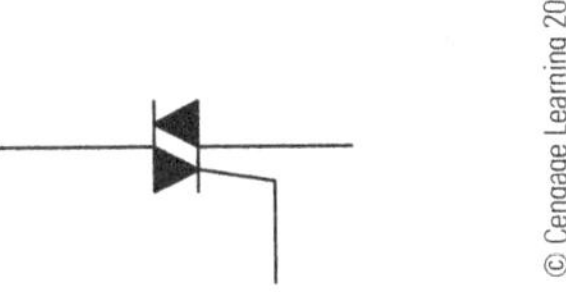

4. Show the connections that would allow you to test the gate function of the remaining portion of the triac in question 1. Be sure to indicate the proper position of the selector switch on the DMM.

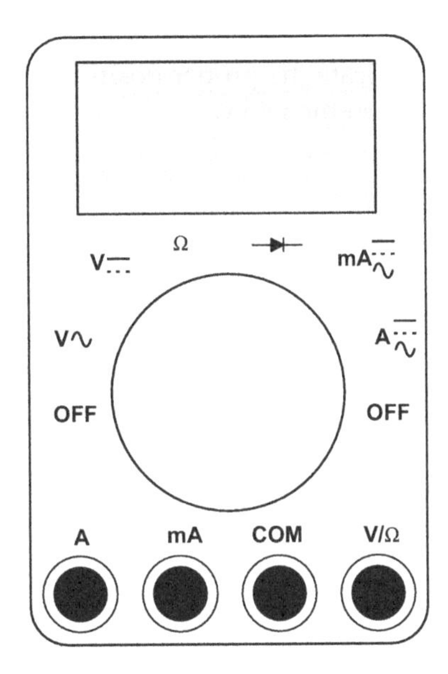

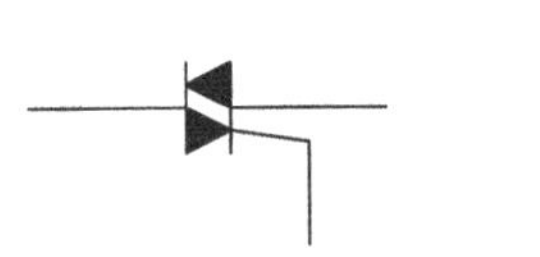

Name: ____________________ Date: ____________________

THE 555 TIMER

1. Correctly number the pins on the 555 timers. Label each pin as to its function.

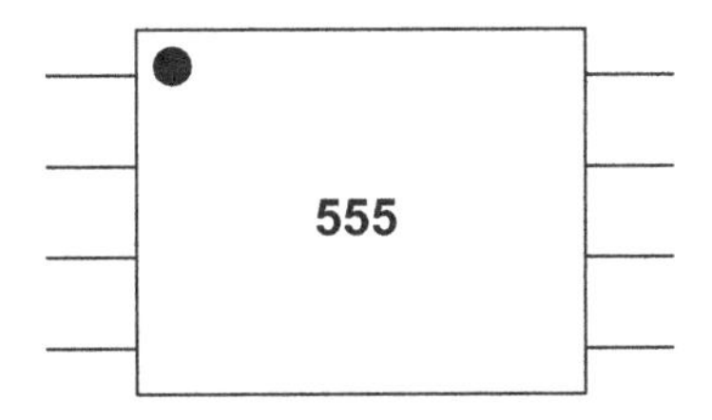

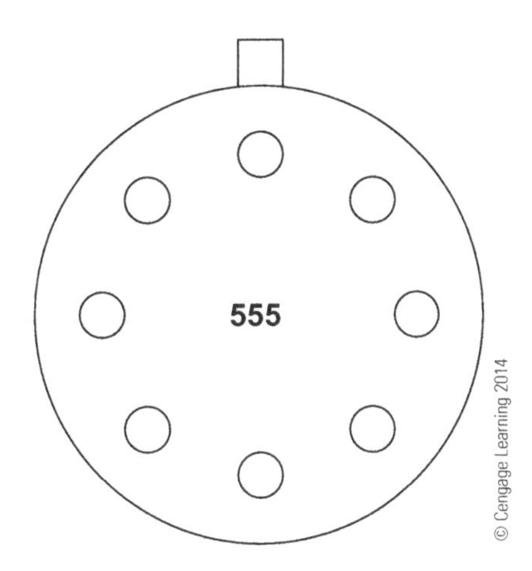

Name: ____________________ Date: ____________________

OPERATIONAL AMPLIFIERS

1. Correctly number the pins on the 741 op-amp. Label each pin as to its function.

Name: ______________________________ Date: ______________________

OP-AMP CIRCUITS

1. Match the circuit to the schematic.

_______ Noninverting amplifier

_______ Inverting amplifier

_______ Buffer amplifier

_______ Difference amplifier

_______ Summing amplifier

_______ Comparator

_______ Integrator

_______ Differentiator

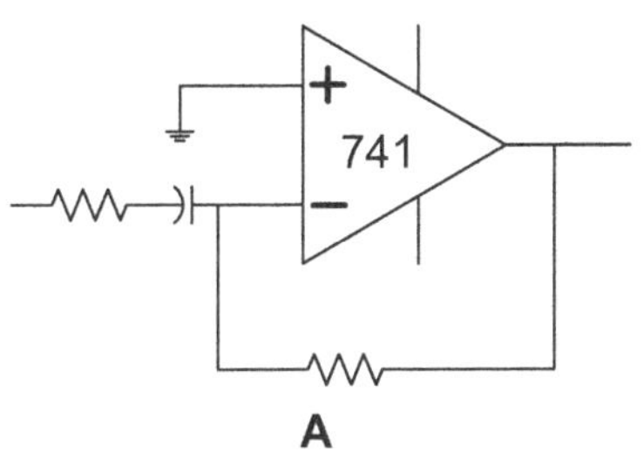

A

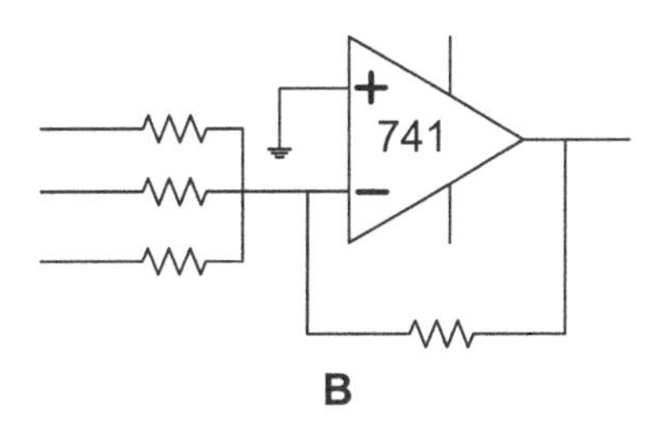

B

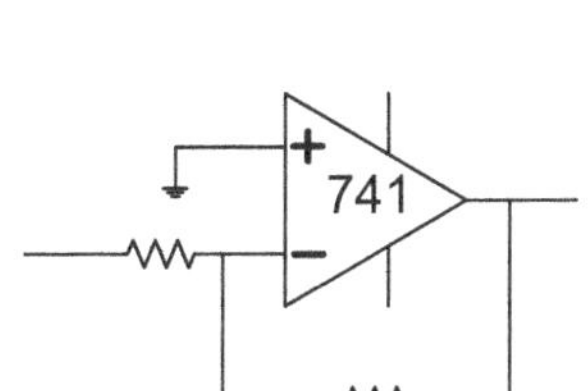

C

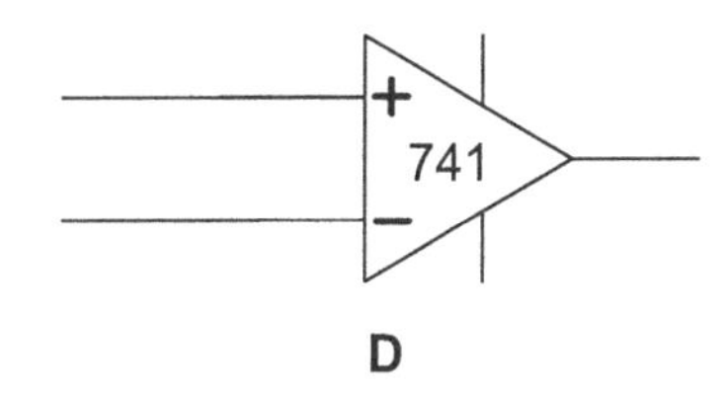

D

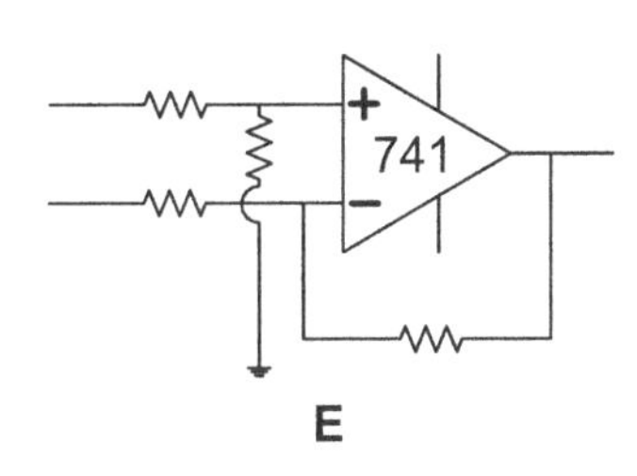

E

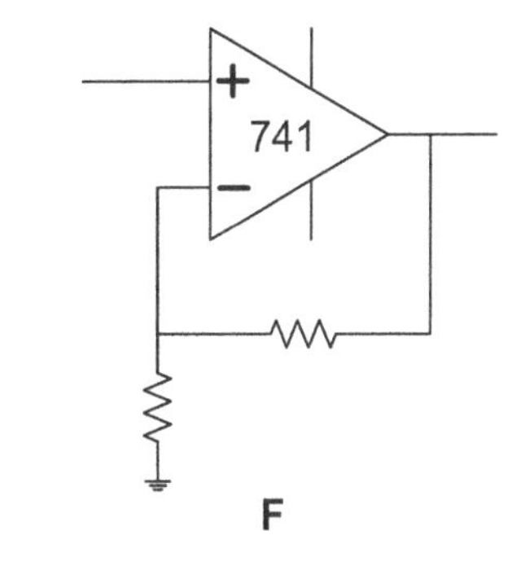

F

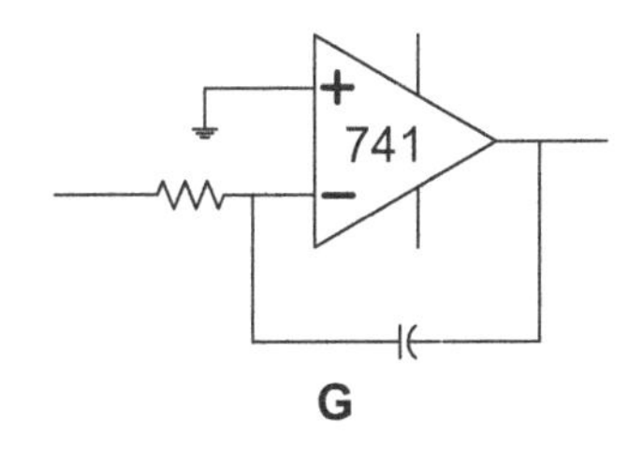

G

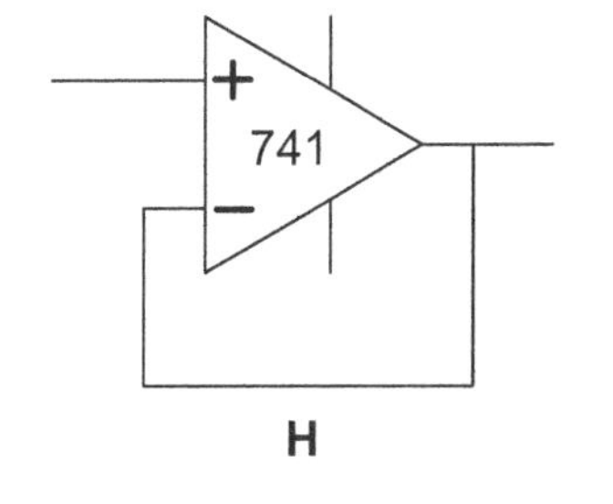

H

Name: ______________________________ Date: ____________________

DIGITAL LOGIC GATES

1. Match the gate to its symbol.

_______ Inverter	_______ OR gate
_______ AND gate	_______ NOR gate
_______ NAND gate	_______ XOR gate

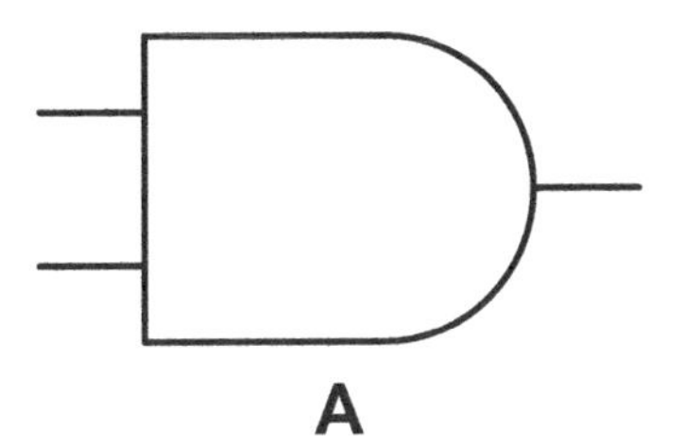

A

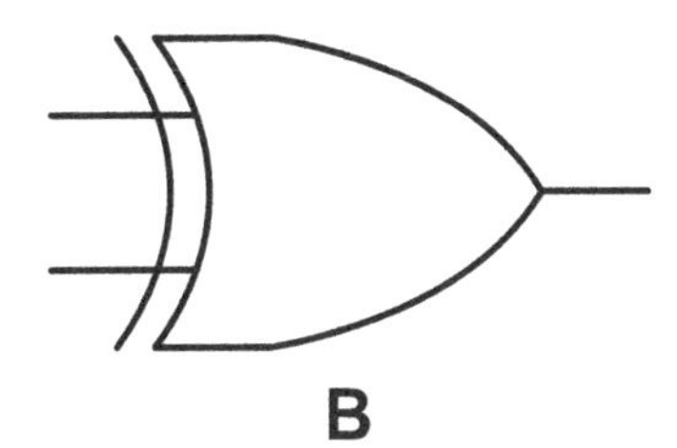

B

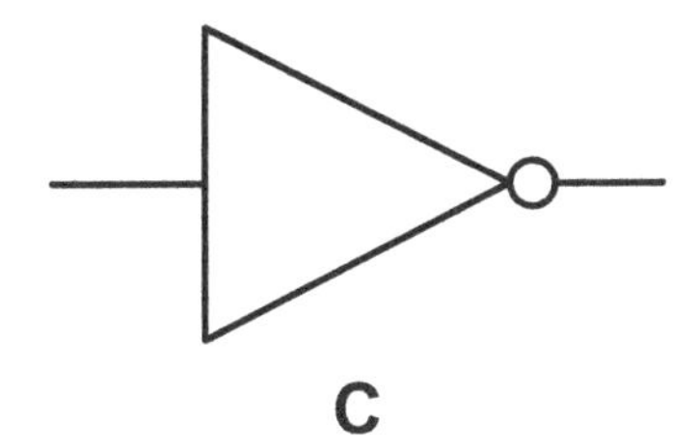

C

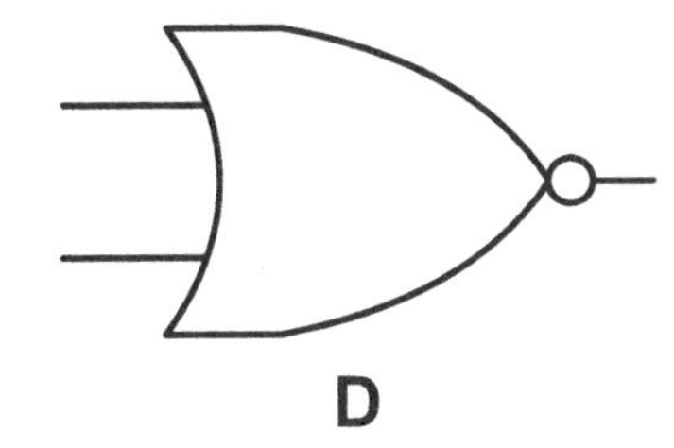

D

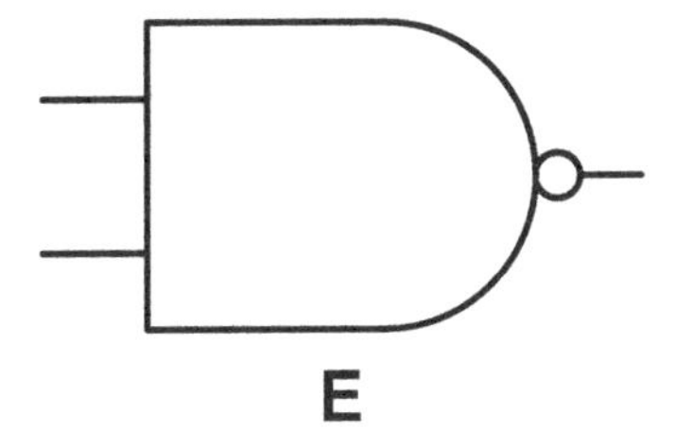

E

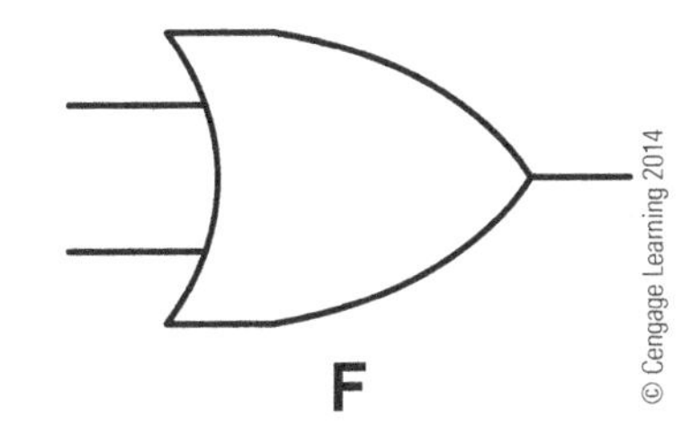

F

Name: ______________________ Date: ______________

TRUTH TABLES

1. Match the digital logic gate to its truth table, found on page 502.

TRUTH TABLE

a.

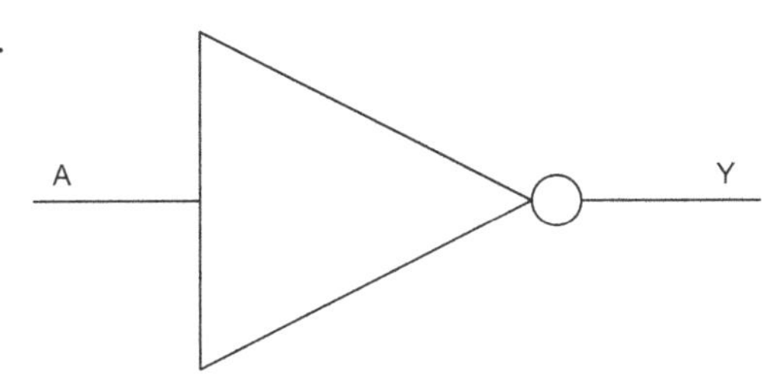

______

b.

______

c.

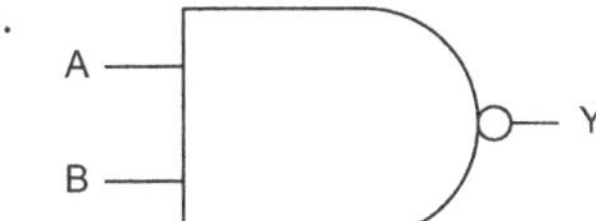

______

d.

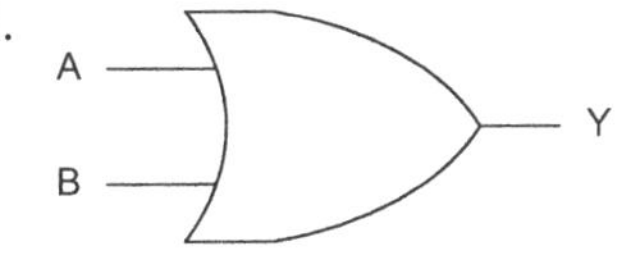

______

e.

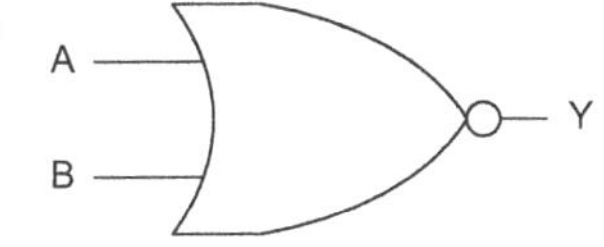

______

f.

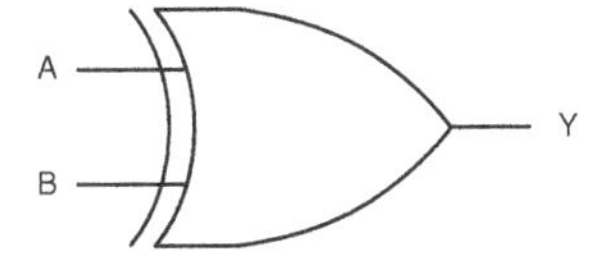

______

TRUTH TABLE A

Inputs		Output
A	B	Y
0	0	0
0	1	0
1	0	0
1	1	1

TRUTH TABLE B

Inputs		Output
A	B	Y
0	0	0
0	1	1
1	0	1
1	1	0

TRUTH TABLE C

Inputs		Output
A	B	Y
0	0	1
0	1	1
1	0	1
1	1	0

TRUTH TABLE D

Input	Output
A	Y
0	1
1	0

TRUTH TABLE E

Inputs		Output
A	B	Y
0	0	1
0	1	0
1	0	0
1	1	0

TRUTH TABLE F

Inputs		Output
A	B	Y
0	0	0
0	1	1
1	0	1
1	1	1

Name: ______________________________ Date: ______________

SWITCHING AMPLIFIER FIELD CURRENT CONTROLLER

1. Given the schematic diagram for a switching amplifier field current controller, use colored pencils, colored markers, or colored highlighters to identify the following areas:

   a. Feedback section

   b. Preamplifier section

   c. Sawtooth generator section

   d. Comparator section

   e. Power amplifier section

(Refer to schematic diagram on page 504.)

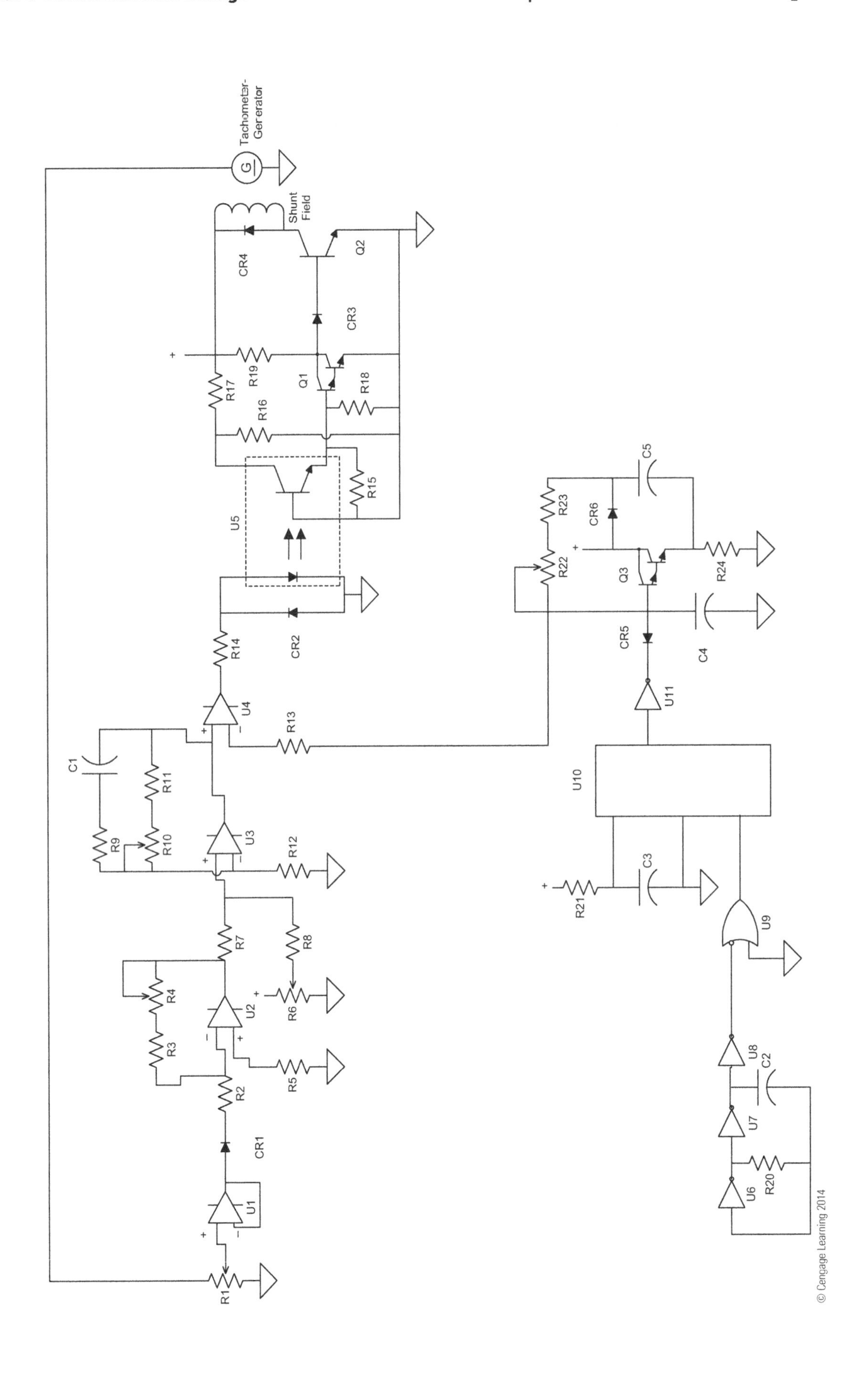

Tachometer-Generator
G
Shunt Field
CR4
Q2
CR3
R17
R19
Q1
R18
R16
R15
U5
CR2
R14
U4
R13
C1
R11
R10
R9
U3
R12
R7
R8
R4
U2
R6
R3
R5
R2
CR1
U1
R1
C5
R23
CR6
R22
Q3
R24
CR5
C4
U11
U10
C3
R21
U9
U8
C2
U7
U6
R20
© Cengage Learning 2014

Name: ______________________________ Date: ____________________

SWITCHING AMPLIFIER FIELD CURRENT CONTROLLER: WAVEFORMS

1. Given the schematic diagram for a switching amplifier field current controller, draw the expected waveforms at the indicated locations.

(Refer to schematic diagram on page 506.)

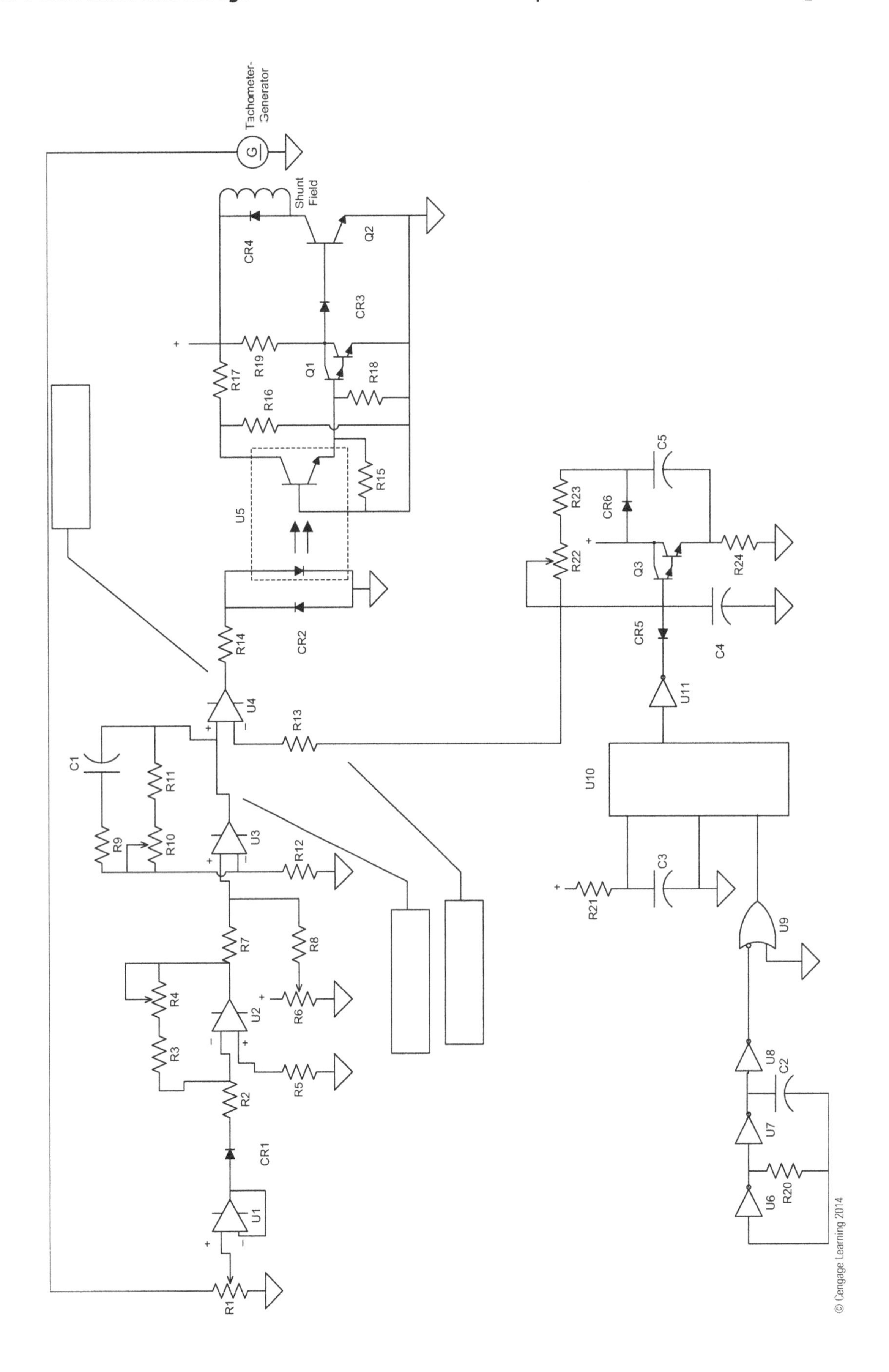

Tachometer-
Generator
G
Shunt
Field
CR4
Q2
CR3
R17
R19
Q1
R18
R16
R15
U5
CR2
R14
U4
R13
C1
R11
R9
R10
U3
R12
R7
R8
R4
U2
R6
R3
R5
R2
CR1
U1
R1
C5
R23
CR6
R22
Q3
R24
CR5
C4
U11
U10
C3
R21
U9
U8
C2
U7
R20
U6
© Cengage Learning 2014

Name: ______________________ Date: ______________

SCR ARMATURE VOLTAGE CONTROLLER

1. Given the schematic diagram for an SCR armature voltage controller, use colored pencils, colored markers, or colored highlighters to identify the following areas:

    a. Null detector, pulse shaper, and sawtooth generator section

    b. Comparator section

    c. Pulse-generator section

    d. Output section

(Refer to schematic diagram on page 508.)

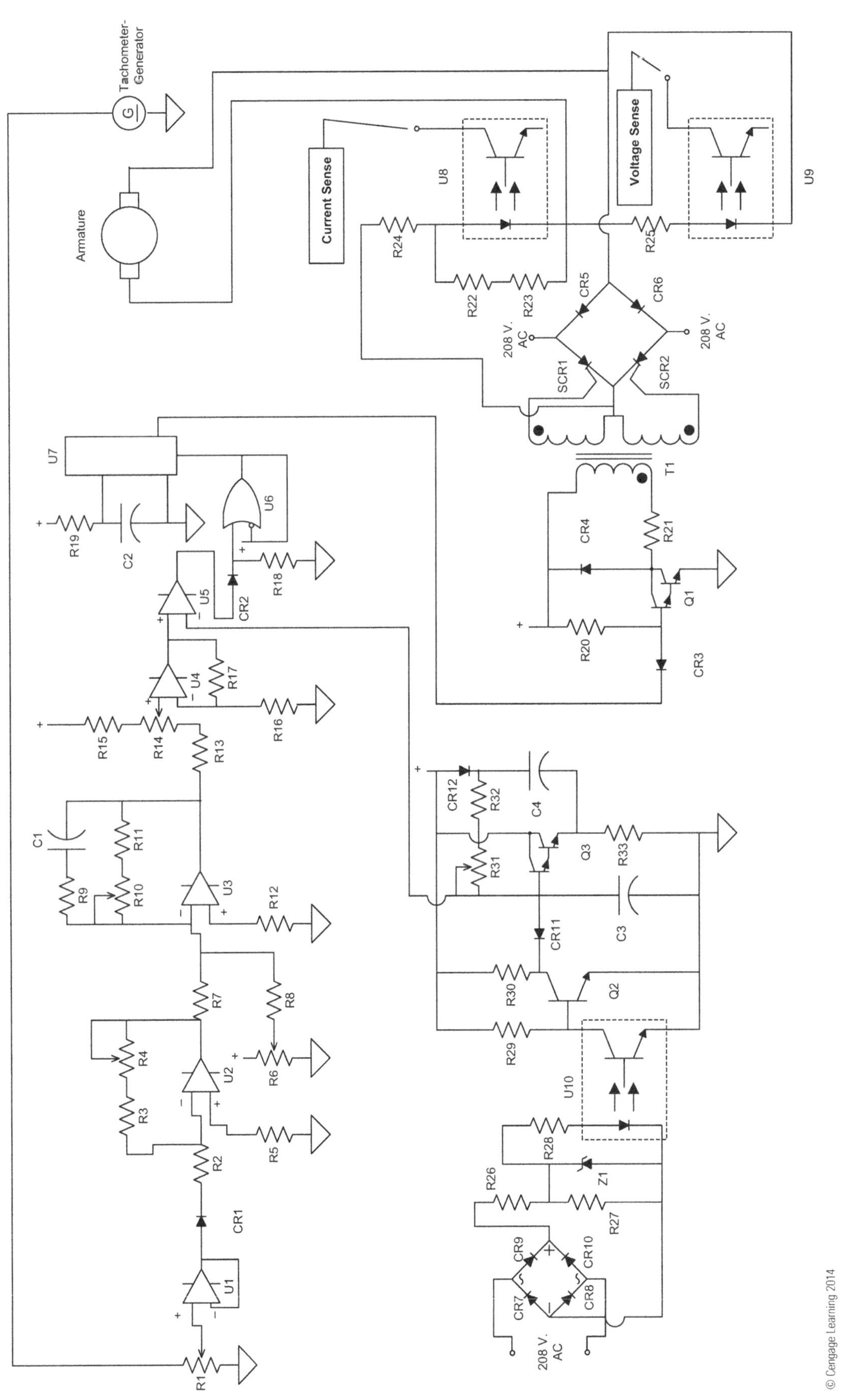
Tachometer-
Generator
G
Armature
Current Sense
Voltage Sense
U8
U9
R24
R25
R22
R23
CR5
CR6
208 V.
AC
208 V.
AC
SCR1
SCR2
T1
U7
U6
R19
C2
R18
CR2
U5
U4
R17
R16
R15
R14
R13
CR4
R21
Q1
R20
CR3
CR12
R32
C4
Q3
R33
R31
C1
R11
R9
R10
U3
R12
CR11
C3
R30
Q2
R7
R8
R4
R3
U2
R6
R29
U10
R5
R2
R28
Z1
R26
R27
CR1
CR9
CR10
U1
CR7
CR8
208 V.
AC
R1

Name: ______________________________ Date: ____________________

SCR ARMATURE VOLTAGE CONTROLLER: WAVEFORMS

1. Given the schematic diagram for an SCR armature voltage controller, draw the expected waveforms at the indicated locations.

(Refer to schematic diagram on page 510.)

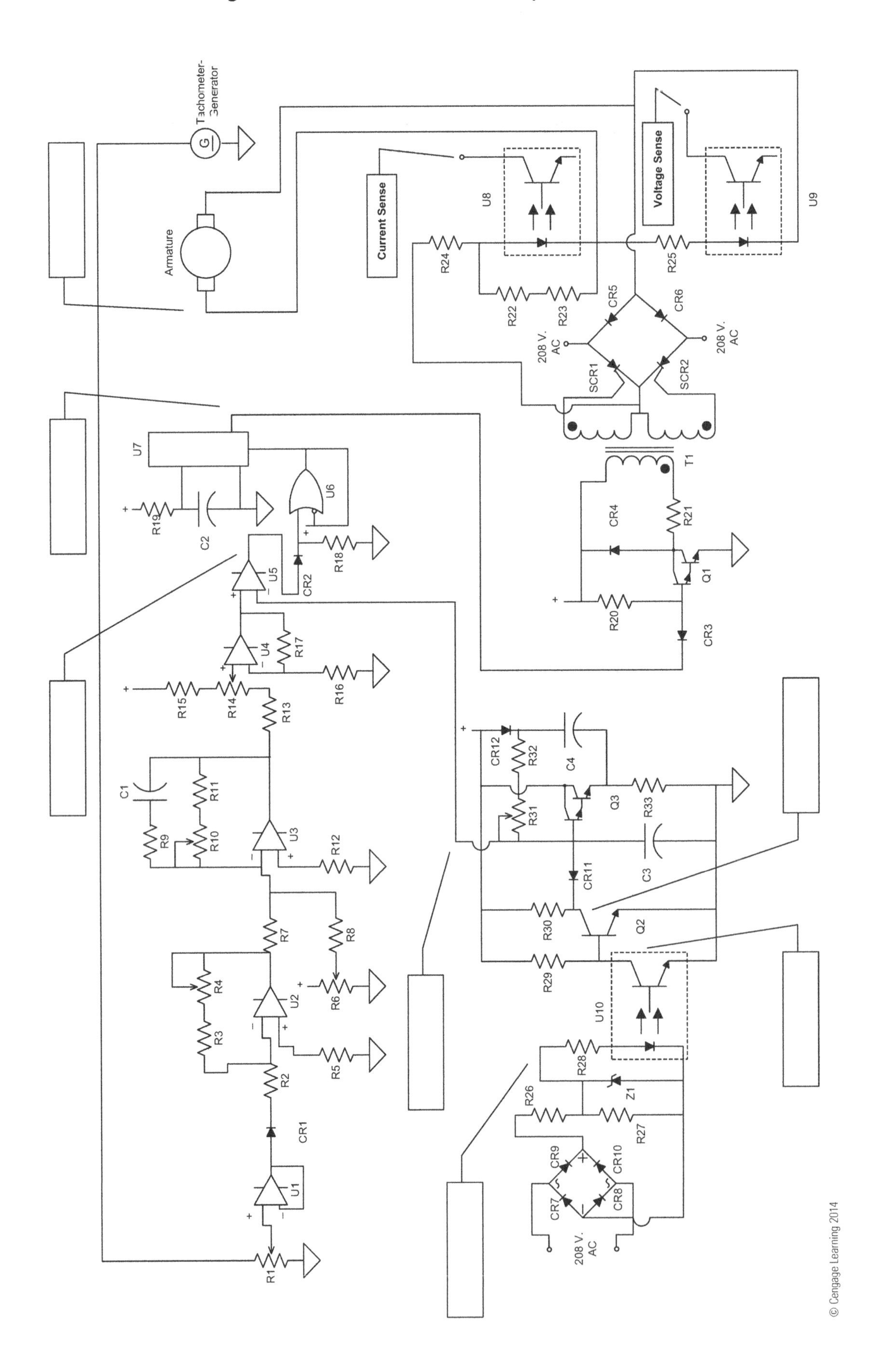
Tachometer-
Generator
Armature
Current Sense
Voltage Sense
U1
U2
U3
U4
U5
U6
U7
U8
U9
U10
208 V.
AC
T1
Q1
Q2
Q3
SCR1
SCR2
CR1
CR2
CR3
CR4
CR5
CR6
CR7
CR8
CR9
CR10
CR11
CR12
Z1
C1
C2
C3
C4
R1
R2
R3
R4
R5
R6
R7
R8
R9
R10
R11
R12
R13
R14
R15
R16
R17
R18
R19
R20
R21
R22
R23
R24
R25
R26
R27
R28
R29
R30
R31
R32
R33

Name: ______________________________ Date: ____________________

CHOPPERS

1. Given the schematic diagrams, identify the buck chopper and the boost chopper.

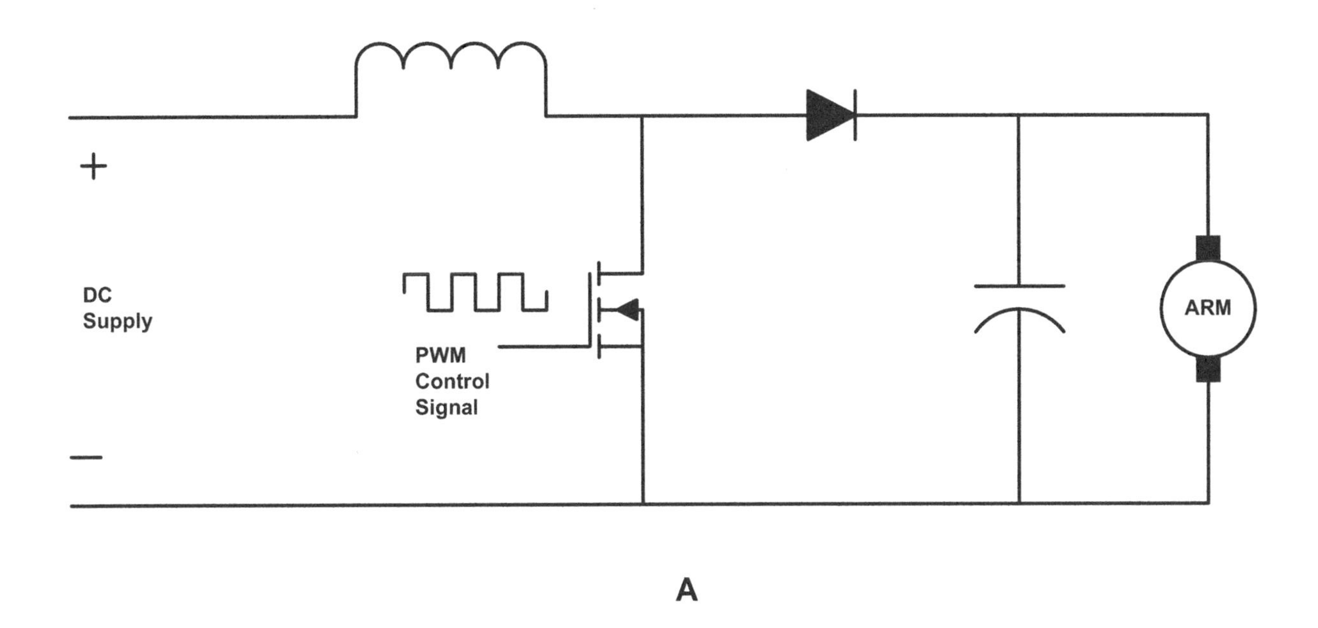

____

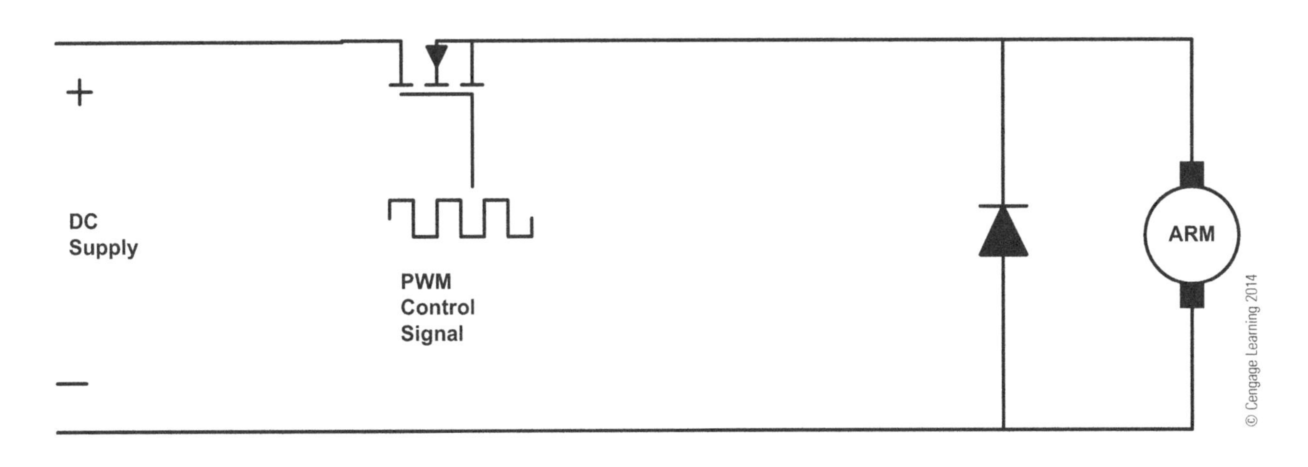

B

____

Name: ______________________________ Date: ______________

FOUR QUADRANTS OF MOTOR OPERATION

1. Given the descriptions of motor operation, place the letter that represents the operating characteristic of the motor in the correct quadrant.

   a. CCW rotation, torque –, motoring reverse

   b. CW rotation, torque +, motoring forward

   c. CCW rotation, torque +, braking reverse

   d. CW rotation, torque –, braking forward

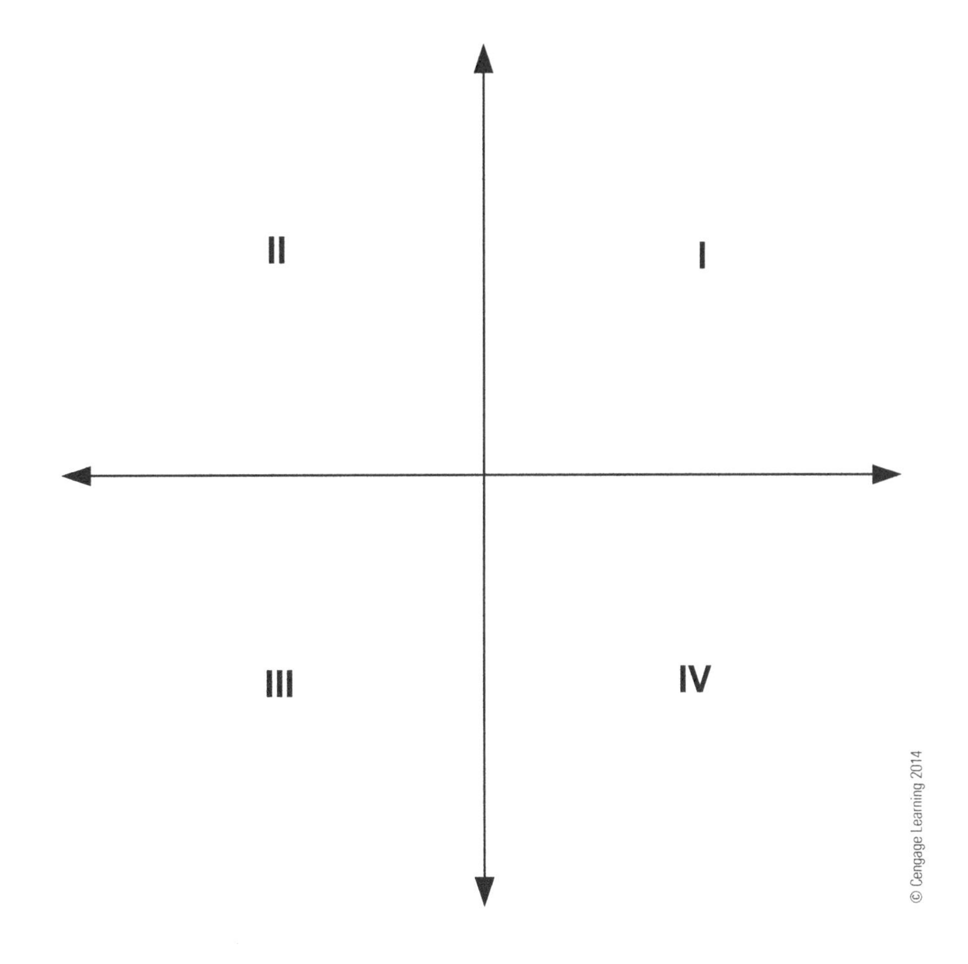

Name: ______________________________ Date: ______________________

VARIABLE VOLTAGE INVERTER

1. Given the schematic diagram of a variable voltage inverter, use colored pencils, colored markers, or colored highlighters to identify the following areas:

   a. Feedback section

   b. Null detector and sawtooth generator section

   c. Comparator section

   d. Pulse-generator section

   e. Output section

(Refer to schematic diagram on page 516.)

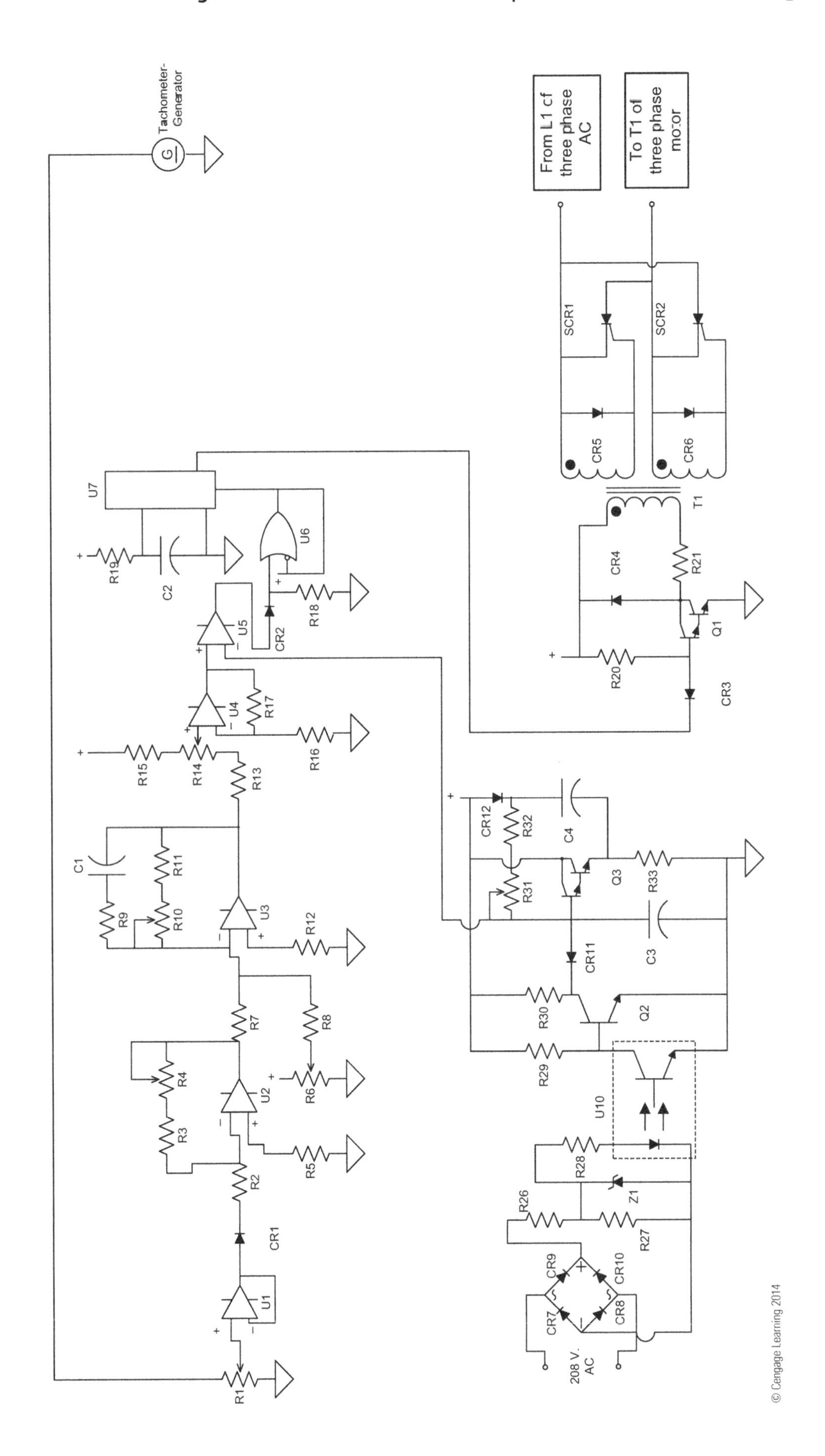
Tachometer-
Generator
From L1 cf
three phase
AC
To T1 of
three phase
mo:or
208 V.
AC
© Cengage Learning 2014

Name: ______________________________ Date: ______________________

VARIABLE VOLTAGE INVERTER: WAVEFORMS

1. Given the schematic diagram for a variable voltage inverter, draw the expected waveforms at the indicated locations.

(Refer to schematic diagram on page 518.)

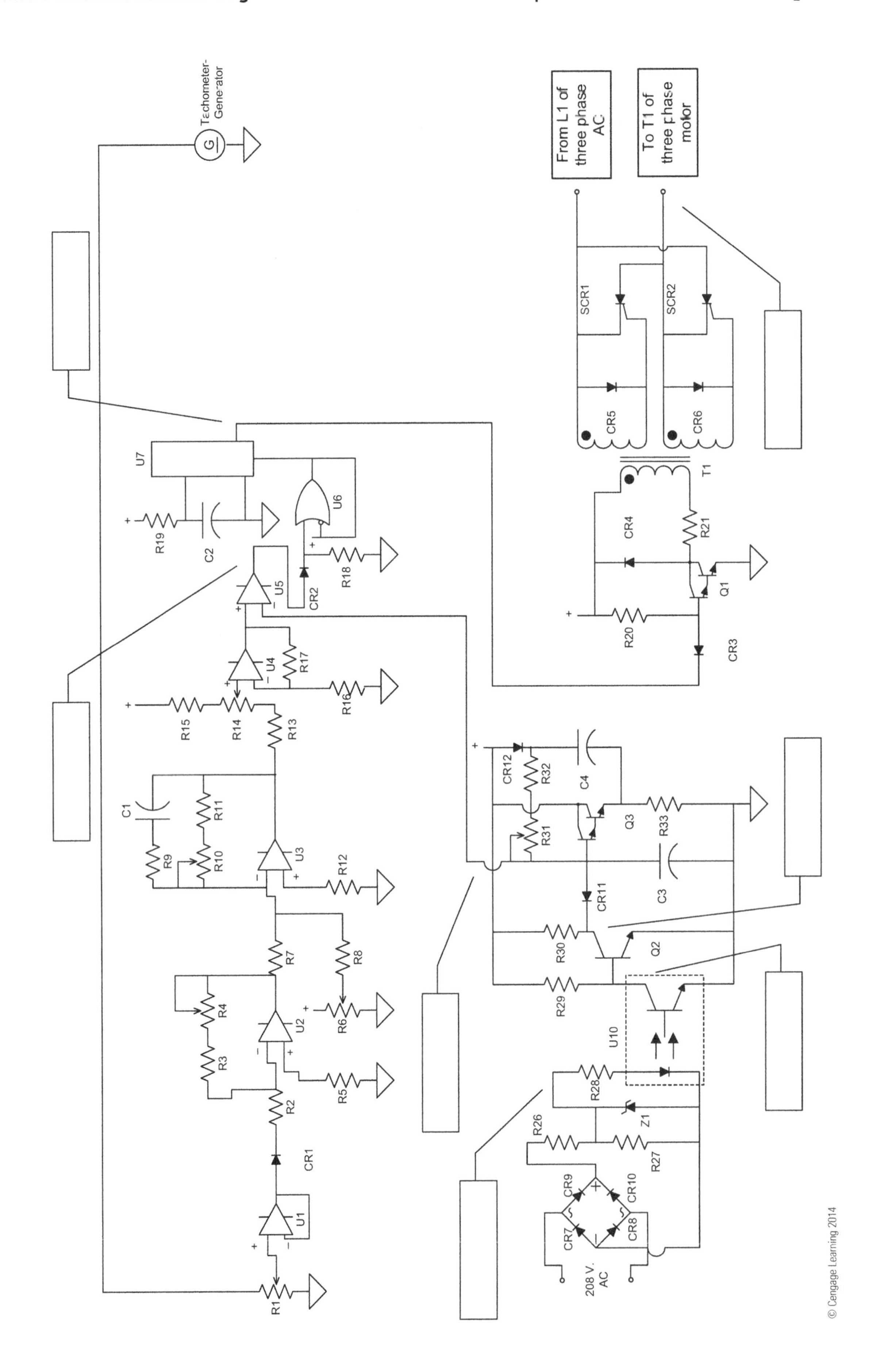
Tachometer-Generator
G
From L1 of three phase AC
To T1 of three phase motor
SCR1
SCR2
CR5
CR6
T1
CR4
R21
Q1
R20
CR3
U7
U6
R19
C2
R18
CR2
U5
U4
R17
R16
R15
R14
R13
C1
R11
R9
R10
U3
R12
R7
R8
R4
R3
U2
R6
R5
R2
CR1
U1
R1
CR12
R32
C4
R31
Q3
R33
C3
CR11
R30
Q2
R29
U10
R28
Z1
R26
R27
CR9
CR10
CR7
CR8
208 V. AC

Name: ______________________________ Date: ____________________

PROGRAMMABLE LOGIC CONTROLLER INPUT/OUTPUT WIRING DIAGRAM

1. Create a programmable logic controller (PLC) input/output (I/O) wiring diagram for the given ladder diagram.

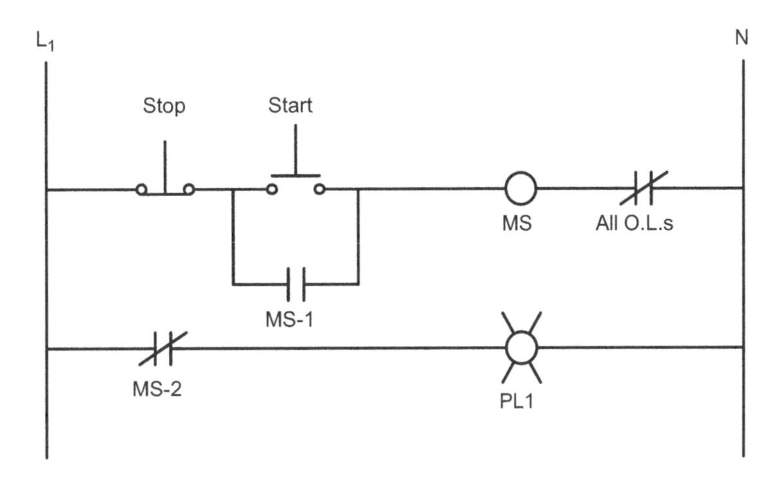

INPUT

A 12345678
F
B 12345678

1 2 3 4 5 6 7 8 9 10 11 12 13 14 15 16 N

OUTPUT

A 12345678
F
B 12345678

1 2 3 4 5 6 7 8 H 11 12 13 14 15 16 17 18 H

2. Create a PLC I/O wiring diagram for the given ladder diagram.

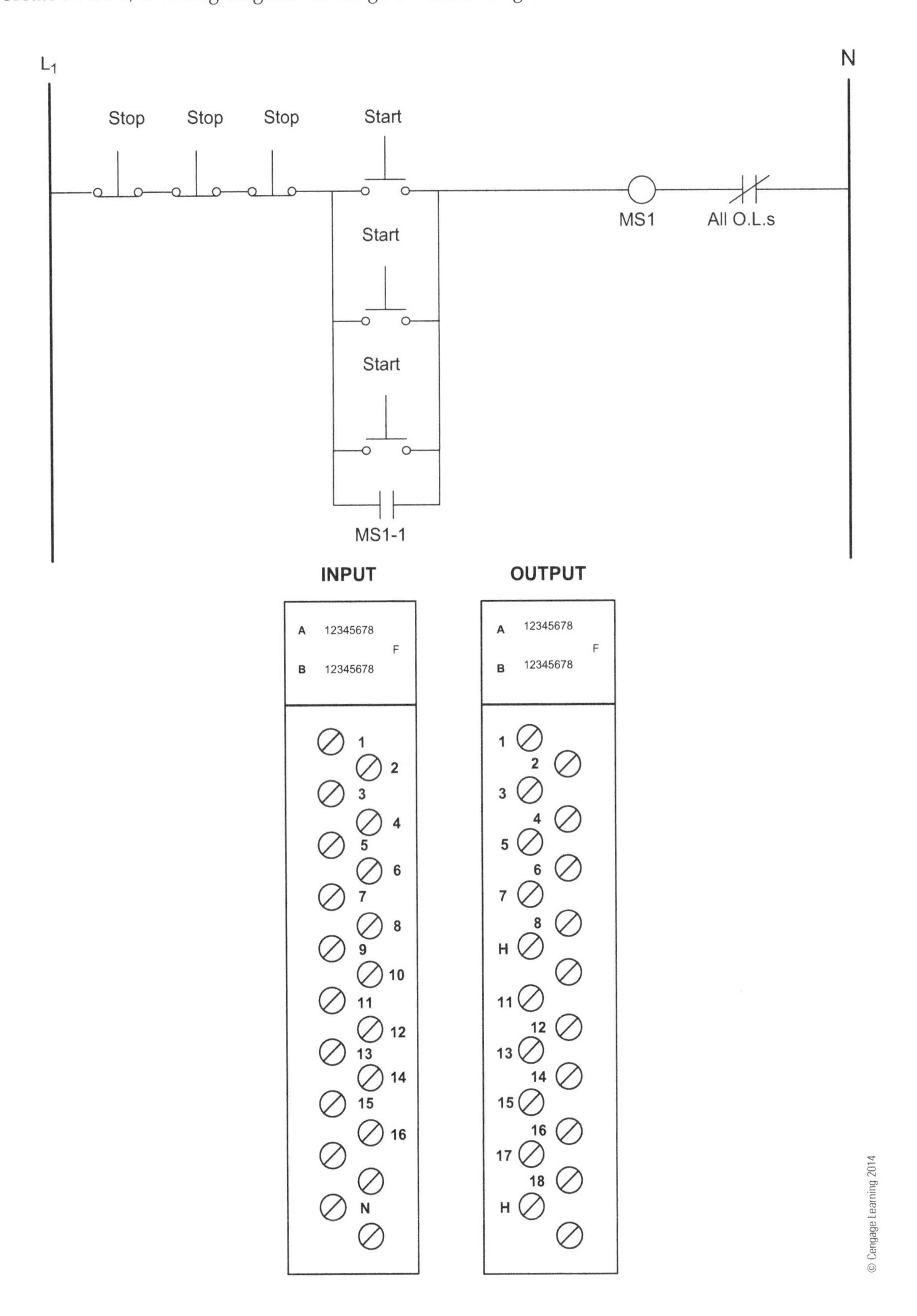

3. Create a PLC I/O wiring diagram for the given ladder diagram.

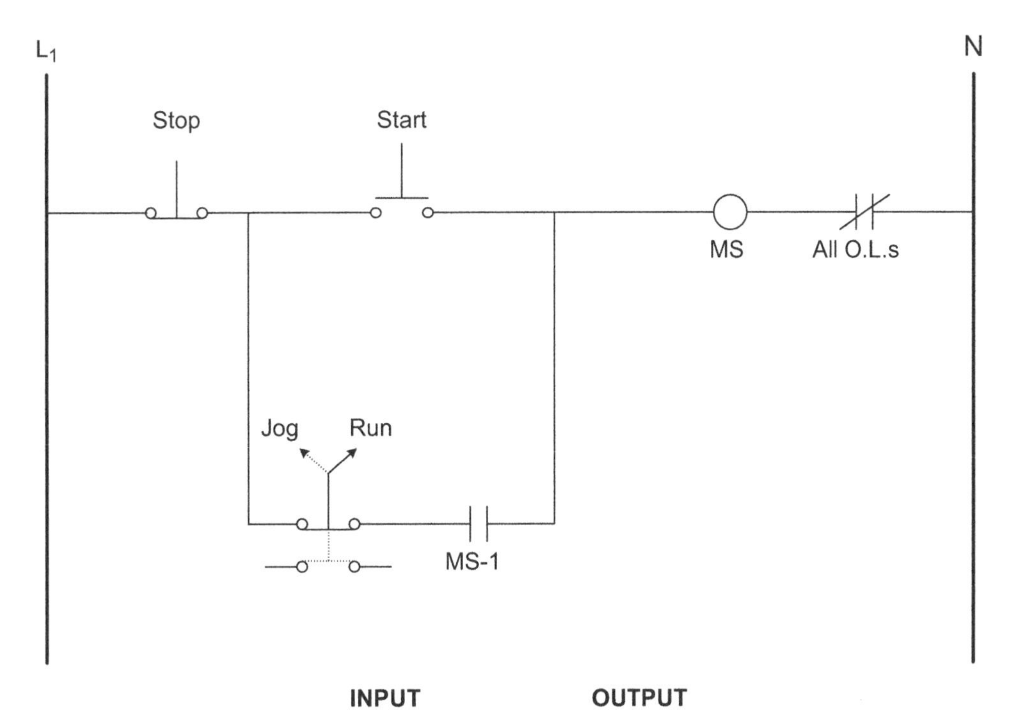

INPUT

A 12345678
F
B 12345678
1 2 3 4 5 6 7 8 9 10 11 12 13 14 15 16
N

OUTPUT

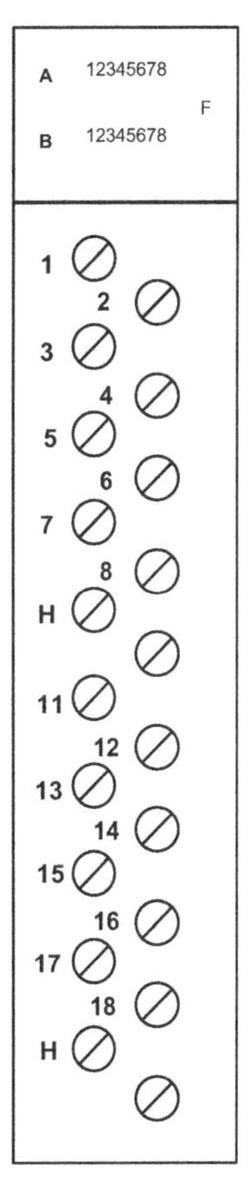

4. Create a PLC I/O wiring diagram for the given ladder diagram.

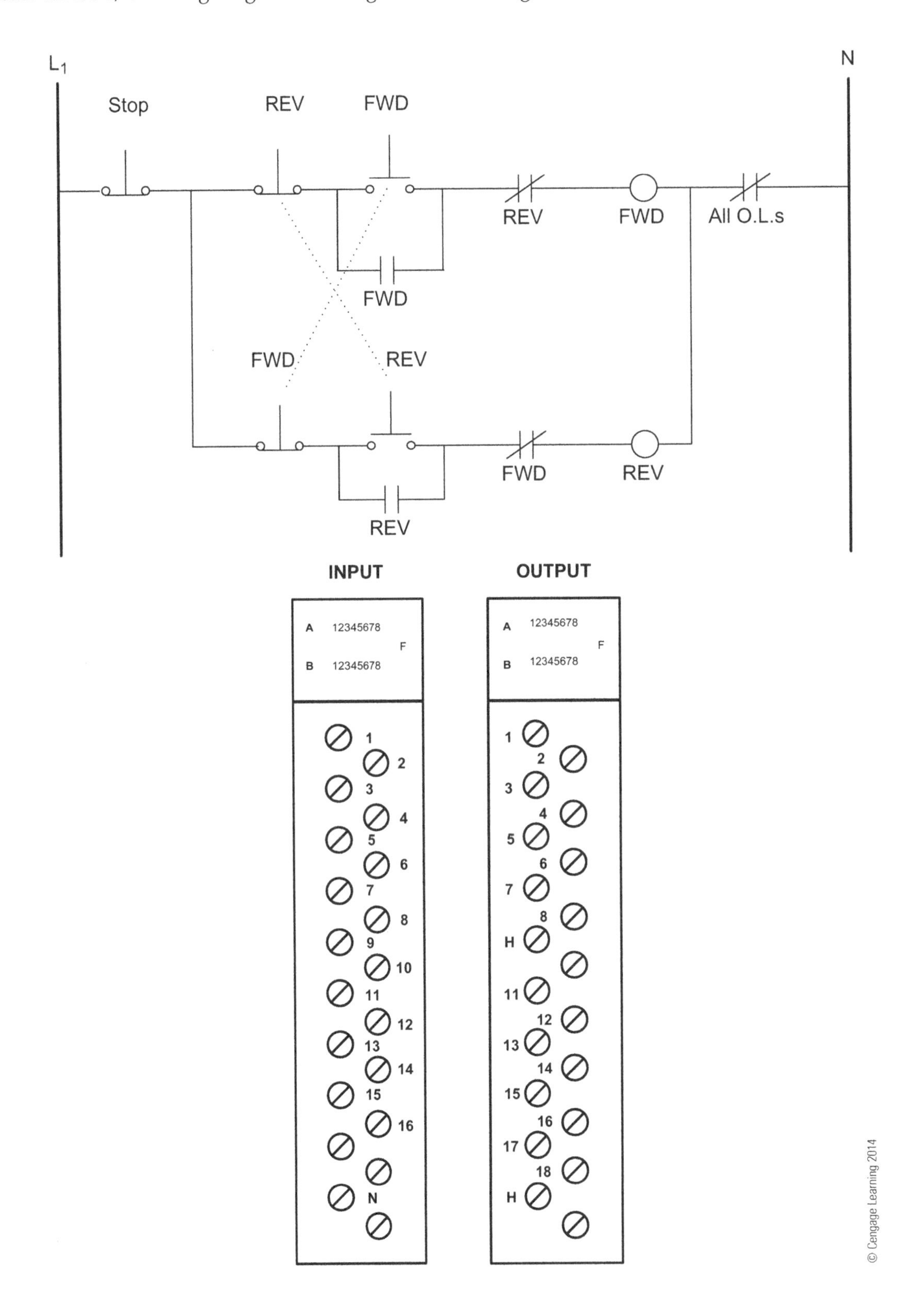

Name: ______________________________ Date: ______________

PLC PROGRAM CONVERSION

1. Create a PLC program for the given ladder diagram.

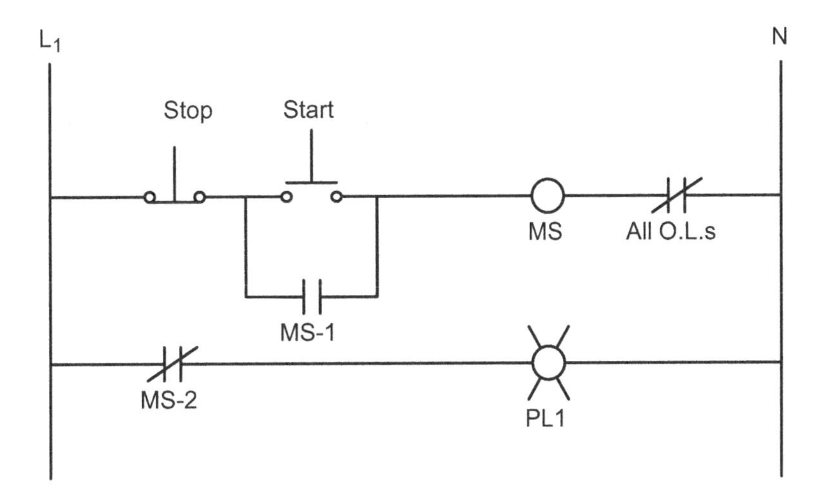

2. Create a PLC program for the given ladder diagram.

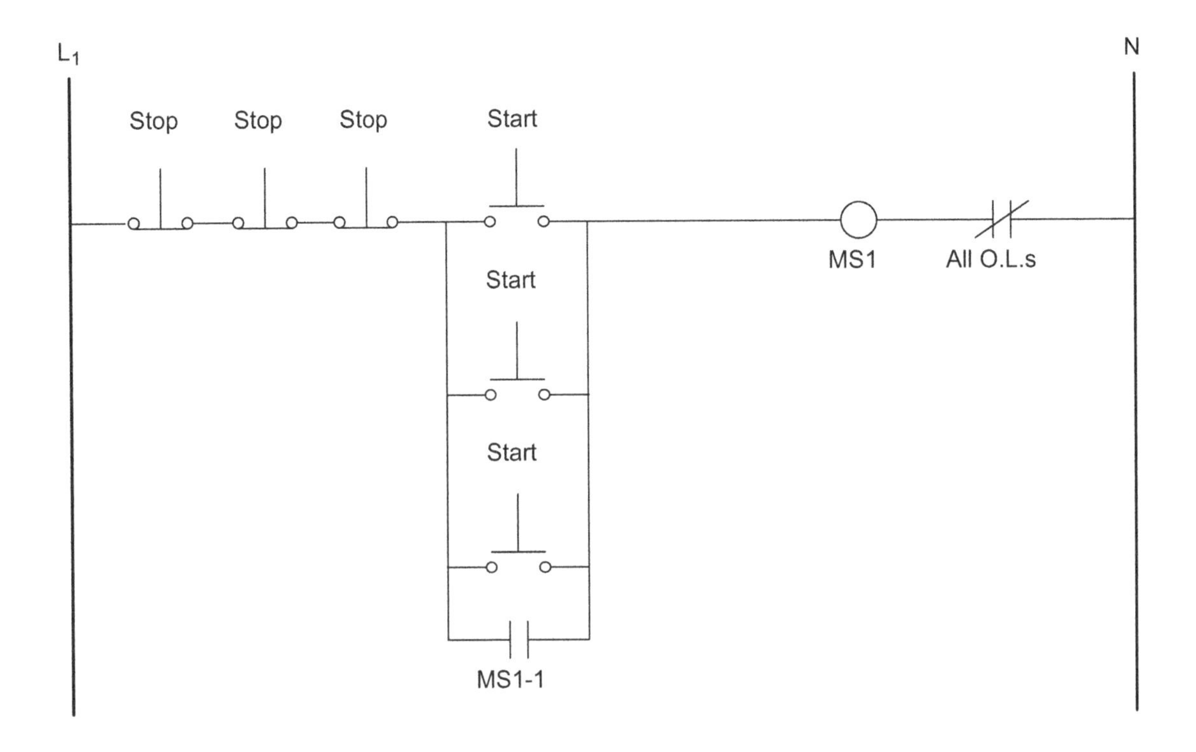

3. Create a PLC program for the given ladder diagram.

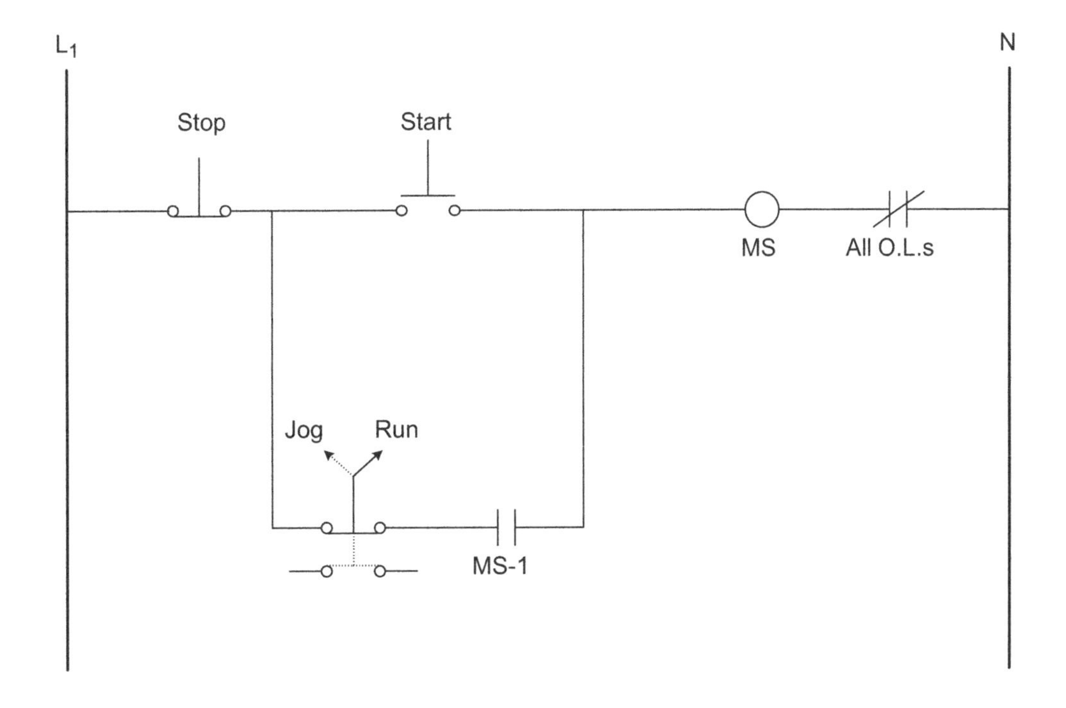

4. Create a PLC program for the given ladder diagram.

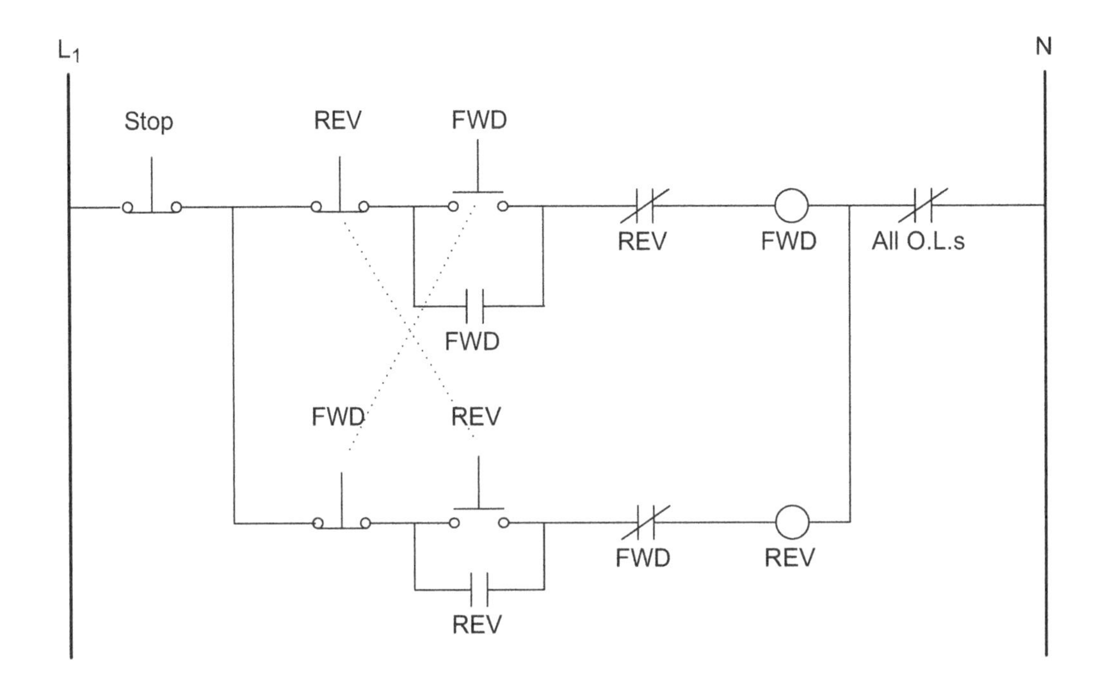

Name: ______________________ Date: ______________

PLC PROJECT 1

1. Develop a PLC program that uses a thermostat to control the operation of a fan.

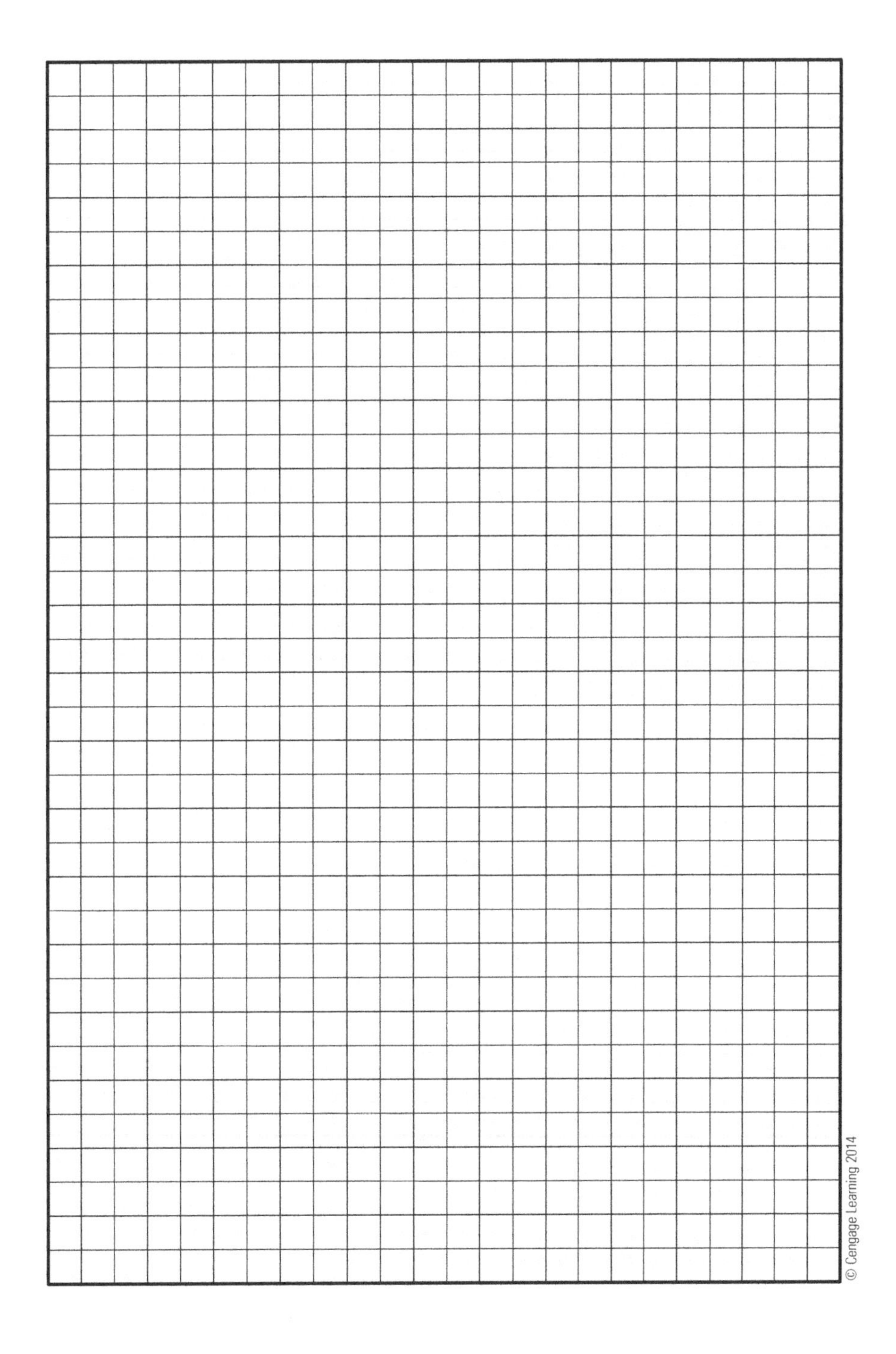

2. Modify the circuit in question 1 to include a red pilot light, which indicates that the fan is not running, and a green pilot light, which indicates that the fan is running.

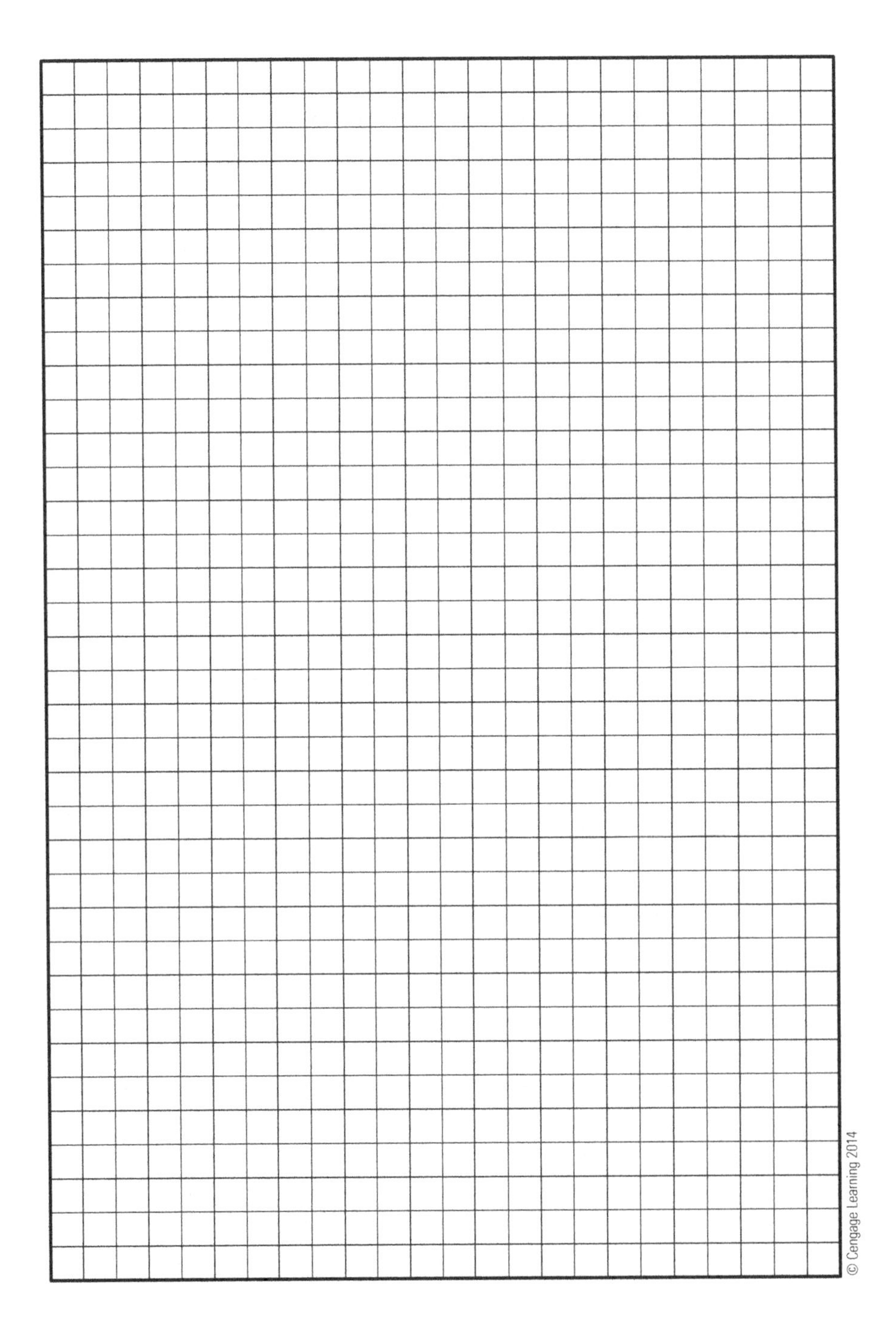

3. Develop the I/O wiring diagram for the circuit in question 2.

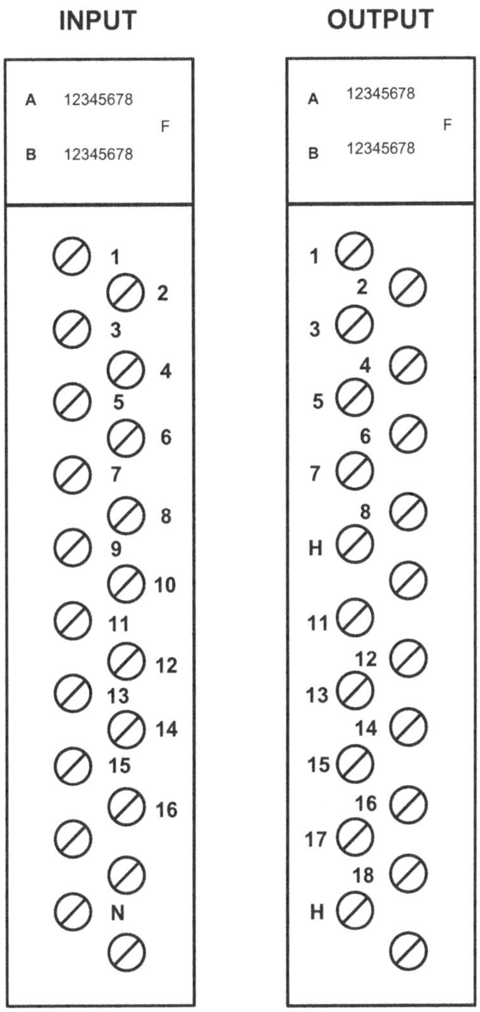

INPUT

A 12345678
F
B 12345678

1
2
3
4
5
6
7
8
9
10
11
12
13
14
15
16
N

OUTPUT

A 12345678
F
B 12345678

1
2
3
4
5
6
7
8
H
11
12
13
14
15
16
17
18
H

Name: ______________________ Date: ______________

PLC PROJECT 2

1. Develop a PLC program that uses a toggle switch to control the operation of a blower.

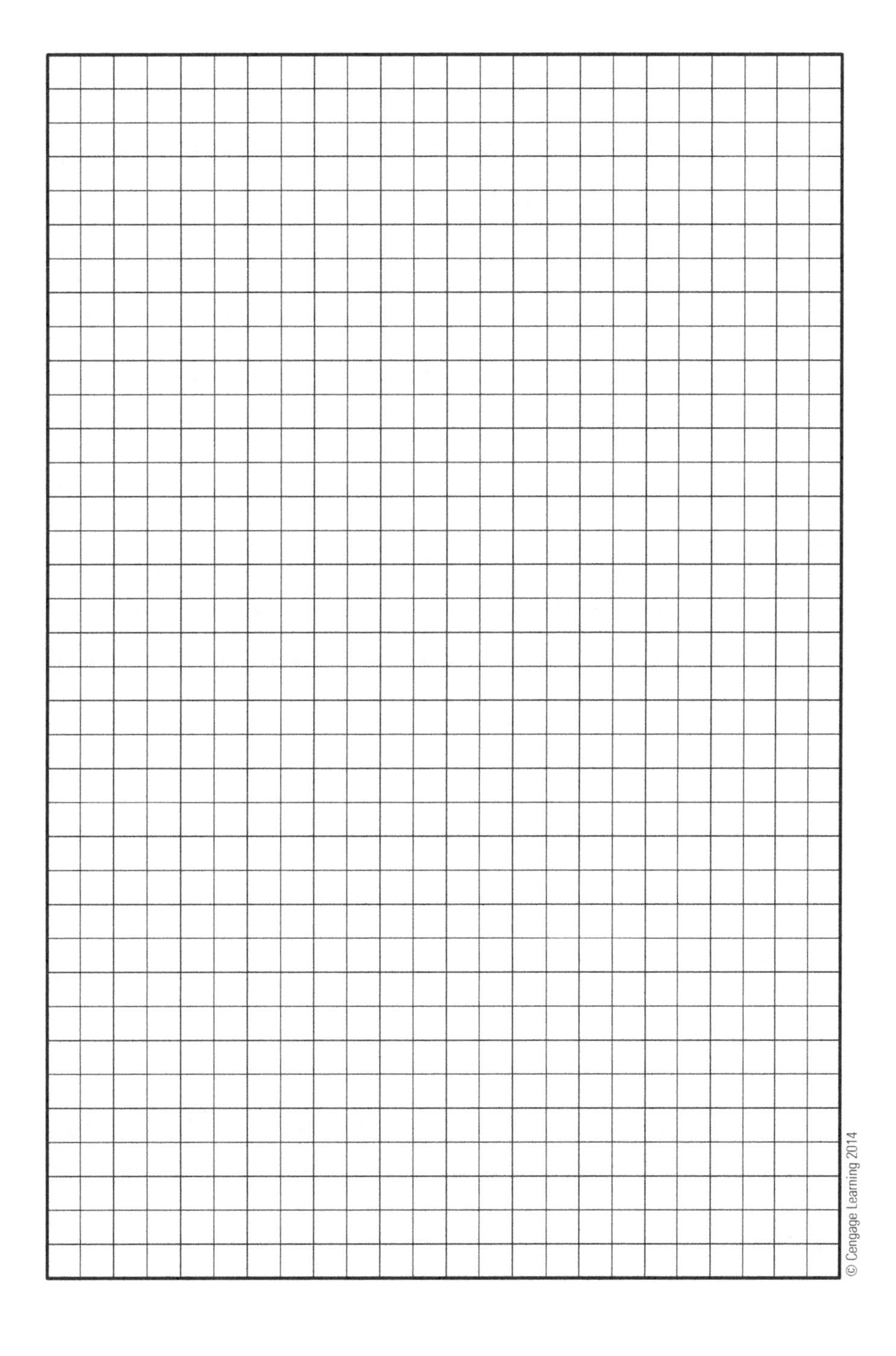

2. Modify the circuit in question 1 to include a flow switch to detect air flow through a filter. Include a red pilot light, which indicates a dirty filter (reduced air flow), and a green pilot light, which indicates normal air flow.

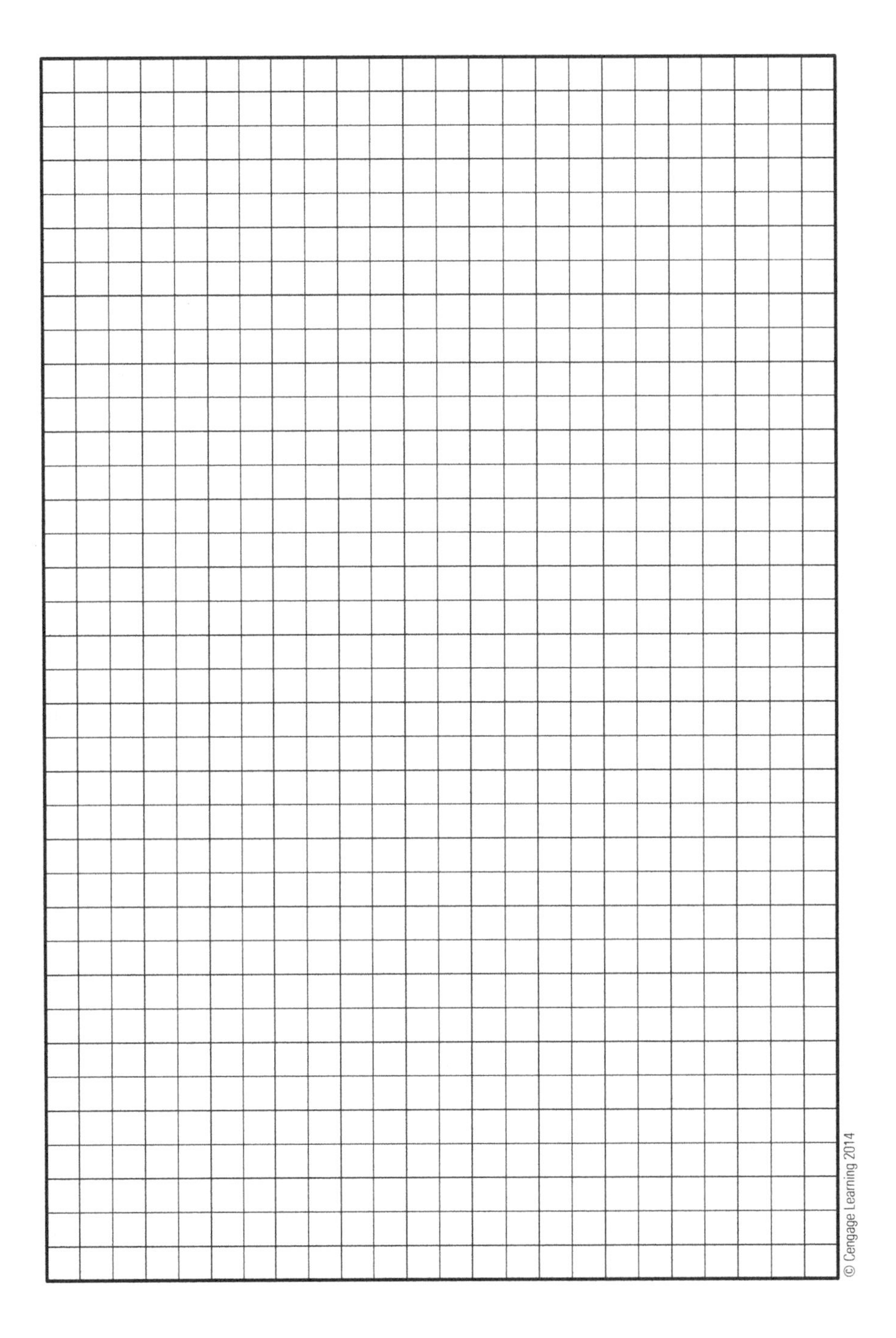

3. Develop the I/O wiring diagram for the circuit in question 2.

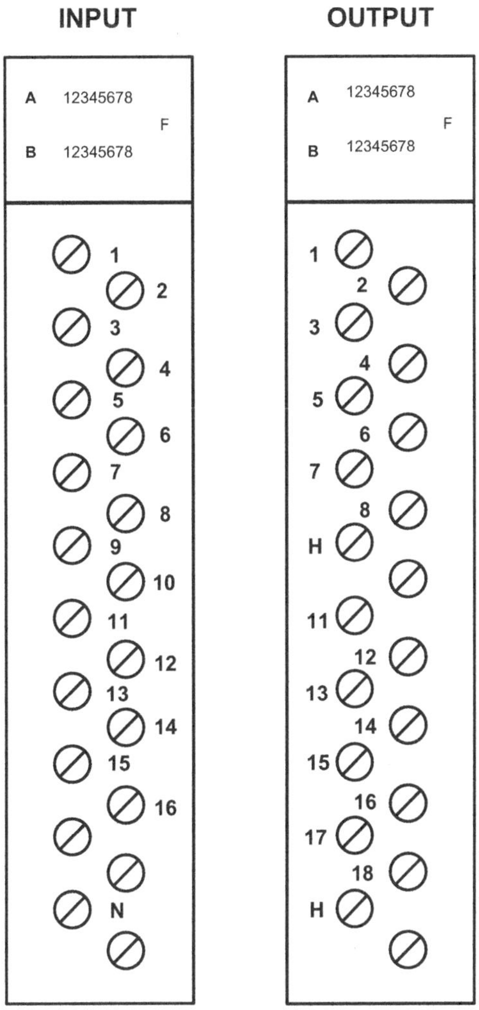

INPUT

A 12345678
F
B 12345678

1 2 3 4 5 6 7 8 9 10 11 12 13 14 15 16 N

OUTPUT

A 12345678
F
B 12345678

1 2 3 4 5 6 7 8 H 11 12 13 14 15 16 17 18 H

Name: ______________________ Date: ______________

PLC PROJECT 3

1. Develop a PLC program that performs as follows:
   a. Pressing and releasing a momentary pushbutton causes a cylinder to advance, even after the pushbutton is released.
   b. When the cylinder strikes a limit switch, the cylinder's motion is stopped.

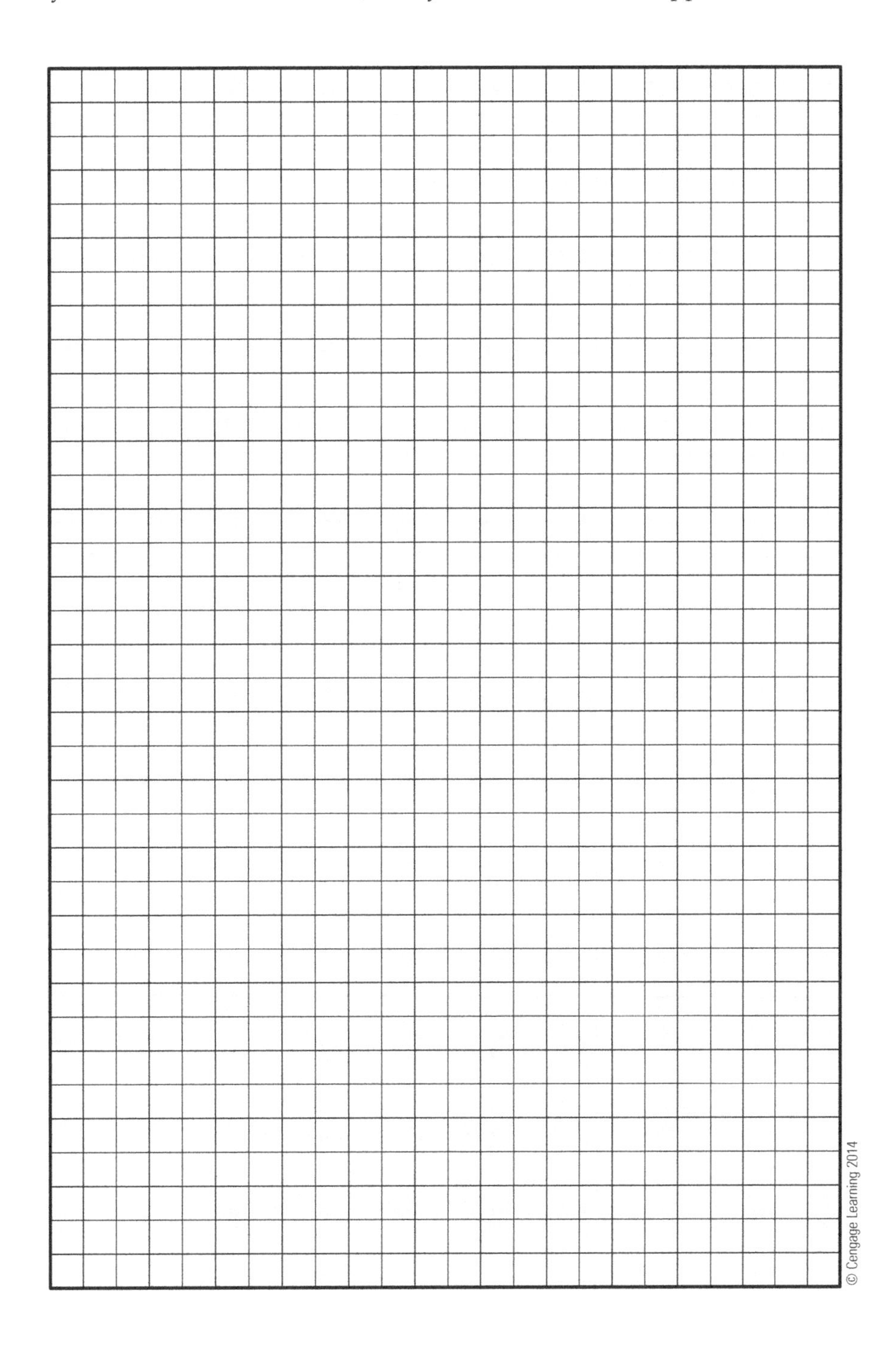

2. Modify the circuit in question 1 to include an alarm that sounds for 5 seconds before the cylinder advances.

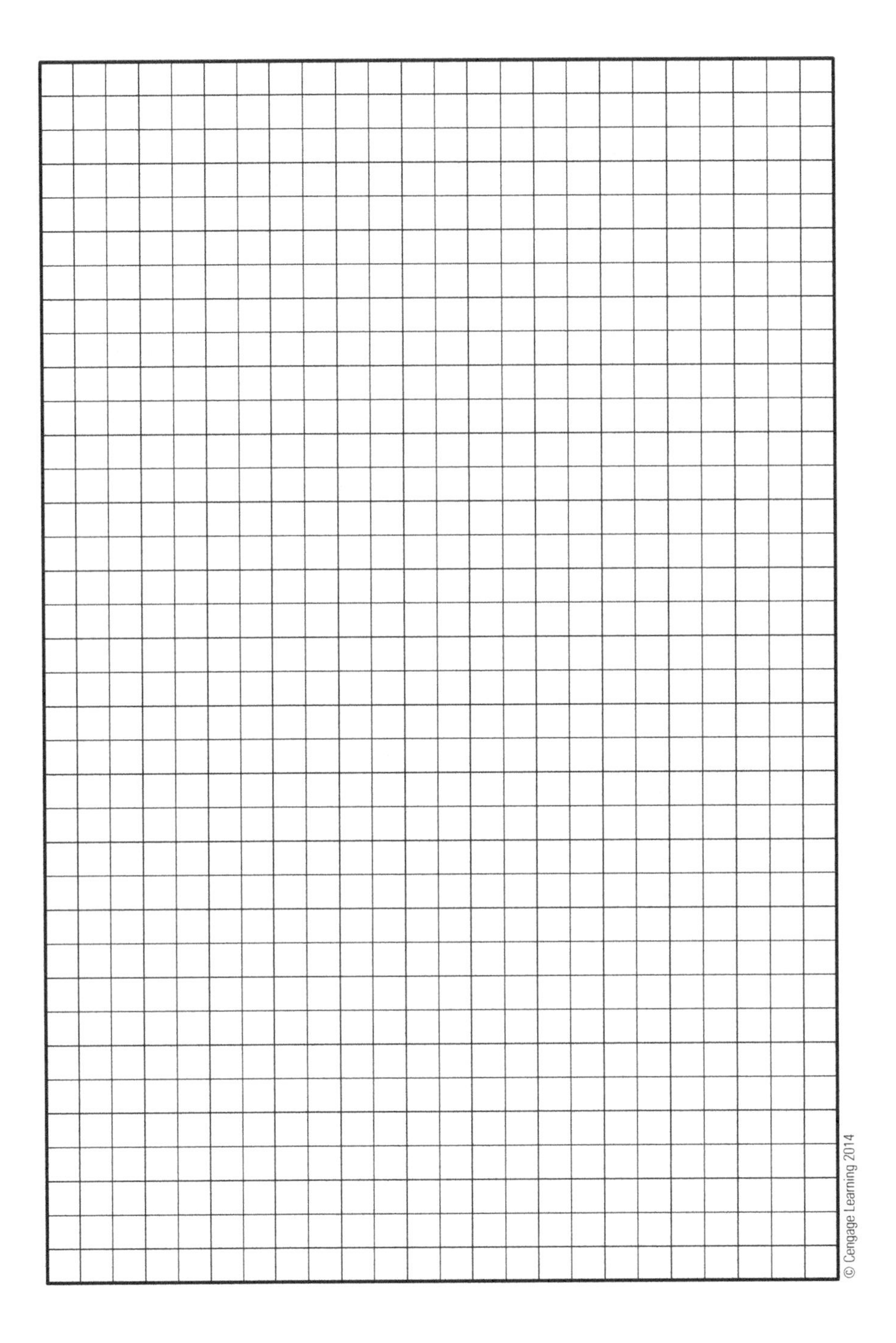

3. Develop the I/O wiring diagram for the circuit in question 2.

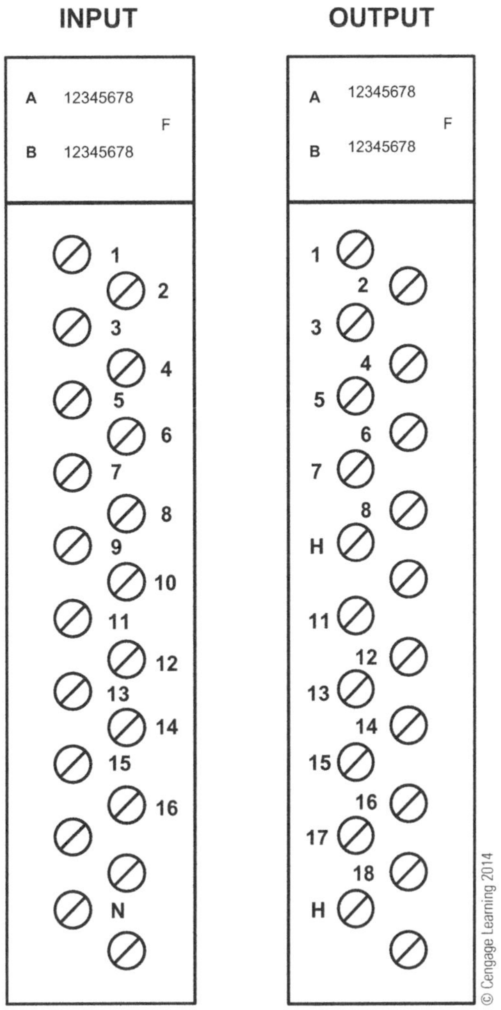

INPUT

A 12345678
F
B 12345678

1 2 3 4 5 6 7 8 9 10 11 12 13 14 15 16 N

OUTPUT

A 12345678
F
B 12345678

1 2 3 4 5 6 7 8 H 11 12 13 14 15 16 17 18 H

Name: ______________________ Date: ______________

PLC PROJECT 4

1. Develop a PLC program that performs as follows:
   a. A conveyor contains three start/stop stations. A start/stop station is mounted at each end of the conveyor and in the middle.
   b. The conveyor must be able to be started or stopped from any one of the three locations.

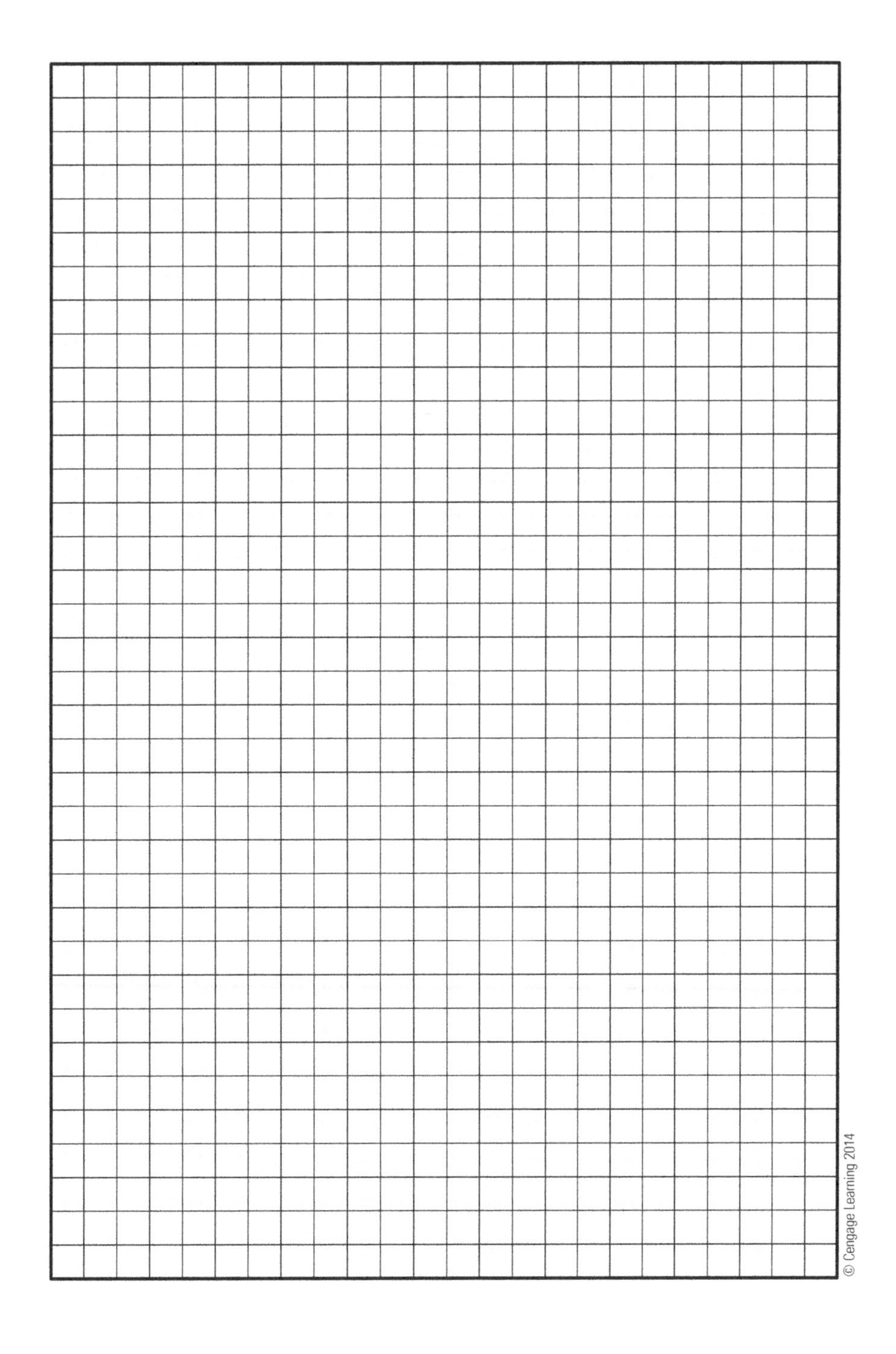

2. Develop the I/O wiring diagram for the circuit in question 1.

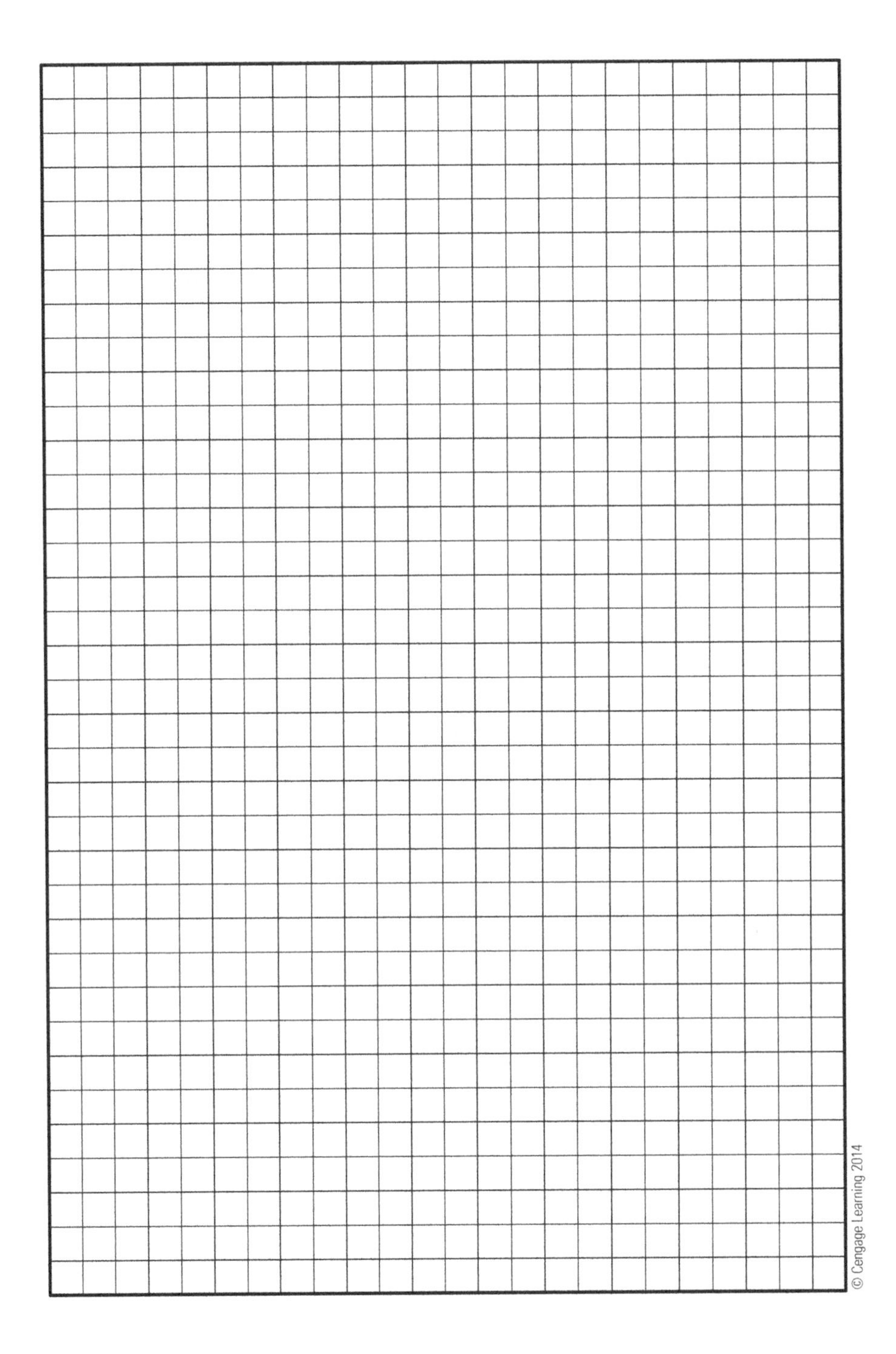

Name: ______________________________ Date: ______________________

PLC PROJECT 5

1. A drill press is started by an operator. To ensure the safety of the operator, the operator must simultaneously press one start pushbutton with the left hand and one pushbutton with the right hand. The pushbuttons are located on opposite sides of the drill table to ensure that the operator's hands are not in the drilling space when the machine is started. The drill press can be stopped by pressing a stop pushbutton located near each start pushbutton. Develop a PLC program that performs this operation.

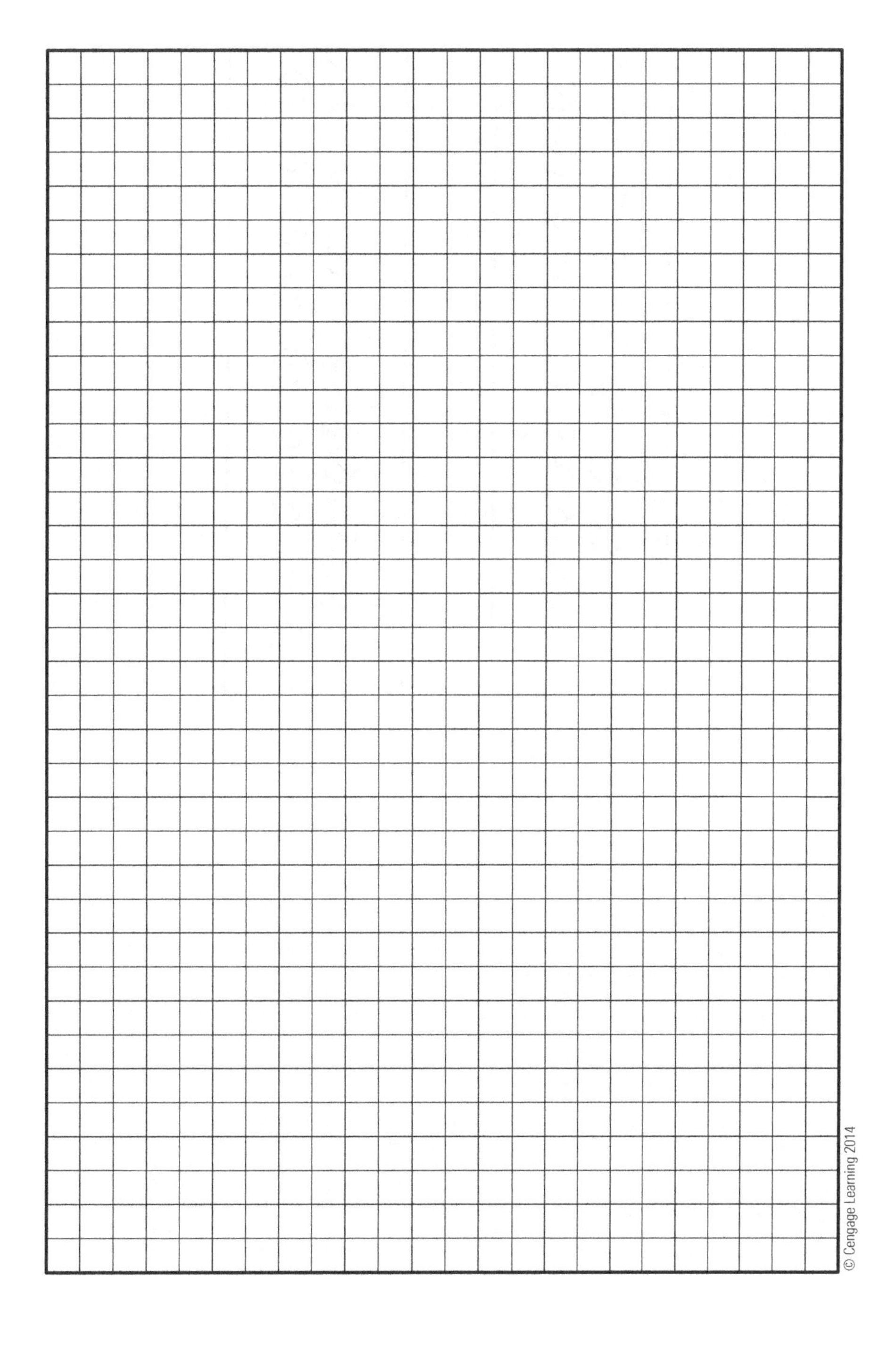

2. Develop the I/O wiring diagram for the circuit in question 1.

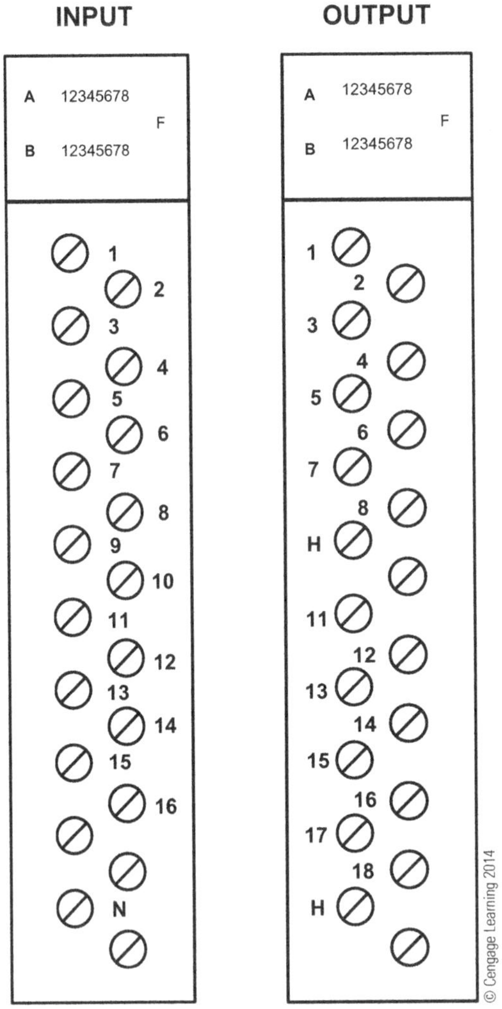

INPUT

A 12345678
F
B 12345678

1
2
3
4
5
6
7
8
9
10
11
12
13
14
15
16
N

OUTPUT

A 12345678
F
B 12345678

1
2
3
4
5
6
7
8
H
11
12
13
14
15
16
17
18
H

Name: ______________________ Date: ______________

PLC PROJECT 6

1. Develop a PLC program that performs as follows:
   a. A momentary start pushbutton is used to cause a cylinder to extend.
   b. When the cylinder extends, it contacts a limit switch, causing the cylinder to retract.
   c. When the cylinder retracts, it contacts a limit switch, causing the cylinder to extend.
   d. The cylinder cycles back and forth until the stop pushbutton is pressed.

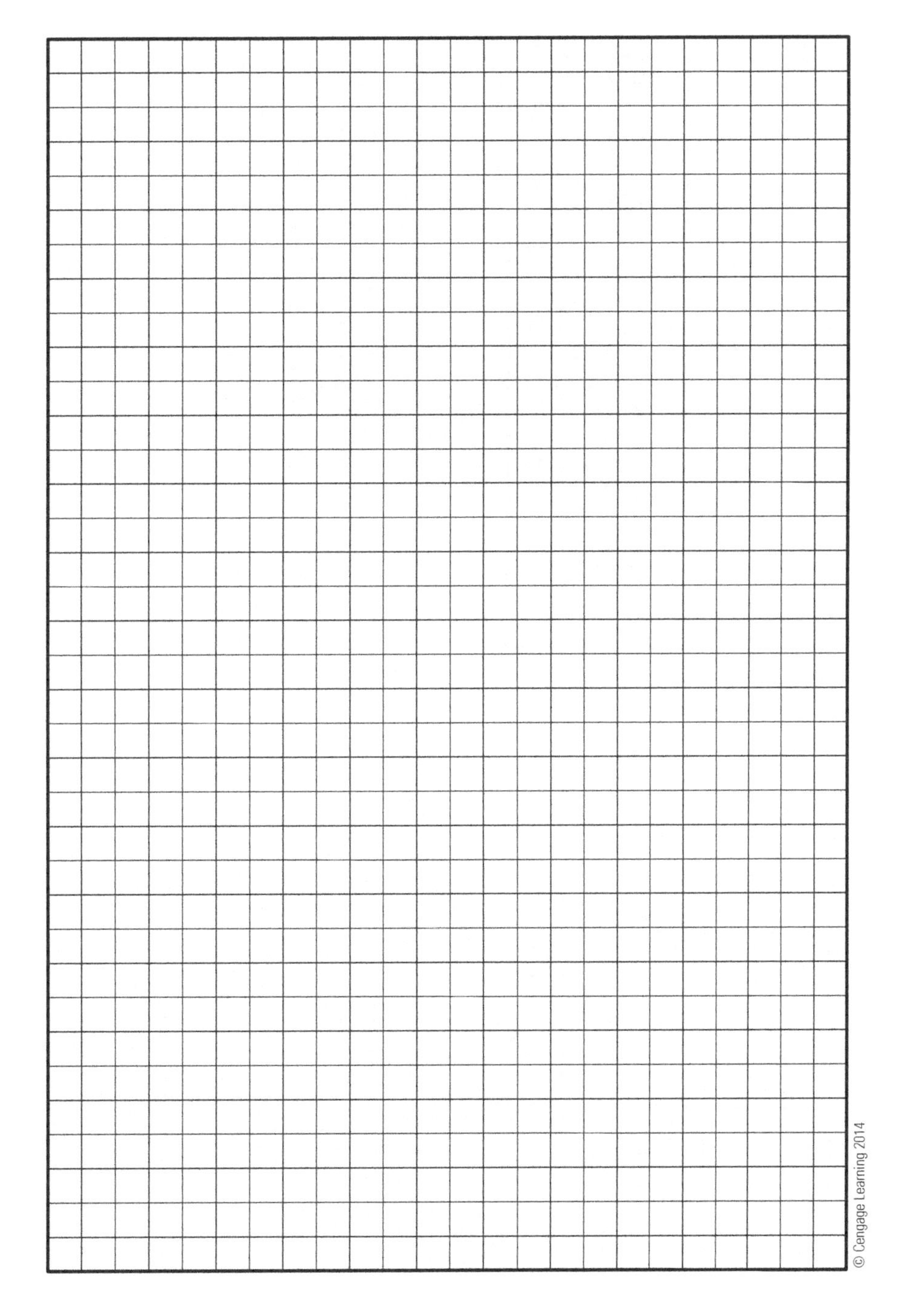

2. Develop the I/O wiring diagram for the circuit in question 1.

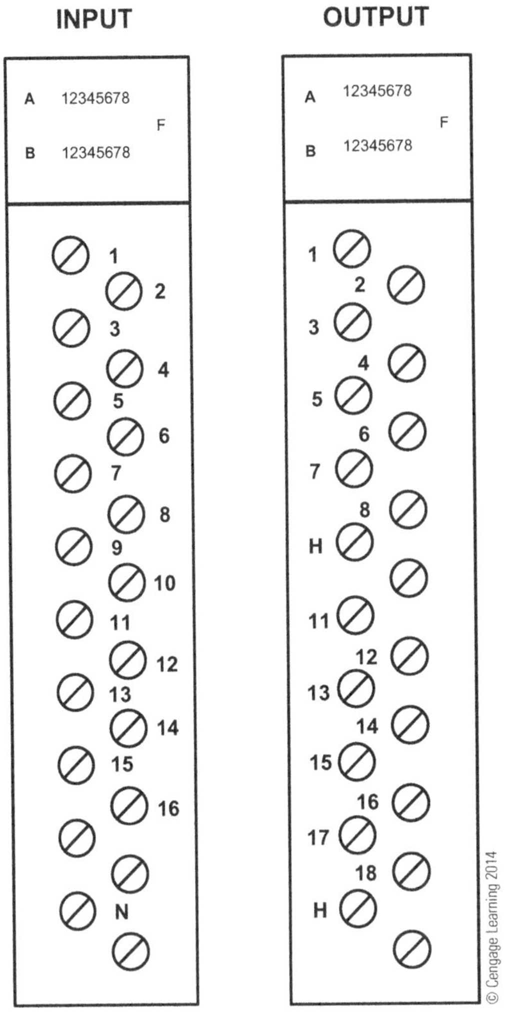

INPUT

OUTPUT

Name: ______________________ Date: ______________

PLC PROJECT 7

1. Develop a PLC program that performs as follows:
   a. Pressing a momentary pushbutton causes a motor to run in the forward direction.
   b. After 10 seconds, the motor stops.
   c. After 5 seconds, the motor runs in the reverse direction.
   d. After 10 seconds, the motor stops.
   e. After 5 seconds, the motor again runs in the forward direction.
   f. The process repeats until the stop pushbutton is pressed.

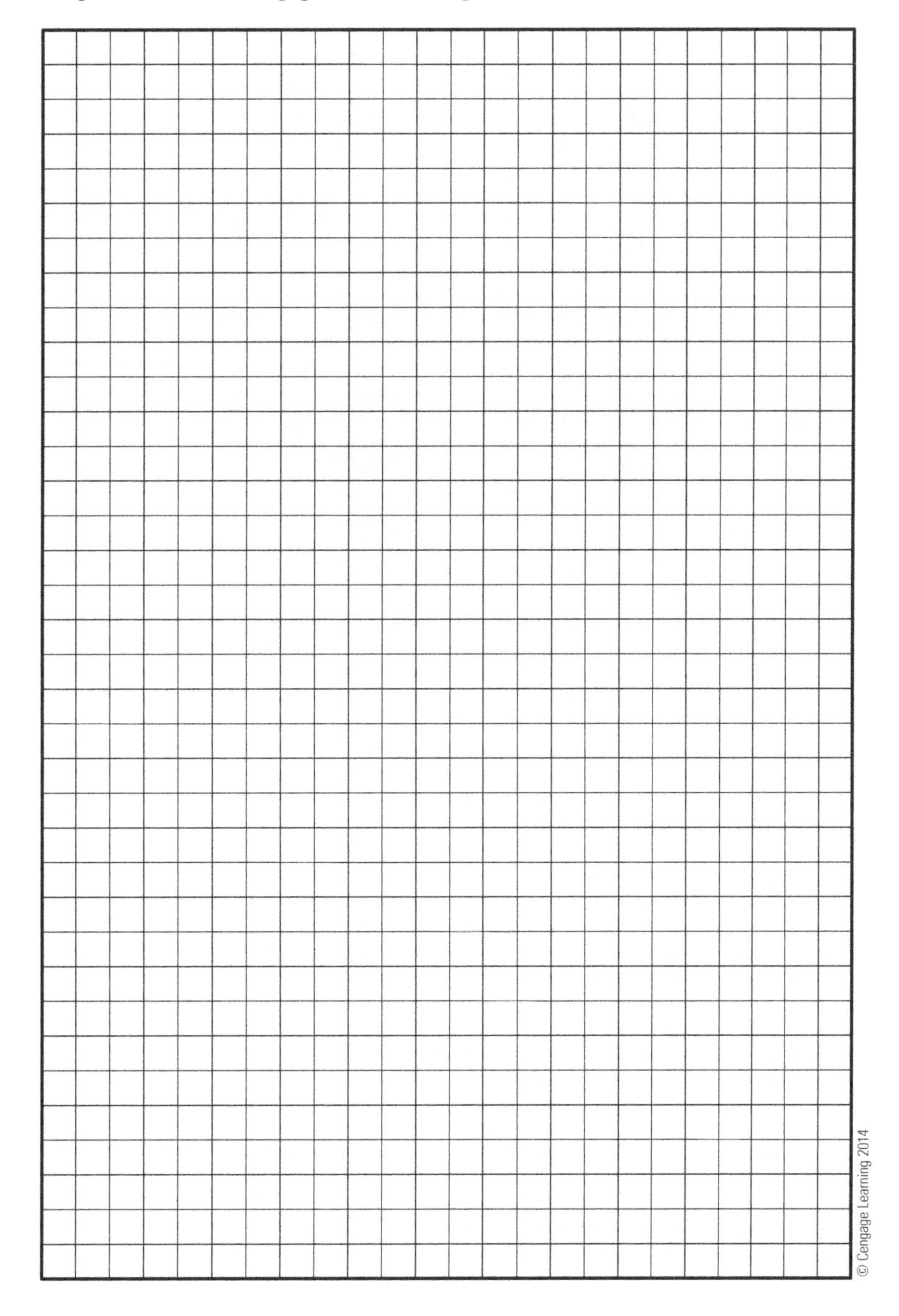

2. Develop the I/O wiring diagram for the circuit in question 1.

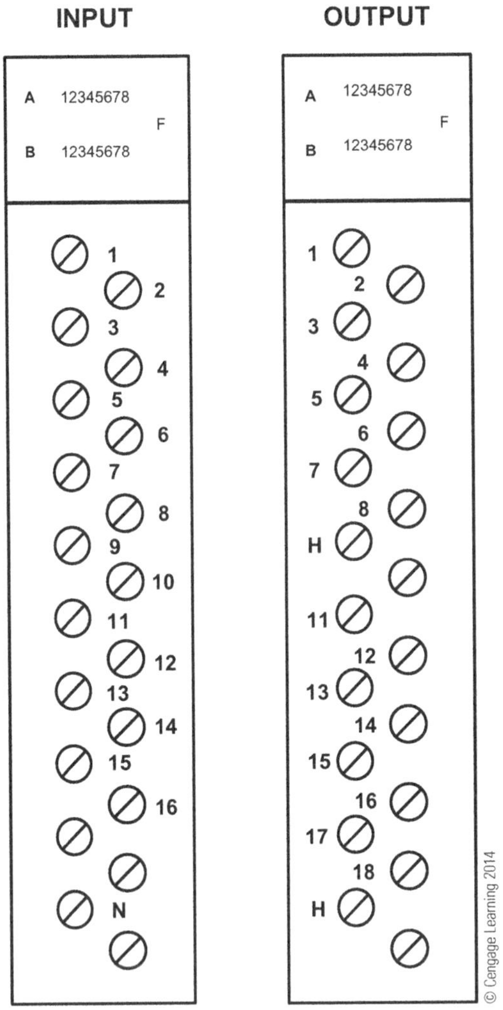

INPUT

A 12345678
F
B 12345678

1 2 3 4 5 6 7 8 9 10 11 12 13 14 15 16 N

OUTPUT

A 12345678
F
B 12345678

1 2 3 4 5 6 7 8 H 11 12 13 14 15 16 17 18 H

Name: ____________________ Date: ____________________

PLC PROJECT 8

1. Develop a PLC program that performs as follows:
   a. Three momentary pushbuttons are used to select slow, medium, or fast speed.
   b. The motor must be started with the slow speed pushbutton.
      i. The motor cannot be started with the medium speed pushbutton.
      ii. The motor cannot be started with the fast speed pushbutton.
   c. Once started in slow speed, the speed of the motor may be changed directly to either medium or fast.
   d. A stop pushbutton is used to stop the motor at any time.

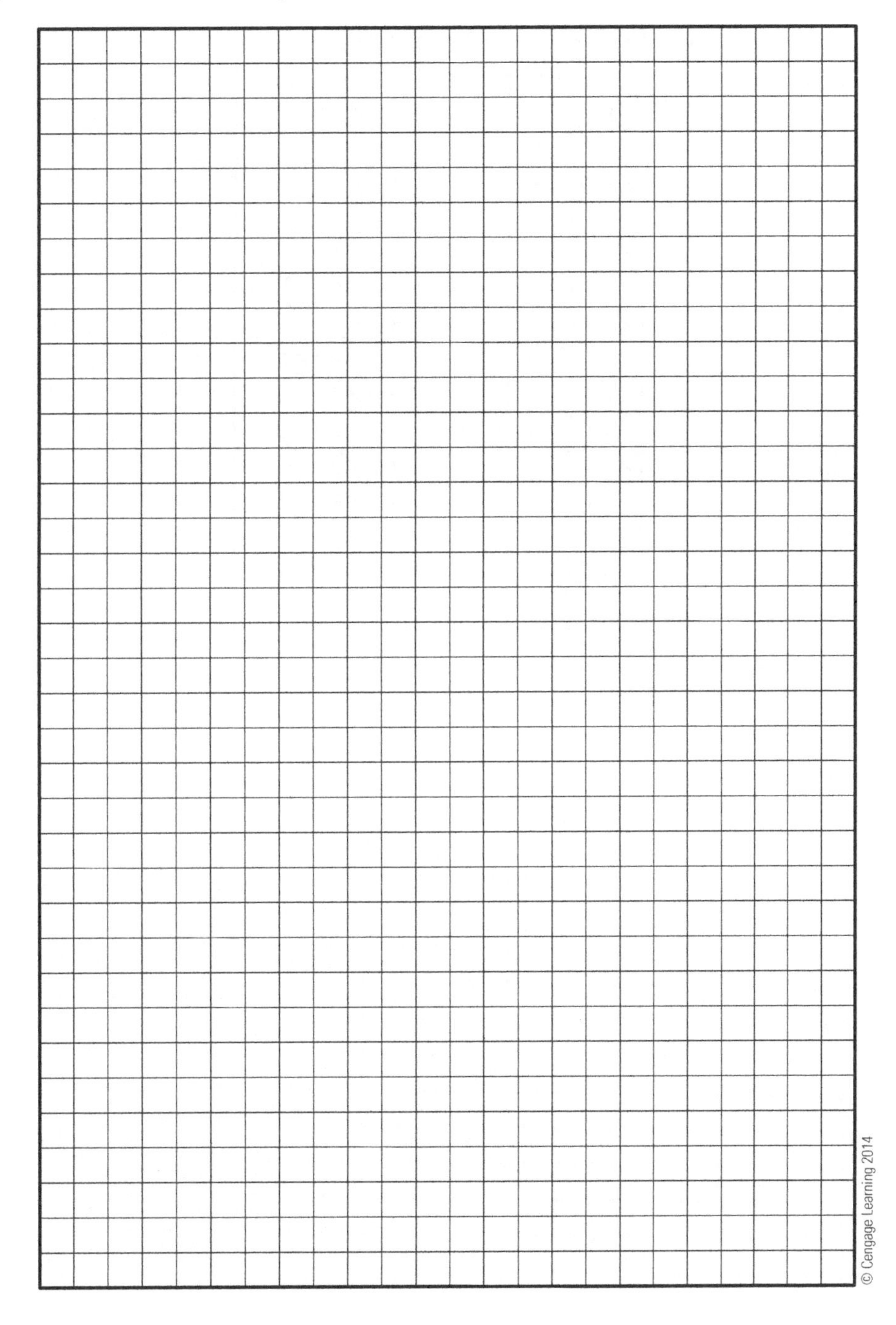

2. Develop the I/O wiring diagram for the circuit in question 1.

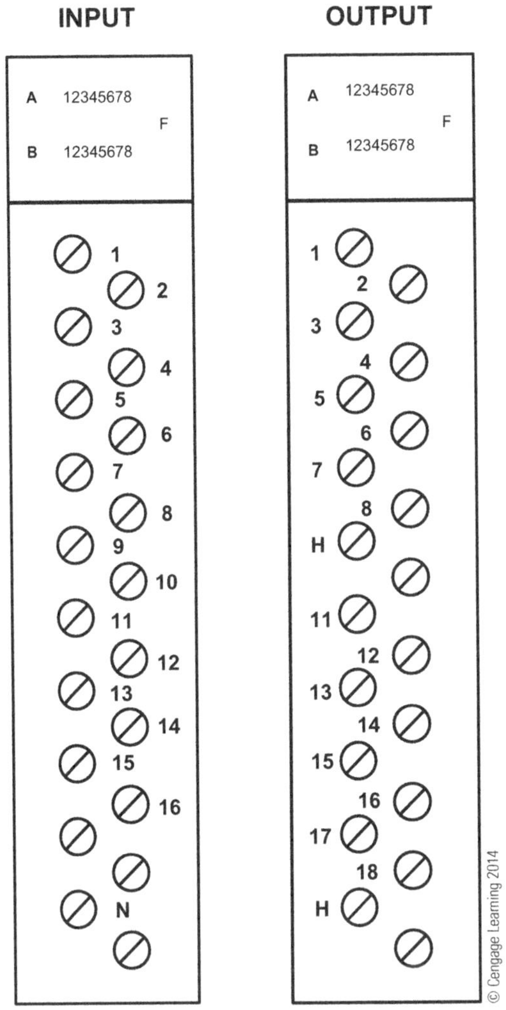

INPUT

OUTPUT

INPUT	OUTPUT
A 12345678	A 12345678
F	F
B 12345678	B 12345678
1	1
2	2
3	3
4	4
5	5
6	6
7	7
8	8
9	H
10	11
11	12
12	13
13	14
14	15
15	16
16	17
N	18
	H

Name: ______________________ Date: ______________

PLC PROJECT 9

1. Develop a PLC program that performs as follows:
   a. A motor is started by pressing a momentary pushbutton switch.
   b. The motor starts in low speed.
   c. After 5 seconds, the motor accelerates to medium speed.
   d. After 5 seconds, the motor accelerates to fast speed.
   e. A stop pushbutton is used to stop the motor at any time.

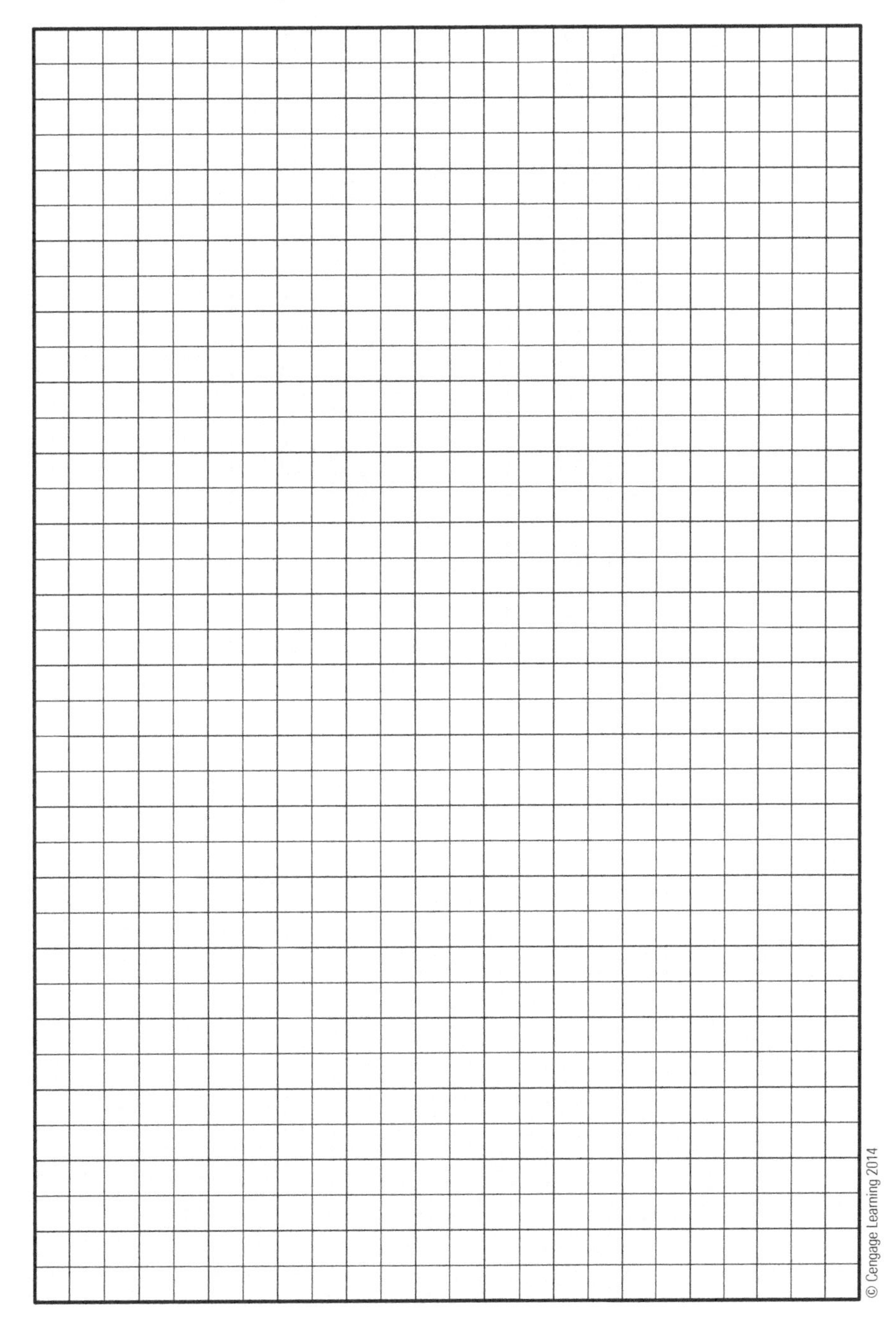

2. Develop the I/O wiring diagram for the circuit in question 1.

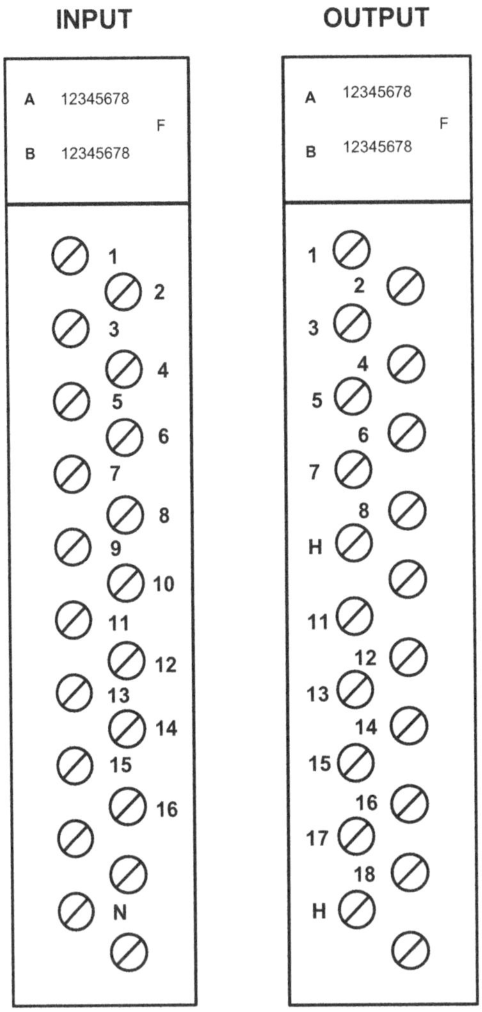

INPUT

A 12345678
F
B 12345678

1 2 3 4 5 6 7 8 9 10 11 12 13 14 15 16 N

OUTPUT

A 12345678
F
B 12345678

1 2 3 4 5 6 7 8 H 11 12 13 14 15 16 17 18 H

Name: ______________________________ Date: ______________________

PLC PROJECT 10

1. Develop a PLC program that performs as follows:
   a. Use three pushbuttons to perform the functions of stop, run, and jog.

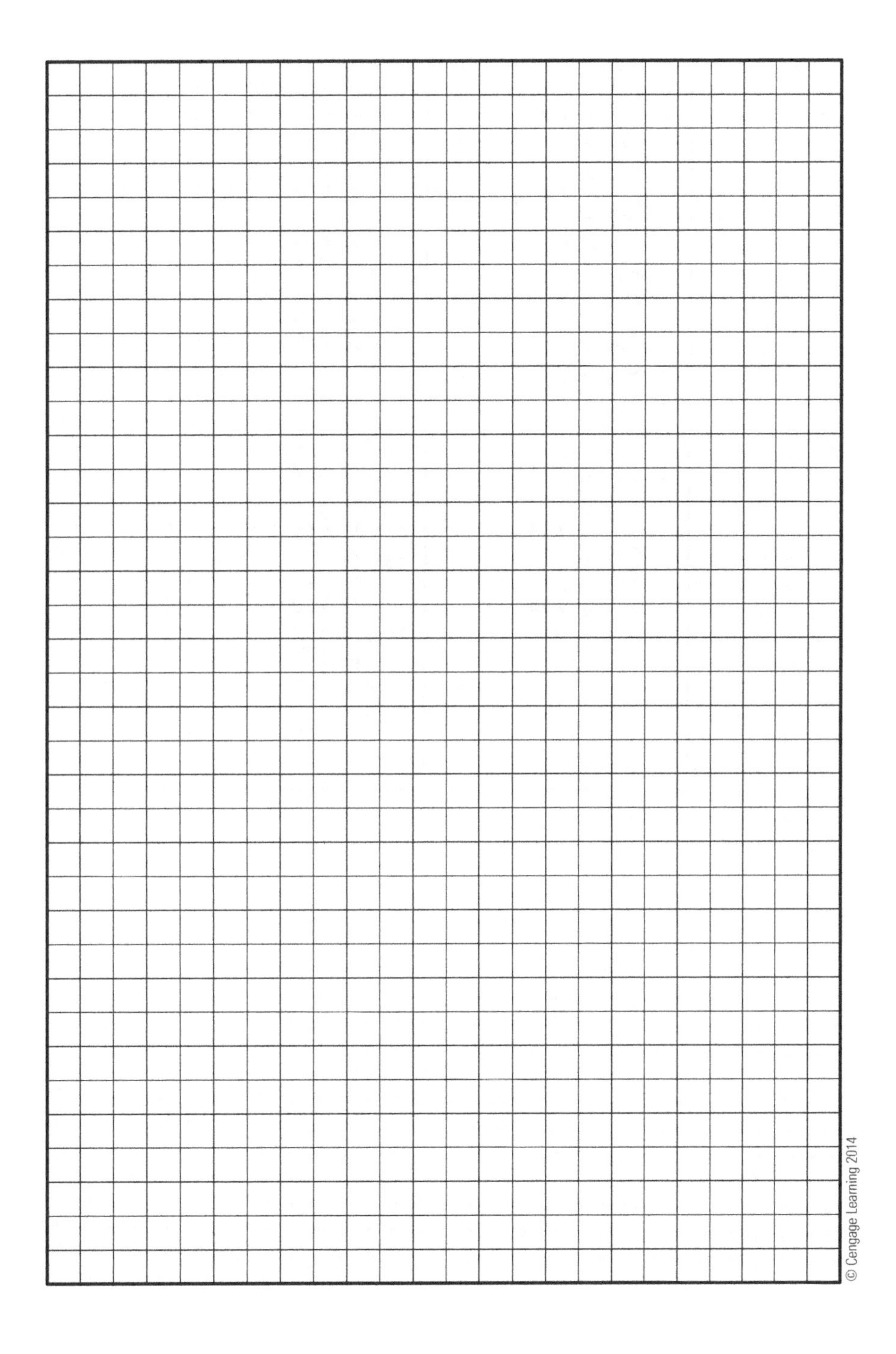

2. Develop the I/O wiring diagram for the circuit in question 1.

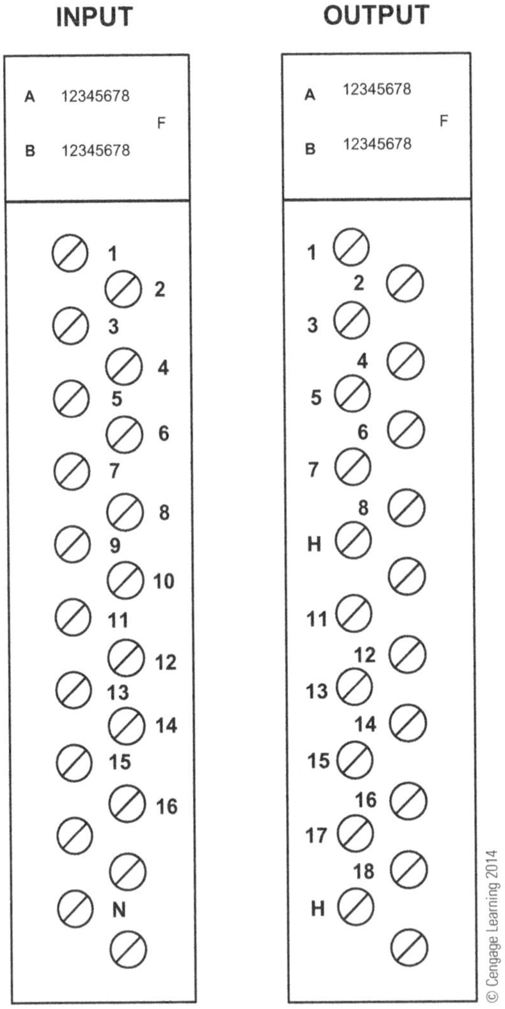

INPUT

A 12345678
F
B 12345678

1 2 3 4 5 6 7 8 9 10 11 12 13 14 15 16 N

OUTPUT

A 12345678
F
B 12345678

1 2 3 4 5 6 7 8 H 11 12 13 14 15 16 17 18 H

Name: ______________________________ Date: ______________

PLC PROJECT 11

1. Develop a PLC program that performs as follows:
   a. Use two pushbuttons (red and green) and a selector switch.
      i. With the selector switch in the "run" position, pressing the green pushbutton causes the motor to run continuously. Pressing the red pushbutton stops the motor.
      ii. With the selector switch in the "jog" position, pressing the green pushbutton jogs the motor.

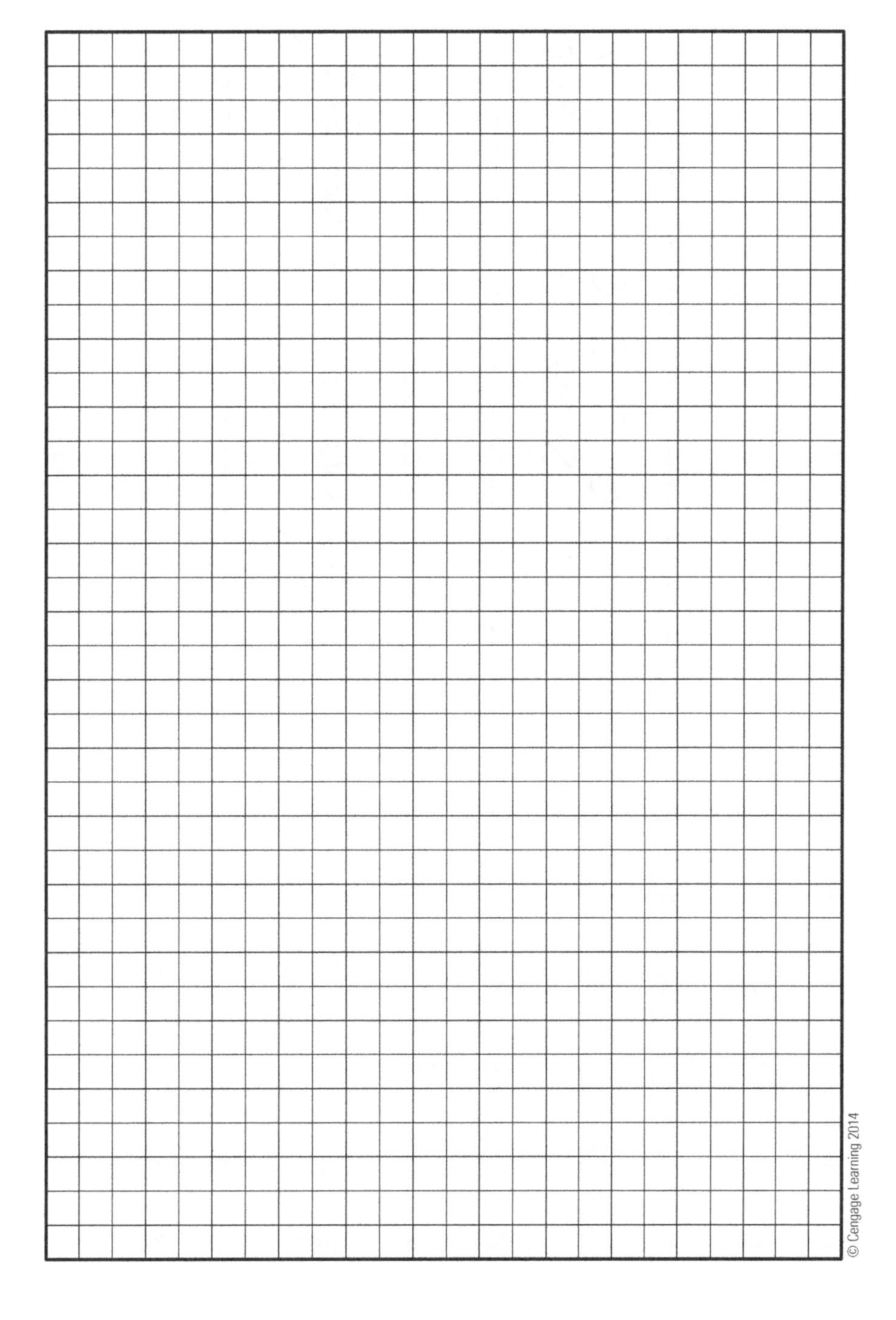

2. Develop the I/O wiring diagram for the circuit in question 1.

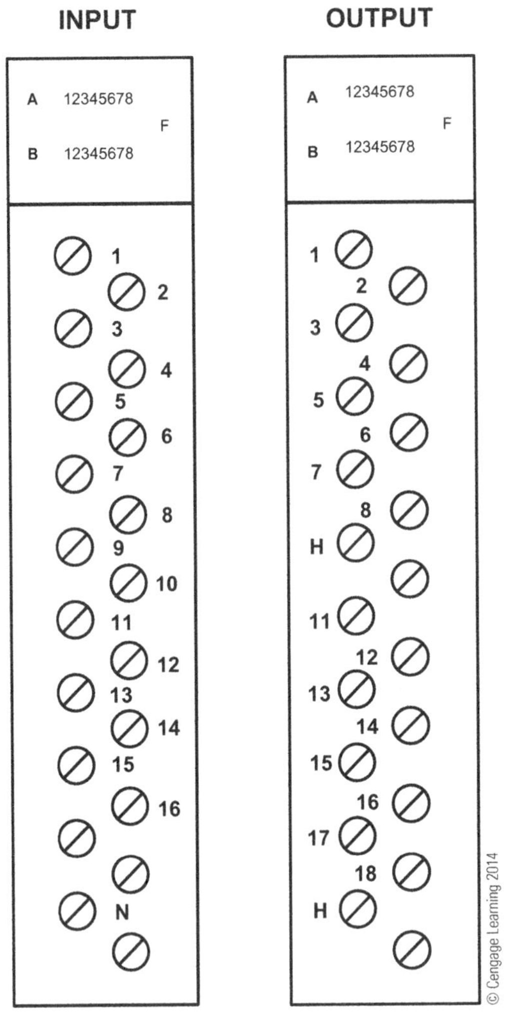

INPUT

A 12345678
F
B 12345678

1
2
3
4
5
6
7
8
9
10
11
12
13
14
15
16
N

OUTPUT

A 12345678
F
B 12345678

1
2
3
4
5
6
7
8
H
11
12
13
14
15
16
17
18
H

Name: ______________________________ Date: ______________________

PLC PROJECT 12

1. Develop a PLC program that performs as follows:
   a. A selector switch is used to select between manual, off, and automatic modes in a residential heating system.
   b. With the selector switch in the off position:
      i. The circuit is de-energized.
   c. With the selector switch in the manual position:
      i. The heater runs continuously.
   d. With the selector switch in the automatic position:
      i. The heater runs whenever a thermostatic switch closes.
      ii. The heater is off whenever a thermostatic switch is open.

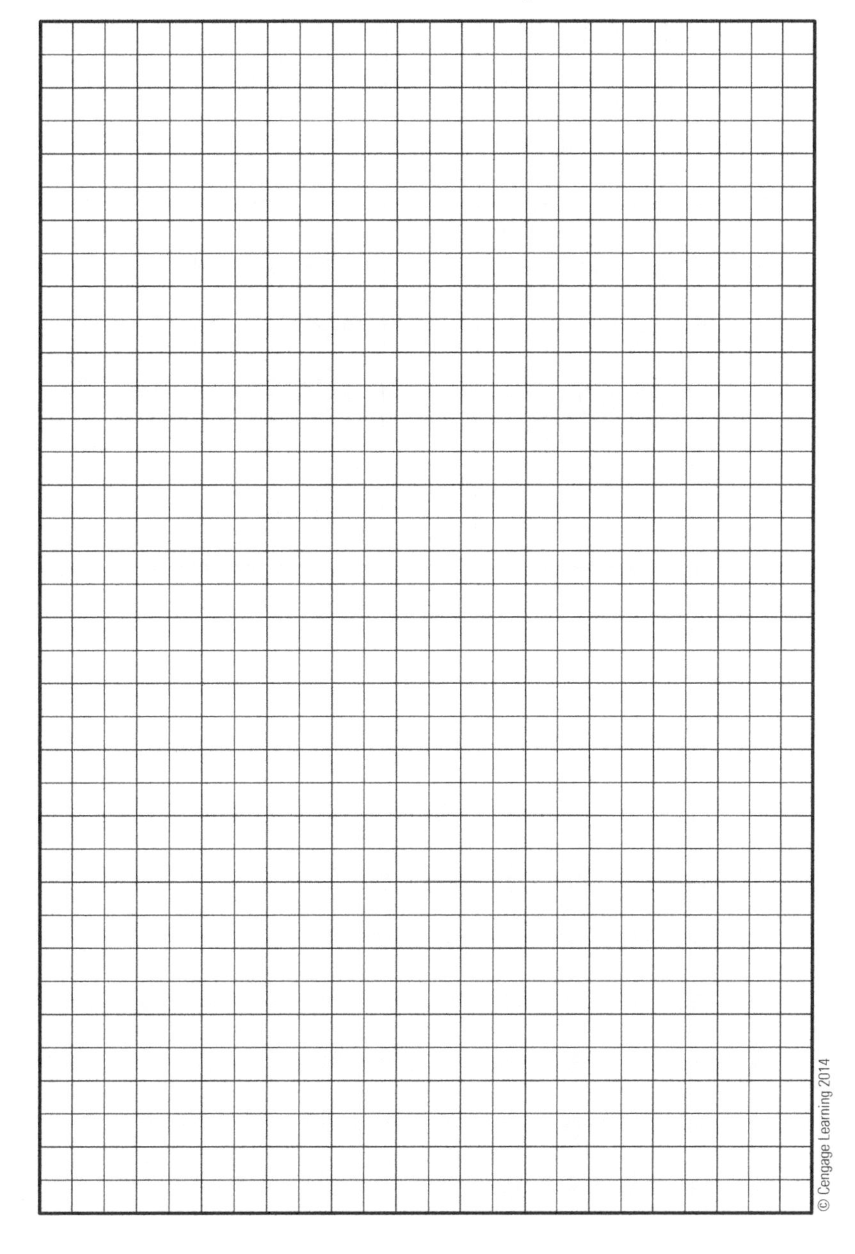

2. Develop the I/O wiring diagram for the circuit in question 1.

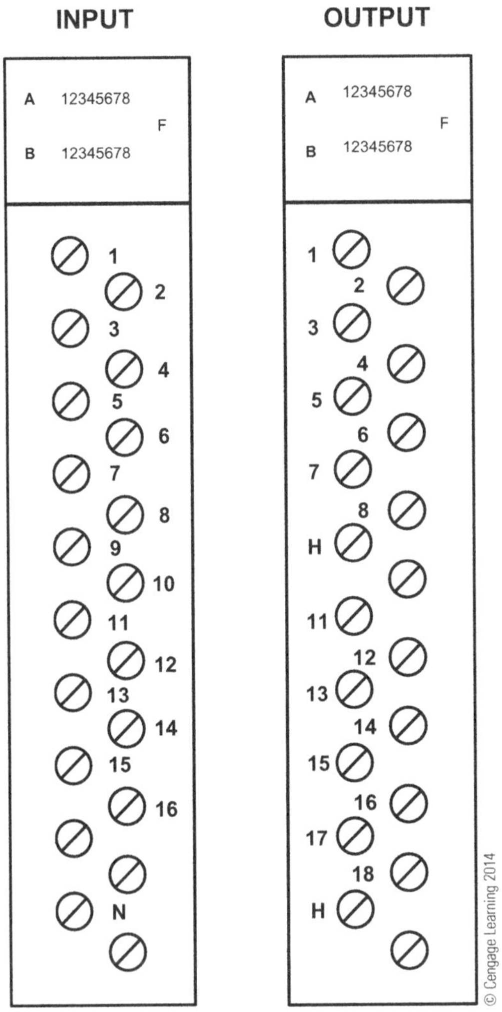

INPUT

A 12345678
F
B 12345678

1
2
3
4
5
6
7
8
9
10
11
12
13
14
15
16
N

OUTPUT

A 12345678
F
B 12345678

1
2
3
4
5
6
7
8
H
11
12
13
14
15
16
17
18
H

3. Modify the circuit in question 1 so that a circulating fan turns on 5 seconds after the heater turns on.
   a. When the heater turns off, the circulating fan continues to run for 5 seconds.

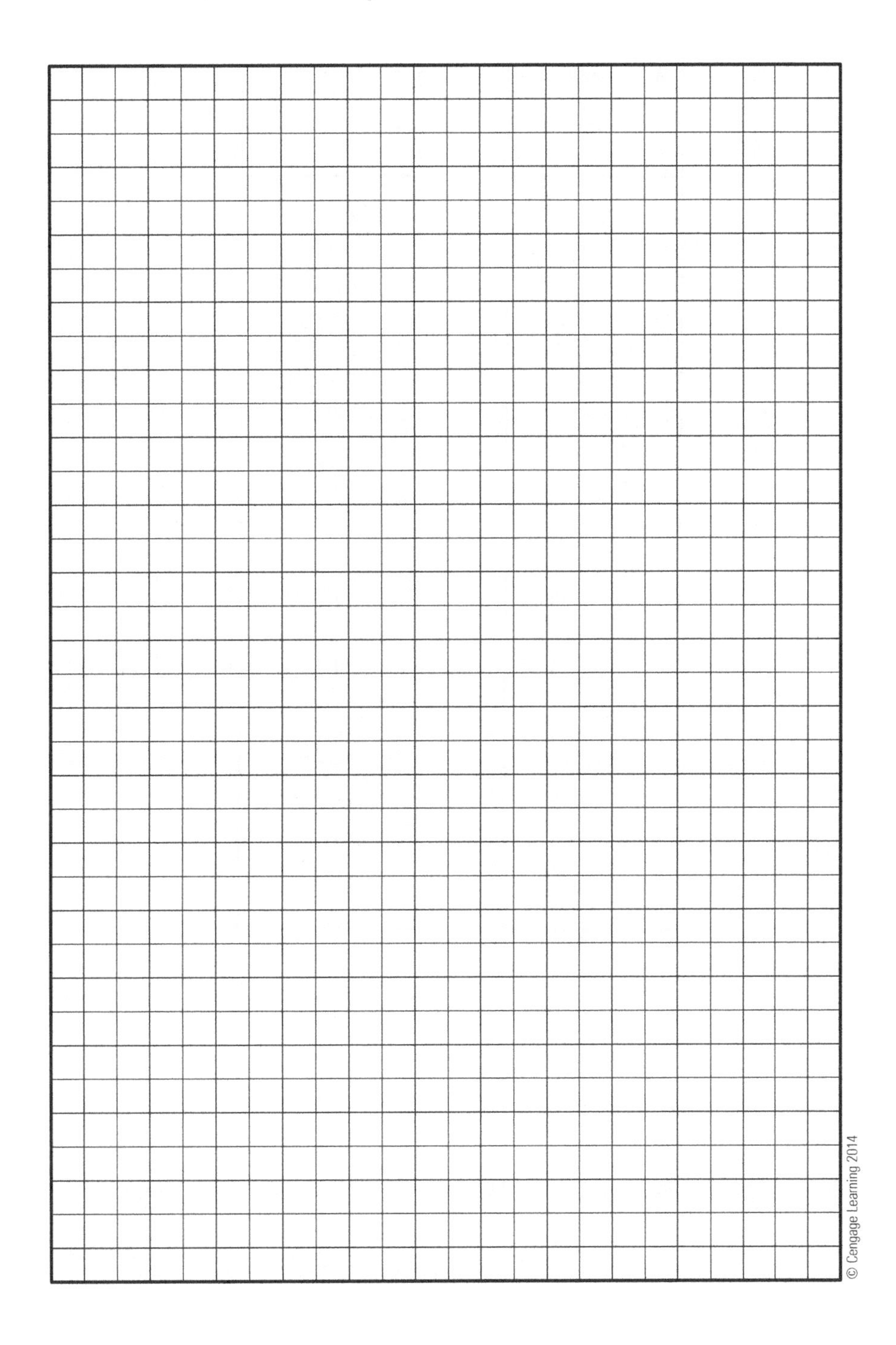

4. Modify the I/O wiring diagram developed in question 2 to match the PLC program developed in question 3.

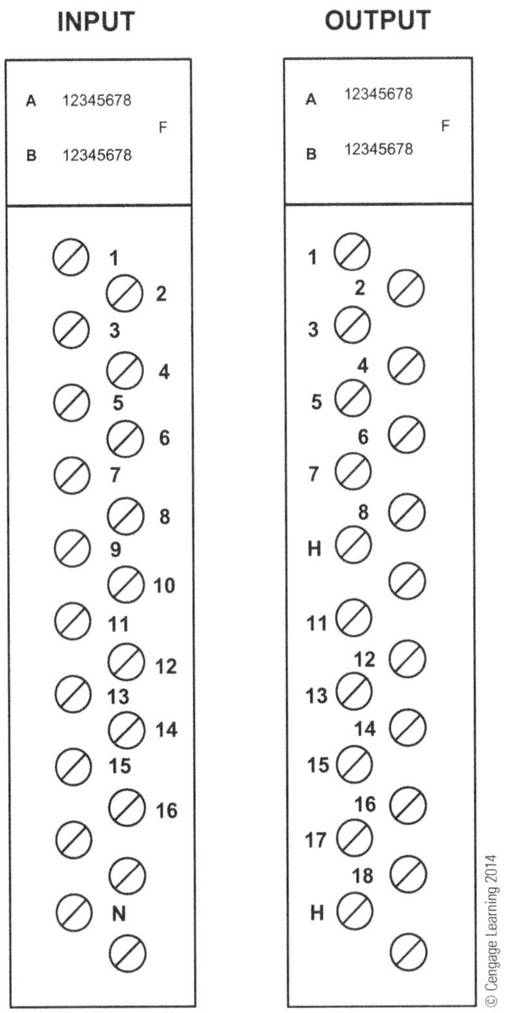

INPUT

OUTPUT

Name: ______________________________ Date: ______________________

PLC PROJECT 13

1. Develop a PLC program that performs as follows:
   a. Pressing a start pushbutton causes three conveyors to start simultaneously.
   b. Pressing a stop pushbutton stops all three conveyors simultaneously.
   c. Should an overload occur on conveyor 1, conveyor 2 and conveyor 3 will not run.
   d. Should an overload occur on conveyor 2, conveyor 1 will run, but conveyor 3 will not run.
   e. Should an overload occur on conveyor 3, conveyor 1 and conveyor 2 will run.

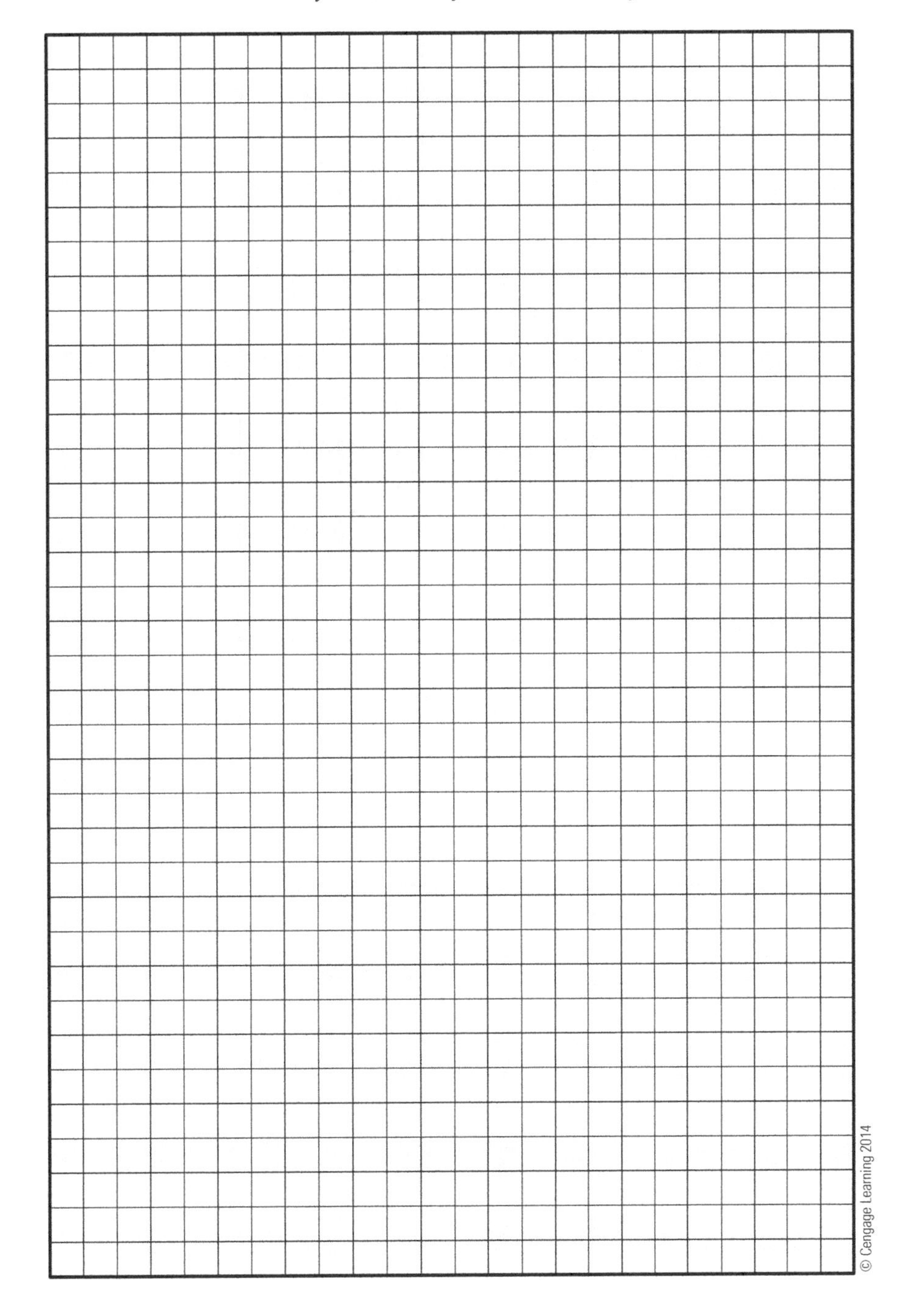

2. Develop the I/O wiring diagram for the circuit in question 1.

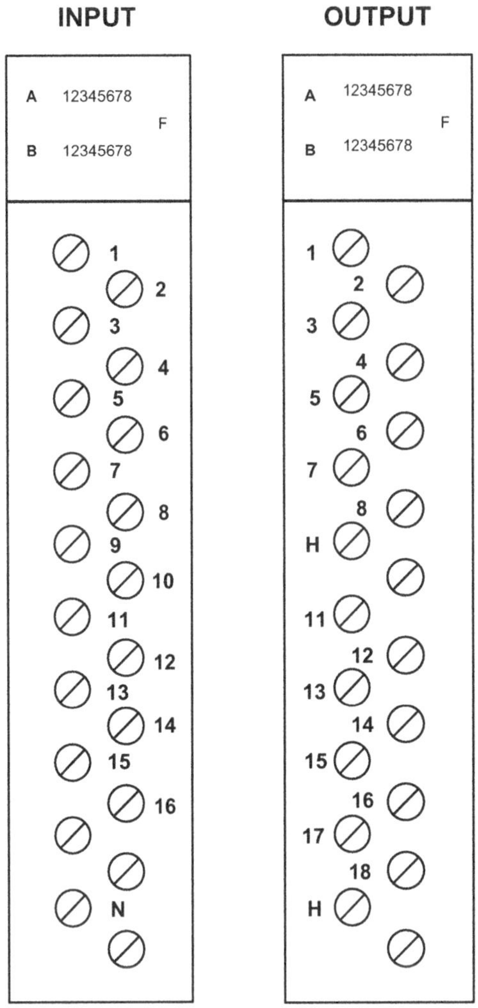

INPUT

A 12345678
F
B 12345678

OUTPUT

A 12345678
F
B 12345678

Name: ______________________ Date: ______________________

PLC PROJECT 14

1. Develop a PLC program for a control circuit that uses a master stop pushbutton and three separate motors, each with its own start/stop station.

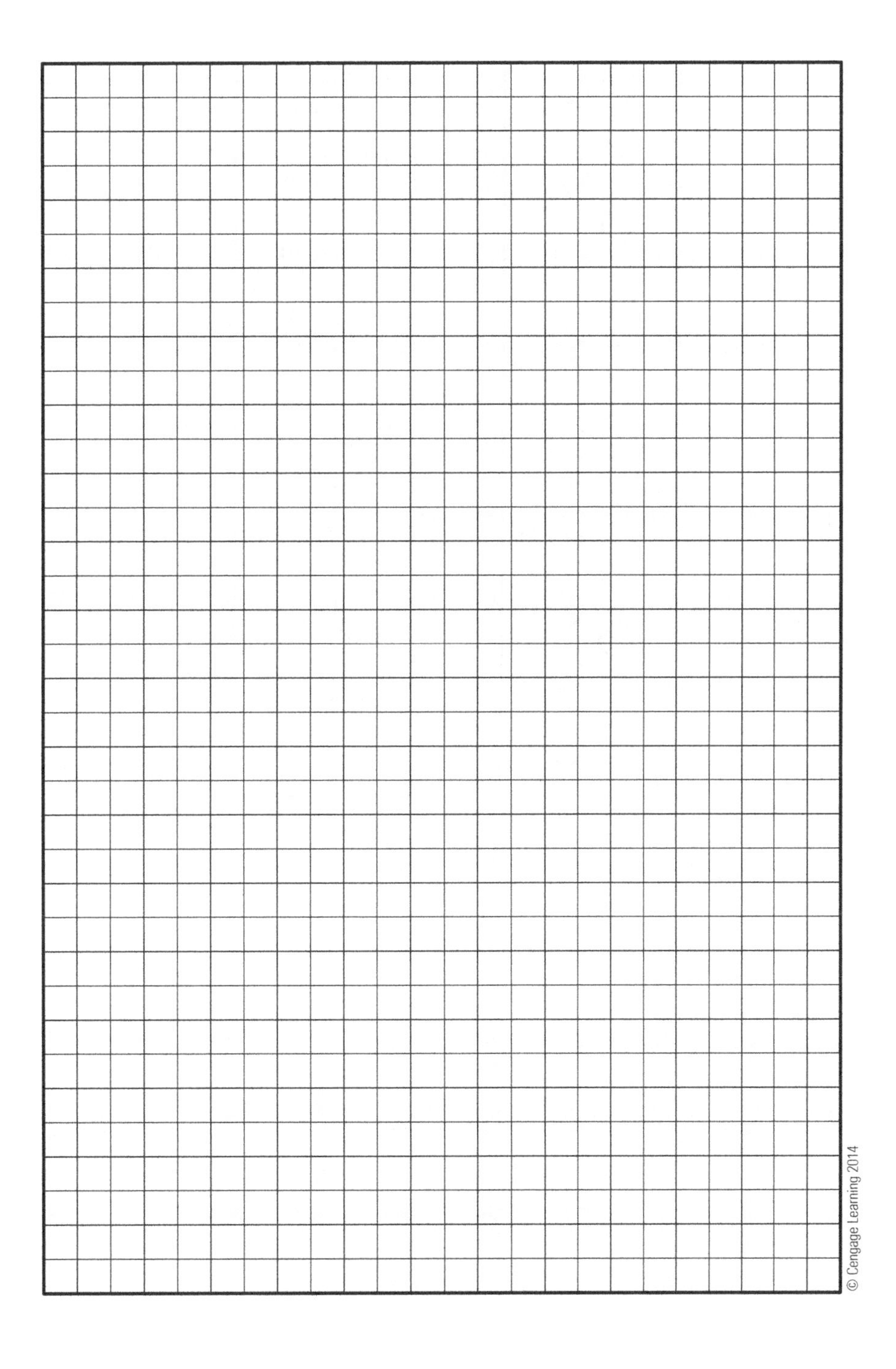

2. Develop the I/O wiring diagram for the circuit in question 1.

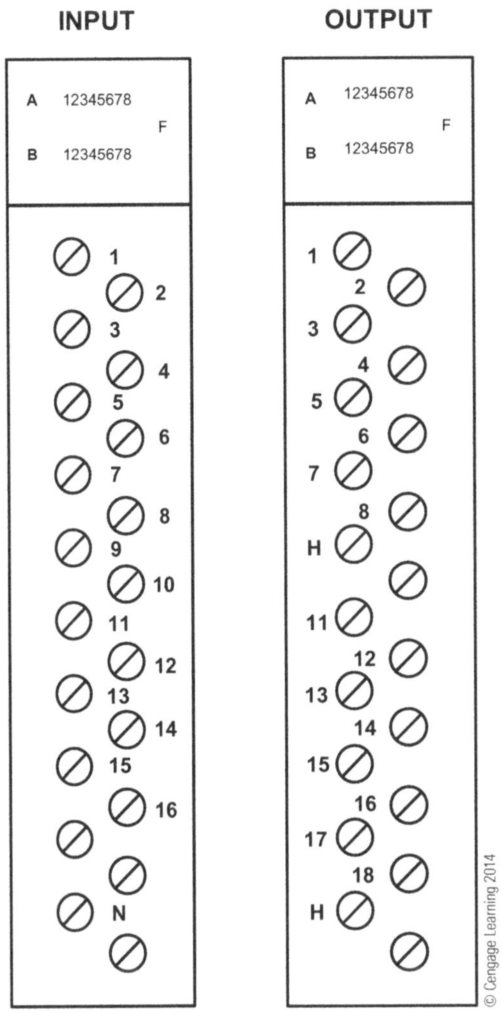

INPUT

OUTPUT

Name: ______________________ Date: ______________

PLC PROJECT 15

1. Develop a PLC program that performs the following:
   a. A start pushbutton is pressed to start a series of three conveyors:
      i. Conveyor 3 must start first.
      ii. 5 seconds after conveyor 3 starts, conveyor 2 starts.
      iii. 5 seconds after conveyor 2 starts, conveyor 1 starts.
   b. If conveyor 3 is not running, conveyor 2 and conveyor 1 will not run.
   c. If conveyor 2 is not running, conveyor 3 may run, but conveyor 1 will not run.
   d. If conveyor 1 is not running, conveyor 2 and conveyor 3 may run.
   e. An overload on any of the conveyors shuts down all conveyors.
   f. Pressing the stop pushbutton stops all conveyors.

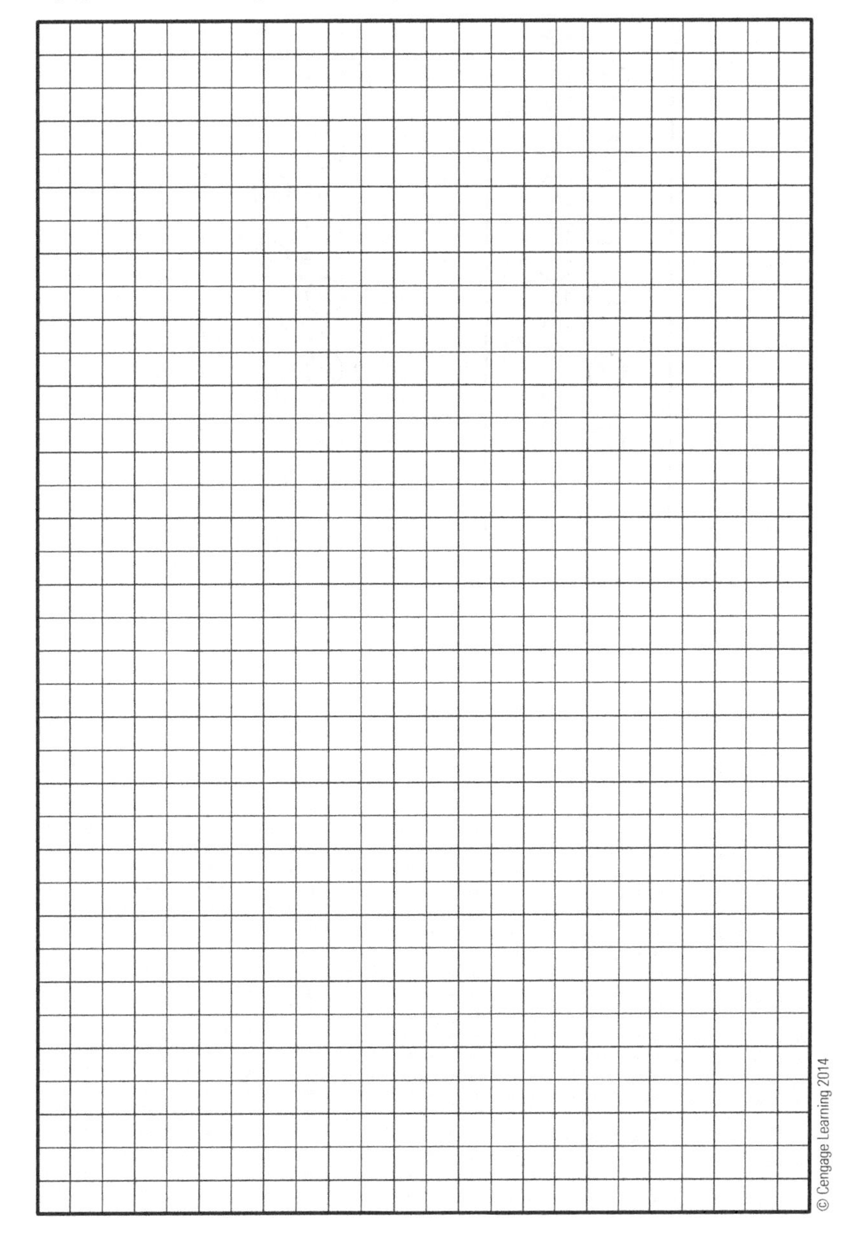

2. Develop the I/O wiring diagram for the circuit in question 1.

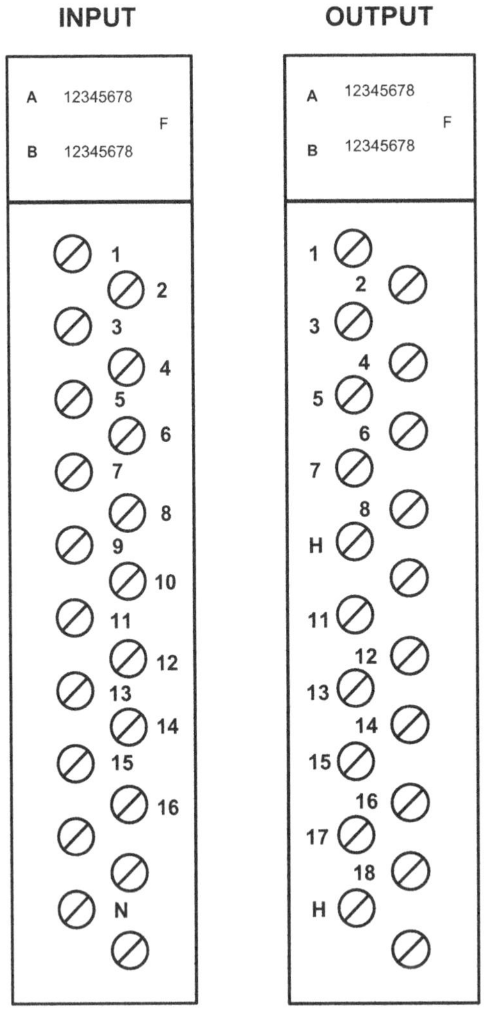

INPUT

A	12345678	
		F
B	12345678	

1, 2, 3, 4, 5, 6, 7, 8, 9, 10, 11, 12, 13, 14, 15, 16, N

OUTPUT

A	12345678	
		F
B	12345678	

1, 2, 3, 4, 5, 6, 7, 8, H, 11, 12, 13, 14, 15, 16, 17, 18, H

Name: ______________________ Date: ______________

PLC PROJECT 16

1. A grinding machine consists of a coolant pump and a grinding motor. Develop a PLC program that performs as follows:
   a. When the start pushbutton is pressed, an alarm sounds for 5 seconds.
   b. After the alarm has sounded, a coolant pump begins to pump coolant to the grinding wheel. A pressure switch detects whether there is sufficient coolant pressure for the operation to continue.
      i. If there is insufficient pressure, the grinding operation cannot go forth.
      ii. If there is sufficient pressure, the grinding wheel begins to turn.
   c. If at any time during the operation a loss of coolant pressure is detected, the grinding operation is stopped.

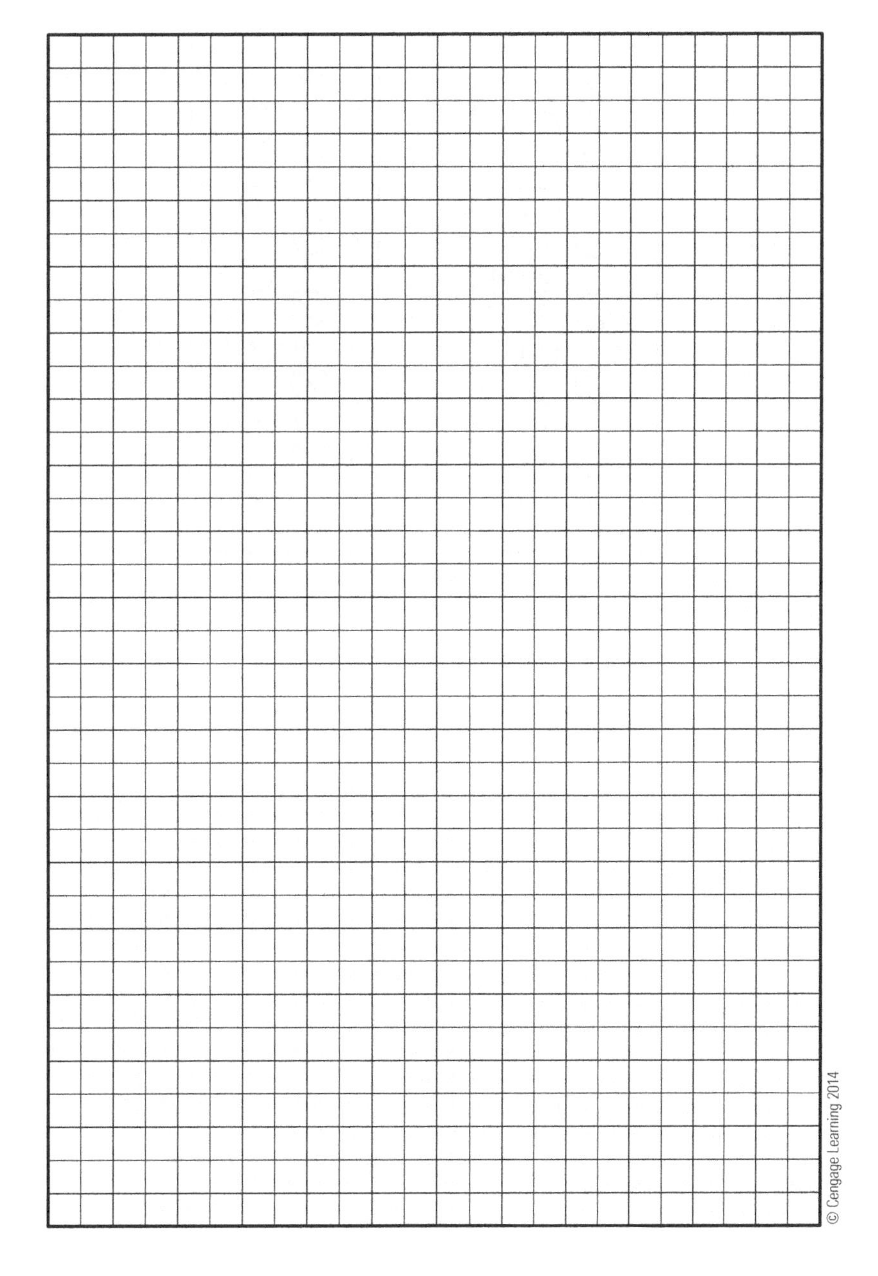

2. Develop the I/O wiring diagram for the circuit in question 1.

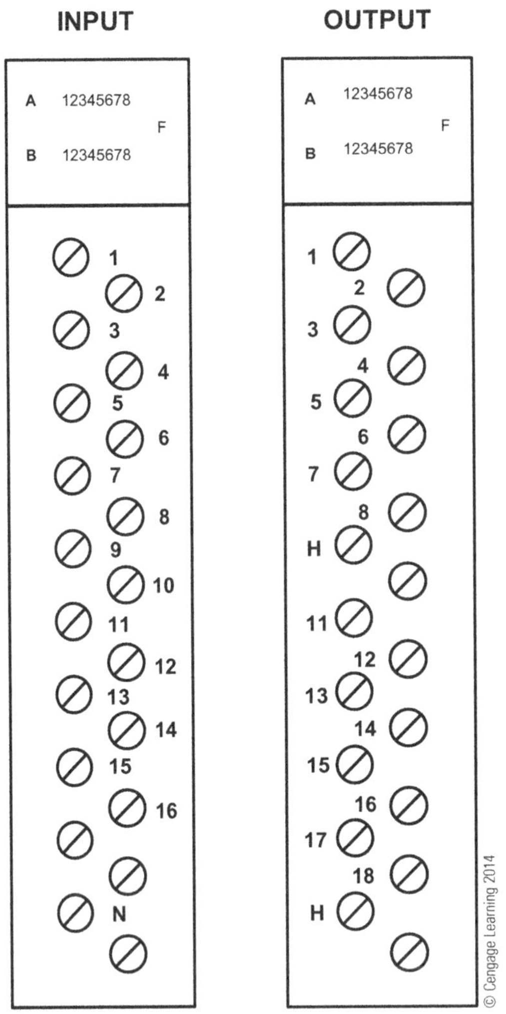

INPUT

A 12345678
F
B 12345678

1 2 3 4 5 6 7 8 9 10 11 12 13 14 15 16 N

OUTPUT

A 12345678
F
B 12345678

1 2 3 4 5 6 7 8 H 11 12 13 14 15 16 17 18 H

Name: ______________________________ Date: ____________________

LIGHTING

1. Using the given components, show how you would connect the ballast, lamp, and starter.

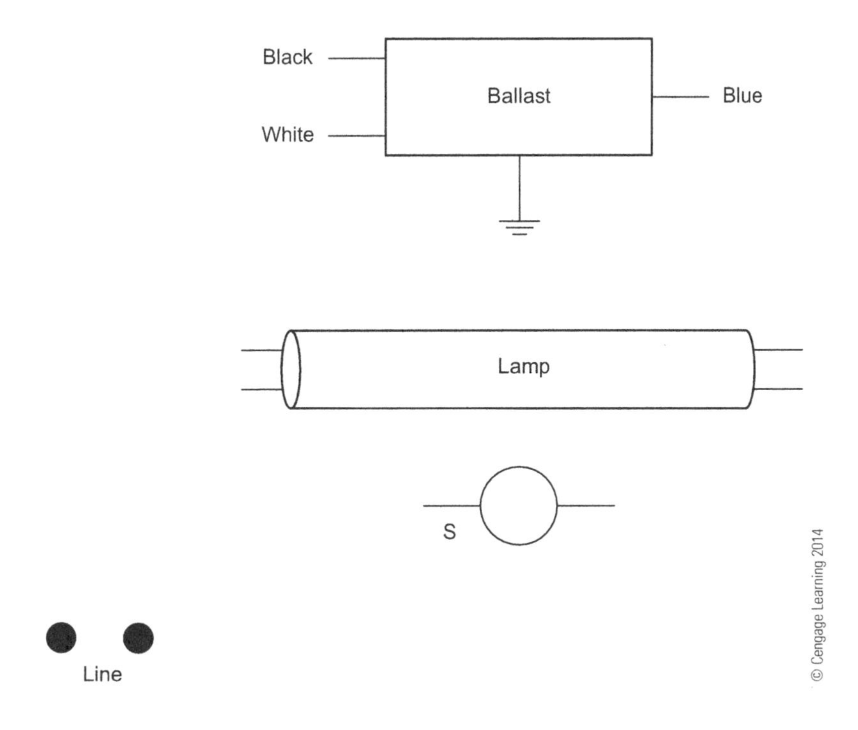

2. Using the given components, show how you would connect the ballast and lamp.

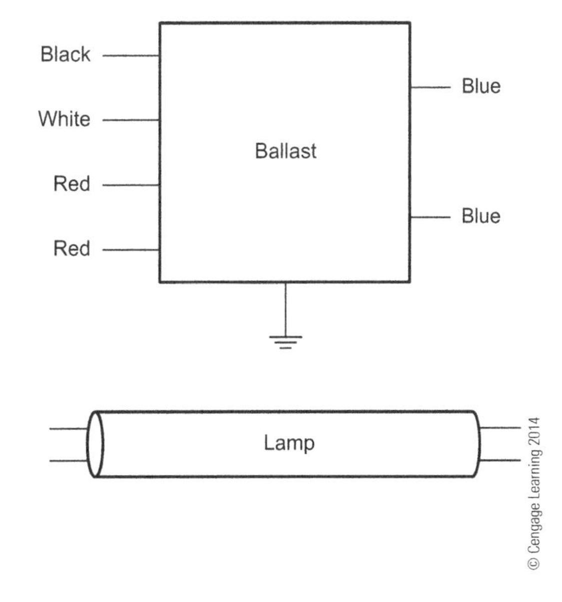

Line

3. Using the given components, show how you would connect the ballast, dimmer, and lamp.

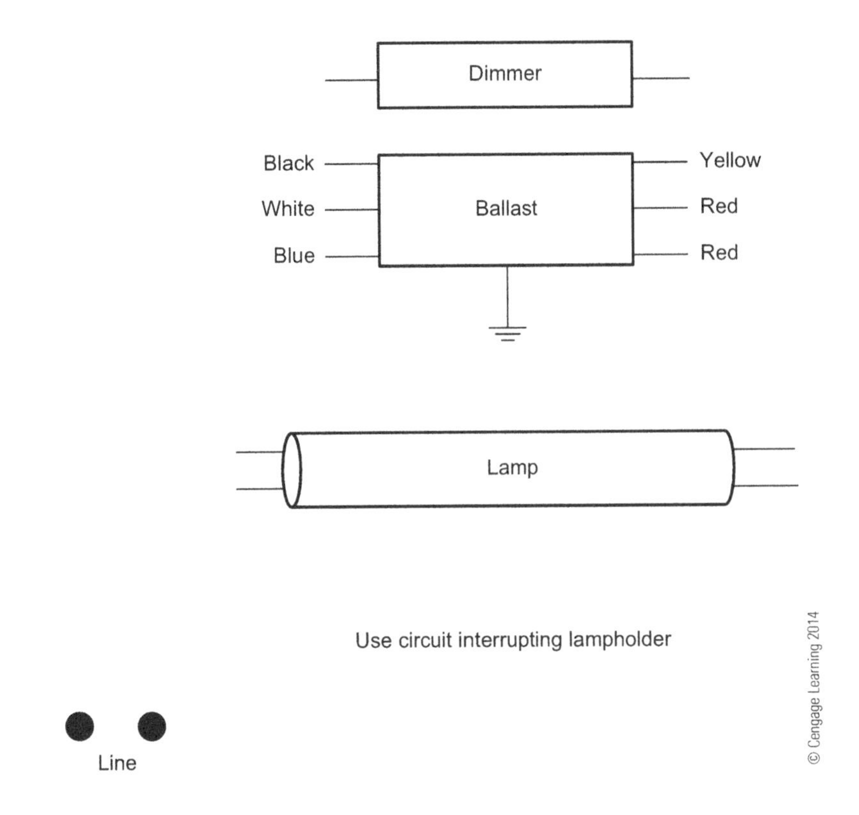

4. Using the given components, show how you would connect the ballast, starters, and lamps.

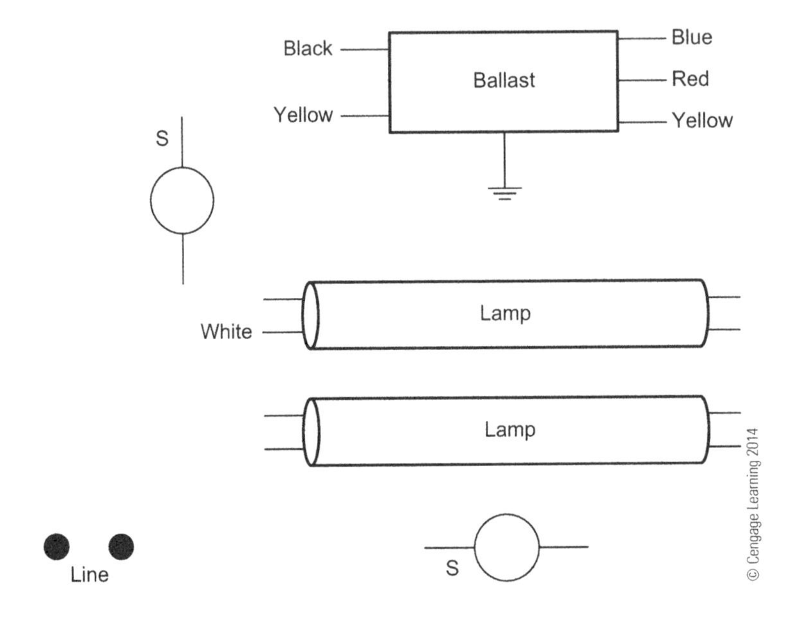

5. Using the given components, show how you would connect the ballast, starters, and lamps.

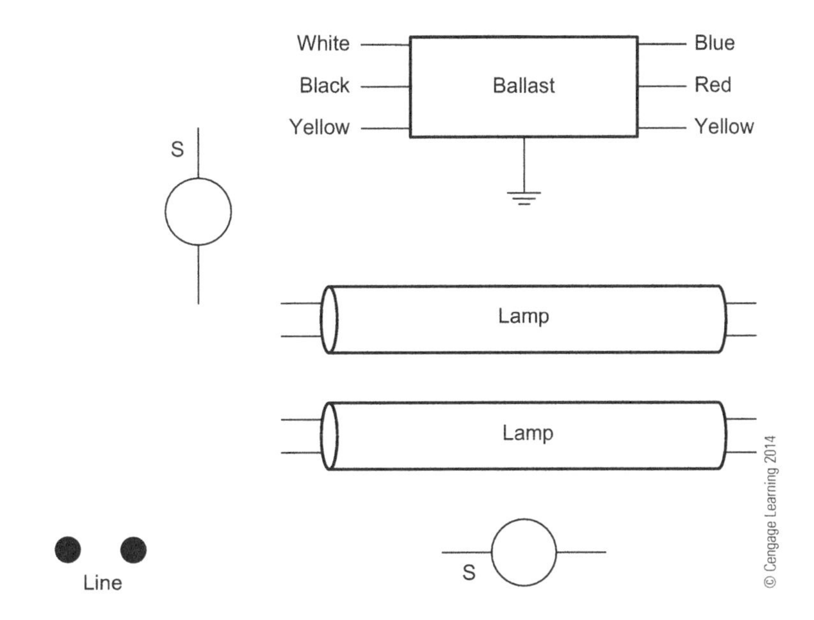

6. Using the given components, show how you would connect the ballast and lamps.

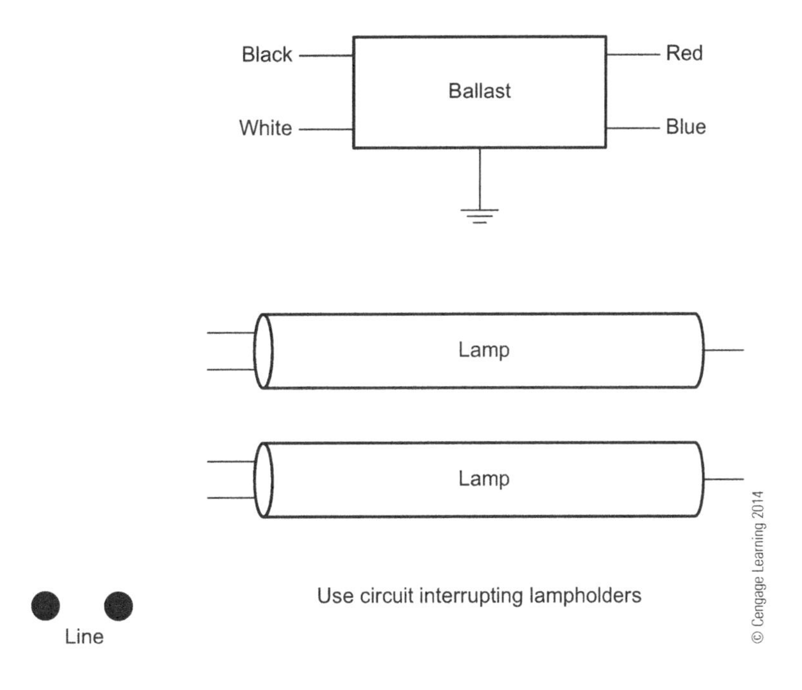

7. Using the given components, show how you would connect the ballast, starters, and lamps.

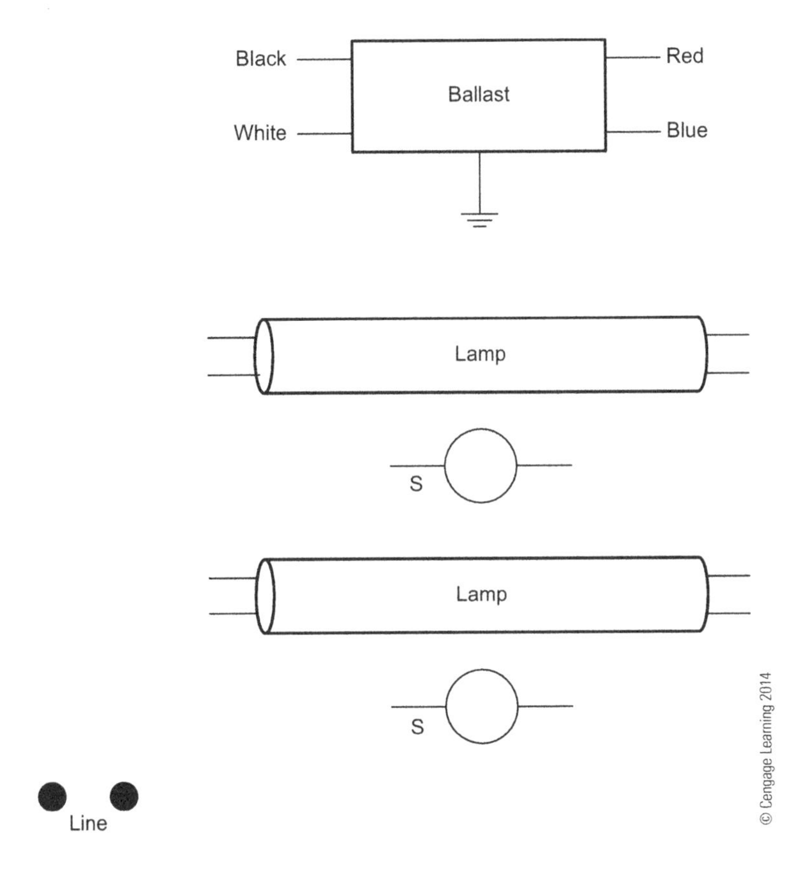

8. Using the given components, show how you would connect the ballast, starters, and lamps.

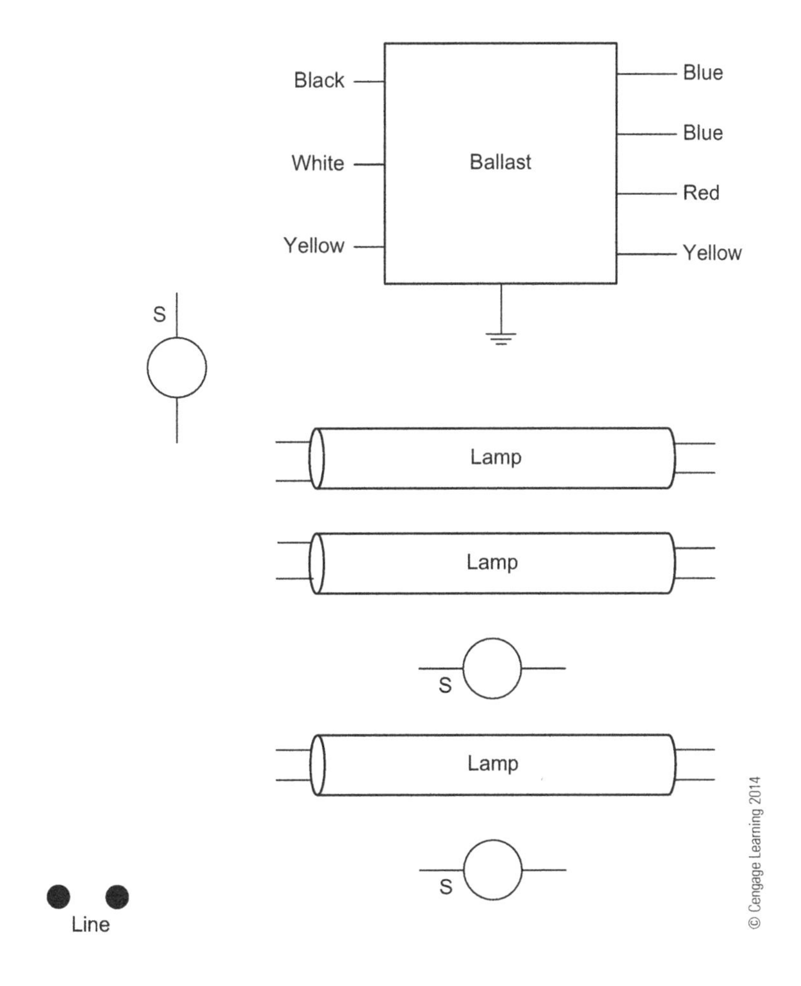

9. Using the given components, show how you would connect the ballast and lamps.

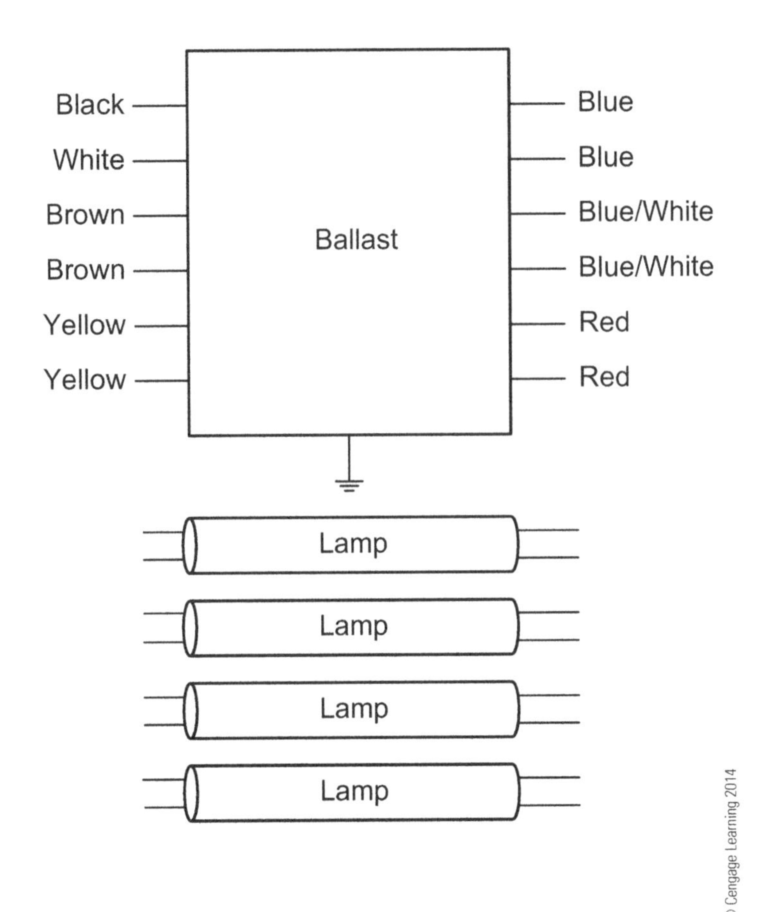

10. Using the given components, show how you would connect the luminaire (fixture).

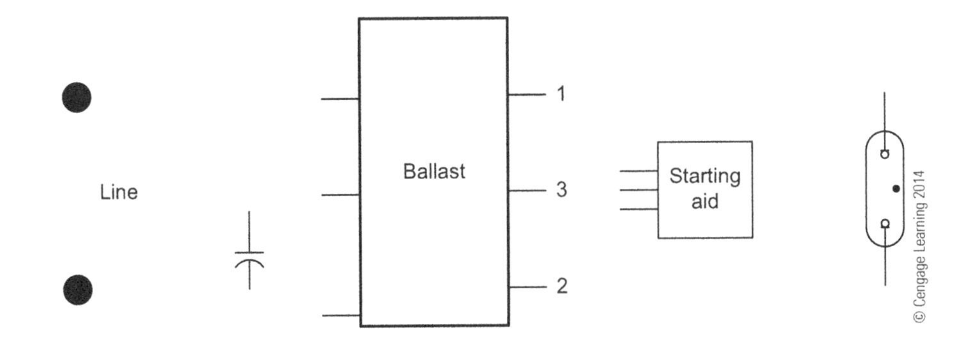

Name: ______________________ Date: ______________

GAS WELDING: THE CYLINDER

1. Label the parts of the cylinder shown in the figure.

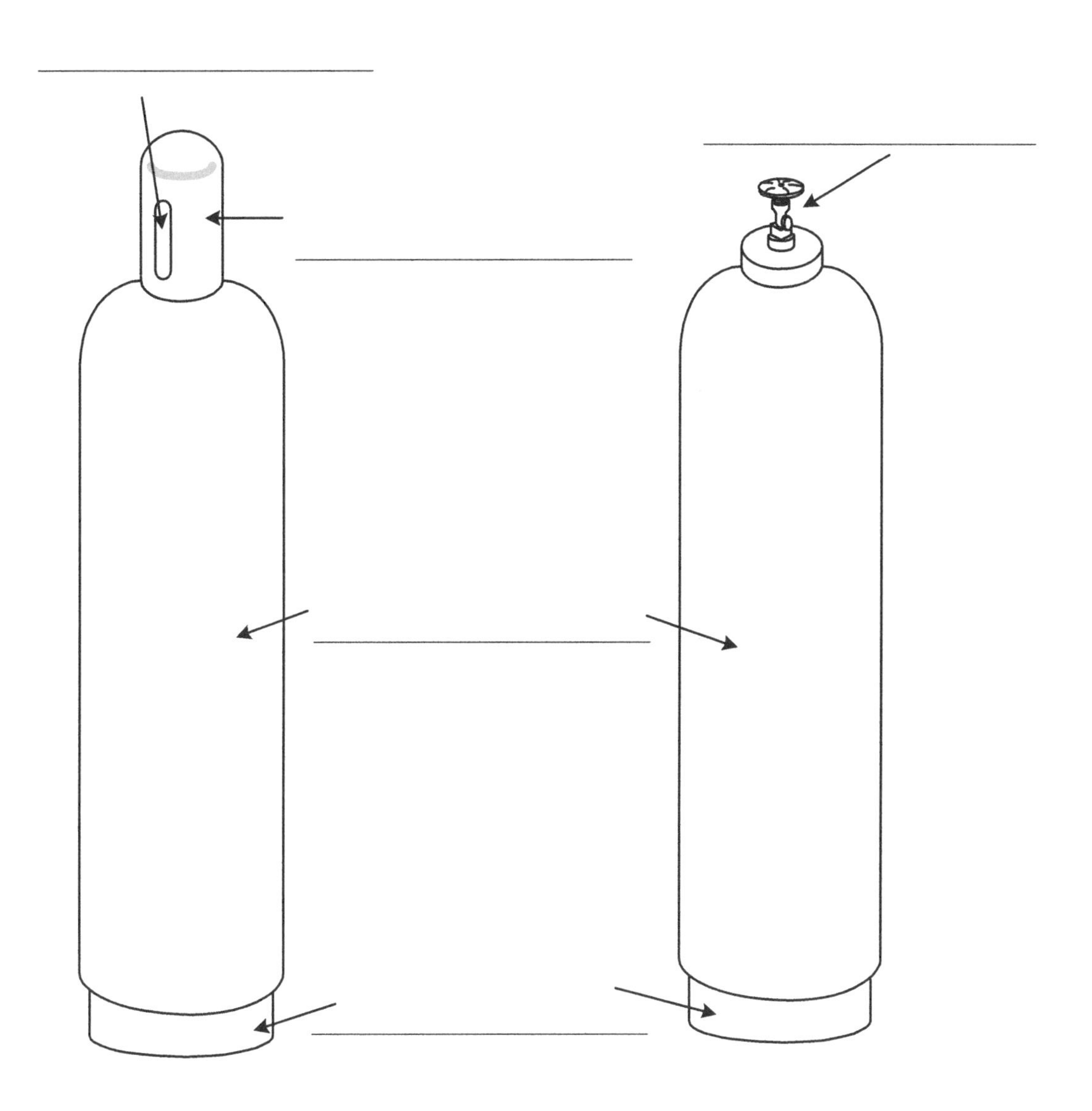

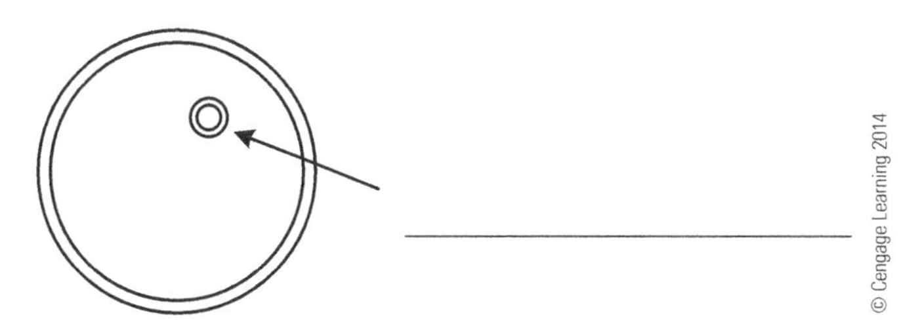

Name: ______________________ Date: ______________

GAS WELDING: THE CUTTING TORCH

1. Label the parts of the cutting torch.

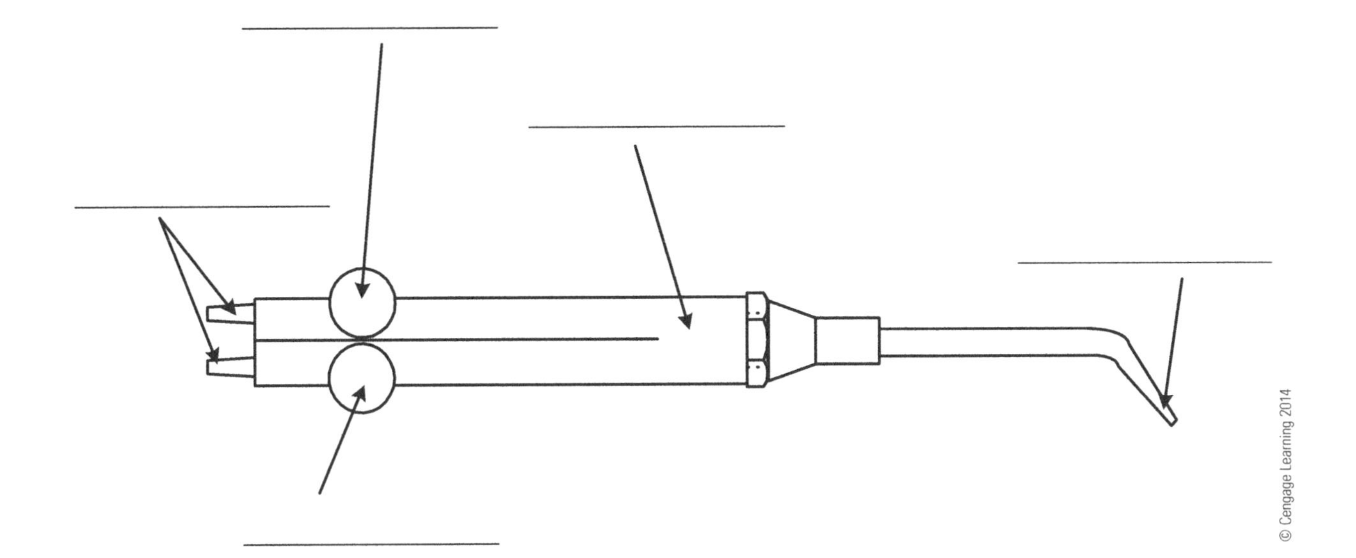

Name: ______________________________ Date: ______________________

GAS WELDING: GAS IDENTIFICATION

1. Identify each component as used in the oxygen system or in the acetylene system.

   A. ____________ A regulator with left-hand threads on it

   B. ____________ A green-colored hose

   C. ____________ An inlet port on the torch handle that is marked with an F

   D. ____________ A fitting collar that has a groove machined into its sides

   E. ____________ A red-colored hose

   F. ____________ A regulator with right-hand threads on it

   G. ____________ A fitting collar that has no grooves machined into its sides

   H. ____________ An inlet port on the torch handle that is marked with an O

Name: ______________________________ Date: ____________________

GAS WELDING: SETUP PROCEDURES

1. Number, in the proper chronological order, the following setup procedures for gas welding.

A. _______ Connect the hoses to the torch handle.

B. _______ Open the oxygen needle valve slowly until a well-defined white cone appears at the tip of the torch.

C. _______ Slowly crack the acetylene cylinder valve to provide a quick burst to blow out debris.

D. _______ Inspect both of the regulators to ensure that the adjustment valves are turned all the way out to close the valve.

E. _______ Connect both of the regulators to their appropriate cylinders.

F. _______ Select the size of the tip that is needed for the job at hand, and locate the working pressure that is used for the chosen tip.

G. _______ Continue to open the acetylene needle valve slowly until all of the smoke leaves the end of the flame.

H. _______ Inspect all of the equipment and threads prior to assembly.

I. _______ Slowly crack the oxygen cylinder valve to provide a quick burst to blow out debris.

J. _______ Begin your work.

K. _______ Check the connections for leaks.

L. _______ Slowly open the oxygen regulator to 3 or 4 psi. Allow the gas to blow through the hose for a couple of seconds, then close the regulator. Repeat the process for the acetylene side.

M. _______ Making sure that the needle valves are closed, insert the tip into the torch handle. Finger-tighten the connecting nut.

N. _______ Adjust the oxygen regulator to the proper working pressure.

O. _______ Connect the reverse-flow check valves to the output of the regulators.

P. _______ Adjust the acetylene regulator to the proper operating pressure.

Q. _______ Point the tip of the torch toward the floor, and ignite the acetylene with the spark lighter.

R. _______ Making sure that the oxygen needle valve is closed on the torch handle, open the acetylene needle valve about ¼ turn.

S. _______ Connect the hoses to the check valves.

T. _______ Stand to the side of the cylinder and slowly open the cylinder valves, remembering to open the acetylene valve only by one complete turn. The oxygen cylinder valve can be completely opened at this time.

Name: ____________________ Date: ____________________

GAS WELDING: SHUTDOWN PROCEDURES

1. Number, in the proper chronological order, the following shutdown procedures for gas welding.

   A. _______ Neatly coil the hose out of the way, and allow the torch tip time to cool. Store the cylinders in a safe place until they are needed again.

   B. _______ Close the oxygen needle valve.

   C. _______ Release both of the adjusting screws on the regulator to close the regulator.

   D. _______ Secure the cylinders with a chain.

   E. _______ Open the acetylene needle valve while watching the gauges on the acetylene regulator. The pressures should drop to zero. Once this happens, close the acetylene needle valve and repeat this procedure for the oxygen side.

   F. _______ Close the acetylene needle valve.

   G. _______ Close both of the cylinder valves.

Name: ______________________________ Date: ______________________

GAS WELDING: FLAME TYPES

1. Match the flame type to the correct graphic.

A

Torch tip
Translucent blue cone
Inner cone

B

Torch tip
Translucent blue cone
Inner cone

________ Neutral flame

________ Carburizing flame

________ Oxidizing flame

C

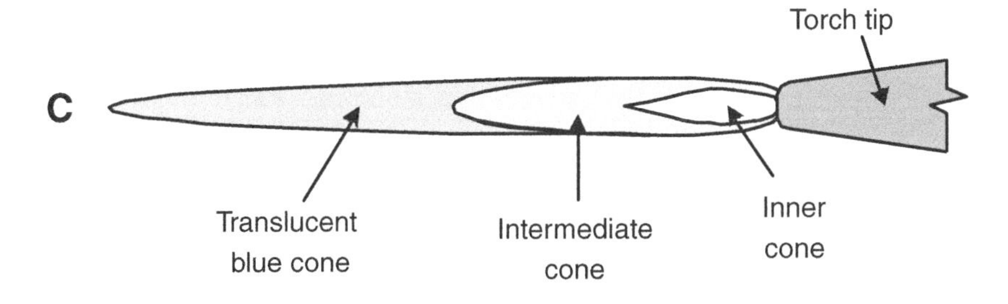

Name: ______________________ Date: ______________

GAS WELDING: WELD TYPES

1. In the spaces that are provided, write the type of weld that is shown in each graphic.

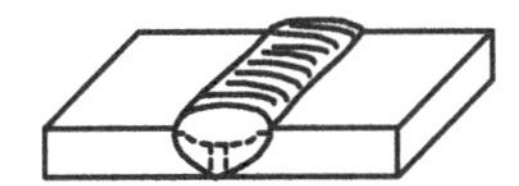

A. ______________________

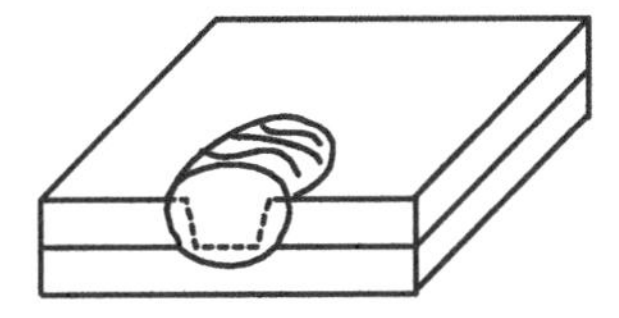

B. ______________________

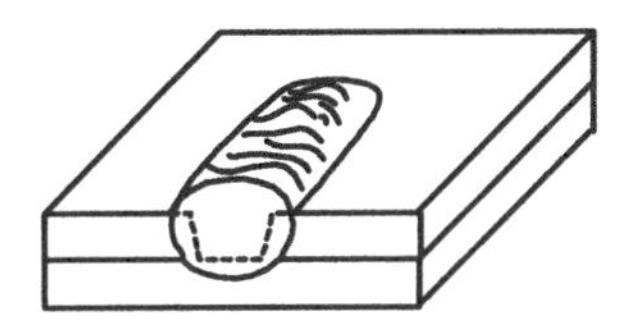

C. ______________________

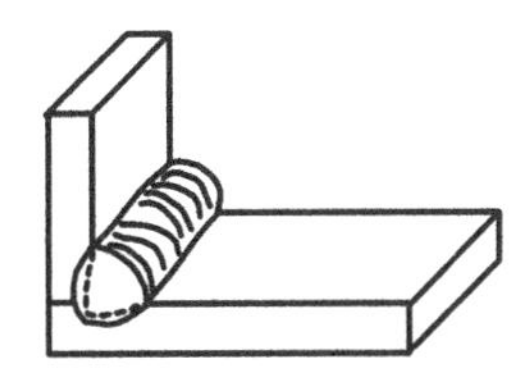

D. ______________________

Name: ____________________ Date: ____________

GAS WELDING: JOINT TYPES

1. In the spaces that are provided, write the type of joint that is shown in each graphic.

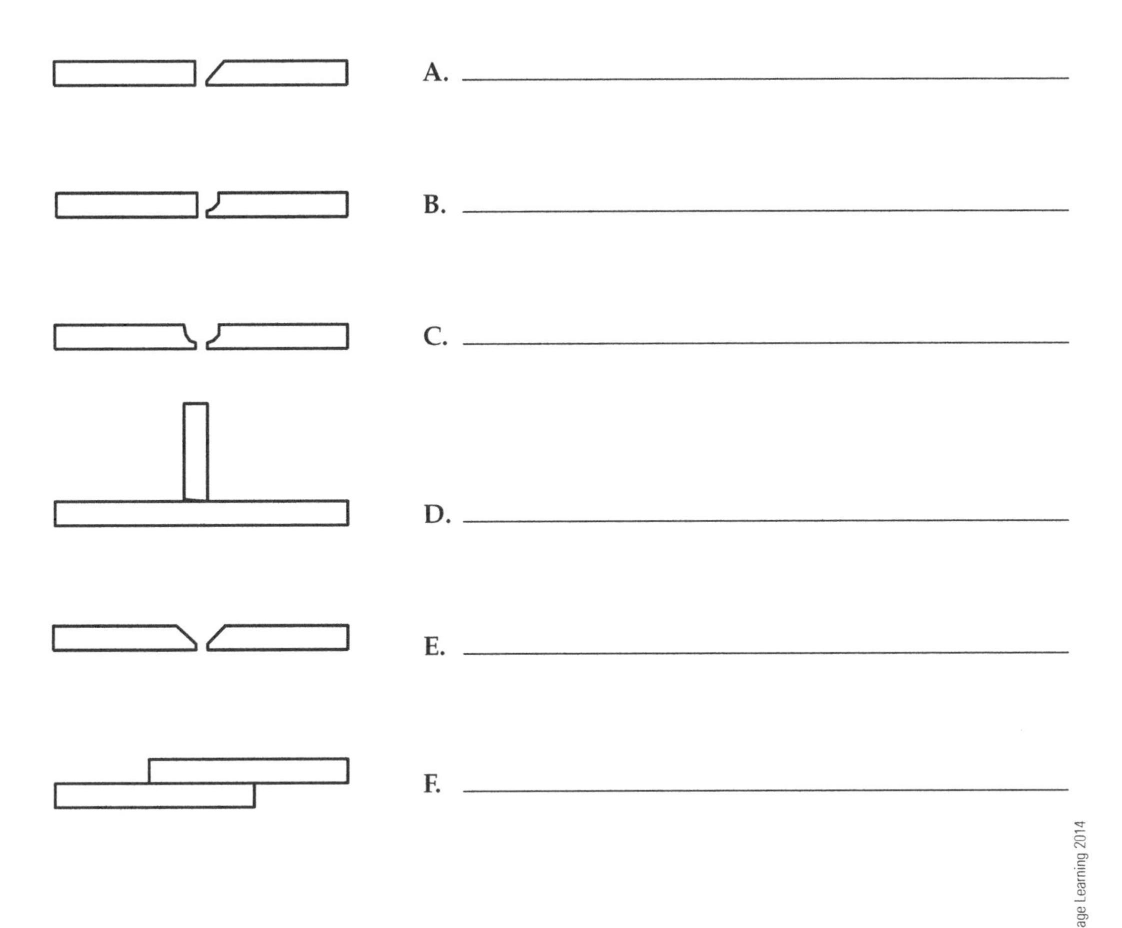

Name: ______________________________ Date: ______________________

GAS WELDING: WELDING POSITIONS

1. Match each definition to the correct welding position.

a. The flat welding position

b. The horizontal position

c. The vertical position

d. The overhead position

1. _______ With this position, the metal to be welded is at a right angle to the floor, and the bead is drawn across the metal from right to left or left to right. It takes a little practice to become proficient at welding in this position because the molten puddle tends to sag as the result of gravity.

2. _______ This position is the most difficult and the most uncomfortable of all the positions. The reason is because the molten metal has a tendency to form drops that fall onto you, the welder. Even many experienced welders find this position a challenge.

3. _______ The material is parallel to the floor as if it were lying on a table or workbench. This is often the most comfortable position for welding.

4. _______ With this position, the metal to be welded is at a right angle to the floor, and the bead is drawn from top to bottom or bottom to top. Most people use the bottom-to-top method because it allows for easier control of the puddle.

Name: ______________________________ Date: ______________

ARC WELDING: SAFETY

1. List all of the personal protective equipment that should be worn during arc welding.

____

____

____

____

____

____

____

Name: ______________________________ Date: ____________________

ARC WELDING: ELECTRICAL QUANTITIES

1. Match each electrical quantity to the correct definition.

   a. Wattage
   b. Current
   c. DC
   d. Resistance
   e. Voltage
   f. AC

   1. _______ The amount of electrical pressure that is in a circuit.
   2. _______ This current changes its direction of current flow every 1⁄120 of a second.
   3. _______ The amount of work that is being done in a given application.
   4. _______ The opposition to current flow.
   5. _______ This type of current is either positive or negative and does not alternate.
   6. _______ The movement of the electrons through a conductor.

Name: ______________________ Date: ______________

ARC WELDING: ARC POLARITY

1. Define, in your own words, each of the acronyms listed.

   a. DCEP ______________________

   ____

   ____

   b. AC ______________________

   ____

   ____

   c. DCEN ______________________

   ____

   ____

Name: ______________________________ Date: ______________

ARC WELDING COMPONENTS

1. Label all of the components that are shown in the figure.

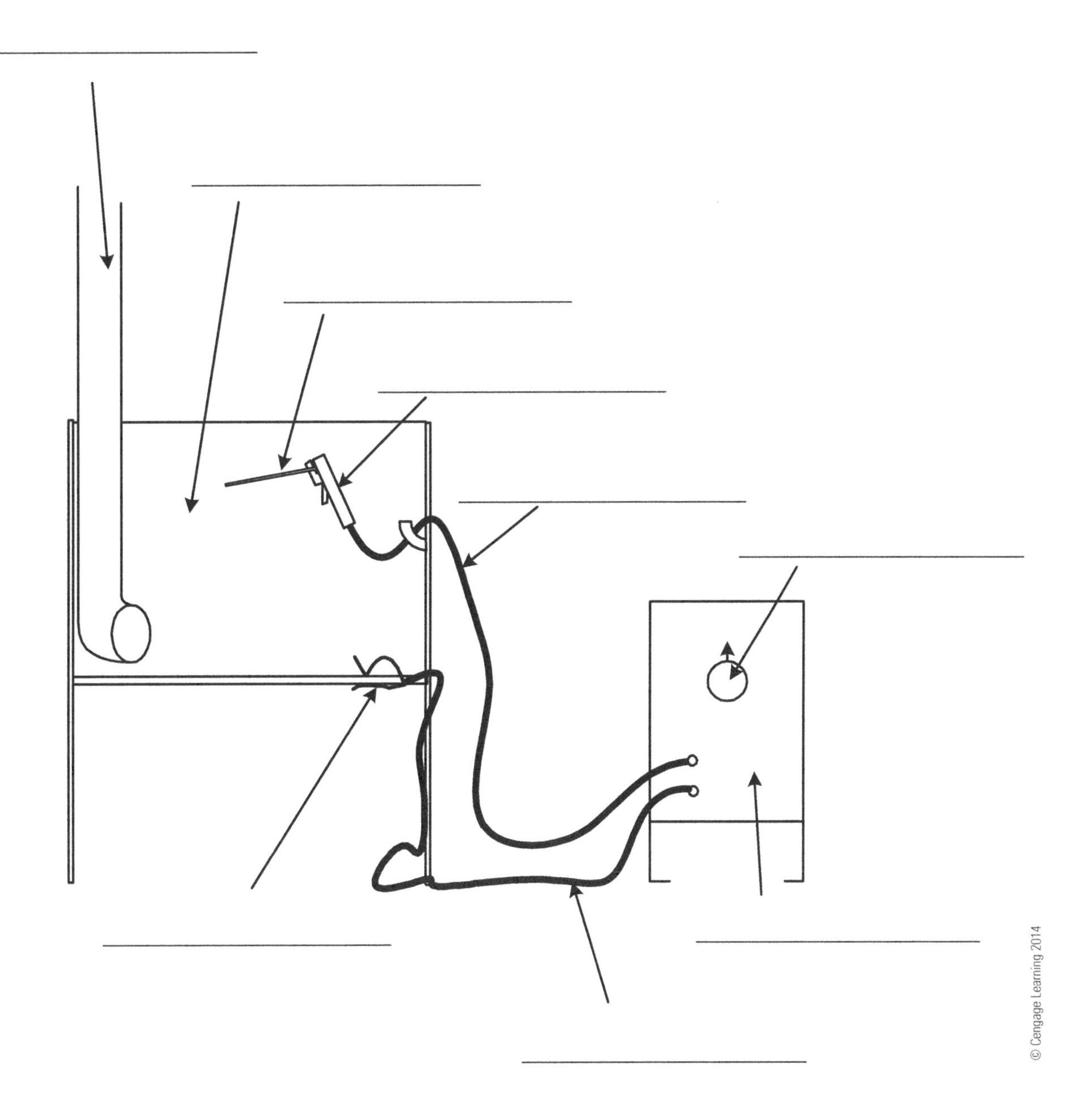

Name: ______________________ Date: ______________

ARC WELDING: ELECTRODE IDENTIFICATION—PART 1

1. Provide all of the information that can be retrieved from each rod identification code that is listed.

   a. E6010 ______________________

   ____

   b. E6018 ______________________

   ____

   c. E7028 ______________________

   ____

   d. E6011 ______________________

   ____

Name: ____________________ Date: ____________________

ARC WELDING: ELECTRODE IDENTIFICATION—PART 2

1. Identify each component of the code in the figure.

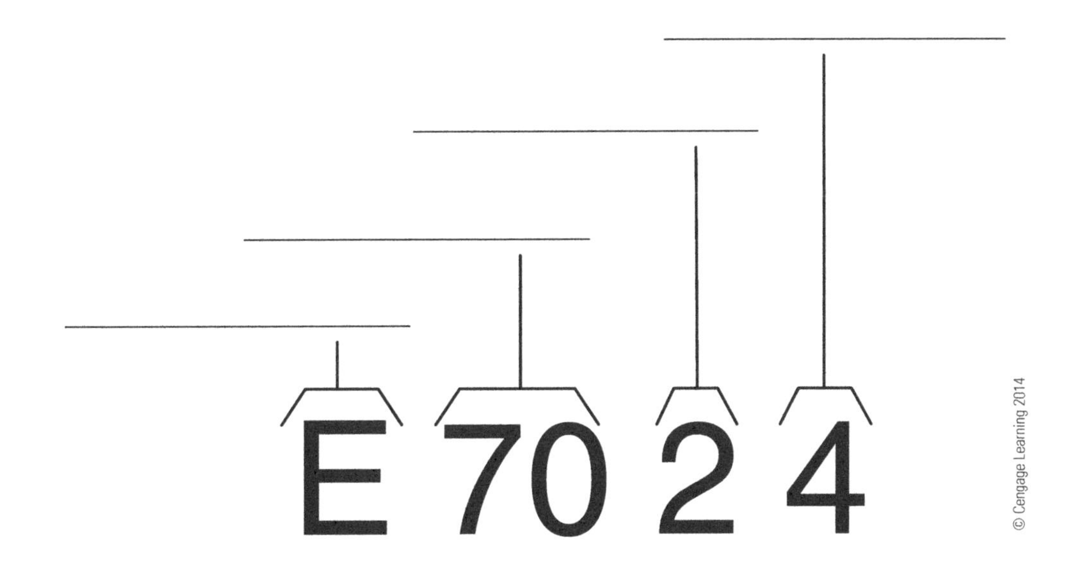

Name: ______________________________ Date: ______________

ARC WELDING: STRIKING THE ARC

1. List, in proper chronological order, the procedures for striking an arc.

____

____

____

____

____

____

____

____

Name: ________________________________ Date: ____________________

ARC WELDING: RUNNING A BEAD

1. Name all of the indicated items in the figure.

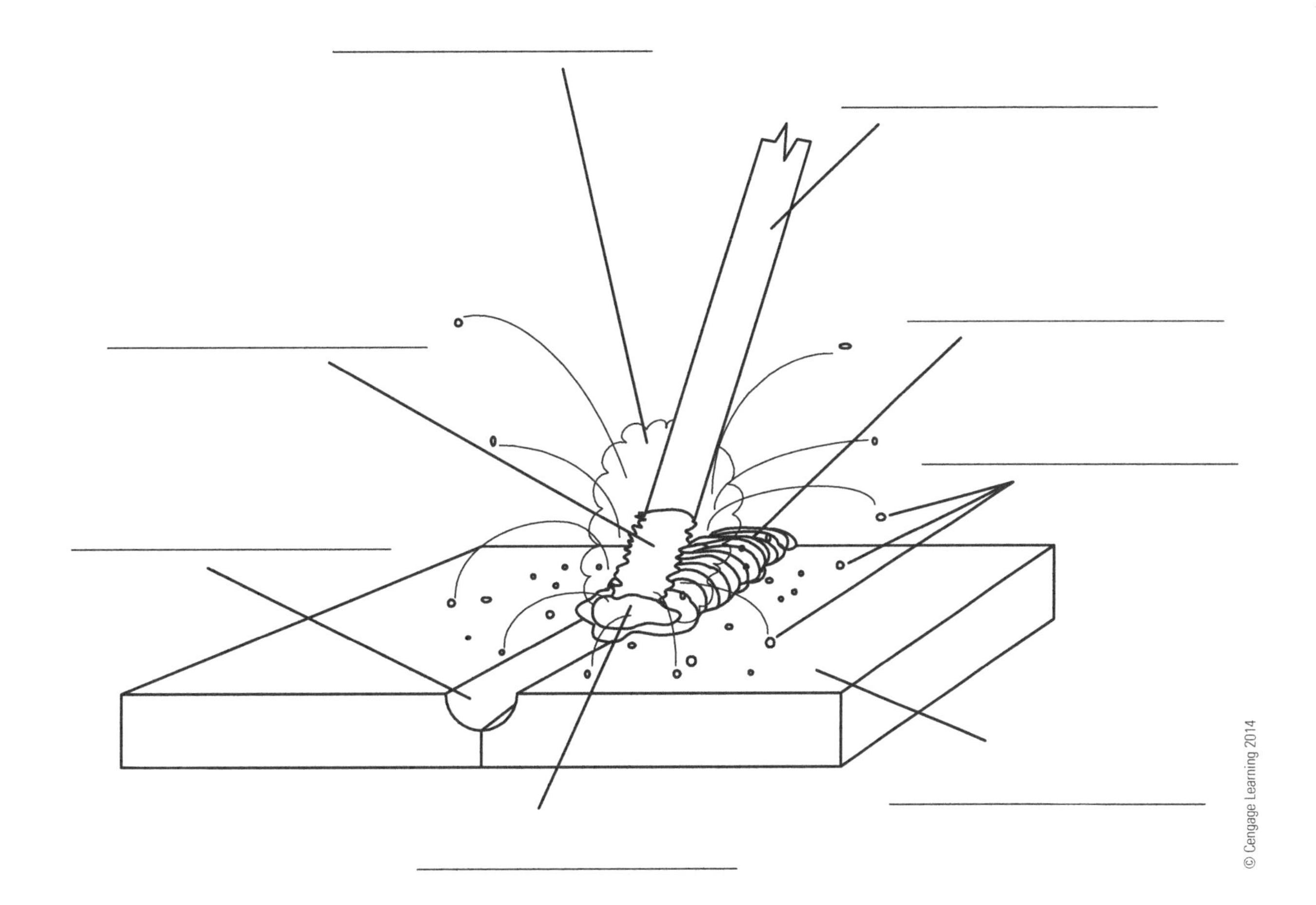

Name: ______________________ Date: ______________

ARC WELDING: WELD TYPES

1. In the spaces that are provided, write the type of weld that is shown in each graphic.

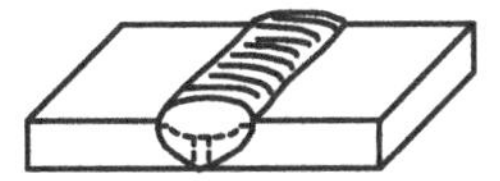

A. ______________________

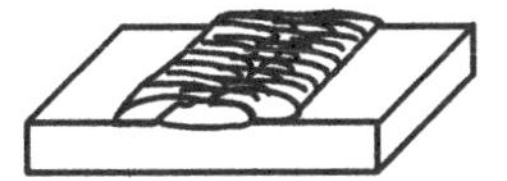

B. ______________________

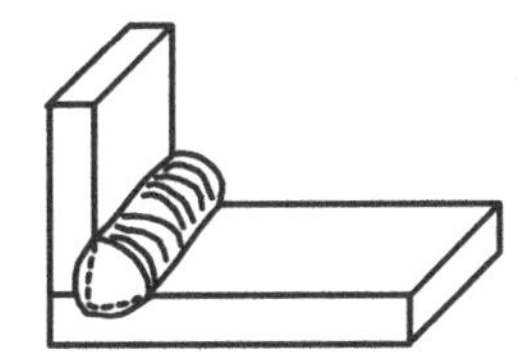

C. ______________________

Name: ______________________________ Date: ____________________

ARC WELDING: JOINT TYPES

1. In the spaces that are provided, write the type of joint that is shown in each graphic.

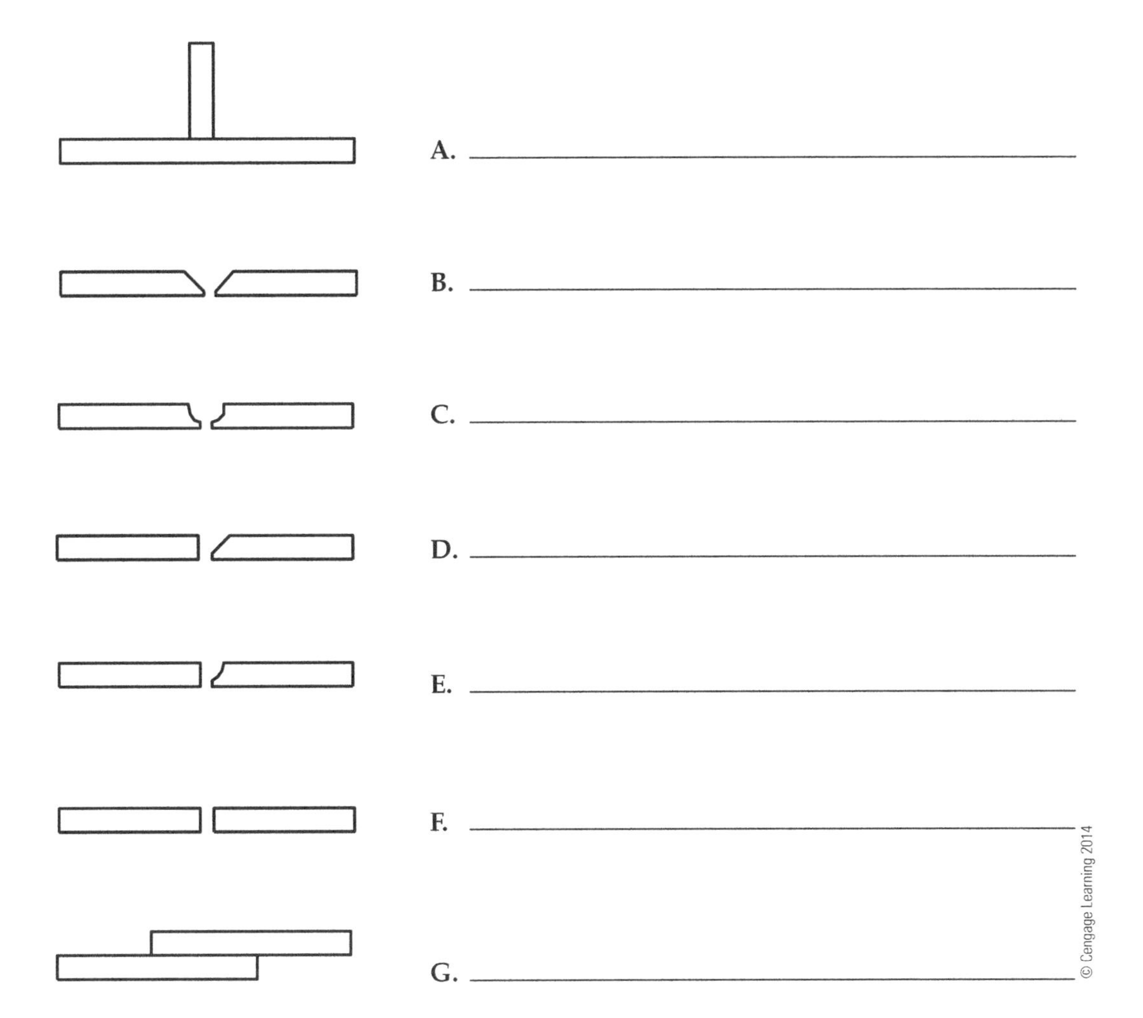

Name: ______________________________ Date: ______________________

ARC WELDING: WELDING POSITIONS—PART 1

1. Match each definition to the correct welding position.

a. The vertical position

b. The horizontal position

c. The overhead position

d. The flat welding position

1. _______ With this position, the metal to be welded is at a right angle to the floor, and the bead is drawn from top to bottom or bottom to top. Most people use the bottom-to-top method because it allows easier control of the puddle.

2. _______ This position is the most difficult and the most uncomfortable of all the positions. Because the molten metal has a tendency to form drops that fall onto you, the welder. Even many experienced welders find this position a challenge.

3. _______ With this position, the metal to be welded is at a right angle to the floor, and the bead is drawn across the metal from right to left or left to right. It takes a little practice to become proficient at welding in this position because the molten puddle tends to sag as a result of gravity.

4. _______ The material is parallel to the floor as if it were lying on a table or workbench. This is often the most comfortable position for welding.

Name: ______________________________ Date: ______________

ARC WELDING: WELDING POSITIONS—PART 2

1. Think of some application in the field that may require a certain welding position, and write a brief description of it. Do this for each welding position. Be creative.

   Flat welding: ______________________________

   Vertical welding: ______________________________

   Horizontal welding: ______________________________

   Overhead welding: ______________________________

Name: ______________________________ Date: ______________

MAINTENANCE LOG—PART 1

1. Study the maintenance log in the figure. What would be the first thing that you, as the preventive maintenance (PM) coordinator, would list as a task during the PM of this machine?

MAINTENANCE LOG

Date	Time	Mach. Identification	Total Downtime	Nature of Breakdown	Work Performed/Maintenance Personnel
3/3/02	4:32 PM	#1 Finish Frame	1 Hour	Machine not running. Notice a lot of heat coming from the right turnstile on the exit end of the machine. After the machine has time to cool, the machine can be restarted. It runs awhile and then stops running again.	Found that the right turnstile needed to be lubricated. Lubricated the turnstile and monitored the machine. Finish frame did not shut down. Released the machine to the operator. Jerry Jackson
3/7/02	8:19 AM	#1 Finish Frame	45 Min.	Oven temperature not getting hot enough.	Only one burner was running. Found faulty high temperature limit switch in the oven. Replaced the limit and fired the oven up to 280 degrees. All is well. Freddy Frazier
3/8/02	2:20 AM	#1 Finish Frame	15 minutes	Machine locked up. Noise coming from right turnstile just before it stopped running.	Noticed some metal filings on the floor below the turnstile. Checked chain on the turnstile. Lubricated the chain & bearing. Runs good now. Ricky Rodgers
3/9/02	7:32 PM	#1 Finish Frame	34 minutes	Yardage counter not running.	Vane switch on main drum roll malfunctioned. Removed the bad vane switch and replaced with a new one. Monitored the counter for about 10 minutes. Released the machine to the operator. Jerry Jackson
3/10/02	10:45 AM	#1 Finish Frame	2 Hrs	Machine shut down by itself. Was running great when a loud screeching started. Seconds later, the machine shut down.	The sprocket on the right turnstile was locked up. Noticed a lot of heat. Applied some grease to the bearing and checked the oiler on the chain. Started the machine up. Ran good. All is well. Freddy Frazier
3/10/02	7:23 PM	#1 Finish Frame	20 minutes	Can't get any fluid down from the mezzanine to the dye tank on the entry end of the frame.	Faulty valve. Replaced Valve and filled dye tank. Monitored the system. Seems to be running OK. Jerry Jackson
3/11/02	5:20 AM	#1 Finish Frame	1 hour 15 minutes	Machine locked up. Right turnstile making a lot of noise.	Noticed more metal filings on the floor below the turnstile. Checked chain on the turnstile again. Lubricated the bearing. Runs good now. Ricky Rodgers

Name: ______________________ Date: ______________

MAINTENANCE LOG—PART 2

1. Create your own maintenance log as if you were the maintenance technician for the scenario described. Use your imagination and write a realistic action that was performed in the Work Performed section.

   The following calls have been made to the maintenance department:

   10:00 A.M.—Roller on entry end of machine will not turn.

   11:26 A.M.—Product counter is not working. Missing a count every now and again. Suspect the sensor, as the operator accidentally hit it.

   12:02 P.M.—Cannot get the roller on the entry end of the machine to turn.

   12:22 P.M.—Coolant motor is not running on the product grinder.

   2:13 P.M.—Oven will not ignite. Fresh air indicator says that the oven is not getting any fresh air from the blower motor on the roof.

   2:55 P.M.—Roller on entry end of the machine is locked up.

(Blank log on following page.)

Maintenance Log

Date	Time	Mach. Identification	Total Downtime	Work Performed/ Maintenance Personnel
Nature of Breakdown:				
Date	Time	Mach. Identification	Total Downtime	Work Performed/ Maintenance Personnel
Nature of Breakdown:				
Date	Time	Mach. Identification	Total Downtime	Work Performed/ Maintenance Personnel
Nature of Breakdown:				
Date	Time	Mach. Identification	Total Downtime	Work Performed/ Maintenance Personnel
Nature of Breakdown:				
Date	Time	Mach. Identification	Total Downtime	Work Performed/ Maintenance Personnel
Nature of Breakdown:				
Date	Time	Mach. Identification	Total Downtime	Work Performed/ Maintenance Personnel
Nature of Breakdown:				
Date	Time	Mach. Identification	Total Downtime	Work Performed/ Maintenance Personnel
Nature of Breakdown:				

Name: ______________________ Date: ______________

PM PLANNING

1. Assume that the machine described in Worksheet 28–2 is scheduled for a PM. Write a detailed plan of the actions that should be taken during the PM.

Name: ______________________________ Date: ______________________

MAINTENANCE LOG

1. Refer to the drawing. Work needs to be performed inside the J-box shown.

   a. What, in detail, should be done prior to beginning any work within the J-box?

   ____

   ____

   ____

   b. What should be done while work is being done in the J-box?

   ____

   ____

   ____

(Drawing located on following page.)

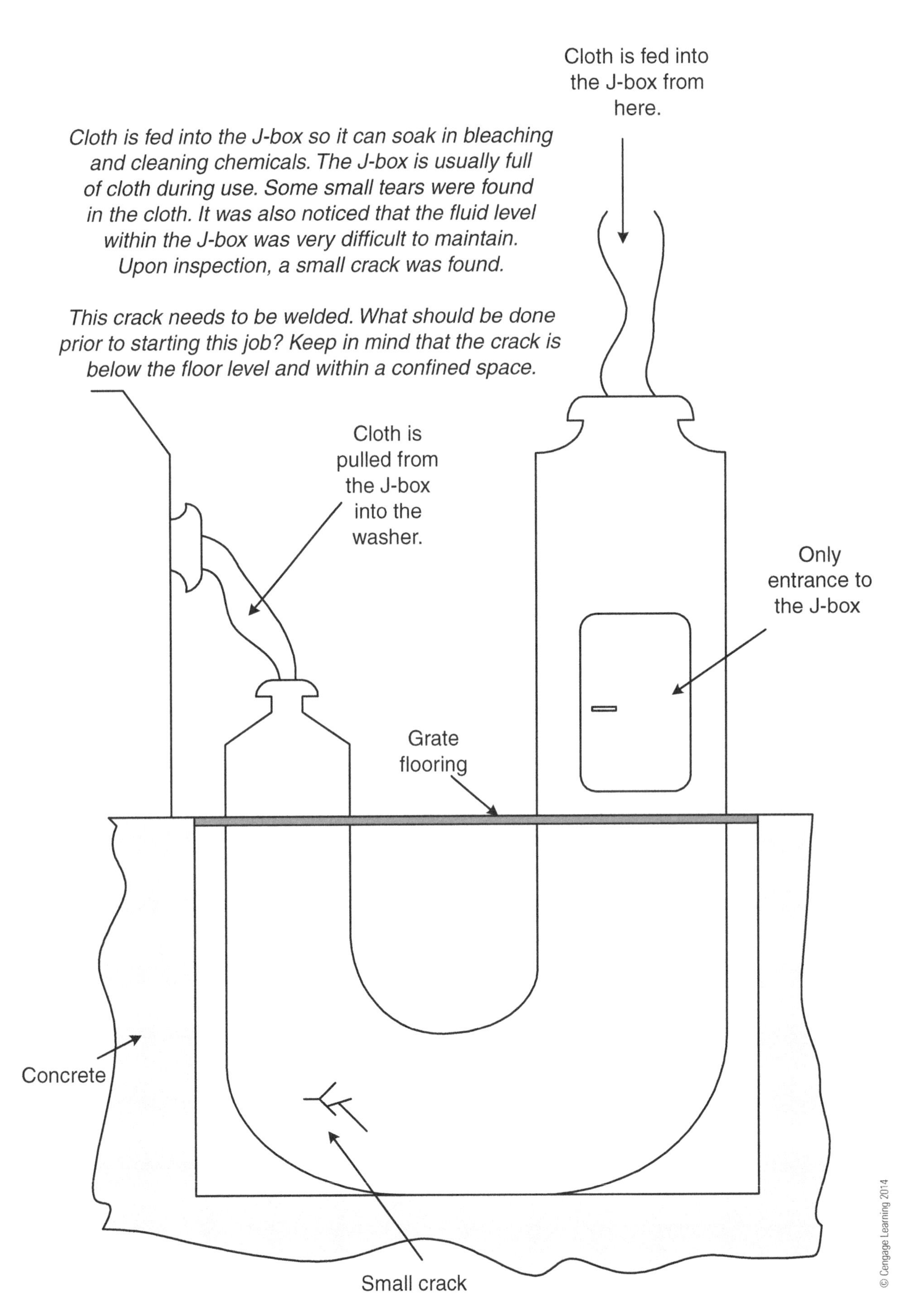
Cloth is fed into the J-box from here.
Cloth is fed into the J-box so it can soak in bleaching and cleaning chemicals. The J-box is usually full of cloth during use. Some small tears were found in the cloth. It was also noticed that the fluid level within the J-box was very difficult to maintain. Upon inspection, a small crack was found.
This crack needs to be welded. What should be done prior to starting this job? Keep in mind that the crack is below the floor level and within a confined space.
Cloth is pulled from the J-box into the washer.
Only entrance to the J-box
Grate flooring
Concrete
Small crack
© Cengage Learning 2014

Name: ______________________ Date: ______________

BEARING FAILURE

1. Match the cause of failure to the symptom. Choose a cause from the list, and write the letter that precedes the cause on the line in front of the description. You can use each cause only once.

a. False Brinell damage
b. Contamination
c. Spalling
d. Improper lubrication
e. Moisture
f. High temperatures
g. Pitting
h. Overloading or excessive thrust
i. Misalignment
j. Fluting

1. _______ No lubrication causes friction and overheating within the bearing. Overlubrication can place internal pressure on the bearing because the rolling elements have to move the excessive amount of lubrication within the bearing, as well as the load.

2. _______ The discoloring of the raceways and the rolling elements indicates this problem. The metal is usually darkened with a bluish-purple coloring where the overheating occurred. It is not uncommon for the bearing to become deformed as the result of excessive internal temperatures.

3. _______ In this condition, one side of the bearing shows more wear than the other. Also, opposing sides may show signs of wear, and there is uneven wear on the rolling element.

4. _______ This occurs mostly when welding currents pass through the bearing.

5. _______ This is an indication of electrical current flow through the bearing. The problem occurs where the contact is made between the outer race and the rolling elements, and the rolling elements and the inner race. The presence of thin lines that are etched into the races makes it easy to recognize this condition.

6. _______ This occurs any time a foreign particle enters the bearing, usually when it is operating in a dirty environment. The damage is usually in the form of deformity, which causes damage to the races and rolling elements as well.

7. _______ With this problem, damage is indicated by marks on the shoulder or upper portions of the inner and outer races. There is also anything from a slight discoloration to heavy galling.

8. _______ This occurs when continuous impacting forces (such as vibrations) are passed from one ring to the other, through the rolling elements, when there is no rotation of the shaft. Indentations are formed on the outer races from the impacts. As the rolling elements begin to rotate, they create heat as they encounter these evenly spaced indentations.

9. _______ Rusting surfaces are an indication of this cause of failure. Oxidation occurs when moisture is present and lubrication is lacking. This is commonly referred to as fretting corrosion. If a bearing has a suitable amount of lubrication, oxidation should not occur even when the bearing is in a moist environment.

10. _______ This problem is seen as the flaking away of metal pieces due to metal fatigue. This occurs when the rolling element and the bearing race begin to flex from application of an excess load. This flexing (distortion) is momentary and repetitive. As the metal begins to fatigue, microscopic fractures begin to appear. This causes the metal to begin flaking.

Name: ______________________________ Date: ______________

GEARBOX FAILURE

1. What is the most common cause of failure in a gearbox?

   ____

   a. Lack of oil
   b. Excess of oil
   c. Contamination
   d. Overwork
   e. None of the above

2. What does it mean when the oil in a gearbox looks metallic?

   ____

   ____

   ____

Name: ______________________ Date: ______________

INSPECTION OF SEALS

1. What is the most important thing to check when inspecting a seal?

____

____

____

2. What should be done if leakage is found in a packing seal?

____

____

____

Name: ______________________________ Date: ____________________

MECHANICAL PMs: EQUIPMENT INFORMATION

1. Identify a piece of equipment and create an Equipment Information record for this item.

   a. Make: ______________________________

   b. Model: ______________________________

   c. Serial number: ______________________________

   d. Manufacturer: ______________________________

   e. Manufacturer contact information: ______________________________

   ____

   ____

   ____

   f. Installation date: ______________________________

   g. Location: ______________________________

   h. Operating voltage: ______________________________

   i. Operating current: ______________________________

   j. Other: ______________________________

Name: ______________________________ Date: ______________

MECHANICAL PMs: INSPECTION CHECKLIST

1. Create an Inspection Checklist for the equipment identified in Worksheet 30-4.

   a. Inspector: ______________________________

   b. Inspection date: ______________________________

   c. Inspection time: ______________________________

   d. Equipment to be inspected: ______________________________

   e. Items needed to perform inspection: ______________________________
   ____
   ____
   ____

   f. Results of inspection: ______________________________
   ____
   ____
   ____

   g. Recommendations: ______________________________
   ____
   ____
   ____

Name: ______________________ Date: ______________

MECHANICAL PMs: REPAIR INFORMATION

1. Create a Repair Information record for the equipment identified in Worksheet 30-4.

   a. Date: ______________________

   b. Time: ______________________

   c. Individual making repairs: ______________________

   d. Operator comments: ______________________

   e. Symptoms: ______________________

   f. Action taken: ______________________

Name: ______________________________ Date: ______________

ELECTRICAL PMs: EQUIPMENT INFORMATION

1. Identify a piece of equipment and create an Equipment Information record for this item.

   a. Make: ______________________________

   b. Model: ______________________________

   c. Serial number: ______________________________

   d. Manufacturer: ______________________________

   e. Manufacturer contact information: ______________________________

   ____

   ____

   ____

   f. Installation date: ______________________________

   g. Location: ______________________________

   h. Operating voltage: ______________________________

   i. Operating current: ______________________________

   j. Other: ______________________________

Name: ______________________ Date: ______________

ELECTRICAL PMs: INSPECTION CHECKLIST

1. Create an Inspection Checklist for the equipment identified in Worksheet 31-1.

   a. Inspector: ______________________

   b. Inspection date: ______________________

   c. Inspection time: ______________________

   d. Equipment to be inspected: ______________________

   e. Items needed to perform inspection: ______________________

   f. Results of inspection: ______________________

   g. Recommendations: ______________________

Name: ______________________ Date: ______________

ELECTRICAL PMs: REPAIR INFORMATION

1. Create a Repair Information record for the equipment identified in Worksheet 31-1.

   a. Date: ______________________

   b. Time: ______________________

   c. Individual making repairs: ______________________

   d. Operator comments: ______________________

   ____

   ____

   ____

   e. Symptoms: ______________________

   ____

   ____

   ____

   f. Action taken: ______________________

   ____

   ____

   ____

Notes

Notes

Notes

Notes

Notes

Notes

Notes